Cell Biology Research Progress

nova
Medicine & Health
New York

Cell Biology Research Progress

Dendritic Spines: An Update
Ignacio González-Burgos, PhD (Editor)
2023. ISBN: 979-8-89113-080-7 (Hardcover)
2023. ISBN: 979-8-89113-138-5 (eBook)

Transcriptomics and Their Importance
Sarah Naiyer, PhD (Editor)
Khalid Hussain Bhat, PhD (Editor)
2022. ISBN: 979-8-88697-044-9 (Hardcover)
2022. ISBN: 979-8-88697-203-0 (eBook)

Endothelial Progenitor Cells in Health and Disease
Amankeldi A. Salybekov, MD, PhD (Editor)
Shuzo Kobayashi, MD, PhD (Editor)
Takayuki Asahara, MD, PhD (Editor)
2022. ISBN: 978-1-68507-794-5 (Hardcover)
2022. ISBN: 978-1-68507-948-2 (eBook)

Dendritic Cell Biology as an Immune Dysfunction in Neoplasia
Lawrence M. Agius (Author)
2022. ISBN: 978-1-68507-486-9 (Softcover)
2022. ISBN: 978-1-68507-537-8 (eBook)

TRiPs Across Epithelial and Endothelial Barriers in Health and Disease
Alexander V. Zholos, PhD, DSc (Editor)
Ganna M. Tolstanova, PhD (Editor)
2021. ISBN: 978-1-68507-020-5 (Hardcover)
2021. ISBN: 978-1-68507-139-4 (eBook)

More information about this series can be found at https://novapublishers.com/product-category/series/cell-biology-research-progress/

Shalini Mani
and Samiksha Wasnik
Editors

Mitophagy and Its Role in Health and Disease

nova
Medicine & Health
New York

DOI: https://doi.org/10.52305/ZKBP1526

NOTICE TO THE READER

The Publisher has taken reasonable care in the preparation of this book but makes no expressed or implied warranty of any kind and assumes no responsibility for any errors or omissions. No liability is assumed for incidental or consequential damages in connection with or arising out of information contained in this book. The Publisher shall not be liable for any special, consequential, or exemplary damages resulting, in whole or in part, from the readers' use of, or reliance upon, this material. Any parts of this book based on government reports are so indicated and copyright is claimed for those parts to the extent applicable to compilations of such works.

Independent verification should be sought for any data, advice or recommendations contained in this book. In addition, no responsibility is assumed by the Publisher for any injury and/or damage to persons or property arising from any methods, products, instructions, ideas or otherwise contained in this publication.

This publication is designed to provide accurate and authoritative information with regards to the subject matter covered herein. It is sold with the clear understanding that the Publisher is not engaged in rendering legal or any other professional services. If legal or any other expert assistance is required, the services of a competent person should be sought. FROM A DECLARATION OF PARTICIPANTS JOINTLY ADOPTED BY A COMMITTEE OF THE AMERICAN BAR ASSOCIATION AND A COMMITTEE OF PUBLISHERS.

Library of Congress Cataloging-in-Publication Data

Names: Mani, Shalini, editor. | Wasnik, Samiksha, editor.
Title: Mitophagy and its role in health and disease / editors Shalini Mani, PhD (Associate Professor, Department of Biotechnology, Jaypee Institute of Information Technology, Noida, Uttar Pradesh), Samiksha Wasnik, PhD (Assistant Professor of Research, Division of Regenerative Medicine, Loma Linda University, Loma Linda, CA USA).
Description: New York : Nova Science Publishers, [2024] | Series: Cell biology research progress | Includes bibliographical references and index. |
Identifiers: LCCN 2024006330 (print) | LCCN 2024006331 (ebook) | ISBN 9798891134706 (hardcover) | ISBN 9798891136113 (adobe pdf)
Subjects: LCSH: Mitochondrial pathology. | Cellular control mechanisms. | Mitochondria. | Cytology. | Cell physiology.
Classification: LCC RB147.5 .M593 2024 (print) | LCC RB147.5 (ebook) | DDC 616.07--dc23/eng/20240313
LC record available at https://lccn.loc.gov/2024006330
LC ebook record available at https://lccn.loc.gov/2024006331

Published by Nova Science Publishers, Inc. † New York

Contents

Preface

In recent years, the study of mitophagy has emerged as a fascinating field of research. Mitophagy, the process of selective degradation and recycling of damaged or dysfunctional mitochondria, holds immense promise in understanding and potentially treating various diseases. From cardiovascular diseases to neurodegenerative disorders and cancer, mitophagy has been implicated in a wide range of medical conditions.

This book aims to provide a comprehensive overview of the current state of knowledge regarding mitophagy and its role in different disease processes. Each chapter delves into a specific aspect of mitophagy, exploring its mechanisms, implications, and potential therapeutic applications.

The first few chapters lay the foundation by providing an overview of mitophagy and its significance in various fields. Aging cardiomyopathy, cardiovascular diseases, neurological disorders, and neurodegenerative diseases are all explored in separate chapters, highlighting the role of mitophagy in these conditions.

The next crucial section of the book is dedicated to unraveling the intricate connection between mitophagy and carcinogenesis, as well as its application in innovative cancer therapies.. Chapters are devoted to understanding the relationship between mitophagy and carcinogenesis and exploring how triggering mitophagy may hold promise in overcoming chemoresistance in cancer therapy.Researchers and medical professionals can gain insights into overcoming chemoresistance and improving treatment outcomes by exploring the potential of triggering mitophagy.

As the book progresses, it delves into the intriguing relationship between mitophagy and aging muscle and its implication in overall aging processes.

Finally, the book explores the potential therapeutic interventions and tools that can be used to modulate mitophagy. From the role of herbs in treating neurodegenerative disorders to the latest therapeutic approaches, this section intends to provide a glimpse into the future of mitophagy research and its clinical applications. The final chapters highlight the latest advancements and breakthroughs. Readers will gain invaluable insights into harnessing the power of mitophagy in stem cells and its implications in lung cancer research.

This book brings together contributions from renowned experts in the field, who have conducted ground-breaking research in mitophagy. Their collective knowledge and expertise aim to offer readers a deep understanding of mitophagy and its vast implications in human health and disease.

Whether you are a student, researcher, or medical professional, the wealth of information presented in this book will undoubtedly expand your understanding of mitophagy and pave the way for further advancements in this exciting study area.

We hope that readers find this book a valuable resource and a stepping stone towards unraveling the intricacies of mitophagy and harnessing its potential for improving human health.

Shalini Mani
Samiksha Wasnik
Editors

Acknowledgments

We sincerely thank the authors for offering to write comprehensive chapters on a tight schedule. This is generally an added responsibility in the hectic work schedules of researchers. We express our earnest gratitude to the reviewers, who provided their critical views for the improvement of the book chapters. We would like to thank reviewers of our book proposal for their suggestions in the framing of the chapters. We also thank Editorial Project Manager, Nova Science Publishers, Inc., whose efforts during the preparation of this book were very useful.

Shalini Mani
Samiksha Wasnik
Editors

Chapter 1

Overview of Mitophagy

Shashi Prakash Singh[1]
Madhu Rani[2]
Annu Devi[3]
Rashmi Kumari[2]
Moshahid Alam Rizvi[2]
and Anil Kumar[4,*]

[1]Pharmacology and Therapeutics, Centre for Genetics and Pharmacology, Roswell Park Comprehensive Cancer Care Buffalo, NY, USA
[2]Genome Biology Laboratory, Department of Biosciences, Jamia Millia Islamia, New Delhi, India
[3]Special Centre for Molecular Medicine, Jawaharlal Nehru University, New Delhi, India
[4]Directorate of Education, Rani Lakshmi Bai Central Agricultural University, Jhansi, Uttar Pradesh, India

Abstract

Mitophagy is an autophagic process that removes irrelevant and impaired mitochondria. Mitophagy is a centric molecular mechanism for mitochondrial quality control, mainly for cellular homeostasis. Deregulation of mitophagy has been linked with the occurrence of metabolic diseases. Therefore, mitophagy is a key target of emerging therapeutic methods in this area.

Keywords: mitophagy, mitochondria, autophagy

Introduction

Mitochondria, the cell's powerhouse, plays a critical role in the quiescence to senescence transition. Mitochondrial dysfunction is a hallmark of aging. Besides generating power (ATP) in cells, mitochondria are involved in calcium metabolism, gluconeogenesis, pentose phosphate pathway, fatty acid metabolism, ROS formation, and initiate apoptosis.

* Corresponding Author's Email: anilkumar.mbge@gmail.com.

In: Mitophagy and Its Role in Health and Disease
Editors: Shalini Mani and Samiksha Wasnik
ISBN: 979-8-89113-470-6

A vast amount of literature is available on the role of mitochondria in inducing apoptosis. The initiation of apoptosis occurs when the outer membrane of mitochondria permeabilizes and releases pro-apoptotic proteins such as "cytochrome C, a second mitochondria-derived activator of caspases (SMAC)/DIABLO, mitochondrial enzyme Endonuclease G (ENDOG), OM1, and Apoptosis-inducing factor (AIF)," leading to cellular death. Mitochondria play a critical role in cellular homeostasis. Therefore, any dysfunctionality in mitochondria in the cell environment is removed by proteolysis and mitophagy. Mitophagy is a unique form of autophagy that regulates and reorganizes mitochondrial components and networks.

This book chapter discusses the general mechanism of mitophagy and its effect on different diseases.

Mitochondrial Quality Control

Like any industry, quality control is an important process. Similarly, the quality control of mitochondria is necessary at the cellular level, as it plays several functions. Stress response factors are activated during cell starvation, leading to "mitochondrial trash" removal (Kornmann 2014). This mitochondrial trash can be damaged, degenerated, or old mitochondria or its components, interfering with the healthy mitochondria. The mitochondrial quality control (MQC) process organizes proteostasis biogenesis, transfer, and mitophagy.

This process of eliminating the damaged mitochondria is known as Mitophagy. In mitophagy, mitochondrial components such as amino acids are degraded, and ATP is conserved. Dysfunctional mitochondria are a fundamental component affecting aging. In aging, genomic instability leads to chronic damage to the DNA and finally causes cellular senescence. The DNA damage response affects energy metabolism, insulin pathway, and attrition of telomeres. This attrition causes activation of p53, which downregulates peroxisome proliferator-activated receptor gamma and alpha and beta (PGC-1α and 1β), leading to dysfunction of oxidative phosphorylation, oxidative stress, gluconeogenesis, and mitochondrial operations (Sahin et al., 2011). The decline in PGC-1α/1β level causes a shift in the metabolic profile of mitochondria and induces apoptosis, finally affecting Senescence(Sen et al., 2011). The senescence caused by mitochondria is a mitochondrial dysfunction-associated senescent response (MiDAS). In MiDAS, cells show a low level of NAD+/NADH ratio, which showed the growth arrest and IL-1b independent and activation of p53, which initiates the Senescence and Senescence-associated secretory phenotype (SASP). The low level of NAD+ may drive the aging process. This phenomenon is found in cultured human cells and progeroid mouse aging models(Wiley et al., 2016). It will be interesting to investigate whether the increase in NAD+ level enhances the anti-aging effect, which needs further investigation.

Mitophagy

Mmitochondrial autophagy, or mitophagy, occurs in eukaryotic cells to selectively degrade damaged or dysfunctional mitochondria (Bratic & Larsson, 2013). This process is vital for maintaining cellular homeostasis, as it helps to eliminate damaged or unnecessary organelles and can be activated in response to cellular stressors. Mitophagy is distinct from macroautophagy, a general non-selective autophagy process. Selective degradation of mitochondria by mitophagy has been observed in various eukaryotes, including yeast, plants, and mammals(McWilliams et al., 2019). Mitophagy occurs when cells recognize damaged or nonfunctional mitochondria and initiate a process of autophagic degradation. This process begins with the recruitment of autophagy-related proteins to the outer membrane of the mitochondrion. This recruitment is mediated by proteins such as Parkin, PINK1, and Atg32, which are known to detect and respond to mitochondrial stress(Saito & Sadoshima, 2015).

Once recruited, these proteins can recruit and activate other autophagic proteins, such as Atg9 and LC3, to form a double membrane structure known as the phagophore(Kim & Lemasters, 2011). This structure is important for the isolation and degradation of the mitochondria. Once the phagophore is formed, the autophagic machinery engulfs and separates the mitochondrion from the rest of the cell. At this point, the outer membrane of the mitochondrion is degraded while the inner membrane remains intact, allowing the contents of the mitochondrion, including the mitochondrial DNA, to be degraded by the lysosomal machinery and removed from the cell. Degradation of mitochondria by mitophagy is essential for various cellular processes, including energy metabolism, cellular stress response, and apoptosis. Mitochondrial autophagy helps to eliminate damaged or excess mitochondria, preventing the accumulation of reactive oxygen species and other harmful by-products (Bakula & Scheibye-Knudsen, 2020; Moreira et al., 2017).

Additionally, mitophagy has been implicated in the regulation of apoptosis as it can help to remove damaged mitochondria and prevent excessive cell death. Lastly, mitophagy can help maintain cellular energy levels, as it helps to recycle the components of mitochondria for energy production. Overall, mitochondrial autophagy is a critical process for maintaining cellular homeostasis. It helps to eliminate damaged or unnecessary mitochondria and can be activated in response to cellular stress. Additionally, it is crucial for regulating energy metabolism, apoptosis, and other cellular processes(Bakula & Scheibye-Knudsen, 2020). As such, it is essential for the proper functioning of eukaryotic cells (Figure 1).

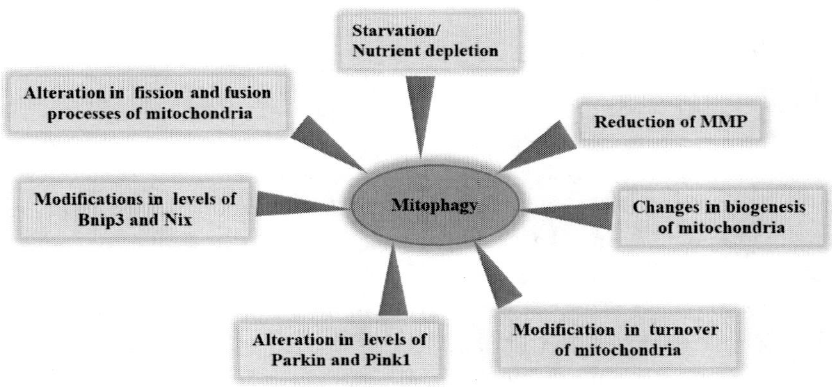

Figure 1. Various molecular mechanisms linked with mitophagy induction include nutrient depletion, impaired mitochondria, modifications in levels of proapoptotic proteins, alteration in mitochondrial membrane potential (MMP), biogenesis of mitochondria, and turnover of mitochondria.

Mechanism of Mitophagy

Mitophagy is a highly conserved process in eukaryotes that clears damaged mitochondria from cells. It is a selective process that is initiated by the recognition of damaged mitochondria by autophagy receptors. These receptors, such as p62 (SQSTM1) and NDP52, recognize ubiquitin-tagged mitochondrial proteins. This recognition triggers the formation of a phagophore, a specialized autophagosome responsible for engulfing the damaged mitochondria (Kataura et al., 2023). Once the phagophore has engulfed the targeted mitochondria, it matures into an autophagosome. The autophagosome then fuses with lysosomes to form an autolysosome, where lysosomal hydrolases degrade the enclosed materials. Autophagy-related proteins such as Atg5, Atg7, and Atg12 facilitate this degradative process. These proteins are responsible for the formation of the autophagosome and the subsequent fusion of the autophagosome with lysosomes (Kataura et al., 2023; Nakamura et al., 2020; Zachari & Ktistakis, 2020).

In addition to the autophagy-related proteins, other factors such as parkin, mitochondrial fission proteins, and mitochondrial dynamics proteins are involved in the process of mitophagy. Parkin is a ubiquitin ligase responsible for tagging damaged mitochondria for degradation by the autophagosome. Mitochondrial fission proteins are responsible for the fragmentation of the mitochondria, which is essential for efficiently clearing damaged mitochondria. Lastly, mitochondrial dynamics proteins like "dynamin-related protein 1 (DRP1) and mitochondrial fission factor (MFF)" are responsible for regulating mitochondrial membrane potential and the fragmentation of mitochondria, which are both essential for the efficient clearance of damaged mitochondria (Gegg et al., 2010; Kageyama et al., 2014) (Figure 1). Once the damaged mitochondria have been degraded, the released components are recycled for the biosynthesis of new mitochondria. This recycling process involves several enzymes, including LON proteases, responsible for the degradation of proteins found in the mitochondrial matrix, and AMP-activated protein kinase, which synthesizes new mitochondria. In summary, mitophagy is essential for maintaining cellular homeostasis in eukaryotes. It is initiated by the recognition of damaged mitochondria by autophagy receptors.

Regulators of Mitophagy

Mitophagy is a process by which defective or damaged mitochondria are selectively removed from the cell. It is essential for maintaining mitochondrial homeostasis and is regulated by various cellular components. In eukaryotes, the B-cell lymphoma-2 (BCL-2) family of proteins primarily controls mitophagy. Bcl-2 proteins are involved in the regulation of apoptosis, autophagy, and mitophagy. They are composed of four subfamilies, namely Bcl-2,-B-cell lymphoma-extra large (Bcl-xL), Bcl-w, and Mcl-1 (Cen et al., 2020; Hollville et al., 2014). The Bcl-2 proteins act as molecular switches, allowing cells to decide whether to undergo autophagy or apoptosis. The Bcl-2 proteins are responsible for initiating the process of mitophagy by tagging mitochondria with the LC3 protein. Once tagged, the mitochondria are identified as damaged or defective and engulfed by autophagosomes. This process is further aided by other autophagy factors such as Beclin-1, Atg14, and Atg5. Beclin-1 is a protein responsible for the initiation of autophagy and is found to interact with Bcl-2 and Bcl-XL proteins. Atg14 and Atg5, on the other hand, are proteins that form part of the autophagosome and are critical for the maturation and fusion of the autophagosome. In addition to the Bcl-2 family, other proteins are known to regulate mitophagy in eukaryotes. These proteins include the p62, Parkin, and Pink1 proteins. The p62 protein functions as an adaptor protein and is known to interact with the LC3 protein and Bcl-2 family proteins. It is also essential for recognizing damaged mitochondria and forming autophagosomes (Bakula & Scheibye-Knudsen, 2020; Zachari & Ktistakis, 2020).

On the other hand, the Parkin and Pink1 proteins are E3 ubiquitin ligases involved in the recognition and degradation of damaged mitochondria. Parkin interacts with Pink1 to initiate the ubiquitination of damaged mitochondria and promote their subsequent destruction. Other proteins involved in the regulation of mitophagy include the LC3-II protein, phosphatidylinositol 3-kinase (PI3K), and protein kinase B (PKB). LC3-II is an isoform of the LC3 protein essential for autophagosome formation. It interacts with the Bcl-2 family proteins and the p62 protein. PI3K also interacts with Bcl-2 family proteins and p62 protein and is essential for the formation of autophagosomes. PKB, conversely, regulates the activities of PI3K and Bcl-2 family proteins (Gureev et al., 2020).

In summary, mitophagy in eukaryotes is regulated by various proteins, including the Bcl-2 family, p62, Parkin, Pink1, LC3-II, PI3K, and PKB. These proteins are essential for the recognition of damaged mitochondria, the formation of autophagosomes, and the subsequent degradation of the mitochondria. The coordinated action of these proteins ensures the maintenance of mitochondrial homeostasis and the efficient removal of damaged mitochondria.

The Physiological and Pathophysiological Role of Mitophagy

Aaberration in mitophagy causes significant ramifications in various diseases such as developmental disorders, neurodegenerative diseases, aging, cardiovascular disorders, and liver diseases. Studies have found that mitochondrial number shift also plays a critical role in these diseases. However, deeper investigation will be needed to elucidate the mechanisms fully. Here we are listing some of the majorly studied mitophagy in different diseases.

Mitophagy in Cancer

The advancement and progression of mitophagy in cancer remain to be explored. The development and spread of cancer are significantly influenced by changes in the functional complexity of mitochondria. To acclimatize to the tumor environment, cancer cells show various types of effects such as aerobic glycolysis, abnormal mitochondrial quality control, Warburg effect, ROS production, modification of cellular redox state, and creation of apoptotic signals. It has been shown that mitophagy contributes to carcinogenesis by either inducing or inhibiting it and that the failure of mitophagy favors carcinogenic stressors and tumors. However, more research needs to be done on the therapeutic potential of focusing on mitophagy. Furthermore, it has been discovered that several proteins, including Parkin, p62/SQSTM1, BNIP3L/NIX, and FUNDC1, are involved in mitophagy and are deactivated in cancers (Panigarhi et al., 2020).

The Parkin gene, a putative tumor suppressor found at chromosome 6 has been linked to juvenile Parkinson's disease. It has been noted to have mutations in numerous cancers, including colorectal, lung, liver, glioma (Maugeri et al.,2015), and breast. On the other hand, in response to a hypoxic situation, BNIP3 and BNIP3L/NIX cause mitophagy-mediated inhibition that plays a crucial role in the progression of various cancers. In addition to being linked to chemotherapy resistance, the reduction of BNIP3 expression in pancreatic cancer has also been linked to poor prognosis.BNIP3 slows the growth of the primary mammary malignancy by inhibiting the accumulation of damaged mitochondria. Another hypoxia-induced mitophagy receptor, FUNDC1, is crucial for controlling the development and spread of malignancies. In a study, it was found that FUNDC1 decreased inflammasome activation and, with the aid of mitophagy, prevented the improvement of hepatocellular carcinoma in mice. Through the mitophagy-mediated selective turnover of mitochondria, autophagy adaptors have been linked to cancer. According to reports, p62/SQSTM1 is over-expressed in renal cell carcinoma. It builds up in pancreatic and hepatic cell pre-carcinogenic circumstances. By promoting mitophagy in the renal tubular epithelial cells, OPTN, another cargo adaptor associated with autophagy, also prevents the activation of the NLRP3 inflammasome. According to reports, OPTN is down-regulated in lung and urothelial cancer.

Development and Ageing

Mitophagy is a vital process in eukaryotes that contributes to development and is involved in aging. It is a form of selective autophagy that explicitly targets mitochondria for degradation and recycling. Mitochondria are multifunctional organelles in eukaryotic cells essential for energy production, metabolism, and signaling. As such, they play a vital role in the development and aging of eukaryotes. Mitophagy ensures that mitochondria are efficiently removed when they become damaged or dysfunctional, thereby preventing them from accumulating and causing damage to the cell. In development, mitophagy maintains proper mitochondrial distribution and number in the cell (Fivenson et al., 2017). To ensure that newly formed cells are supplied with the necessary number and type of mitochondria, cells must be able to remove old and damaged mitochondria efficiently.

Additionally, mitophagy is involved in the differentiation of cells. Mitochondria play an essential role in cell differentiation, providing the energy necessary for the cell to specialize into its distinct form. By ensuring that mitochondria are of the appropriate type and number, mitophagy facilitates cell differentiation. In aging, mitophagy is essential for maintaining mitochondrial health and preventing the accumulation of damaged or dysfunctional mitochondria. As cells age, their mitochondria become increasingly damaged leading to a decrease in their function. Suppose these damaged mitochondria are allowed to accumulate; in that case, they can become a significant source of oxidative stress and damage to the cell. Mitophagy selectively removes damaged mitochondria from the cell to maintain mitochondrial health, preserving the cell's metabolic activity and preventing the accumulation of oxidative stress (Liu et al. 2020; Moreira et al., 2017).

In conclusion, mitophagy is essential for the development and aging of eukaryotes. In development, it ensures the proper distribution and number of mitochondria in newly formed cells and lipid cell differentiation. In aging, it helps to maintain mitochondrial health and prevents the accumulation of damaged mitochondria that can contribute to oxidative stress and damage to the cell. It is an essential process for proper eukaryotic development and aging.

Mitophagy in Neurodegenerative Diseases

Mitophagy is a process of autophagy that specifically targets mitochondria for degradation. It is a highly regulated and essential cellular process that works as an efficient way to monitor and maintain mitochondrial health. Mitophagy is vital in regulating mitochondrial homeostasis and critical in preventing neurodegenerative disorders. It is believed that defects in the mitophagy process may contribute to the development of various neurodegenerative disorders. "Alzheimer's disease (AD), Parkinson's disease (PD), and Huntington's disease (HD)" are among the most common neurodegenerative disorders. The pathogenesis of these diseases is complex and not fully understood. Still, recent studies have suggested that an impaired mitophagy process may be involved in developing these diseases (Fivenson et al., 2017; Moreira et al., 2017). Mitochondrial dysfunction has been linked to the pathogenesis of these diseases and other neurodegenerative disorders, and it has been suggested that impaired mitophagy may be responsible for this dysfunction. Perturbations in the mitophagy process have been identified as a possible underlying cause of many neurodegenerative disorders. Mitochondrial fragmentation decreased mitochondrial membrane potential, and mitochondrial leakages are some indicators of mitochondrial dysfunction linked to impaired mitophagy.

In addition, an increase in damaged mitochondria has been observed in patients with neurodegenerative disorders, suggesting that an impaired mitophagy process may be responsible for accumulating these damaged organelles. The role of mitophagy in the pathogenesis of neurodegenerative diseases is still being explored (Cai & Jeong, 2020; De Gaetano et al., 2021). Recent studies have suggested that increased levels of mitophagy may be beneficial in preventing these diseases. For example, in animal models of PD, increased levels of mitophagy have been observed to reduce the levels of reactive oxygen species (ROS) and improve neuronal survival. In addition, increased levels of mitophagy have been found to improve mitochondrial function, reduce mitochondrial fragmentation, and protect against neuronal cell death in animal models of AD (Lin et al., 2021; Youle & Narendra, 2010).

In conclusion, mitophagy is critical in maintaining mitochondrial homeostasis and preventing neurodegenerative disorders. Impaired mitophagy has been linked to the pathogenesis of several neurodegenerative diseases, suggesting that increased levels of mitophagy may be beneficial in preventing these diseases. Further research is needed to fully understand the role of mitophagy in the pathogenesis of neurodegenerative disorders and to develop effective therapeutic strategies to prevent or treat these diseases.

Mitophagy in Diabetes/Obesity

Mitophagy is a process that involves selective removal of dysfunctional mitochondria from the cell. It is a critical process for maintaining cellular homeostasis. It involves various physiological conditions like aging, cancer, and metabolic diseases. Diabetes is one of the most common metabolic diseases and represents a significant public health problem worldwide. Mitochondrial dysfunction is associated with diabetes, and evidence suggests impaired mitophagy may play a role in developing diabetic complications. Mitochondrial dysfunction is believed to be one of the significant pathophysiological factors in the development of diabetes. Several studies have demonstrated that mitochondrial dysfunction is associated with increased oxidative stress and insulin resistance (Zhang et al., 2018). Oxidative stress has been implicated in the development of diabetes by contributing to the formation of advanced glycation end products and disrupting the insulin signaling pathway. In addition, mitochondrial dysfunction can lead to insulin resistance by impairing glucose uptake and metabolism. Mitochondrial dysfunction is also believed to be associated with increased inflammation, which can further contribute to insulin resistance and diabetes. In diabetes, impaired mitophagy has been linked to increased oxidative stress and inflammation. Oxidative stress is believed to be a significant factor in developing diabetic complications. It is thought that impaired mitophagy leads to the accumulation of damaged mitochondria, leading to increased oxidative stress due to the release of reactive oxygen species. Increased oxidative stress causes further damage to the mitochondria, resulting in a vicious cycle that can eventually lead to cell death (Tong et al., 2021; Zhang et al., 2018). The increased inflammation associated with diabetes is also believed to be linked to impaired mitophagy, as it can accumulate damaged mitochondria and increase oxidative stress. The effects of impaired mitophagy on diabetic complications are not entirely understood. However, impaired mitophagy is thought to lead to an accumulation of dysfunctional mitochondria, which can contribute to diabetic complications such as cardiovascular disease, retinopathy, and nephropathy.

Mitochondrial dysfunction is also believed to be involved in the development of hepatic steatosis, a major cause of morbidity and mortality in diabetes. In addition, mitochondrial dysfunction and impaired mitophagy have been linked to the development of metabolic syndrome, a cluster of metabolic abnormalities associated with an increased risk of cardiovascular disease in diabetes. Mitophagy is essential for maintaining cellular homeostasis and therefore plays a vital role in the development of diabetes and its complications. Understanding the molecular pathways that regulate mitophagy in diabetes is critical for developing effective therapies to prevent and treat diabetic complications. Several drugs have been developed that can modulate mitophagy and reduce oxidative stress in diabetes, and further research is needed to determine their efficacy in treating diabetes and its complications.

Mitophagy in Liver Diseases

Mitophagy, a type of autophagy, is a process by which dysfunctional mitochondria are selectively degraded and recycled to maintain cellular homeostasis. Mitophagy is an essential process that occurs under both physiological and pathological conditions. Its importance is highlighted by the observation that defects in mitophagy are associated with various pathologies, including liver diseases. The liver is an organ that plays a significant role in gluconeogenesis, lipid metabolism, and detoxification (Ke, 2020). As a result, it is vulnerable to various metabolic and oxidative stresses, and mitochondria are the primary source of energy for the cells of the liver. When these cells are exposed to disease-causing agents, the mitochondria may become dysfunctional, leading to oxidative stress, apoptosis, and necrosis. Mitophagy is a cellular mechanism for removing damaged mitochondria and other cellular components, thereby protecting the liver from further damage. Mitophagy is critical in maintaining cellular homeostasis and recovering from liver disease injury. In particular, it plays a vital role in resolving the inflammatory response, significantly contributing to the pathogenesis of liver disease. In chronic liver disease, mitophagy has been shown to reduce the accumulation of dysfunctional mitochondria, reduce oxidative stress, and restore mitochondrial function.

In some cases, mitophagy may also help prevent disease progression by promoting the clearance of toxins and preventing the accumulation of toxic metabolites. In addition to its protective functions, mitophagy has also been implicated in the pathogenesis of some liver diseases. For example, it has been suggested that impaired mitophagy contributes to the development of nonalcoholic fatty liver disease (NAFLD), a condition characterized by the accumulation of lipids in the liver. In this condition, dysfunctional mitochondria cannot properly metabolize lipids, accumulating toxic metabolites, which can further exacerbate the disease process (Aryapour& Kietzmann, 2022; Williams et al., 2015). Mitophagy is also implicated in the pathogenesis of hepatotoxicity, an adverse reaction to drugs or other toxic agents. In this setting, mitophagy is believed to play a role in the clearance of the toxic agent and preventing further damage.

Additionally, it has been suggested that mitophagy may be involved in the pathogenesis of hepatic fibrosis, a condition characterized by the excessive accumulation of extracellular matrix proteins in the liver. In summary, mitophagy plays a critical role in the maintenance of cellular homeostasis and recovery from injury in the setting of liver diseases. It has been implicated in the pathogenesis of various liver diseases, including NAFLD, hepatotoxicity, and hepatic fibrosis. Further research is needed to understand mitophagy's role in liver disease and to develop therapeutic strategies to target it (Figure 2).

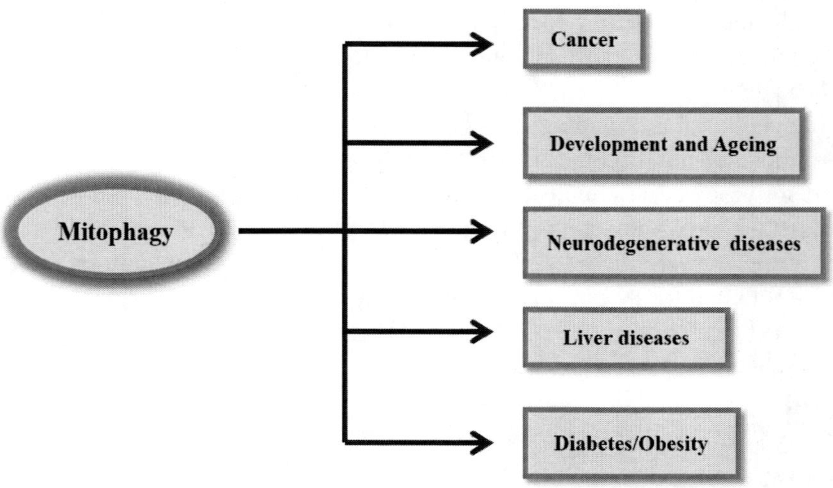

Figure 2. The role of mitophagy in physiological (development and aging) and pathophysiological conditions like cancer, neurodegenerative diseases, liver diseases, and diabetes/obesity.

Mitophagy Targeting Drugs and Their Role in Therapeutics

Mitophagy, which translates to "self-eating," is a form of "autophagy" whereby damaged or dysfunctional mitochondria are selectively degraded and recycled by lysosomes as part of the normal "mitochondrial quality control" mechanisms. This process is essential for the maintenance of normal mitochondrial function and homeostasis, and its dysregulation has been implicated in a wide range of neurodegenerative diseases, such as "Parkinson's disease, Alzheimer's disease, and Huntington's disease." As such, developing drugs targeting mitophagy has become an increasingly attractive therapeutic strategy for treating these diseases. One of the most promising approaches is to target the proteins involved in the regulation of mitophagy. These include proteins of the PINK1/Parkin pathway, a ubiquitin-dependent pathway primarily responsible for recognizing and degrading dysfunctional mitochondria.

The PINK1/Parkin pathway is activated in response to mitochondrial damage. It is thought to be the most crucial pathway for regulating mitophagy (Gegg et al., 2010). Drugs targeting the proteins of this pathway have been developed, and these drugs are effective in inducing mitophagy and protecting against mitochondrial dysfunction in preclinical models. In addition to targeting the PINK1/Parkin pathway, drugs have also been developed to target other proteins involved in the regulation of mitophagy. For example, the protein Beclin-1 is a crucial regulator of autophagy and is required for mitophagy induction. Drugs targeting Beclin-1 and its upstream regulators, such as the Bcl-2 family and the mammalian target of the rapamycin (mTOR) pathway, have been developed and are currently being explored as potential mitophagy-targeting drugs. In addition to targeting proteins involved in the regulation of mitophagy, drugs targeting the mitochondrial fission and fusion proteins have also been developed (Hollville et al., 2014). Mitochondrial fission and fusion are essential processes for maintaining normal mitochondrial morphology and function, and their dysregulation has been implicated in a wide range of diseases. Drugs targeting mitochondrial fission and fusion proteins have been developed and shown to effectively induce mitophagy and restore mitochondrial morphology and function in preclinical models. Finally, drugs targeting the proteins involved in the transport and release of mitochondrial components, such as mitofusins, have also been developed. These drugs have effectively induced mitochondrial degradation and restored mitochondrial morphology and function in preclinical models. Developing drugs targeting mitophagy has become an increasingly attractive therapeutic strategy for treating neurodegenerative diseases. Such drugs can potentially restore mitochondrial function and homeostasis, thus providing a promising approach to treating these debilitating diseases. Therefore, further research is needed to identify and develop mitophagy-targeting drugs and explore their potential for treating neurodegenerative diseases.

Conclusion

Recent discoveries in mitochondrial biology showed that mitophagy is a central process in regulating organismal health and disease. Here we have listed the extensive role of mitophagy in different diseases. Mitophagy induction can be helpful in some diseases while opposite in others; thus, a thorough investigation must be done to understand and target mitophagy. Mitophagy interacts with and influences ROS production and aging; therefore, it is crucial in mitochondrial quality and quality control. This delicate balance of mitophagy must be studied to understand, target, and harness mitophagy.

References

Aryapour, E., & Kietzmann, T. (2022). Mitochondria, mitophagy, and the role of deubiquitinases as novel therapeutic targets in liver pathology. *Journal of Cellular Biochemistry*, *123*(10), 1634–1646.

Bakula, D., & Scheibye-Knudsen, M. (2020). MitophAging: Mitophagy in Aging and Disease. *Frontiers in Cell and Developmental Biology*, *8*, 239.

Bratic, A., & Larsson, N. G. (2013). The role of mitochondria in aging. *The Journal of Clinical Investigation*, *123*(3), 951–957.

Cai, Q., & Jeong, Y. Y. (2020). Mitophagy in Alzheimer's Disease and Other Age-Related Neurodegenerative Diseases. *Cells*, *9*(1).

Cen, X., Chen, Y., Xu, X., Wu, R., He, F., Zhao, Q., Sun, Q., Yi, C., Wu, J., Najafov, A., & Xia, H. (2020). Pharmacological targeting of MCL-1 promotes mitophagy and improves disease pathologies in an Alzheimer's disease mouse model. *Nature Communications 2020 11:1*, *11*(1), 1–13.

De Gaetano, A., Gibellini, L., Zanini, G., Nasi, M., Cossarizza, A., & Pinti, M. (2021). Mitophagy and Oxidative Stress: The Role of Aging. *Antioxidants*, *10*(5).

Fivenson, E. M., Lautrup, S., Sun, N., Scheibye-Knudsen, M., Stevnsner, T., Nilsen, H., Bohr, V. A., & Fang, E. F. (2017). Mitophagy in neurodegeneration and aging. *Neurochemistry International*, *109*, 202–209.

Gegg, M. E., Cooper, J. M., Chau, K. Y., Rojo, M., Schapira, A. H. V., & Taanman, J. W. (2010). Mitofusin 1 and mitofusin 2 are ubiquitinated in a PINK1/parkin-dependent manner upon induction of mitophagy. *Human Molecular Genetics*, *19*(24), 4861.

Gureev, A. P., Sadovnikova, I. S., Starkova, N. N., Starkov, A. A., & Popov, V. N. (2020). p62-Nrf2-p62 Mitophagy Regulatory Loop as a Target for Preventive Therapy of Neurodegenerative Diseases. *Brain Sciences*, *10*(11), 1–14.

Hollville, E., Carroll, R. G., Cullen, S. P., & Martin, S. J. (2014). Bcl-2 family proteins participate in mitochondrial quality control by regulating parkin/PINK1-dependent mitophagy. *Molecular Cell*, *55*(3), 451–466.

Kageyama, Y., Hoshijima, M., Seo, K., Bedja, D., Sysa-Shah, P., Andrabi, S. A., Chen, W., Höke, A., Dawson, V. L., Dawson, T. M., Gabrielson, K., Kass, D. A., Iijima, M., &Sesaki, H. (2014). Parkin-independent mitophagy requires Drp1 and maintains the integrity of the mammalian heart and brain. *The EMBO Journal*, *33*(23), 2798–2813.

Kataura, T., Otten, E. G., Rabanal-Ruiz, Y., Adriaenssens, E., Urselli, F., Scialo, F., Fan, L., Smith, G. R., Dawson, W. M., Chen, X., Yue, W. W., Bronowska, A. K., Carroll, B., Martens, S., Lazarou, M., & Korolchuk, V. I. (2023). NDP52 acts as a redox sensor in PINK1/Parkin-mediated mitophagy. *The EMBO Journal*, *42*(5), e111372.

Ke, P. Y. (2020). Mitophagy in the Pathogenesis of Liver Diseases. *Cells 2020, Vol. 9, Page 831*, *9*(4), 831.

Kim, I., & Lemasters, J. J. (2011). Mitochondrial degradation by autophagy (mitophagy) in GFP-LC3 transgenic hepatocytes during nutrient deprivation. *American Journal of Physiology - Cell Physiology*, *300*(2), C308.

Kornmann, B. (2014). Quality control in mitochondria: use it, break it, fix it, trash it. *F1000Prime Reports*, *6*.

Lin, Q., Chen, J., Gu, L., Dan, X., Zhang, C., & Yang, Y. (2021). New insights into mitophagy and stem cells. *Stem Cell Research and Therapy*, *12*(1), 1–14.

Liu, L., Liao, X., Wu, H., Li, Y., Zhu, Y., & Chen, Q. (2020). Mitophagy and its contribution to metabolic and aging-associated disorders. *Antioxidants and Redox Signaling*, *32*(12), 906–927.

Maugeri, G., D'Amico, A. G., Magro, G., Salvatorelli, L., Barbagallo, G. M., Saccone, S., Drago, F., Cavallaro, S., & D'Agata, V. (2015). Expression profile of parkin isoforms in human gliomas. International journal of oncology, 47(4), 1282–1292.

McWilliams, T. G., Prescott, A. R., Villarejo-Zori, B., Ball, G., Boya, P., & Ganley, I. G. (2019). A comparative map of macroautophagy and mitophagy in the vertebrate eye. *Autophagy*, *15*(7), 1296.

Moreira, O. C., Estébanez, B., Martínez-Florez, S., De Paz, J. A., Cuevas, M. J., & González-Gallego, J. (2017). Mitochondrial Function and Mitophagy in the Elderly: Effects of Exercise. *Oxidative Medicine and Cellular Longevity*, *2017*.

Nakamura, M., Verboon, J. M., Prentiss, C. L., & Parkhurst, S. M. (2020). The kinesin-like protein Pavarotti functions noncanonically to regulate actin dynamics. *Journal of Cell Biology*, *219*(9).

Panigrahi, D. P., Praharaj, P. P., Bhol, C. S., Mahapatra, K. K., Patra, S., Behera, B. P., Mishra, S. R., & Bhutia, S. K. (2020). The emerging, multifaceted role of mitophagy in cancer and cancer therapeutics. Seminars in cancer biology, 66, 45–58.

Sahin, E., Colla, S., Liesa, M., Moslehi, J., Müller, F. L., Guo, M., Cooper, M., Kotton, D., Fabian, A. J., Walkey, C., Maser, R. S., Tonon, G., Foerster, F., Xiong, R., Wang, Y. A., Shukla, S. A., Jaskelioff, M., Martin, E. S., Heffernan, T. P.,DePinho, R. A. (2011). Telomere dysfunction induces metabolic and mitochondrial compromise. *Nature 2011 470:7334*, *470*(7334), 359–365.

Saito, T., & Sadoshima, J. (2015). Molecular Mechanisms of Mitochondrial Autophagy/Mitophagy in the Heart. *Circulation Research*, *116*(8), 1477–1490.

Sen, N., Satija, Y. K., & Das, S. (2011). PGC-1α, a Key Modulator of p53, Promotes Cell Survival upon Metabolic Stress. *Molecular Cell*, *44*(4), 621–634. 4.

Tong, M., Saito, T., Zhai, P., Oka, S. I., Mizushima, W., Nakamura, M., Ikeda, S., Shirakabe, A., & Sadoshima, J. (2021). Alternative Mitophagy Protects the Heart Against Obesity-Associated Cardiomyopathy. *Circulation Research*, *129*(12), 1105–1121.

Wiley, C. D., Velarde, M. C., Lecot, P., Liu, S., Sarnoski, E. A., Freund, A., Shirakawa, K., Lim, H. W., Davis, S. S., Ramanathan, A., Gerencser, A. A., Verdin, E., & Campisi, J. (2016). Mitochondrial dysfunction induces senescence with a distinct secretory phenotype. *Cell Metabolism*, *23*(2), 303–314.

Williams, J. A., Ni, H. M., Ding, Y., & Ding, W. X. (2015). Parkin regulates mitophagy and mitochondrial function to protect against alcohol-induced liver injury and steatosis in mice. *American Journal of Physiology - Gastrointestinal and Liver Physiology*, *309*(5), G324–G340.

Youle, R. J., & Narendra, D. P. (2010). Mechanisms of mitophagy. *Nature Reviews Molecular Cell Biology 2011 12:1*, *12*(1), 9–14. Zachari, M., & Ktistakis, N. T. (2020). Mammalian Mitophagosome Formation: A Focus on the Early Signals and Steps. *Frontiers in Cell and Developmental Biology*, *8*, 171.

Zhang, Y., Sowers, J. R., & Ren, J. (2018). Targeting autophagy in obesity: from pathophysiology to management. *Nature Reviews Endocrinology 2018 14:6*, *14*(6), 356–376.

Chapter 2

Aging Cardiomyopathy: The Role of Mitophagy

Dinesh D. Anandani[1]
Pranali Chandurkar[1]
Nitu L. Wankhede[1]
Shubhada V. Mangrulkar[1]
Manjit B. Chauhan[1]
Brijesh G. Taksande[1]
Aman B. Upaganlawar[2]
Milind J. Umekar[1]
and Mayur B. Kale[*]
[1]Smt. KishoritaiBhoyar College of Pharmacy, Kamptee, Nagpur, Maharashtra, India
[2]SNJB's ShrimanSureshdada Jain College of Pharmacy, Neminagar, Chandwad, Nashik, Maharashtra, India

Abstract

The essential organelles known as mitochondria, manage the balance between the cellular energy production and cell death. As a result, maintaining proper cellular functions depends on removing damaged mitochondria by autophagy, also known as mitophagy. In reality, it has recently been demonstrated that mitophagy is critical for various activities, including tissue damage brought on by ischemia, drugs, or other conditions. Inducing general autophagy and processing damaged mitochondria for selective autophagic detection are two processes necessary for removing damaged mitochondria via autophagy. Recent developments in mitophagy research have revealed that mitochondrial activation is regulated by either the Pink1-Parkin signaling route, the Drp1-mediated pathway, the Ubiquitin-independent pathway, or the mitophagy receptors BNIP3, NIX/BNIP3L, FKBP8, FUNDC1, and Bcl2L13. This chapter reviews what we now know about the mechanisms underlying mitophagy and how they relate to aging cardiomyopathy.

Keywords: mitophagy, cardiomyopathy, autophagy, aging, mitochondria

[*] Corresponding Author's Email: mayur.kale28@gmail.com.

In: Mitophagy and Its Role in Health and Disease
Editors: Shalini Mani and Samiksha Wasnik
ISBN: 979-8-89113-470-6

Introduction

Organ failure and loss of tissue homeostasis are inevitable consequences of aging over time (Biala et al., 2015). The relationship of the aging process with dysfunctional organelles, which results in oxidative stress, protein misfolding, and cell death, as well as a sharp decline in cellular quality control mechanisms, has been revealed through research on the biology of aging (Buchberger et al., 2010; Denzel et al., 2014; Finkel et al., 2007). The most numerous organelles in cardiomyocytes are mitochondria, which are essential for normal cell activity.

To maintain mitochondrial homeostasis, cells engage in selective autophagy, known as mitophagy, in which damaged or dysfunctional mitochondria are destroyed (Nah et al., 2016; Shirakabe et al., 2016). Mitophagy was initially postulated by Lemasters et al. in 2005. Since then, mitophagy has been linked to several illnesses (Lemasters et al., 2005). The heart heavily relies on mitochondrial metabolism to meet its vast energy requirements (Taegtmeyer et al., 1994). Mitochondrial function and integrity must always be maintained for appropriate cell physiology and survival, especially for cells that demand a lot of energy, such as cardiomyocytes. In cardiomyocytes, mitochondria comprise about 30% of the total cell volume and generate a tonne of ATP through oxidative phosphorylation to keep the cells' contractile activity active (Vasquez-Trincado et al., 2016). According to its metabolic state, a mitochondrion's shape can vary significantly (Eisner et al., 2018). These morphological variations also require specialized equipment. For instance, neighboring mitochondria can fuse to extend the mitochondria. The outer mitochondrial membrane (OMM) is fused by mitofusins (MFN) 1 and 2 (Santel and Fuller, 2001; Chen et al., 2003; Eura et al., 2003; Wankhede et al., 2021), and the inner mitochondrial membrane (IMM) is fused by optic atrophy protein 1 (OPA1) (Santel and Fuller, 2001; Chen et al., 2003; Olichon et al., 2003 Eura et al., 2003; Meeusen et al., 2006; Umare et al., 2021). For mitochondrial fragmentation to take place, dynamin-related protein 1 (DRP1), which is encoded by the mouse and human genes DNM1L and DNM1L, respectively, must be active (Pitts et al., 1999; Smirnova et al., 2001). In mitochondria, DRP1 translocates and creates spiral-shaped structures that squeeze the organelle until it separates (Smirnova et al., 2001; Upaganlawar et al., 2022; Ingerman et al., 2005). Inhibiting mitochondrial fission with a dominant-negative version of DRP1 prevents pancreatic cells from undergoing mitophagy, which occurs in mammalian cells in conjunction with mitochondrial fission (Twig and Shirihai, 2011). Moreover, mouse embryonic fibroblasts with defective fusion due to Opa1 or Mfn2 deletion exhibit increased mitophagy in response to fasting (Rambold et al., 2011). In tissues that rely predominantly on mitochondrial metabolism, such as the heart, it is also simpler to understand how defective mitophagy may result in accumulated oxidative damage to the mitochondrial circuitry. Mitophagy is a process that most cell types in the body, including those in the heart, must go through, which increases the likelihood that this mechanism will be used to mitigate the effects of cardiac aging (Shirakabe et al., 2016a).

Several routes allow for the regulation of circulatory activity by mitophagy. Changes in autophagy and mitophagy are considered crucial for the aging process throughout time (Liang et al., 2020). This chapter discusses the current understanding of autophagy and mitophagy in aging, along with age-related changes in these processes that contribute to heart aging and may cause cardiomyopathy and other aging-related cardiovascular illnesses.

Mitochondrial Connectivites

Cycles of fusion and fission within the mitochondrial network alter mitochondrial transport, size, shape, and turnover (Chan et al., 2012). These actions collectively are known as mitochondrial dynamics, and they directly affect cell viability, proliferation, and ATP generation, especially in organs with high metabolic demand such as the heart (de Carvalho et et al., 2017; Chenet al., 2010; Ferreira et al., 2019). Furthermore, metabolites, solutes, proteins, DNA, and membranes are constantly exchanged between fission and fusion in the mitochondria (Schrepfer et al., 2016). Because of this, disruption of mitochondrial fission/fusion balance often causes morphological and functional abnormalities inside the mitochondria, which are strongly linked to several chronic ailments, particularly cardiac diseases (Campos et al., 2016, Song et al., 2017).

Mitochondrial Fusion

Three GTPases, including the ocular atrophy factor 1 (Opa1) on the mitochondria's inner membrane and the mitofusins 1 and 2 (Mfn1 and Mfn2, respectively), are responsible for controlling the fusion of two mitochondria (Song et al., 2009).

Mitofusins-Mediated Mitochondrial Fusion

Mitofusins feature a GTPase domain that hydrolyzes GTP for oligomerization and fusing of outer mitochondrial membranes to neighboring mitochondria (Koshiba et al., 2004). Mitofusins need membrane curvature for binding, and phospholipase D-dependent cardiolipin hydrolysis increases this curvature (Choi et al., 2006; DeVay et al., 2009). A comparable mechanism is also required for inner membrane fusion, which is initiated by Opa1 and coordinated by several other proteins, including prohibitins (Merkwirth et al., 2008; Song et al., 2009). The inner mitochondrial membrane must interact with cardiolipin to fuse, resulting in interconnected and elongated mitochondria (Ban et al., 2017). Mitofusins are essential in regulating numerous cellular processes, such as protein turnover, apoptosis, calcium homeostasis, biogenesis, and mitochondrial abundance and structure (Bach et al., 2003; Chen et al., 2005). (Leboucher et al., 2012). In cardiac physiology, mitofusins are required for the heart's normal and pathological morphology and function. Eliminating mitofusins promotes mitochondrial fragmentation in cultured myocytes (Chen et al., 2010) and the whole heart (Song et al., 2015), resulting in mouse embryonic mortality. Conditional deletion of MFN1 under basal conditions results in cardiac mitochondrial fragmentation without affecting mitochondrial bioenergetics or heart function (Papanicolaou et al., 2012). The distinct cardiac abnormalities of MFN1 and MFN2 knockout animals could be attributed to their dissimilar biochemical properties and functions in mitochondrial dynamics. For instance, mitochondria that contain MFN1 are successfully connected in a GTP-dependent manner (but not mitochondria that carry MFN2) (Ishihara et al., 2004). Moreover, Mfn1 hydrolyzes molecules more quickly and with less affinity for GTP than Mfn2 (Ishihara et al., 2004). Mice with left ventricular failure after MFN1/MFN2 in the heart are knocked out, which severely disrupts the sarcomere structure of the cardiomyocytes (Hall et al., 2016; Song et al., 2015).

Optic Atrophy Factor 1 (Opa1) Mediated Mitochondrial Fusion

Opa1 influences the organization of the cristae, the number of mitochondria, and the control of bioenergetic efficiency (Del Dotto et al., 2017; Ishihara et al., 2006). The ATP-dependent zinc metalloprotease Yme1l (Anand et al., 2014) or the metallo-endopeptidase Oma1 (Ehses et al., 2009; Head et al., 2009) can cleave Opa1 with enough force to produce opposing phenotypes both at rest and under stress. Longer forms produced by Yme1l's constitutive cleavage of Opa1 are necessary to initiate mitochondrial fusion and control mitochondrial crista dynamics (Anand et al., 2014; Hu et al., 2020). However, the short forms generated when Oma1 cleaves Opa1 in response to stress cause mitochondrial fission and cell death. (Kale et al., 2020; Anand et al., 2014; Korwitz et al., 2016). Opa1 levels are lower in failing human hearts, which is related to the development of smaller mitochondria (Chen et al., 2009). OPA1 mutant mice display anomalies in the left ventricle, a build-up of fragmented and dysfunctional cardiac mitochondria, and severe oxidative stress (homozygous OPA1 mutation is embryonic lethal) (Chen et al., 2012). Moreover, Yme1l and Oma1 modify Opa1 post-translationally by constitutive and inducible proteolytic cleavage, which is crucial for cardiac physiology (Wai et al., 2015). Age-related cardiomyopathy and heart failure occur in mice with YMEL1L1 cardiac deletion because of shortened forms of Opa1 and fragmented mitochondrial buildup (Wai et al., 2015).

Mitochondrial Fission

A mitochondrion splitting into two smaller mitochondria is known as mitochondrial fission, and it is crucial for controlling mitochondrial quality and bioenergetics under both normal and stressed conditions (Figure 1).

The mitochondrial protein Fis1, the mitochondrial dynamics proteins of 49 kDa (MiD49) and 51 kDa (MiD51) just at the outer mitochondrial membrane, the mitochondrial fission factor (Mff) (Otera et al., 2010), and others interact with the cytosolic protein Drp1 (Loson et al., 2013; Palmer et al., 2011). Drp1 translocation to mitochondria is triggered by various cardiomyocyte events, including b-adrenergic stimulation and ischemia/reperfusion injury (Jhun et al., 2018; Disatnik et al., 2013). Following translocation and binding to the proteins mentioned above at constriction sites, Drp1 polymerizes around the organelle, changing its conformation to constrict the outer and inner membranes through GTP hydrolysis. This causes mitochondria to split, forming smaller, spherical mitochondria (Disatnik et al. 2013; Ingerman et al., 2005; Mears et al., 2011). Because damaged and fragmented mitochondria cannot be reintegrated into the mitochondrial network, autophagy is used to remove them.Drp1 contributes to quality control in mitochondria by facilitating cardiac autophagy-mediated mitochondrial clearance, also known as mitophagy (Ikeda et al., 2015; Song et al., 2015). Our results demonstrate a critical role for Drp1 in maintaining cardiac biochemical balance and contractility properties in physiological settings. Under pathological conditions such as myocardial infarction, a pro-fission phenotype coupled with impaired mitophagy causes an accumulation of fragmented and dysfunctional mitochondria, promoting the spread of damage and heart degeneration over time (Disatnik et al., 2013). Cardiovascular ablation of DNM1L results in a decrease in mitochondrial membrane potential, activation of mPTP, reduced mitochondrial respiration, disruption of mitophagy, cardiomyocyte necrosis, and aging cardiomyopathy in mice (Song et al., 2015; Ishihara et al., 2015; Twig et al., 2008). (Song et al., 2015; Ikeda et al., 2015, Song et al., 2017).

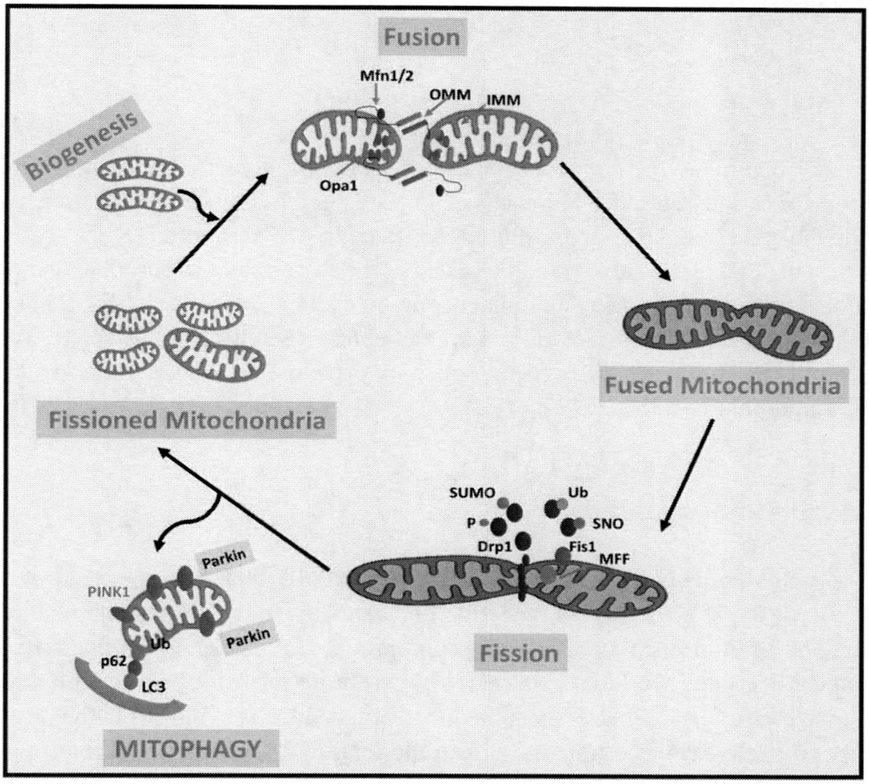

Figure 1. Mitochondrial fusion and fission.
(P:Phosphate group;Ub: Ubiquitin; LC3: Light Chain 3 IMM: Inner mitochondrial membrane; OMM: Outer mitochondrial membrane; Drp1: Dynamin-related protein; Opa1:Ocular atrophy factor 1 (Opa1); MFN: Mitofusin; PINK1: PTEN-induced putative kinase-1; Fis-1: Mitochondrial fission protein 1; MFF: Mitochondrial fission factor; SUMO: Small Ubiquitin-related modifier protein; SNO: Storms Compromise Protein).

Mitophagy

Deffective mitochondria are targeted for removal during the mitophagy process, which is related to the fusion and fission of mitochondria (Palikaras et al., 2018). The complete opposite of mitochondrial biogenesis or production is mitophagy. Mitochondrial biogenesis and mitophagy act together to modulate cellular mitochondrial content and function. They play a crucial role in regulating signaling and metabolism as a result. Mitophagy's underlying molecular pathways have only lately been identified. A decline in mitochondrial membrane potential was initially thought to be the catalyst for the degradation of mitochondria during mitophagy (Youle and Narendra, 2011). Mitophagy in the heart can be triggered by hypoxia and situations where excessive reactive oxygen species (ROS) are formed, such as during reperfusion (Doblado et al., 2021). The accumulation of damaged mitochondria, loss of myocytes, and dysfunctional contraction are all effects of abnormal mitophagy. According to Vasquez-Trincado et al. (2016), dynamic mitochondrial processes, mitophagy, incorrect clearance of damaged mitochondria, and stimulation of the inflammatory response are all factors in heart aging.

Experimental Approaches for Assessing Mitophagy

Klionsky adds that many methods must be used to conclusively show that autophagy is altered in response to various stimuli (Klionsky et al., 2009). To summarize electron microscopy imaging is the gold standard for morphological analysis of structures relevant to autophagy (Eskelinen et al., 2011). It is also possible to detect the recruitment of autophagosomal vesicles to the mitochondria by measuring autophagosome-resident proteins, such as LC3 or autophagy receptors, in preparations that are rich in mitochondria (Kubli et al., 2013; B. Wang et al., 2018a). Similarly, fluorescently tagged probes or proteins have been used to identify cells undergoing mitophagy.

Cardiac Aging and Mitophagy

According to Sussman et al. (2004) and Lakatta et al. (2003), the aging process presents itself in the heart through a variety of genetic, biochemical, and histological factors, such as decreasing DNA synthesis, increased cell proliferation, and enlarging cardiac size. While the early neonatal heart and the fetus experience hyperplastic growth of cardiac myocytes, adults experience hypertrophic cardiac expansion due to reduced DNA synthesis in the postnatal heart following cell cycle exit. The transition from the fetal to adult heart and from hypertrophied myocardium to heart failure have both been linked to changes in genetic and metabolic reprogramming (Kajstura et al., 1996; Olivetti et al., 1994; Swynghedauw et al., 1995).

Mitophagy Receptors

The mitochondrial proteins found in the OMM can guide mitochondria to autophagosomes. NIX/BNIP3L, FUNDC1, BNIP3, Bcl2L13, FKBP8, and prohibitin-2 (PHB2) are a few of the mitophagy receptors that have been identified thus far (Wei et al., 2017; Bhujabal et al., 2017; Murakawa et al., 2015; Liu et al., 2012; Novak et al., 2010; Hanna et al., 2012; Umare et al., 2021). These proteins are part of the mitochondria's integrated membrane and interact with the cytoplasm. One exception is PHB2, which is found in the inner mitochondrial membrane. PHB2 stimulates the evacuation of residual mitochondria following an outer membrane break (Wei et al., 2017).LIRs enable the mitophagy receptors to bind to LC3 on the autophagosome membrane, even without ubiquitin or adaptor proteins. A phospholipid called cardiolipin has mitophagy receptor-like properties. Cardiolipin separates damaged mitochondria, where it interacts with LC3 to encourage mitophagy. Cardiolipin is positioned on the inner mitochondrial membrane (Chu et al., 2013). Once the outer mitochondrial membrane has ruptured, cardiolipin may be sufficient to ensure the inner mitochondrial compartment undergoes mitophagy, much like PHB2. It is unknown how most of them have been activated to cause mitochondria to undergo mitophagy, despite all being shown to be mitophagy receptors. Although it is known that these proteins play additional roles, it is not apparent how they switch between the two. The physiological mechanisms regulating the activation of the two distinct mitophagy pathways are currently poorly understood and are the focus of extensive investigation. Recent research has revealed that PINK1/Parkin-mediated mitophagy is more critical for basal mitophagy and plays a minor role in stress adaptation and repair(McWilliams et al., 2018 Lee et al., 2018; Kubli et al., 2013; Song et al., 2015).In other studies, mitophagy receptors have been recognized as crucial regulators of programmable mitophagy for development or differentiation (Lampert et al., 2019; Esteban-Martinez et al., 2017; Sandoval et al., 2008). Hence, the two distinct mitophagy routes appear to play various roles in the cell, although more investigation is likely needed. Moreover, the two mitophagy mechanisms interact (Zhang et al., 2016; Gao et al., 2015; Wankhede et al., 2022). For instance, the protein phosphatase PGAM5 dephosphorylates FUNDC1, which enhances its association with LC3 (Chen et al. 2014). Moreover, PHB2 inhibits PARL to speed up PINK1 processing, while PGAM5 consolidates PINK1 on the OMM to increase PINK1/Parkin-mediated mitophagy (Yan et al., 2019). These two routes work well together and can slightly counterbalance one another's drawbacks.

Mechanisms Mediating Mitophagy in the Heart

Parkin-Dependent Mitophagy

The two most well-researched processes of mitophagy in cardiomyocytes are the cytosolic E3 ubiquitin ligase Parkin-induced putative kinase-1 (PINK1) and the mitochondrial membrane kinase PTEN (Kitada et al., 1998; Valente et al., 2004). Parkin, a protein that ubiquitinates proteins of the mitochondrial outer membrane such as MFN1, MFN2, and voltage-dependent anion channel (VDAC), is attracted from the cytosol to the injured or malfunctioning mitochondria (Valente et al., 2004). Only depolarized mitochondria can accumulate PINK1 because it is unstable and rapidly degrades in healthy mitochondria but is stabilized there (Matsuda et al., 2010). MFN2 is phosphorylated by PINK1, which enables Parkin to go toward the mitochondria (Chen et al., 2013). Phosphorylated ubiquitin, which PINK1 phosphorylates, activates Parkin E3 ubiquitin ligase activity (Koyano et al., 2014; Kane et al., 2014; Kale et al., 2022). Cardiomyopathy results from the absence of MFN2 in cardiomyocytes, inhibiting Parkin from being recruited to repair damaged mitochondria by depolarization (Chen et al., 2013).

Role of Drp1 in Mitophagy in The Heart

Drp1 is a small GTPase linked to dynamin and involved in mitochondrial fission. It has been determined that Dnm1, a yeast homolog of Drp1, is a crucial mediator of mitophagy in yeast

(Kanki et al., 2009), but it is unknown whether Drp1 is also necessary for mitophagy in mammalian hearts. In contrast to other tissues, the heart and brain express Drp1 in higher amounts (Ishihara et al., 2015). Moreover, postnatal cardiac-specific downregulation of Drp1 in mice results in dilated cardiomyopathy and early mortality by causing the condition. (Ikeda et al., 2015; Kageyama et al., 2014; Ishihara et al., 2015; Song et al., 2015; Marde et al., 2021). Drp1 may be crucial in controlling mitophagy in the heart, but this is debatable. According to Song et al., mitophagy is overactive in mouse hearts lacking Drp1. According to Cahill et al., the C452F mutation in Drp1 causes dilated cardiomyopathy with abnormal mitochondrial morphology and poor mitophagy (2015). In a clinical study, a newborn girl with a Drp1 heterozygous mutation (A395D) demonstrated a failure during mitochondrial fission, as well as chronically elevated lactate and plasma fatty acids with a highly long chain(Waterham et al., 2007).

Light Chain 3 (LC3) Mediated Mitophagy

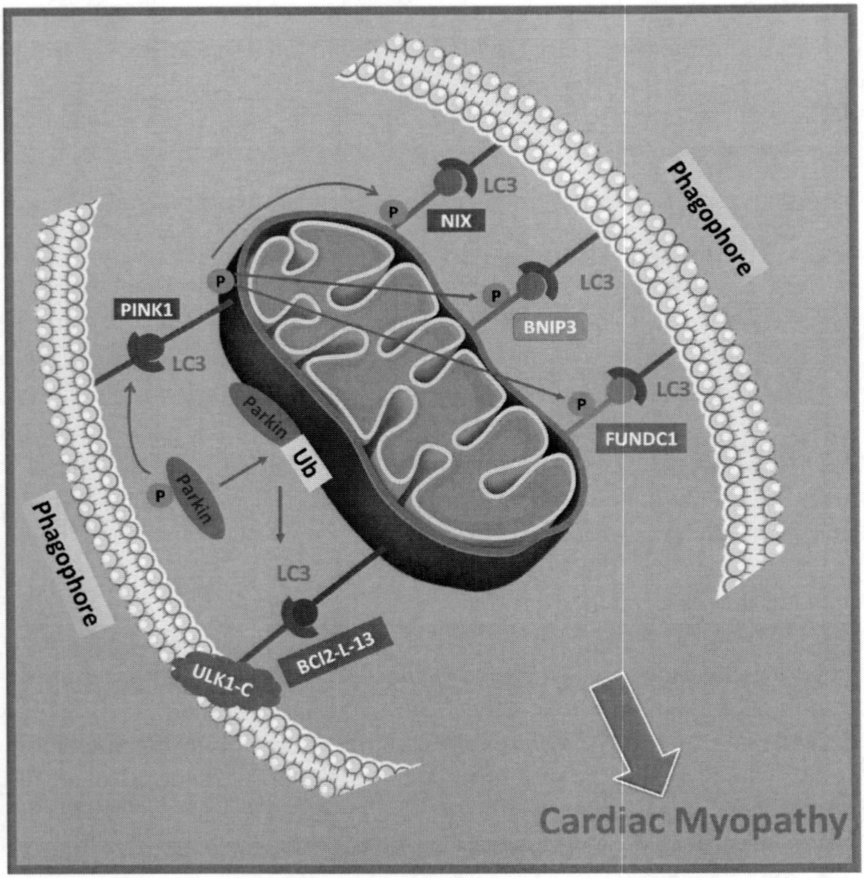

Figure 2. Mechanisms mediating mitophagy in the heart.
(BNIP3:BCL2 interacting protein 3; FUNDC1: FUN 14 domain-containing protein; ULK1-c: Unc-51 like autophagy activating kinase; LC3: Light chain 3; P: Phosphate group; PINK1: PTEN-induced putative kinase-1).

Ubiquitination is a crucial step in all types of selective autophagy (Yin et al., 2020). A well-established model for linking cargo with the ready, autophagy-generated membrane is the recruitment of protein 1 LC3 (MAP1LC3) associated with microtubules by cargo-bound receptors via LIR (Ney et al., 2015) (Figure 2). In this scenario, receptors are either a component of the cargo or are drawn to it by ubiquitination. More autophagy-related proteins must be acquired, and scaffold proteins are necessary (Eickhorst et al., 2020). Ubiquitin-bound adaptor proteins in mitochondria are crucial for delivering depolarized mitochondria to the perinuclear region via the microtubule after OMM remodeling brought on by the degradation of ubiquitinated proteins (Eickhorst et al., 2020). The interaction between LC3 and adaptor proteins facilitates the sequestration of damaged mitochondria into autophagosomes (Wei et al., 2015). Damaged mitochondria are destroyed due to the interaction between autophagosomes and lysosomes. Optineurin, trans-activating transcriptional regulatory protein of HTLV-1-binding protein 1, sequestosome-1 (p62), nuclear domain 10 protein 52, and neighbor of breast cancer LC3 can recognize and recruit cargo-binding receptors (LC3 adaptors) via their ubiquitin-binding domain to polyubiquitinated substrates on the mitochondria (Gatica et al., 2018).

Ubiquitin-Independent Mitophagy

According to an earlier study, PINK1-deficient cells undergo autophagy and mitophagy (Y. Zhang et al., 2017). Without ubiquitination, the LC3 adaptor can also eliminate damaged mitochondria. Directly detecting damaged mitochondria, these adaptors change the subcellular position or link proteins to enable the transportation of damaged mitochondria to autophagosomes. The best examples implicated in mitochondrial elimination are the BCL2/adenovirus E1B 19 kDa-interacting protein 3-like (NIX/BNIP3L) and BCL2/adenovirus E1B 19 kDa-interacting protein 3 (BNIP3) pathways (Zhang and Ney, 2009; Upaganlawar et al., 2021). It has been shown that BNIP3 and NIX induce mitophagy in response to hypoxia, as well as mitochondrial depolarization and fusion with cell membranes (Zhang and Ney, 2009). The cytoplasmic N-terminals of NIX and BNIP3 activate mitophagy through interactions with LC3-related molecules. Direct binding of BNIP3 stabilizes the protein and boosts Parkin recruitment (Dorn et al., 2008).

Additional Parkin-Independent Mitophagy

There are numerous parkin-independent mechanisms, including receptor-mediated pathways. For instance, the OMM-localized protein FUNDC1 can induce mitophagy in mammalian cells by attracting MAP1LC3B/LC3B via its LIR motif (Zhang, W et al., 2021). FUNDC1 is phosphorylated by various kinases that control its activity (Wu et al., 2014). Dephosphorylation by PGAM5 is required for LC3 activation and interaction, which results in mitophagy (Chen et al., 2014). In response to damaged mitochondrial potential, often localized in the IMM and OMM, choline dehydrogenase (CHDH) accumulates in the OMM (Yao et al., 2021). The CHDHp62-LC3 complex, implicated in mitophagy, is created when CHDH binds to p62. Importantly, via phosphorylation and activation of TBK1, the food deprivation sensor monophosphate-activated protein kinase (AMPK) may also encourage Parkin-independent mitophagy (Herzig and Shaw et al., 2018).

Novel Regulatory Pathways

LC3-independent mitophagy, which is physiologically linked, has also been observed in recent studies. By generating a Rab9, Unc51-like kinase 1 (ULK1), and DRP1 protein complex, the Rab9-mediated autophagosome causes mitophagy (Saito et al., 2019). Furthermore, novel mitophagy routes are still being discovered at a rapid pace. For example, autophagy may be influenced by SMAD-specific E3 (Petrasek et al., 2019). Coiled-coil myosin-like BCL2-interacting protein (BECN1)/Beclin1 is an autophagy-related protein that engages with Parkin but is not necessary for Parkin to move to the mitochondria. (Sun et al., 2018) At this time, further research being conducted has linked mitophagy to ER stress via the specialized ER-mitochondrial contact areas (MAMs), which regulate Ca2+ fluxes and apoptosis (Kumar and Mait et al., 2021; Luan et al., 2021). MAMs have a well-established function in mitophagy due to their necessity for the beginning of autophagy and multiple autophagy-related proteins in MAMs (Luan et al., 2021). It has been discovered that PINK1 and Beclin1 relocalize in MAMs in response to mitophagy stimulation, promoting the interaction of mitochondria with ER and the formation of the auto-phagosome(Sun et al., 2018).

Other Mechanisms Mediating Mitophagy in the Heart

Damaged mitochondria are ingested by autophagosomes when they come into contact with LC3- or LC3-related proteins, autophagy adaptors, or receptors. Autophagy adaptors like p62/SQSTM1, Tax1BP1, NBR1, NDP52, and optineurin possess ubiquitin-binding domains through which they recognize damaged mitochondria ubiquitinated through either Parkin-dependent (Huang et al., 2011) or independent mechanism, mitochondrial receptors such as NIX/BNIP3L, mitochondrial pro-apoptotic BH3-only domain protein (BNIP3), and FUNDC1 are situated inside (Lazarou et al., 2015; Yang et al., 2015). When pathological cardiac hypertrophy occurs, NIX is raised (Gálvez et al., 2006). When there is hypoxia, BNIP3 increases in the myocardium (Regula et al., 2002). BNIP3 and NIX function as mitophagy regulators in adult hearts (Dorn et al., 2010). Hypoxia causes mitophagy in HeLa cells when FUNDC1, a crucial mitochondrial outer-membrane protein, interacts with LC3 (Liu et al., 2012). A Ser/Thr phosphatase called phosphoglycerate mutase family member 5 (PGAM5) dephosphorylates FUNDC1 at Serine 13 and triggers mitophagy (Chen et al., 2014).PGAM5 is necessary for controlling PINK1-mediated mitophagy in the heart (Lu et al., 2016). The relationship between PINK1 and FUNDC1 is currently unknown. Additionally activated, ULK1 moves to the mitochondria, which engages in FUNDC1 and phosphorylates it at Serine 17 to enhance its interaction with LC3 in MEF cells. ULK1 is a Ser/Thr kinase required for early autophagosome formation (Wuet al., 2014). Nevertheless, it is not yet known whether the heart has a similar pathway. Bcl2-L-13, a human mitophagy receptor similar to Atg32 in yeast (Kanki et al., 2009; Okamoto et al., 2009), has been identified (Murakawa et al., 2015). Bcl2-

L-13 was shown to bind to LC3 in HEK293A cells and fragment mitochondria without the aid of Drp1 or Parkin (Murakawa et al., 2015), offering a fresh look at the mitophagy process.

Manifestation of the Aging Heart

Aging hearts have different histological and biochemical features. (Sussman et al., 2004; Lakatta et al., 2003). Increases in apoptosis and necrosis, myocyte nuclei proliferation, increased myocyte volume, and connective tissue accumulation are frequently seen in the myocardium of elderly animals (Kajstura et al., 1996; Olivetti et al., 1994; Swynghedauw et al., 1995). Ectopic-galactosidase activity associated with senescence and lipofuscin, dense masses with autofluorescence composed of accumulated lipid and worn-out proteins, are also increased. Moreover, aging impacts the quantity of several molecules, including the tumor suppressors p16INK4 and p19ARF (Krishnamurthy et al., 2004; Satyanarayana et al., 2004). Aged cardiomyocytes can compensate less through compensatory hypertrophy and proliferation in response to increased strain. (Levine et al., 2002; Sussman et al., 2004; Kajstura et al., 2000; Takahashi et al., 1992; Nadal-Ginard et al., 2003). As a result, ventricular remodeling does not reduce the wall tension that results from aging. (1990; Capasso et al.).Moreover, in aging hearts, the activation of cell defense systems such as antioxidant and heat shock proteins is decreased in response to pathogenic stressors. Mariani et al., 2000; Edwards et al., 2004; Takahashi et al., 1992). The greatest therapeutic approaches should prevent cell death and accumulation of senescent myocytes to combat the detrimental effects of aging.

Conclusion

The importance of mitochondrial activity in cardiac performance is evident, as it is a highly ATP-demanding tissue. As a result, removing damaged mitochondria is critical for heart cell function. The results presented here suggest that mitophagy has an ambiguous effect on heart function because reduced and increased mitochondrial elimination results in some circulatory impairment. Mitophagy can be deficient or overactive in the case of cardiomyopathy. This idea may directly affect the treatment of aging cardiomyopathy and other cardiovascular diseases. It suggests that maintaining a healthy mitochondrial network in heart cells may be a better therapeutic goal than encouraging the regeneration or removal of damaged mitochondria under various stress circumstances.

References

Anand, R., Wai, T., Baker, M. J., Kladt, N., Schauss, A. C., Rugarli, E. and Langer, T. (2014). The i-AAA protease YME1L and OMA1 cleave OPA1 to balance mitochondrial fusion and fission. *Journal of Cell Biology*. 204(6), pp.919-929.

Bach, D., Pich, S., Soriano, F. X., Vega, N., Baumgartner, B., Oriola, J., Daugaard, J. R., Lloberas, J., Camps, M., Zierath, J. R. and Rabasa-Lhoret, R. (2003). Mitofusin-2 determines mitochondrial network architecture and mitochondrial metabolism: a novel regulatory mechanism altered in obesity. *Journal of Biological Chemistry*. 278(19), pp.17190-17197.

Ban, T., Ishihara, T., Kohno, H., Saita, S., Ichimura, A., Maenaka, K., Oka, T., Mihara, K. and Ishihara, N. (2017). Molecular basis of selective mitochondrial fusion by heterotypic action between OPA1 and cardiolipin. *Nature cell biology*. 19(7), pp.856-863.

Bhujabal, Z., Birgisdottir, Å. B., Sjøttem, E., Brenne, H. B., Øvervatn, A., Habisov, S., Kirkin, V., Lamark, T. and Johansen, T. (2017). FKBP8 recruits LC3A to mediate Parkin-independent mitophagy. *EMBO reports*. 18(6), pp.947-961.

Biala, A. K., Dhingra, R. and Kirshenbaum, L. A. (2015). Mitochondrial dynamics: orchestrating the journey to advanced age. *Journal of molecular and cellular cardiology*. 83, pp.37-43.

Bravo-San Pedro, J. M., Kroemer, G. and Galluzzi, L. (2017). Autophagy and mitophagy in cardiovascular disease. *Circulation research*. 120(11), pp.1812-1824.

Buchberger, A., Bukau, B. and Sommer, T. (2010). Protein quality control in the cytosol and the endoplasmic reticulum: brothers in arms. *Molecular cell*. 40(2), pp.238-252.

Cahill, T. J., Leo, V., Kelly, M., Stockenhuber, A., Kennedy, N. W., Bao, L., Cereghetti, G., Harper, A. R., Czibik, G., Lao, C. and Bellahcene, M. (2015). Resistance of dynamin-related protein 1 oligomers to disassembly impairs mitophagy, resulting in myocardial inflammation and heart failure. *Journal of Biological Chemistry*. 290(43), pp.25907-25919.

Campos, J. C., Bozi, L. H., Bechara, L. R., Lima, V. M. and Ferreira, J. C. (2016). Mitochondrial quality control in cardiac diseases. *Frontiers in physiology*. 7, p.479.

Cavellini, L., Meurisse, J., Findinier, J., Erpapazoglou, Z., Belgareh-Touzé, N., Weissman, A. M. and Cohen, M. M. (2017). An ubiquitin-dependent balance between mitofusin turnover and fatty acids desaturation regulates mitochondrial fusion. *Nature communications*. 8(1), pp.1-15.

Chan, D. C. (2012). Fusion and fission: interlinked processes critical for mitochondrial health. *Annual review of genetics*. 46, pp.265-287.

Chen, G., Han, Z., Feng, D. U., Chen, Y., Chen, L., Wu, H., Huang, L., Zhou, C., Cai, X., Fu, C. and Duan, L. (2014). A regulatory signaling loop comprising the PGAM5 phosphatase and CK2 controls receptor-mediated mitophagy. *Molecular cell*. 54(3), pp.362-377.

Chen, H., Chomyn, A. and Chan, D. C. (2005). Disruption of fusion results in mitochondrial heterogeneity and dysfunction. *Journal of Biological Chemistry*. 280(28), pp.26185-26192.

Chen, H., Detmer, S. A., Ewald, A. J., Griffin, E. E., Fraser, S. E. and Chan, D. C. (2003). Mitofusins Mfn1 and Mfn2 coordinately regulate mitochondrial fusion and are essential for embryonic development. *The Journal of Cell Biology*. 160(2), pp.189-200.

Chen, H., Vermulst, M., Wang, Y. E., Chomyn, A., Prolla, T. A., McCaffery, J. M. and Chan, D. C. (2010). Mitochondrial fusion is required for mtDNA stability in skeletal muscle and tolerance of mtDNA mutations. *Cell*. 141(2), pp.280-289.

Chen, L., Gong, Q., Stice, J. P. and Knowlton, A. A. (2009). Mitochondrial OPA1, apoptosis, and heart failure. *Cardiovascular research*. 84(1), pp.91-99.

Chen, L., Liu, T., Tran, A., Lu, X., Tomilov, A. A., Davies, V., Cortopassi, G., Chiamvimonvat, N., Bers, D. M., Votruba, M. and Knowlton, A. A. (2012). OPA 1 Mutation and Late-Onset Cardiomyopathy: Mitochondrial Dysfunction and mtDNA Instability. *Journal of the American Heart Association*. 1(5), p.e003012.

Chen, Y. and Dorn, G. W. (2013). PINK1-phosphorylated mitofusin 2 is a Parkin receptor for culling damaged mitochondria. *Science*. 340(6131), pp.471-475.

Choi, S. Y., Huang, P., Jenkins, G. M., Chan, D. C., Schiller, J. and Frohman, M. A. (2006). A common lipid links Mfn-mediated mitochondrial fusion and SNARE-regulated exocytosis. *Nature cell biology*. 8(11), pp.1255-1262.

Chu, C. T., Ji, J., Dagda, R. K., Jiang, J. F., Tyurina, Y. Y., Kapralov, A. A., Tyurin, V. A., Yanamala, N., Shrivastava, I. H., Mohammadyani, D. and Qiang Wang, K. Z. (2013). Cardiolipin externalization to the outer mitochondrial membrane acts as an elimination signal for mitophagy in neuronal cells. *Nature cell biology*. 15(10), pp.1197-1205.

Cuervo, A. M., Bergamini, E., Brunk, U. T., Dröge, W., Ffrench, M. and Terman, A. (2005). Autophagy and aging: the importance of maintaining" clean" cells. *Autophagy*. 1(3), pp.131-140.

De Brito, O. M. and Scorrano, L. (2008). Mitofusin 2 tethers endoplasmic reticulum to mitochondria. *Nature*. 456(7222), pp.605-610.

de Carvalho, A. E. T. S., Bassaneze, V., Forni, M. F., Keusseyan, A. A., Kowaltowski, A. J. and Krieger, J. E. (2017). Early postnatal cardiomyocyte proliferation requires high oxidative energy metabolism. *Scientific reports*. 7(1), pp.1-11.

Del Dotto, V., Mishra, P., Vidoni, S., Fogazza, M., Maresca, A., Caporali, L., McCaffery, J. M., Cappelletti, M., Baruffini, E., Lenaers, G. and Chan, D. (2017). OPA1 isoforms in the hierarchical organization of mitochondrial functions. *Cell reports*. 19(12), pp.2557-2571.

Denzel, M. S., Storm, N. J., Gutschmidt, A., Baddi, R., Hinze, Y., Jarosch, E., Sommer, T., Hoppe, T. and Antebi, A. (2014). Hexosamine pathway metabolites enhance protein quality control and prolong life. *Cell*. 156(6), pp.1167-1178.

DeVay, R. M., Dominguez-Ramirez, L., Lackner, L. L., Hoppins, S., Stahlberg, H. and Nunnari, J. (2009). Coassembly of Mgm1 isoforms requires cardiolipin and mediates mitochondrial inner membrane fusion. *Journal of Cell Biology*. 186(6), pp.793-803.

Disatnik, M. H., Ferreira, J. C., Campos, J. C., Gomes, K. S., Dourado, P. M., Qi, X. and Mochly-Rosen, D. (2013). Acute inhibition of excessive mitochondrial fission after myocardial infarction prevents long-term cardiac dysfunction. *Journal of the American Heart Association*. *2*(5), p.e000461.

Doblado, L., Lueck, C., Rey, C., Samhan-Arias, A. K., Prieto, I., Stacchiotti, A. and Monsalve, M. (2021). Mitophagy in human diseases. *International journal of molecular sciences*, *22*(8), p.3903.

Dorn, G. W. and Kirshenbaum, L. A. (2008). Cardiac reanimation: targeting cardiomyocyte death by BNIP3 and NIX/BNIP3L. *Oncogene*, *27*(1), pp.S158-S167.

Dorn, G. W., (2010). Mitochondrial pruning by Nix and BNip3: an essential function for cardiac-expressed death factors. *Journal of cardiovascular translational research*, *3*(4), pp.374-383.

Edwards, M. G., Sarkar, D., Klopp, R., Morrow, J. D., Weindruch, R. and Prolla, T. A., (2004). Impairment of the transcriptional responses to oxidative stress in the heart of aged C57BL/6 mice. *Annals of the New York Academy of Sciences*, *1019*(1), pp.85-95.

Ehses, S., Raschke, I., Mancuso, G., Bernacchia, A., Geimer, S., Tondera, D., Martinou, J. C., Westermann, B., Rugarli, E. I. and Langer, T. (2009). Regulation of OPA1 processing and mitochondrial fusion by m-AAA protease isoenzymes and OMA1. *Journal of Cell Biology*, *187*(7), pp.1023-1036.

Eickhorst, C., Licheva, M. and Kraft, C. (2020). Scaffold proteins in bulk and selective autophagy. *Progress in molecular biology and translational science*, *172*, pp.15-35.

Eisner, V., Picard, M. and Hajnóczky, G. (2018). Mitochondrial dynamics in adaptive and maladaptive cellular stress responses. *Nature Cell Biology*, *20*(7), pp.755-765.

Eskelinen, E. L., Reggiori, F., Baba, M., Kovács, A. L. and Seglen, P. O., (2011). Seeing is believing: the impact of electron microscopy on autophagy research. *Autophagy*, *7*(9), pp.935-956.

Esteban-Martínez, L., Sierra-Filardi, E., McGreal, R. S., Salazar-Roa, M., Mariño, G., Seco, E., Durand, S., Enot, D., Graña, O., Malumbres, M. and Cvekl, A. (2017). Programmed mitophagy is essential for the glycolytic switch during cell differentiation. *The EMBO journal*, *36*(12), pp.1688-1706.

Eura, Y., Ishihara, N., Yokota, S. and Mihara, K., (2003). Two mitofusin proteins, mammalian homologues of FZO, with distinct functions are both required for mitochondrial fusion. *Journal of Biochemistry*, *134*(3), pp.333-344.

Ferreira, J. C., Campos, J. C., Qvit, N., Qi, X., Bozi, L. H., Bechara, L. R., Lima, V. M., Queliconi, B. B., Disatnik, M. H., Dourado, P. M. and Kowaltowski, A. J. (2019). A selective inhibitor of mitofusin 1-βIIPKC association improves heart failure outcome in rats. *Nature Communications*, *10*(1), pp.1-14.

Finkel, T., Serrano, M. and Blasco, M. A. (2007). The common biology of cancer and aging. *Nature*, *448*(7155), pp.767-774.

Gálvez, A. S., Brunskill, E. W., Marreez, Y., Benner, B. J., Regula, K. M., Kirschenbaum, L. A. and Dorn, G. W. (2006). Distinct pathways regulate proapoptotic Nix and BNip3 in cardiac stress. *Journal of Biological Chemistry*, *281*(3), pp.1442-1448.

Gao, F., Chen, D., Si, J., Hu, Q., Qin, Z., Fang, M. and Wang, G. (2015). The mitochondrial protein BNIP3L is the substrate of PARK2 and mediates mitophagy in PINK1/PARK2 pathway. *Human molecular genetics*, *24*(9), pp.2528-2538.

Gatica, D., Lahiri, V. and Klionsky, D. J. (2018). Cargo recognition and degradation by selective autophagy. *Nature Cell Biology*, *20*(3), pp.233-242.

Hall, A. R., Burke, N., Dongworth, R. K., Kalkhoran, S. B., Dyson, A., Vicencio, J. M., Dorn, G. W., Yellon, D. M. and Hausenloy, D. J. (2016). Hearts deficient in both Mfn1 and Mfn2 are protected against acute myocardial infarction. *Cell death & disease*, *7*(5), pp.e2238-e2238.

Hanna, R. A., Quinsay, M. N., Orogo, A. M., Giang, K., Rikka, S. and Gustafsson, Å. B. (2012). Microtubule-associated protein 1 light chain 3 (LC3) interacts with Bnip3 protein to selectively remove endoplasmic reticulum and mitochondria via autophagy. *Journal of Biological Chemistry*, *287*(23), pp.19094-19104.

Head, B., Griparic, L., Amiri, M., Gandre-Babbe, S. and van der Bliek, A. M. (2009). Inducible proteolytic inactivation of OPA1 mediated by the OMA1 protease in mammalian cells. *Journal of Cell Biology*, *187*(7), pp.959-966.

Hu, C., Shu, L., Huang, X., Yu, J., Gong, L., Yang, M., Wu, Z., Gao, Z., Zhao, Y., Chen, L. and Song, Z. (2020). OPA1 and MICOS Regulate mitochondrial crista dynamics and formation. *Cell death & disease*, *11*(10), pp.1-17.

Huang, C., Andres, A. M., Ratliff, E. P., Hernandez, G., Lee, P. and Gottlieb, R. A. (2011). Preconditioning involves selective mitophagy mediated by Parkin and p62/SQSTM1. *PloS one*, *6*(6), p.e20975.

Ikeda, Y., Shirakabe, A., Maejima, Y., Zhai, P., Sciarretta, S., Toli, J., Nomura, M., Mihara, K., Egashira, K., Ohishi, M. and Abdellatif, M. (2015). Endogenous Drp1 mediates mitochondrial autophagy and protects the heart against energy stress. *Circulation Research*, *116*(2), pp.264-278.

Ingerman, E., Perkins, E. M., Marino, M., Mears, J. A., McCaffery, J. M., Hinshaw, J. E. and Nunnari, J. (2005). Dnml forms spirals that are structurally tailored to fit mitochondria. *The Journal of Cell Biology*, *170*(7), pp.1021-1027.

Ishihara, N., Eura, Y. and Mihara, K. (2004). Mitofusin 1 and 2 play distinct roles in mitochondrial fusion reactions via GTPase activity. *Journal of cell science*, *117*(26), pp.6535-6546.

Ishihara, N., Fujita, Y., Oka, T. and Mihara, K. (2006). Regulation of mitochondrial morphology through proteolytic cleavage of OPA1. *The EMBO journal*, *25*(13), pp.2966-2977.

Ishihara, T., Ban-Ishihara, R., Maeda, M., Matsunaga, Y., Ichimura, A., Kyogoku, S., Aoki, H., Katada, S., Nakada, K., Nomura, M. and Mizushima, N. (2015). Dynamics of mitochondrial DNA nucleoids regulated by mitochondrial fission is essential for maintenance of homogeneously active mitochondria during neonatal heart development. *Molecular and cellular biology*, *35*(1), pp.211-223.

Jhun, B. S., O-Uchi, J., Adaniya, S. M., Mancini, T. J., Cao, J. L., King, M. E., Landi, A. K., Ma, H., Shin, M., Yang, D. and Xu, X. (2018). Protein kinase D activation induces mitochondrial fragmentation and dysfunction in cardiomyocytes. *The Journal of Physiology*, *596*(5), pp.827-855.

Kageyama, Y., Hoshijima, M., Seo, K., Bedja, D., Sysa-Shah, P., Andrabi, S. A., Chen, W., Höke, A., Dawson, V. L., Dawson, T. M. and Gabrielson, K. (2014). Parkin-independent mitophagy requires D rp1 and maintains the integrity of mammalian heart and brain. *The EMBO journal*, *33*(23), pp.2798-2813.

Kajstura, J., Cheng, W., Sarangarajan, R., Li, P., Li, B., Nitahara, J. A., Chapnick, S., Reiss, K., Olivetti, G., and Anversa, P. (1996). Necrotic and apoptotic myocyte cell death in the aging heart of Fischer 344 rats. *American Journal of Physiology-Heart and Circulatory Physiology*, *271*(3), pp.H1215-H1228.

Kajstura, J., Pertoldi, B., Leri, A., Beltrami, C. A., Deptala, A., Darzynkiewicz, Z. and Anversa, P. (2000). Telomere shortening is an in vivo marker of myocyte replication and aging. *The American Journal of pathology*, *156*(3), pp.813-819.

Kale, M., Nimje, N., Aglawe, M. M., Umekar, M., Taksande, B. and Kotagale, N. (2020). Agmatine modulates anxiety and depression-like behaviour in diabetic insulin-resistant rats. *Brain Research*, *1747*, p.147045.

Kale, M. B., Bajaj, K., Umare, M., Wankhede, N. L., Taksande, B. G., Umekar, M. J. and Upaganlawar, A. (2022). Exercise and nutraceuticals: eminent approach for diabetic neuropathy. *Current Molecular Pharmacology*, *15*(1), pp.108-128.

Kane, L. A., Lazarou, M., Fogel, A. I., Li, Y., Yamano, K., Sarraf, S. A., Banerjee, S. and Youle, R. J. (2014). PINK1 phosphorylates ubiquitin to activate Parkin E3 ubiquitin ligase activity. *Journal of Cell Biology*, *205*(2), pp.143-153.

Kanki, T., Wang, K., Cao, Y., Baba, M. and Klionsky, D. J. (2009). Atg32 is a mitochondrial protein that confers selectivity during mitophagy. *Developmental cell*, *17*(1), pp.98-109.

Kanki, T., Wang, K. E., Baba, M., Bartholomew, C. R., Lynch-Day, M. A., Du, Z., Geng, J., Mao, K., Yang, Z., Yen, W. L. and Klionsky, D. J. (2009). A genomic screen for yeast mutants defective in selective mitochondria autophagy. *Molecular biology of the cell*, *20*(22), pp.4730-4738.

Kitada, T., Asakawa, S., Hattori, N., Matsumine, H., Yamamura, Y., Minoshima, S., Yokochi, M., Mizuno, Y. and Shimizu, N. (1998). Mutations in the parkin gene cause autosomal recessive juvenile parkinsonism. *nature*, *392*(6676), pp.605-608.

Korwitz, A., Merkwirth, C., Richter-Dennerlein, R., Tröder, S. E., Sprenger, H. G., Quirós, P. M., López-Otín, C., Rugarli, E. I. and Langer, T. (2016). Loss of OMA1 delays neurodegeneration by preventing stress-induced OPA1 processing in mitochondria. *Journal of Cell Biology*, *212*(2), pp.157-166.

Koshiba, T., Detmer, S. A., Kaiser, J. T., Chen, H., McCaffery, J. M. and Chan, D. C. (2004). Structural basis of mitochondrial tethering by mitofusin complexes. *Science*, *305*(5685), pp.858-862.

Koyano, F., Okatsu, K., Kosako, H., Tamura, Y., Go, E., Kimura, M., Kimura, Y., Tsuchiya, H., Yoshihara, H., Hirokawa, T. and Endo, T. (2014). Ubiquitin is phosphorylated by PINK1 to activate parkin. *Nature*, *510*(7503), pp.162-166.

Krishnamurthy, J., Torrice, C., Ramsey, M. R., Kovalev, G. I., Al-Regaiey, K., Su, L. and Sharpless, N. E. (2004). Ink4a/Arf expression is a biomarker of aging. *The Journal of Clinical Investigation*, *114*(9), pp.1299-1307.

Kubli, D. A., Zhang, X., Lee, Y., Hanna, R. A., Quinsay, M. N., Nguyen, C. K., Jimenez, R., Petrosyan, S., Murphy, A. N. and Gustafsson, Å. B. (2013). Parkin Protein Deficiency Exacerbates Cardiac Injury and Reduces Survival following Myocardial Infarction. *Journal of Biological Chemistry*, *288*(2), pp.915-926.

Kumar, V. and Maity, S. (2021). ER, stress-sensor proteins and ER-mitochondrial crosstalk—signaling beyond (ER) stress response. *Biomolecules*, *11*(2), p.173.

Lakatta, E. G. and Levy, D. (2003). Arterial and cardiac aging: major shareholders in cardiovascular disease enterprises: Part I: aging arteries: a "set up" for vascular disease. *Circulation*, *107*(1), pp.139-146.

Lakatta, E. G. (2003). Arterial and cardiac aging: major shareholders in cardiovascular disease enterprises: Part III: cellular and molecular clues to heart and arterial aging. *Circulation*, *107*(3), pp.490-497.

Lampert, M. A., Orogo, A. M., Najor, R. H., Hammerling, B. C., Leon, L. J., Wang, B. J., Kim, T., Sussman, M. A. and Gustafsson, Å. B. (2019). BNIP3L/NIX and FUNDC1-mediated mitophagy is required for mitochondrial network remodeling during cardiac progenitor cell differentiation. *Autophagy*, *15*(7), pp.1182-1198.

Lazarou, M., Sliter, D. A., Kane, L. A., Sarraf, S. A., Wang, C., Burman, J. L., Sideris, D. P., Fogel, A. I. and Youle, R. J. (2015). The ubiquitin kinase PINK1 recruits autophagy receptors to induce mitophagy. *Nature*, *524*(7565), pp.309-314.

Leboucher, G. P., Tsai, Y. C., Yang, M., Shaw, K. C., Zhou, M., Veenstra, T. D., Glickman, M. H. and Weissman, A. M. (2012). Stress-induced phosphorylation and proteasomal degradation of mitofusin 2 facilitates mitochondrial fragmentation and apoptosis. *Molecular cell*, *47*(4), pp.547-557.

Lee, J. J., Sanchez-Martinez, A., Zarate, A. M., Benincá, C., Mayor, U., Clague, M. J. and Whitworth, A. J. (2018). Basal mitophagy is widespread in Drosophila but minimally affected by loss of Pink1 or parkin. *Journal of Cell Biology*, *217*(5), pp.1613-1622.

Lemasters, J. J. (2005). Selective mitochondrial autophagy, or mitophagy, as a targeted defense against oxidative stress, mitochondrial dysfunction, and aging. *Rejuvenation research*, *8*(1), pp.3-5.

Liang, W. J. and Gustafsson, Å. B. (2020). The aging heart: mitophagy at the center of rejuvenation. *Frontiers in Cardiovascular Medicine*, *7*, p.18.

Liu, L., Feng, D., Chen, G., Chen, M., Zheng, Q., Song, P., Ma, Q., Zhu, C., Wang, R., Qi, W. and Huang, L. (2012). Mitochondrial outer-membrane protein FUNDC1 mediates hypoxia-induced mitophagy in mammalian cells. *Nature Cell Biology*, *14*(2), pp.177-185.

Losón, O. C., Song, Z., Chen, H. and Chan, D. C. (2013). Fis1, Mff, MiD49, and MiD51 mediate Drp1 recruitment in mitochondrial fission. *Molecular biology of the cell*, *24*(5), pp.659-667.

Lu, W., Sun, J., Yoon, J. S., Zhang, Y., Zheng, L., Murphy, E., Mattson, M. P. and Lenardo, M. J. (2016). Mitochondrial protein PGAM5 regulates mitophagic protection against cell necroptosis. *PloS one*, *11*(1), p.e0147792.

Macri, C., Wang, F., Tasset, I., Schall, N., Page, N., Briand, J. P., Cuervo, A. M. and Muller, S. (2015). Modulation of deregulated chaperone-mediated autophagy by a phosphopeptide. *Autophagy*, *11*(3), pp.472-486.

Marde, V. S., Tiwari, P. L., Wankhede, N. L., Taksande, B. G., Upaganlawar, A. B., Umekar, M. J. and Kale, M. B. (2021). Neurodegenerative disorders associated with genes of mitochondria. *Future Journal of Pharmaceutical Sciences*, *7*(1), pp.1-8.

Matsuda, N., Sato, S., Shiba, K., Okatsu, K., Saisho, K., Gautier, C. A., Sou, Y. S., Saiki, S., Kawajiri, S., Sato, F. and Kimura, M. (2010). PINK1 stabilized by mitochondrial depolarization recruits Parkin to damaged mitochondria and activates latent Parkin for mitophagy. *Journal of Cell Biology*, *189*(2), pp.211-221.

McWilliams, T. G., Prescott, A. R., Montava-Garriga, L., Ball, G., Singh, F., Barini, E., Muqit, M. M., Brooks, S. P. and Ganley, I. G. (2018). Basal mitophagy occurs independently of PINK1 in mouse tissues of high metabolic demand. *Cell metabolism*, *27*(2), pp.439-449.

Mears, J. A., Lackner, L. L., Fang, S., Ingerman, E., Nunnari, J. and Hinshaw, J. E. (2011). Conformational changes in Dnm1 support a contractile mechanism for mitochondrial fission. *Nature Structural & molecular biology*, *18*(1), pp.20-26.

Meeusen, S., DeVay, R., Block, J., Cassidy-Stone, A., Wayson, S., McCaffery, J. M. and Nunnari, J. (2006). Mitochondrial inner-membrane fusion and crista maintenance requires the dynamin-related GTPase Mgm1. *Cell*, *127*(2), pp.383-395.

Merkwirth, C., Dargazanli, S., Tatsuta, T., Geimer, S., Löwer, B., Wunderlich, F. T., von Kleist-Retzow, J. C., Waisman, A., Westermann, B. and Langer, T. (2008). Prohibitins control cell proliferation and apoptosis by regulating OPA1-dependent cristae morphogenesis in mitochondria. *Genes & development*, *22*(4), pp.476-488.

Morales, Pablo E.; Arias-Durán, Carla; Ávalos-Guajardo, Yáreni; Aedo, Geraldine; Verdejo, Hugo E.; Parra, Valentina; Lavandero, Sergio. (2019). Emerging role of mitophagy in cardiovascular physiology and pathology. *Molecular Aspects of Medicine,* 71:100822.

Mozaffarian, D., Benjamin, E. J., Go, A. S., Arnett, D. K., Blaha, M. J., Cushman, M., De Ferranti, S., Després, J. P., Fullerton, H. J., Howard, V. J. and Huffman, M. D. (2015). Heart disease and stroke statistics—2015 update: a report from the American Heart Association. *circulation*, *131*(4), pp.e29-e322.

Mozdy, A. D., McCaffery, J. M. and Shaw, J. M. (2000). Dnm1p GTPase-mediated mitochondrial fission is a multi-step process requiring the novel integral membrane component Fis1p. *The Journal of Cell Biology*, *151*(2), pp.367-380.

Murakawa, T., Yamaguchi, O., Hashimoto, A., Hikoso, S., Takeda, T., Oka, T., Yasui, H., Ueda, H., Akazawa, Y., Nakayama, H. and Taneike, M. (2015). Bcl-2-like protein 13 is a mammalian Atg32 homologue that mediates mitophagy and mitochondrial fragmentation. *Nature Communications*, *6*(1), pp.1-14.

Nah, J., Fernández, Á. F., Kitsis, R. N., Levine, B. and Sadoshima, J. (2016). Does autophagy mediate cardiac myocyte death during stress. *Circulation Research*, *119*(8), pp.893-895.

Nah, J., Miyamoto, S. and Sadoshima, J. (2011). Mitophagy as a protective mechanism against myocardial stress. *Comprehensive Physiology*, *7*(4), pp.1407-1424.

Narendra, D. P., Jin, S. M., Tanaka, A., Suen, D. F., Gautier, C. A., Shen, J., Cookson, M. R. and Youle, R. J. (2010). PINK1 is selectively stabilized on impaired mitochondria to activate Parkin. *PLoS biology*, *8*(1), p.e1000298.

Ney, P. A. (2015). Mitochondrial autophagy: Origins, significance, and role of BNIP3 and NIX. *Biochimica et Biophysica Acta (BBA)-Molecular Cell Research*, *1853*(10), pp.2775-2783.

Novak, I., Kirkin, V., McEwan, D. G., Zhang, J., Wild, P., Rozenknop, A., Rogov, V., Löhr, F., Popovic, D., Occhipinti, A. and Reichert, A. S. (2010). Nix is a selective autophagy receptor for mitochondrial clearance. *EMBO reports*, *11*(1), pp.45-51.

Okamoto, K., Kondo-Okamoto, N. and Ohsumi, Y. (2009). Mitochondria-anchored receptor Atg32 mediates degradation of mitochondria via selective autophagy. *Developmental cell*, *17*(1), pp.87-97.

Olichon, A., Baricault, L., Gas, N., Guillou, E., Valette, A., Belenguer, P. and Lenaers, G., (2003). Loss of OPA1 perturbates the mitochondrial inner membrane structure and integrity, leading to cytochrome c release and apoptosis. *Journal of Biological Chemistry*, *278*(10), pp.7743-7746.

Olivetti, G., Melissari, M., Balbi, T., Quaini, F., Sonnenblick, E. H. and Anversa, P. (1994). Myocyte nuclear and possible cellular hyperplasia contribute to ventricular remodeling in the hypertrophic senescent heart in humans. *Journal of the American College of Cardiology*, *24*(1), pp.140-149.

Otera, H., Wang, C., Cleland, M. M., Setoguchi, K., Yokota, S., Youle, R. J. and Mihara, K., (2010). Mff is an essential factor for mitochondrial recruitment of Drp1 during mitochondrial fission in mammalian cells. *Journal of Cell Biology*, *191*(6), pp.1141-1158.

Palikaras, K., Lionaki, E. and Tavernarakis, N. (2018). Mechanisms of mitophagy in cellular homeostasis, physiology and pathology. *Nature Cell Biology*, *20*(9), pp.1013-1022.

Palmer, C. S., Osellame, L. D., Laine, D., Koutsopoulos, O. S., Frazier, A. E. and Ryan, M. T. (2011). MiD49 and MiD51, new components of the mitochondrial fission machinery. *EMBO reports*, *12*(6), pp.565-573.

Papanicolaou, K. N., Kikuchi, R., Ngoh, G. A., Coughlan, K. A., Dominguez, I., Stanley, W. C. and Walsh, K. (2012). Mitofusins 1 and 2 are essential for postnatal metabolic remodeling in heart. *Circulation Research*, *111*(8), pp.1012-1026.

Papanicolaou, K. N., Ngoh, G. A., Dabkowski, E. R., O'Connell, K. A., Ribeiro Jr, R. F., Stanley, W. C. and Walsh, K. (2012). Cardiomyocyte deletion of mitofusin-1 leads to mitochondrial fragmentation and improves tolerance to ROS-induced mitochondrial dysfunction and cell death. *American Journal of Physiology-Heart and Circulatory Physiology*, *302*(1), pp.H167-H179.

Pitts, K. R., Yoon, Y., Krueger, E. W. and McNiven, M. A. (1999). The dynamin-like protein DLP1 is essential for normal distribution and morphology of the endoplasmic reticulum and mitochondria in mammalian cells. *Molecular biology of the cell*, *10*(12), pp.4403-4417.

Rambold, A. S., Kostelecky, B., Elia, N. and Lippincott-Schwartz, J. (2011). Tubular network formation protects mitochondria from autophagosomal degradation during nutrient starvation. *Proceedings of the National Academy of Sciences*, *108*(25), pp.10190-10195.

Regula, K. M., Ens, K. and Kirshenbaum, L. A. (2002). Inducible expression of BNIP3 provokes mitochondrial defects and hypoxia-mediated cell death of ventricular myocytes. *Circulation Research*, *91*(3), pp.226-231.

Saito, T., Nah, J., Oka, S. I., Mukai, R., Monden, Y., Maejima, Y., Ikeda, Y., Sciarretta, S., Liu, T., Li, H. and Baljinnyam, E. (2019). An alternative mitophagy pathway mediated by Rab9 protects the heart against ischemia. *The Journal of Clinical Investigation*, *129*(2), pp.802-819.

Sandoval, H., Thiagarajan, P., Dasgupta, S. K., Schumacher, A., Prchal, J. T., Chen, M. and Wang, J. (2008). Essential role for Nix in autophagic maturation of erythroid cells. *Nature*, *454*(7201), pp.232-235.

Santel, A. and Fuller, M. T. (2001). Control of mitochondrial morphology by a human mitofusin. *Journal of cell science*, *114*(5), pp.867-874.

Satyanarayana, A. and Rudolph, K. L. (2004). p16 and ARF: activation of teenage proteins in old age. *The Journal of Clinical Investigation*, *114*(9), pp.1237-1240.

Schrepfer, E. and Scorrano, L. (2016). Mitofusins, from mitochondria to metabolism. *Molecular cell*, *61*(5), pp.683-694.

Shirakabe, A., Ikeda, Y., Sciarretta, S., Zablocki, D. K. and Sadoshima, J. (2016). Aging and autophagy in the heart. *Circulation Research*, *118*(10), pp.1563-1576.

Smirnova, E., Griparic, L., Shurland, D. L. and Van Der Bliek, A. M. (2001). Dynamin-related protein Drp1 is required for mitochondrial division in mammalian cells. *Molecular biology of the cell*, *12*(8), pp.2245-2256.

Song, M., Franco, A., Fleischer, J. A., Zhang, L. and Dorn II, G. W. (2017). Abrogating mitochondrial dynamics in mouse hearts accelerates mitochondrial senescence. *Cell metabolism*, *26*(6), pp.872-883.

Song, M., Gong, G., Burelle, Y., Gustafsson, Å. B., Kitsis, R. N., Matkovich, S. J. and Dorn, G. W. (2015). Interdependence of Parkin-mediated mitophagy and mitochondrial fission in adult mouse hearts. *Circulation Research*, *117*(4), pp.346-351.

Song, M., Mihara, K., Chen, Y., Scorrano, L. and Dorn II, G. W. (2015). Mitochondrial fission and fusion factors reciprocally orchestrate mitophagic culling in mouse hearts and cultured fibroblasts. *Cell metabolism*, *21*(2), pp.273-286.

Song, Z., Ghochani, M., McCaffery, J. M., Frey, T. G. and Chan, D. C. (2009). Mitofusins and OPA1 mediate sequential steps in mitochondrial membrane fusion. *Molecular biology of the cell*, *20*(15), pp.3525-3532.

Sun, Y., Yao, X., Zhang, Q. J., Zhu, M., Liu, Z. P., Ci, B., Xie, Y., Carlson, D., Rothermel, B. A., Sun, Y. and Levine, B. (2018). Beclin-1-dependent autophagy protects the heart during sepsis. *Circulation*, *138*(20), pp.2247-2262.

Sussman, M. A. and Anuersa, P. (2004). Myocardial aging and senescence: where have the stem cells gone?. *Annual review of physiology*, *66*, p.29.

Swynghedauw, B., Besse, S., Assayag, P., Carré, F., Chevalier, B., Charlemagne, D., Delcayre, C., Hardouin, S., Heymes, C. and Moalic, J. M. (1995). Molecular and cellular biology of the senescent hypertrophied and failing heart. *The American Journal of cardiology*, *76*(13), pp.2D-7D.

Taegtmeyer, H. (1994). Energy metabolism of the heart: From basic concepts to clinical applications applications. *Current problems in cardiology*, *19*(2), pp.61-113.

Takahashi, T., Schunkert, H., Isoyama, S., Wei, J. Y., Nadal-Ginard, B., Grossman, W. and Izumo, S. (1992). Age-related differences in the expression of proto-oncogene and contractile protein genes in response to pressure overload in the rat myocardium. *The Journal of Clinical Investigation*, *89*(3), pp.939-946.

Twig, G. and Shirihai, O. S. (2011). The interplay between mitochondrial dynamics and mitophagy. *Antioxidants & redox signaling*, *14*(10), pp.1939-1951.

Twig, G., Elorza, A., Molina, A. J., Mohamed, H., Wikstrom, J. D., Walzer, G., Stiles, L., Haigh, S. E., Katz, S., Las, G. and Alroy, J. (2008). Fission and selective fusion govern mitochondrial segregation and elimination by autophagy. *The EMBO journal*, *27*(2), pp.433-446.

Umare, M. D., Chimthanawala, N. M., Trivedi, R. V., Wankhede, N. L., Umekar, M. J. and Kale, M. B. (2021). Neuropsychiatric Disorders: A Pharmacovigilance Perspective. *Bull. Env. Pharmacol. Life Sci*, *10*, pp.141-155.

Umare, M. D., Wankhede, N. L., Bajaj, K. K., Trivedi, R. V., Taksande, B. G., Umekar, M. J., Mahore, J. G. and Kale, M. B. (2021). Interweaving of reactive oxygen species and major neurological and psychiatric disorders. In *Annales PharmaceutiquesFrançaises*. Elsevier Masson.

Upaganlawar, A., Upasani, C. and Kale, M. B. (2022). Medicinal Potential of Fenugreek in Neuropathy and Neuroinflammation Associated Disorders. In *Fenugreek: Traditional and Modern Medicinal Uses* (pp. 211-229). CRC Press.

Upaganlawar, A. B., Wankhede, N. L., Kale, M. B., Umare, M. D., Sehgal, A., Singh, S., Bhatia, S., Al-Harrasi, A., Najda, A., Nurzyńska-Wierdak, R. and Bungau, S. (2021). Interweaving epilepsy and neurodegeneration: Vitamin E as a treatment approach. *Biomedicine & Pharmacotherapy*, *143*, p.112146.

Valente, E. M., Abou-Sleiman, P. M., Caputo, V., Muqit, M. M., Harvey, K., Gispert, S., Ali, Z., Del Turco, D., Bentivoglio, A. R., Healy, D. G. and Albanese, A. (2004). Hereditary early-onset Parkinson's disease caused by mutations in PINK1. *Science*, *304*(5674), pp.1158-1160.

Vásquez-Trincado, C., García-Carvajal, I., Pennanen, C., Parra, V., Hill, J. A., Rothermel, B. A. and Lavandero, S. (2016). Mitochondrial dynamics, mitophagy and cardiovascular disease. *The Journal of Physiology*, *594*(3), pp.509-525.

Wai, T., García-Prieto, J., Baker, M. J., Merkwirth, C., Benit, P., Rustin, P., Rupérez, F. J., Barbas, C., Ibañez, B. and Langer, T. (2015). Imbalanced OPA1 processing and mitochondrial fragmentation cause heart failure in mice. *Science*, *350*(6265), p.aad0116.

Wankhede, N. L., Kale, M. B., Upaganlawar, A. B., Taksande, B. G., Umekar, M. J., Behl, T., Abdellatif, A. A., Bhaskaran, P. M., Dachani, S. R., Sehgal, A. and Singh, S. (2022). Involvement of molecular chaperone in protein-misfolding brain diseases. *Biomedicine & Pharmacotherapy*, *147*, p.112647.

Wankhede, N. L., Samrit, S. R., Umare, M. D., Bajaj, K. K., Trivedi, R. V., Umekar, M. J. and Kale, M. B. (2021). Biomarkers of cancer: A Comprehensive Review. *Advances in Bioresearch Adv. Biores.,* Vol. 12 (1) January 2021: 221-233.

Wei, Y., Chiang, W. C., Sumpter Jr, R., Mishra, P. and Levine, B. (2017). Prohibitin 2 is an inner mitochondrial membrane mitophagy receptor. *Cell*, *168*(1-2), pp.224-238.

Xu X., Kobayashi S., Chen K., Timm D., Volden P., Huang Y., Gulick J., Yue Z., Robbins J., Epstein P. N., Q. Liang. (2013). Diminished autophagy limits cardiac injury in mouse models of type 1 diabetes. *J Biol Chem*. 288:18077-18092.

Yan, C., Gong, L., Chen, L., Xu, M., Abou-Hamdan, H., Tang, M., Désaubry, L. and Song, Z. (2020). PHB2 (prohibitin 2) promotes PINK1-PRKN/Parkin-dependent mitophagy by the PARL-PGAM5-PINK1 axis. *Autophagy*, *16*(3), pp.419-434.

Yang, K. C., Ma, X., Liu, H., Murphy, J., Barger, P. M., Mann, D. L. and Diwan, A., 2015. Tumor necrosis factor receptor–associated factor 2 mediates mitochondrial autophagy. *Circulation: Heart Failure*, *8*(1), pp.175-187.

Yin, Z., Popelka, H., Lei, Y., Yang, Y. and Klionsky, D. J. (2020). The roles of ubiquitin in mediating autophagy. *Cells*, *9*(9), p.2025.

Yoon, Y., Krueger, E. W., Oswald, B. J. and McNiven, M. A. (2003). The mitochondrial protein hFis1 regulates mitochondrial fission in mammalian cells through an interaction with the dynamin-like protein DLP1. *Molecular and cellular biology*, *23*(15), pp.5409-5420.

Youle, R. J. and Narendra, D. P. (2011). Mechanisms of mitophagy. *Nature Reviews Molecular cell biology*, *12*(1), pp.9-14.

Zhang, T., Xue, L., Li, L., Tang, C., Wan, Z., Wang, R., Tan, J., Tan, Y., Han, H., Tian, R. and Billiar, T. R. (2016). BNIP3 protein suppresses PINK1 kinase proteolytic cleavage to promote mitophagy. *Journal of Biological Chemistry*, *291*(41), pp.21616-21629.

Zhang, Y., Nguyen, D. T., Olzomer, E. M., Poon, G. P., Cole, N. J., Puvanendran, A., Phillips, B. R. and Hesselson, D. (2017). Rescue of Pink1 deficiency by stress-dependent activation of autophagy. *Cell Chemical Biology*, *24*(4), pp.471-480.

Chapter 3

Mitophagy in Cardiovascular Diseases

Saboor Khan[1]
Aditi Kaushik[1]
and Shalini Mani*
[1]Department of Biotechnology, Jaypee Institute of Information Technology, Noida, Uttar Pradesh, India

Abstract

Mitochondria are essential organelles responsible for producing energy within cells, and their proper function is critical for maintaining cellular homeostasis. However, mitochondrial dysfunction can accumulate damaged mitochondria and produce harmful reactive oxygen species (ROS), contributing to various diseases, including neurodegenerative disorders, cancer, and cardiovascular disease. Mitochondrial dysfunction is implicated in the pathogenesis of cardiovascular diseases such as heart failure, ischemia-reperfusion injury, and atherosclerosis. Mitophagy plays a critical role in maintaining mitochondrial function and preventing the accumulation of damaged mitochondria in cardiovascular cells. Several studies have demonstrated the importance of mitophagy in cardiovascular diseases. For example, impaired mitophagy has been implicated in the development of heart failure. In a mouse model of heart failure, impaired mitophagy was associated with decreased cardiac function and increased cardiac cell death.

Similarly, impaired mitophagy has also been implicated in the development of atherosclerosis. In a mouse model of atherosclerosis, impaired mitophagy was associated with increased plaque formation and inflammation. On the other hand, enhancing mitophagy is protective against cardiovascular diseases. For example, in a mouse model of ischemia-reperfusion injury, enhancing mitophagy was associated with reduced cardiac cell death and improved cardiac function. Overall, these studies suggest that mitophagy plays an essential role in the pathogenesis of cardiovascular diseases and that enhancing mitophagy may be a potential therapeutic strategy for these conditions.

Keywords: ischemia-reperfusion injury, heart failure, atherosclerosis, mitochondria, mitophagy

---

* Corresponding Author's Email: shalini.mani@jiit.ac.in.

In: Mitophagy and Its Role in Health and Disease
Editors: Shalini Mani and Samiksha Wasnik
ISBN: 979-8-89113-470-6

Introduction

Cardiovascular diseases (CVD) significantly contribute to global disease burden and mortality rates, with non-infectious diseases such as CVD becoming increasingly prevalent in the 20th century as infectious disease-related deaths decreased (Joseph et al., 2017). CVD encompasses various maladies related to the circulatory system, including atherosclerosis, peripheral artery disease, heart failure, and arrhythmias (Nichols M et al., 2014).

The incidence of CVD is expected to continue to increase, making prevention and treatment crucial (Yusuf et al., 2014).

Mitochondria are essential organelles found in cardiomyocytes that provide energy for metabolism and play a role in cell death, proliferation, signal transduction, and calcium homeostasis (Bonora et al., 2018), (Galluzzi et al., 2012), (López-Crisosto et al., 2017). Recent research has shown that mitophagy, the process by which cells remove damaged mitochondria, significantly impacts the pathophysiology of CVD (Dorn et al., 2015), (Lesnefsky et al., 2016), (Liang & Kobayashi, 2016), (Moyzis et al., 2015), (Saito & Sadoshima, 2015), (Shi et al., 2018), (Shirihai et al., 2015), (Sun et al., 2016), (Wu et al., 2018), (Tong & Sadoshima, 2016), (Woodall & Gustafsson, 2018). Studies in laboratory animals have demonstrated that deleting mitophagy-related genes can lead to the spontaneous development of cardio-degenerative disorders (Pedro et al., 2017b).

Mitophagy in Cardiovascular Diseases

Atherosclerosis

Atherosclerosis, a chronic inflammatory disease common in industrialized nations, results from the interaction among various factors such as lipid accumulation, vascular smooth muscle cell (VSMC) proliferation, and inflammation, leading to notable narrowing of arteries due to plaque build-up inside the artery lumen (Liu et al., 2015). The irregular growth of VSMCs in the neointima and foam cell generation contributes to the advancement of atherosclerotic lesions. The immense death of macrophage- or VSMC-derived foam cells within advanced atherosclerotic lesions leads to plaque instability (Yamada et al., 2018).

Vascular EC's high expression of dynamin-related protein 1 (Drp1) incites pro-inflammatory factors and increases blood vessel inflammation by affecting mitochondrial conversion and morphological changes (Santovito et al., 2020), (Ikeda et al., 2015). Macrophage polarization is linked to mitochondrial function and metabolism.

Oxidized low-density lipoprotein (ox-LDL) is an atherosclerosis agent that creates atherosclerosis models. In vascular smooth muscle cells, ox-LDL disorganizes mitochondrial structure, increases ROS levels, and induces mitochondrial dysfunction (Kattoor et al., 2017).

Mitochondria deterioration with age leads to mitochondrial dysfunction, which causes increased ROS production, reduced ATP synthesis, enhanced lipid peroxidation, and imbalance in cellular calcium homeostasis (Narendra & Youle, 2011). Mitochondrial dysfunction and AS are interconnected to oxidative stress, abnormal ATP synthesis, extensive cellular damage, and apoptosis (Chistiakov et al., 2017).

Ischemic Heart Disease

Ischemic heart disease refers to damage to the myocardium resulting from an imbalance between myocardial demand and coronary blood flow and is a significant threat to human health. Ischemia causes injury and death of myocardial cells (Anzell et al., 2017) by reducing or eliminating the availability of O_2 for oxidative metabolic reactions. Reduced or absent oxidative phosphorylation causes intracellular acidosis and rapid depletion of intracellular creatine phosphate, which, along with a simultaneous rise in intracellular phosphate, stimulates glycolysis and lactate production. The accumulation of lactate and the hydrolysis of ATP cause further reductions in intracellular pH. Both reduced pH and increased Pi are molecular factors responsible for the immediate downregulation of contractility, which leads to akinesia of the ischemic zone (Cobbe & Poole-Wilson, 1980).

In a study, ischemia and glucose stripping caused significant depletion in ATP levels coinciding with the upregulation of autophagy-dependent-adenosine monophosphate-activated protein kinase (AMPK) and the retardation of the mammalian target of rapamycin (mTOR) pathway (Chen et al., 2013), (Z. Li et al., 2015). In conclusion, transgenic mice expressing dominant-negative AMPK show a limited induction of autophagy following ischemia (Matsui et al., 2007).

During ischemia, there is no change in the total tissue Ca2+, but the mitochondrial Ca2+ increases, indicating a reallocation of intracellular Ca2+ except for the massive net accumulation during reperfusion after extended ischemia. Unregulated Ca2+ entry may occur through leaks or membrane faults caused by the aggregation of toxic metabolites, such as lysophospholipids, oxyradicals, and/or mechanical factors (Ferrari et al., 1986), (Murphy et al., 2019). The previously described relationship between mitochondrial calcium transport and ATP synthesis may be a key factor for cell survival during postischemic reperfusion (Ferrari R et al., 1989).

Hypertension

The accumulation of dysfunctional mitochondria due to malfunctioning mitophagy can lead to various pathologies (Palikaras et al., 2017). Hypertension has several risk factors, such as insufficient energy supply, oxidative damage, abnormal signal transduction, genetic mutations, and kinetic imbalance (Dikalov & Mikalova, 2019). It has been reported that myocardial hypertrophy caused by hypertension can change the function, phenotype, and energy metabolism of cardiomyocytes. These variations are reflected in the processes such as using metabolic substrates, ETC function, and ATP synthesis (Neubauer, 2007). ROS play a significant role in regulating the vascular structure and tension remodeling, and their increase or decrease may lead to vasoconstriction, smooth muscle cell damage, vascular inflammation and remodeling, and aggravation of hypertension (Togliatto et al., 2017). Hypertension is linked with increased mitochondrial fragmentation and is suggested to affect mitochondrial dynamics, energy metabolism, and function in cardiomyocytes (Lu et al., 2007), (Lahera et al., 2017).

Arrhythmia

Arrhythmia, a common feature of cardiovascular disease, can occur alone or in conjunction with other heart conditions. In a healthy heart, coordinated electrical activity ensures proper cardiac function. However, when this activity fails, it can lead to arrhythmia (Santulli et al., 2014), a condition in which the electrical signals that regulate heartbeat malfunction, which can be fatal in some cases (Waldmann et al., 2016).

Mitochondria play a role in arrhythmia development because they are involved in ATP and ROS production. Reducing ATP levels and increased ROS generation disrupt intracellular ion balance (Yang et al., 2014), leading to distinctiveness in cardiac action potentials. This vicious cycle is based on variations in ATP and ROS levels that induce electrophysiological changes and ultimately lead to arrhythmia (Yang et al., 2014). Proper mitophagy can help prevent damage to harmful substances and maintain mitochondrial homeostasis in cells, thereby preventing cellular disturbances (Li et al., 2021).

Heart Failure

Cardiac mitochondria are crucial in generating energy through oxidative phosphorylation, and are essential for heart function (Billia et al., 2011). However, when they malfunction, they contribute to myocyte loss and the development of heart failure (Cahill et al., 2015). To ensure a network of functional mitochondria, myocytes have quality control mechanisms in place (Shires & Gustafsson, 2015). Mitophagy decreases with aging or disease, leading to inadequate elimination of damaged mitochondria and increased ROS and peroxide levels. This results in the oxidative damage of cardiomyocyte mitochondrial proteins, lipids, and DNA, contributing to heart failure. These findings are consistent with the clinical presentation of heart failure in elderly patients and those with advanced heart disease (Shi R et al., 2018), (Dai et al., 2014), (Qiu et al., 2018).

Cardiomyopathies

Cardiomyopathy (CM) is a heart disorder that affects the heart muscle's structure and function and occurs independently of other identifiable cardiovascular diseases, such as coronary artery disease (CAD), uncontrolled hypertension (HT), and congenital heart disorders. The frequency of cardiomyopathy in the global population is 3%, but it rises to 12% in diabetic patients, leading to an increased risk of heart failure and mortality (Dandamudi et al., 2014).

In diabetes, mitochondria are the primary source of intracellular reactive oxygen species (ROS) and are particularly vulnerable to oxidative injury (Nishikawa et al., 2000), (Kristal et al., 1997). Disfigured mitochondria can increase ROS manufacturing further through a process called ROS-induced ROS release (Kristal et al., 1997), (Brady et al., 2006), (Taneike et al., 2010) and may set off the leakage of pro-death factors, leading to cardiomyocyte death (Cai et al., 2002), (Ghosh et al., 2005), (Malhotra et al., 2005), (Shen et al., 2006), (Frustaci et al., 2000). Increased mitochondrial manufacturing of ROS and impaired mitochondrial structures and operations are commonly observed in both type 1 and type 2 diabetic patients' hearts.

Table 1 summarizes the literature showing the efefct of mitochondrial dysfunction in cardiovascular problems.

Table 1. Summary of literature describing impact of mitochondrial dysfunction in cardiovascular diseases

Pathology	Impact Of Mitochondrial Dysfunction	References
Atherosclerosis	Deletion of antioxidant systems in mice led to elevated mitochondrial reaction species (ROS) and accelerated atherosclerosis, suppression of mitochondrial Trx2 in endothelial cells causes endothelial dysfunction, excess ROS production can contribute to atherogenesis, xanthine oxidase can promote atherosclerosis by stimulating LOX-1 and CD-36.	(Kattoor et al., 2017)
	Melatonin induces mitophagy thereby reducing the ROS production through the Sirt3/FOXO3a/Parkin signaling pathway	(Ma et al., 2018)
Ischemic heart disease	Affects the activation of autophagic pathway and degradation of misfolded proteins, rapamycin has been shown to attenuate polyQ aggregate formation and decrease toxicity in cell and mouse models of Huntington's disease suggesting a protective effect in ischemic heart disease.	(Kouroku et al., 2006)
	Dysfunctional mitochondria can lead to programmed cell death or the production of reactive oxygen species	(Liu et al., 2012)
	Mitochondrial dysfunction can lead to impaired energy production in the heart, Impaired mitochondrial function can disrupt cellular signaling pathways in the heart	(Foglio et al., 2016)
	FUNDC1-mediated mitophagy protects the heart from ischemia-reperfusion (I/R) injury	(Zhang et al., 2017)
	Promotion of mitophagy through PEDF/PEDF-R/PA/DAG/PKC-α/ULK1/FUNDC1 pathway can protect hypoxic cardiomyocytes from cellular DNA damage and programmed cell death.	(Li et al., 2018)
Cardiomyopathies and heart failure	BNIP3 is a key regulator of mitochondrial function and cell death in ventricular myocytes during hypoxia, increased BNIP3 expression is detected in postnatal ventricular myocytes and adult rat hearts subjected to hypoxia.	(Regula et al., 2002)
	Inhibition of mitophagy by cytosolic p53 exacerbates mitochondrial compromise and heart failure	(Hoshino et al., 2013)
	Mitochondrial dysfunction in the heart can lead to cardiac damage and dysfunction, reactive oxygen species produced by dysfunctional mitochondria can cause further damage.	(Kubli et al., 2013)
	Mitochondrial dysfunction can cause a decreased NAD/NADH ratio, decreased NAD/NADH ratio can lead to inactivation of sirtuin deacetylases	(Xu et al., 2015)
	Mitochondrial dysfunction can induce hypertrophy and fibrosis in the heart, both unopposed fission and unopposed fusion of mitochondria can cause cardiac dysfunction	(Ikeda et al., 2015)
	Activation of sterile myocardial inflammation is observed as a result of mitochondrial dysfunction.	(Cahill et al., 2015)
	Mitochondrial dysfunction can lead to declining contractility of cardiomyocytes, it can contribute to the development of heart failure	(S. Wang et al., 2018)
	Disruption of energy homeostasis and mitochondrial dysfunction cause cardiac hypertrophy, mitochondrial abnormalities include irregular arrangement, swelling, and reduced density	(B. Wang et al., 2018)
	Impaired mitochondrial function can result in lipid accumulation in the heart, it can exacerbate diastolic dysfunction and induce systolic dysfunction.	(Tong et al., 2019)

Table 1. (Continued)

Pathology	Impact Of Mitochondrial Dysfunction	References
Hypertension	Coexistence of obesity and renovascular hypertension impairs mitochondrial biogenesis and function, mitochondrial dysfunction is associated with tissue remodeling and left ventricular diastolic dysfunction	(X. Zhang et al., 2015)
	Spermidine feeding delays the progression to heart failure in rats with mitochondrial dysfunction.	(Eisenberg et al., 2016)
	Mitochondrial dysfunction in hypertension leads to impaired angiogenic capacity of late EPCs, deficient CXCR4/JAK2/SIRT5 signaling contributes to mitochondrial dysfunction, capillary rarefaction in hypertension is associated with the decline in angiogenic capacity, late EPC mitochondria could be a potential target for treating hypertension-related microvascular loss.	(Yu et al., 2019)
Arrhythmia	Thiol oxidation of ryanodine receptors (RyR2s) by mitochondria-derived reactive oxygen species (mito-ROS) contributes to defective Ca2+ homeostasis in aging rabbit hearts.	(Murphy et al., 2019)
Peripheral vascular disease	The accumulation of LC3 and lack of co-localization with LAMP2 in PAD fibers suggests impaired clearance of damaged mitochondria.	(White et al., 2016)

Conclusion

Maintaining a delicate balance between eliminating and replenishing damaged mitochondria with new ones is crucial for heart function because mitochondria play a vital role in it. Mitophagy, which helps maintain cellular stability, can potentially reduce or prevent certain diseases. Various factors influence the complex systems involved in this process and are interconnected. Several processes trigger mitophagy throughout the cellular life cycle, including growth, development, aging, and death. Altered levels of mitophagy could impact the development of cardiovascular disease. Therefore, targeting mitophagy could be a potential therapy for cardiovascular disease, but further research is necessary to determine whether different substances that induce mitophagy have corresponding effects to advance the development of clinical diagnostics and treatments (Morciano et al., 2020).

References

Anzell, A. R., Maizy, R., Przyklenk, K., & Sanderson, T. H. (2017). Mitochondrial Quality Control and Disease: Insights into Ischemia-Reperfusion Injury. *Molecular Neurobiology,* 55(3), 2547–2564.

Billia, F., Hauck, L., Konecny, F., Rao, V., Shen, J., & Mak, T. W. (2011). PTEN-inducible kinase 1 (PINK1) /Park6 is indispensable for normal heart function. *Proceedings of the National Academy of Sciences of the United States of America, 108*(23), 9572–9577.

Brady, N. R., Hamacher-Brady, A., Westerhoff, H. V., & Gottlieb, R. A. (2006). A wave of reactive oxygen species (ROS)-Induced ROS release in a sea of excitable mitochondria. *Antioxidants & Redox Signaling, 8*(9–10), 1651–1665.

Bonora, M., Wieckowski, M. R., Sinclair, D. A., Kroemer, G., Pinton, P., & Galluzzi, L. (2018). Targeting mitochondria for cardiovascular disorders: therapeutic potential and obstacles. *Nature Reviews Cardiology, 16*(1), 33–55.

Cahill, T. J., Leo, V. C., Kelly, M., Stockenhuber, A., Kennedy, N. W., Bao, L., Cereghetti, G. M., Harper, A., Czibik, G., Liao, C., Bellahcene, M., Steeples, V., Ghaffari, S., Yavari, A., Mayer, A., Poulton, J., Djp,

F., Scorrano, L., Hettiarachchi, N. T., Ashrafian, H. (2015). Resistance of dynamin-related protein 1 oligomers to disassembly impairs mitophagy, resulting in myocardial inflammation and heart failure. *Journal of Biological Chemistry*, *290*(43), 25907–25919.

Cai, L., Li, W., Wang, G., Guo, L., Jiang, Y., & Kang, Y. J. (2002). Hyperglycemia-Induced apoptosis in mouse myocardium. *Diabetes*, *51*(6), 1938–1948.

Chen, K., Kobayashi, S., Xu, X., Viollet, B., & Liang, Q. (2013). AMP activated protein kinase is indispensable for myocardial adaptation to caloric restriction in mice. *PLOS ONE*, *8*(3), e59682.

Chistiakov, D. A., Shkurat, T. P., Melnichenko, A. A., Grechko, A. V., & Orekhov, A. N. (2017). The role of mitochondrial dysfunction in cardiovascular disease: A brief review. *Annals of Medicine*, *50*(2), 121–127.

Cobbe, S. M., & Poole-Wilson, P. A. (1980). The time of onset and severity of acidosis in myocardial ischaemia. *Journal of Molecular and Cellular Cardiology*, *12*(8), 745–760.

Dai, D. F., Karunadharma, P., Chiao, Y. A., Basisty, N., Crispin, D. A., Hsieh, E. J., Chen, T., Gu, H., Djukovic, D., Raftery, D., Beyer, R. P., MacCoss, M. J., & Rabinovitch, P. S. (2014). Altered proteome turnover and remodeling by short-term caloric restriction or rapamycin rejuvenate the aging heart. *Aging Cell*, *13*(3), 529–539.

Dandamudi, S., Slusser, J. P., Mahoney, D. W., Redfield, M. M., Rodeheffer, R. J., & Chen, H. H. (2014). The Prevalence of Diabetic Cardiomyopathy: A Population-Based Study in Olmsted County, Minnesota. *Journal of Cardiac Failure*, *20*(5), 304–309.

Dikalov, S., & Dikalova, A. (2019). Crosstalk between mitochondrial hyperacetylation and oxidative stress in vascular dysfunction and hypertension. *Antioxidants & Redox Signaling*, *31*(10), 710–721.

Dorn, G. W., Vega, R. B., & Kelly, D. P. (2015). Mitochondrial biogenesis and dynamics in the developing and diseased heart. *Genes & Development*, *29*(19), 1981–1991.

Eisenberg, T., Abdellatif, M., Schroeder, S., Primeßnig, U., Stekovic, S., Pendl, T., Harger, A., Schipke, J., Zimmermann, A., Schmidt, A., Tong, M., Ruckenstuhl, C., Dammbrueck, C., Gross, A. S., Herbst, V., Magnes, C., Trausinger, G., Narath, S. H., Meinitzer, A., Madeo, F. (2016). Cardioprotection and lifespan extension by the natural polyamine spermidine. *Nature Medicine,* 22(12), 1428–1438.

Ferrari, R., Ceconi, C., Curello, S., Cargnoni, A., Agnoletti, G., Gm, B., &Visioli, O. (1986). Intracellular effects of myocardial ischaemia and reperfusion : role of calcium and oxygen. *Pub Med, 7 Suppl A*, 3–12.

Ferrari, R., Boffa, G. M., Ceconi, C., Curello, S., Boraso, A., Ghielmi, S., & Cargnoni, A. (1989). Effect of D-600 on ischemic and reperfused rabbit myocardium: relation with timing and modality of administration. *Basic research in cardiology,* 84(6), 606–622.

Foglio, E., Puddighinu, G., Germani, A., Russo, M. A., & Limana, F. (2016). HMGB1 inhibits apoptosis following MI and induces autophagy via MTORC1 inhibition. *Journal of Cellular Physiology*, 232(5), 1135–1143.

Frustaci, A., Kajstura, J., Chimenti, C., Jakoniuk, I., Leri, A., Maseri, A., Nadal-Ginard, B., &Anversa, P. (2000). Myocardial cell death in human diabetes. *Circulation Research,* 87(12), 1123–1132.

Galluzzi, L., Kepp, O., Trojel-Hansen, C., & Kroemer, G. (2012). Mitochondrial control of cellular life, stress, and death. *Circulation Research*, *111*(9), 1198–1207.

Ghosh, S., Pulinilkunnil, T., Yuen, G., Kewalramani, G., Ding, A., Qi, D., Abrahani, A., & Rodrigues, B. (2005). Cardiomyocyte apoptosis induced by short-term diabetes requires mitochondrial GSH depletion. *American Journal of Physiology-heart and Circulatory Physiology*, *289*(2), H768–H776.

Hoshino, A., Mita, Y., Okawa, Y., Ariyoshi, M., Iwai-Kanai, E., Ueyama, T., Ikeda, K., Ogata, T., & Matoba, S. (2013). Cytosolic p53 inhibits Parkin-mediated mitophagy and promotes mitochondrial dysfunction in the mouse heart. *Nature Communications,* 4(1). https://doi.org/10.1038/ncomms3308.

Ikeda, Y., Shirakabe, A., Maejima, Y., Zhai, P., Sciarretta, S., Toli, J., Nomura, M., Mihara, K., Egashira, K., Ohishi, M., Abdellatif, M., &Sadoshima, J. (2015). Endogenous DRP1 mediates mitochondrial autophagy and protects the heart against energy stress. *Circulation Research*, *116*(2), 264–278.

Joseph, P., Leong, D. P., McKee, M., Anand, S. S., Schwalm, J., Teo, K., Mente, A., & Yusuf, S. (2017). Reducing the Global Burden of Cardiovascular Disease, Part 1. *Circulation Research*, *121*(6), 677–694.

Kattoor, A. J., Pothineni, N. V., Palagiri, D., & Mehta, J. L. (2017). Oxidative stress in atherosclerosis. *Current Atherosclerosis Reports*, *19*(11).

Kristal, B. S., Jackson, C. T., Chung, H., Matsuda, M., Nguyen, H., & Yu, B. P. (1997). Defects at center P underlie Diabetes-Associated mitochondrial dysfunction. *Free Radical Biology and Medicine,* 22(5), 823–833.

Kouroku, Y., Fujita, E., Tanida, I., Ueno, T., Isoai, A., Kumagai, H., Ogawa, S., Kaufman, R. J., Kominami, E., & Momoi, T. (2006). ER stress (PERK/eIF2α phosphorylation) mediates the polyglutamine-induced LC3 conversion, an essential step for autophagy formation. *Cell Death & Differentiation,* 14(2), 230–239

Kubli, D. A., Zhang, X., Lee, Y., Hanna, R., Quinsay, M. N., Nguyen, C. K., Jimenez, R. E., Petrosyan, S., Murphy, A. N., & Gustafsson, Å. B. (2013). Parkin Protein Deficiency Exacerbates Cardiac Injury and Reduces Survival following Myocardial Infarction. *Journal of Biological Chemistry*, 288(2), 915–926.

Lahera, V., De Las Heras, N., López-Farré, A., Manucha, W., & Ferder, L. (2017). Role of mitochondrial dysfunction in hypertension and obesity. *Current Hypertension Reports*, *19*(2).

Lesnefsky, E. J., Chen, Q., & Hoppel, C. L. (2016). Mitochondrial metabolism in aging heart. *Circulation Research*, *118*(10), 1593–1611.

Li, Y., Liu, Z., Zhang, Y., Zhao, Q., Wang, X., Lu, P., Zhang, H., Wang, Z., Dong, H., & Zhang, Z. (2018). PEDF protects cardiomyocytes by promoting FUNDC1mediated mitophagy via PEDF-R under hypoxic condition. *International Journal of Molecular Medicine.* https://doi.org/10.3892/ijmm.2018.3536.

Li, Y., Meng, W., Hou, Y., Li, D., Wang, X., Wu, K., Sun, S., Liu, H., Li, X., Lin, F., & Zhao, G. (2021). Dual role of mitophagy in cardiovascular diseases. *Journal of Cardiovascular Pharmacology*, *78*(1), e30–e39.

Li, Z., Song, Y., Liu, L., Hou, N., An, X., Zhan, D., Li, Y., Zhou, L., Li, P., Yu, L., Xia, J., Zhang, Y., Wang, J., & Yang, X. (2015). MiR -199a impairs autophagy and induces cardiac hypertrophy through mTOR activation. *Cell Death & Differentiation*, *24*(7), 1205–1213.

Liang, Q., & Kobayashi, S. (2016). Mitochondrial quality control in the diabetic heart. *Journal of Molecular and Cellular Cardiology*, *95*, 57–69.

Liu, L., Feng, D., Guo, C., Chen, M., Zheng, Q., Song, P., Ma, Q., Zhu, C., Wang, R., Qi, W., Huang, L., Xue, P., Li, B., Wang, X., Jin, H., Wang, J., Yang, F., Liu, P., Zhu, Y., . . . Chen, Q. (2012). Mitochondrial outer-membrane protein FUNDC1 mediates hypoxia-induced mitophagy in mammalian cells. *Nature Cell Biology,* 14(2), 177–185.

Liu, H., Cao, Y., Tong, T., Shi, J., Zhang, Y., Yang, Y., & Liu, C. (2015). Autophagy in atherosclerosis. *Chinese Medical Journal*, *128*(1), 69–74.

López-Crisosto, C., Pennanen, C., Vásquez-Trincado, C., Morales, P. E., Bravo-Sagua, R., Quest, A. F. G., Chiong, M., & Lavandero, S. (2017). Sarcoplasmic reticulum–mitochondria communication in cardiovascular pathophysiology. *Nature Reviews Cardiology*, *14*(6), 342–360.

Lu, F., Moore, X., Gao, X., Dart, A. M., Lim, Y. L., & Du, X. (2007). Down-regulation of mitofusin-2 expression in cardiac hypertrophy in vitro and in vivo. *Life Sciences*, *80*(23), 2154–2160.

Ma, S., Chen, J., Feng, J., Zhang, R., Fan, M., Han, D., Li, X., Li, C., Ren, J., Wang, Y., & Cao, F. (2018). Melatonin ameliorates the progression of atherosclerosis via mitophagy activation and NLRP3 inflammasome inhibition. *Oxidative Medicine and Cellular Longevity,* 2018, 1–12.

Malhotra, A., Begley, R., Kang, B., Rana, I., Liu, J., Yang, G., Mochly-Rosen, D., & Meggs, L. G. (2005). PKC-ε-dependent survival signals in diabetic hearts. *American Journal of Physiology-heart and Circulatory Physiology*, *289*(4), H1343–H1350.

Matsui, Y., Takagi, H., Qu, X., Abdellatif, M., Sakoda, H., Asano, T., Levine, B., &Sadoshima, J. (2007). Distinct roles of autophagy in the heart during ischemia and reperfusion. *Circulation Research*, *100*(6), 914–922.

Morciano, G., Patergnani, S., Bonora, M., Pedriali, G., Tarocco, A., Bouhamida, E., Marchi, S., Ancora, G., Anania, G., Wieckowski, M. R., Giorgi, C., &Pinton, P. (2020). Mitophagy in cardiovascular diseases. *Journal of Clinical Medicine*, *9*(3), 892.

Moyzis, A. G., Sadoshima, J., & Gustafsson, Å. B. (2015). Mending a broken heart: the role of mitophagy in cardioprotection. *American Journal of Physiology-heart and Circulatory Physiology*, *308*(3), H183–H192.

Murphy, K. R., Baggett, B., Cooper, L. N., Lu, Y., O-Uchi, J., Sedivy, J. M., Terentyev, D., & Koren, G. (2019). Enhancing autophagy diminishes aberrant CA2+ homeostasis and arrhythmogenesis in aging rabbit hearts. *Frontiers in Physiology*, *10*.

Narendra, D. P., &Youle, R. J. (2011). Targeting mitochondrial dysfunction: Role for PINK1 and Parkin in mitochondrial quality control. *Antioxidants & Redox Signaling*, *14*(10), 1929–1938. Neubauer, S. (2007). The Failing Heart — An Engine Out of Fuel. *The New England Journal of Medicine*, *356*(11), 1140–1151.

Nichols, M., Townsend, N., Scarborough, P., & Rayner, M. (2014). Cardiovascular disease in Europe 2014 : epidemiological update. *European Heart Journal*, *35*(42), 2950–2959.

Palikaras, K., Daskalaki, I., Markaki, M., &Tavernarakis, N. (2017). Mitophagy and age-related pathologies : Development of new therapeutics by targeting mitochondrial turnover. *Pharmacology & Therapeutics*, *178*, 157–174.

Pedro, J. M. B., Kroemer, G., & Galluzzi, L. (2017b). Autophagy and mitophagy in cardiovascular disease. *Circulation Research*, *120*(11), 1812–1824.

Qiu, Z., Hu, Y., Geng, Y., Wu, H., Bo, R., Shi, J., Wang, J., & Wang, H. (2018). Xin Fu Kang oral liquid inhibits excessive myocardial mitophagy in a rat model of advanced heart failure. *PubMed*, *10*(10), 3198–3210.

Regula, K. M., Ens, K., & Kirshenbaum, L. A. (2002). Inducible expression of BNIP3 provokes mitochondrial defects and Hypoxia-Mediated cell death of ventricular myocytes. *Circulation Research,* 91(3), 226–231.

Saito, T., &Sadoshima, J. (2015). Molecular mechanisms of mitochondrial Autophagy/Mitophagy in the heart. *Circulation Research*, *116*(8), 1477–1490.

Santulli, G., Iaccarino, G., De Luca, N., Trimarco, B., & Condorelli, G. (2014). Atrial fibrillation and microRNAs. *Frontiers in Physiology*, *5*.

Shires, S. E., & Gustafsson, Å. B. (2015). Mitophagy and heart failure. *Journal of Molecular Medicine*, *93*(3), 253–262.

Shen, X., Zheng, S., Metreveli, N., & Epstein, P. N. (2006). Protection of cardiac mitochondria by overexpression of MNSOD reduces diabetic cardiomyopathy. *Diabetes*, *55*(3), 798–805.

Shi, R., Guberman, M., & Kirshenbaum, L. A. (2018). Mitochondrial quality control : The role of mitophagy in aging. *Trends in Cardiovascular Medicine*, *28*(4), 246–260.

Shirihai, O. S., Song, M., & Dorn, G. W. (2015). How mitochondrial dynamism orchestrates mitophagy. *Circulation Research*, *116*(11), 1835–1849.

Sun, N., Youle, R. J., & Finkel, T. (2016). The mitochondrial basis of aging. *Molecular Cell*, *61*(5), 654–666.

Nishikawa, T., Edelstein, D., Du, X., Yamagishi, S., Matsumura, T., Kaneda, Y., Yorek, M. A., Beebe, D. A., Oates, P. J., Hammes, H., Giardino, I., & Brownlee, M. (2000). Normalizing mitochondrial superoxide production blocks three pathways of hyperglycaemic damage. *Nature*, *404*(6779), 787–790.

Taneike, M., Yamaguchi, O., Nakai, A., Hikoso, S., Takeda, T., Mizote, I., Oka, T., Tamai, T., Oyabu, J., Murakawa, T., Nishida, K., Shimizu, T., Hori, M., Komuro, I., Shirasawa, T., Mizushima, N., & Otsu, K. (2010). Inhibition of autophagy in the heart induces age-related cardiomyopathy. *Autophagy*, *6*(5), 600–606.

Togliatto, G., Lombardo, G., & Brizzi, M. F. (2017). The future challenge of reactive oxygen species (ROS) in hypertension : from bench to bed side. *International Journal of Molecular Sciences*, *18*(9), 1988. Tong, M., & Sadoshima, J. (2016). Mitochondrial autophagy in cardiomyopathy. *Current Opinion in Genetics & Development*, *38*, 8–15.

Tong, M., Saito, T., Zhai, P., Oka, S., Mizushima, W., Nakamura, M., Ikeda, S., Shirakabe, A., & Sadoshima, J. (2019). Mitophagy is essential for maintaining cardiac function during high fat Diet-Induced Diabetic cardiomyopathy. *Circulation Research,* 124(9), 1360–1371.

Waldmann, V., & Marijon, E. (2016). Troubles du rythme cardiaque : diagnostic et prise en charge [Cardiac arrhythmias: Diagnosis and management]. *La Revue de medecine interne,* 37(9), 608–615.

Wang, B., Nie, J., Wu, L., Hu, Y., Wen, Z., Dong, L., Zou, M., Chen, C., & Wang, D. W. (2018). AMPKA2 protects against the development of heart failure by enhancing mitophagy via PINK1 phosphorylation. *Circulation Research,* 122(5), 712–729. .

Wang, S., Zhao, Z., Feng, X., Cheng, Z., Xiong, Z., Wang, T., Lin, J., Zhang, M., Hu, J., Yang, F., Reiter, R. J., Wang, H., & Sun, D. (2018). Melatonin activates Parkin translocation and rescues the impaired mitophagy activity of diabetic cardiomyopathy through Mst1 inhibition. *Journal of Cellular and Molecular Medicine,* 22(10), 5132–5144.

White, S.H., M., Ferrucci, L., Tian, L., Zhao, L., Gao, Y., Kibbe, M. R., Criqui, M. H., Leeuwenburgh, C., & Peterson, C. A. (2016). Walking performance is positively correlated to calf muscle fiber size in peripheral artery disease subjects, but fibers show aberrant mitophagy: an observational study. *Journal of Translational Medicine,* 14(1).

Woodall, B. P., & Gustafsson, Å. B. (2018). Autophagy-A key pathway for cardiac health and longevity. *Acta Physiologica,* 223(4), e13074.

Wu, M., Yiang, G. T., Liao, W. T., Tsai, A. P., Cheng, Y. L., Cheng, P., Li, C., & Li, C. J. (2018). Current mechanistic concepts in Ischemia and reperfusion injury. *Cellular Physiology and Biochemistry, 46*(4), 1650–1667.

Xu, W., Barrientos, T., Mao, L., Rockman, H. A., Sauve, A. A., & Andrews, N. C. (2015). Lethal cardiomyopathy in mice lacking transferrin receptor in the heart. *Cell Reports,* 13(3), 533–545.

Yamada, T., Murata, D., Adachi, Y., Itoh, K., Kameoka, S., Igarashi, A., Kato, T., Araki, Y., Huganir, R. L., Dawson, T. M., Yanagawa, T., Okamoto, K., Iijima, M., &Sesaki, H. (2018). Mitochondrial stasis reveals P62-Mediated ubiquitination in Parkin-Independent mitophagy and mitigates nonalcoholic fatty liver disease. *Cell Metabolism, 28*(4), 588-604.e5.

Yang, K., Bonini, M. G., & Dudley, S. C. (2014). Mitochondria and arrhythmias. *Free Radical Biology and Medicine, 71*, 351–361.

Yu, B., Zhi, H., Zhang, X., Liang, J., He, J., Su, C., Xia, W., Zhang, G., & Tao, J. (2019). Mitochondrial dysfunction-mediated decline in angiogenic capacity of endothelial progenitor cells is associated with capillary rarefaction in patients with hypertension via downregulation of CXCR4/JAK2/SIRT5 signaling. *EBioMedicine,* 42, 64–75.

Yusuf, S., Rangarajan, S., Teo, K., Islam, S., Li, W., Liu, L., Jiang, B., Lou, Q., Lu, F., Liu, T., Liu, Y., Zhang, S., Mony, P., Swaminathan, S., Mohan, V., Gupta, R., Kumar, R., Vijayakumar, K., Lear, S. A., . . . Dagenais, G. R. (2014). Cardiovascular risk and events in 17 Low-, Middle-, and High-Income countries. *The New England Journal of Medicine, 371*(9), 818–827.

Zhang, W., Siraj, S., Zhang, R., & Chen, Q. (2017). Mitophagy receptor FUNDC1 regulates mitochondrial homeostasis and protects the heart from I/R injury. *Autophagy,* 13(6), 1080–1081. https://doi.org/10.1080/15548627.2017.1300224.

Zhang, X., Li, Z. L., Eirin, A., Ebrahimi, B., Pawar, A. S., Zhu, X. Y., Lerman, A., & Lerman, L. O. (2015). Cardiac metabolic alterations in hypertensive obese pigs. *Hypertension,* 66(2), 430–436. https://doi.org/10.1161/hypertensionaha.115.05478.

Chapter 4

Mitophagy in Neurological Disorders

Aishani Kulshreshtha[1]
Prashant Kaushik[1]
Avinav Agarwal[1]
Harshita Saxena[1]
Rupesh Kumar[1]
and Shazia Haider[2,*]
[1]Department of Biotechnology, Jaypee Institute of Information Technology, Noida, Uttar Pradesh, India
[2]Department of Biosciences, Jamia Millia University, New Delhi, India

Abstract

Proteins, organelles, aggregates, and other items found in the intracellular cytosol are transported to lysosomes for destruction via autophagy, a catabolic process. The selective autophagic breakdown of mitochondria is a kind of autophagy known as mitophagy. Mitophagy can eliminate stressed or damaged mitochondria, which is crucial for cellular health. We are beginning to comprehend the processes of mitophagy, mainly the functions of ubiquitin-dependent and receptor-dependent signals on injured mitochondria in initiating mitophagy, thanks to genetics, cell biology, and proteomics methods. Many neurodegenerative illnesses have been linked to mitochondrial and mitophagy dysfunction. Maintaining control over the quality and quantity of mitochondria in neurons requires the precise and appropriate destruction of defective mitochondria via mitophagy. This chapter aims to provide an overview of mitophagy's mechanisms in higher organisms and its functions in the pathogenesis of neurodegenerative diseases. The multifactorial nature of neurodegenerative diseases and the phenotypic variations observed across patients warrant the development of new therapeutic strategies; thus, focusing efforts toward a greater understanding of the mechanisms underlying mitophagy dysfunction in neurodegenerative disorders is a primary concern.

Keywords: mitophagy, neurodegenerative disease, oxidative stress, neuroinflammation

* Corresponding Author's Email: shaziahaider786@gmail.com.

In: Mitophagy and Its Role in Health and Disease
Editors: Shalini Mani and Samiksha Wasnik
ISBN: 979-8-89113-470-6

Introduction

The degradation of mitochondria comes under mitophagy, a subset of autophagy that has seen one of the most significant recent advancements in research.

Over the decades, efforts have been concentrated on elucidating various molecules underlying this process and its function under pathological or normal conditions (Martinez-Vicente 2017).

The primary energy source for neurons' high energy requirements is mitochondrial metabolism. As a result, the mitochondrial network must function perfectly. Furthermore, due to the distinct shape of neurons, including their axons and dendrites, mitochondria must be transported through axonal transport to distant regions of the cell to meet the energy needs at those locations. As a result, axonal trafficking flaws can affect neurons. Finally, cellular division cannot dilute components accumulated inside the cell because neurons are post-mitotic. There is a need for a sufficient "cleaning system" to dispose of any mitochondria that might be damaged or capable of inducing cell death. Given that mitochondria are crucial for maintaining neuronal viability, any dysfunction in mitochondria could result in neuronal death, which explains why many neurological disorders have been linked to dysregulation in mitochondria (Martinez-Vicente 2017).

Mitophagy thus plays a pivotal role in the etiology of neurological diseases. Mitophagy can prevent the build-up of defective mitochondria in cases of infective protein entities, mutations, stress, or injury. Hence, mitophagy may mediate neuroprotective effects in neurodegenerative diseases or acute brain injuries. Dysfunction of mitochondria shares significant overlap with various characteristic features of neurological disorders at a molecular level, such as oxidative stress, neuroinflammation, and accumulation of misfolded protein aggregates. Here we explore the link between mitophagy and neurological disorders and the various mechanisms through which mitophagy dysfunction may contribute to their pathogenesis.

The Role of Parkin/PINK1 Mediated Pathway in Neurological Discorderas

Although there is strong evidence for PINK1/Parkin mitophagy in cell lines, it has been disputed whether this pathway occurs in neurons, and its physiological significance is still up for debate (Cummins and Gotz 2018).

In practically all studies on PINK1/Parkin-dependent mitophagy, the uncoupler CCCP was used to trigger the process in non-neuronal cells overexpressing Parkin. This uncoupler produces a quick and significant reduction in the mitochondrial membrane potential across the whole mitochondrial pool. As these conditions (non-neuronal cells, overexpressed Parkin, and prolonged CCCP treatment) are not physiological, they may not correctly depict the type of mitophagy that occurs in neurons under normal conditions (Grenier, McLelland, and Fon 2013).

Neurons have a unique bioenergetic metabolism that differs from non-neuronal cells, which have a more glycolytic metabolism. Neurons mainly rely on oxidative phosphorylation to produce ATP. The discrepancies in how neurons react to CCCP and their failure to recruit parkin could be explained by this distinction. HeLa cells were forced to rely on mitochondrial respiration rather than glycolysis in an investigation (Van Laar et al. 2011) since Parkin translocation could not be seen in primary neurons, simulating how neurons produce energy. This made it more difficult for the cells to carry out Parkin translocation. This supported the hypothesis that part of the variations in Parkin translocation between neurons and non-neuronal cell types may be due to their bioenergetic differences.

However, subsequent research revealed that various triggers can induce Parkitrans location in cultured neurons. For example, authors (Cai et al. 2012) found that incubating cortical neurons with CCCP might, in fact, cause Parkin translocation, but only after at least 18 h, which is around 14–16 h longer than is required for the majority of cell lines. Parkin translocation also required the co-culture of glia and neurons and the addition of apoptosis inhibitors to the cultures. As a result, it was realized that the well-known PINK1/parkin-dependent pathway might be activated in response to severe mitochondrial damage, but given the small amounts of endogenous parkin present in neurons, this would probably only occur occasionally (Martinez-Vicente 2017). Eliminating mitochondria with modest malfunction can also be handled by alternative processes, such as CL-dependent mitophagy (Chu et al. 2013). The Parkin/PINK1 route and Cardiolipin-mediated pathways involved in neuronal mitophagy are depicted in Figure 1. The mechanisms involved are covered in more detail in the ensuing sections.

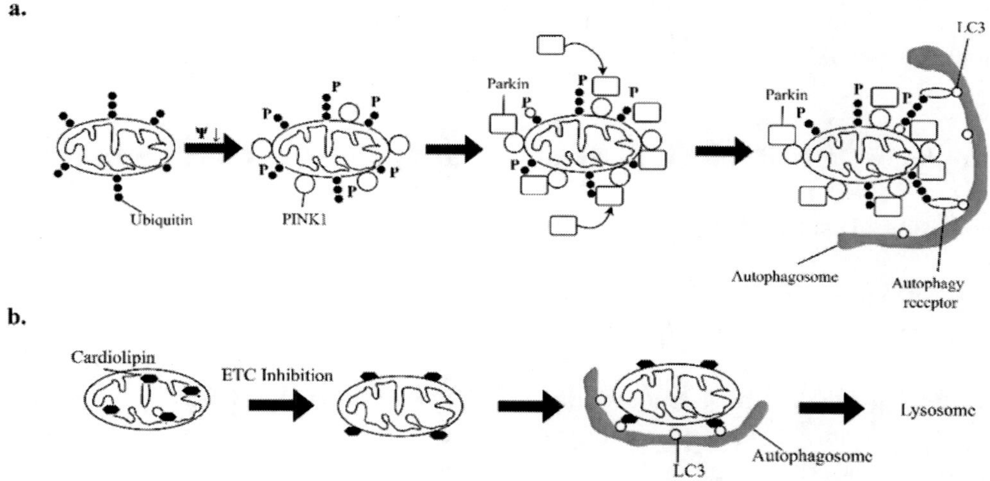

Figure 1. Neuronal mitophagy pathway. a. The Parkin/PINK route: 1. On their outer membrane, healthy mitochondria often have some basal ubiquitin moieties. 2. PINK1 import into mitochondria is stopped as a result of mitochondrial damage that decreases the inner membrane potential (ψ). Instead, the protein phosphorylates ubiquitin and accumulates on the outer mitochondrial membrane. 3. Parkin is both recruited from the cytosol and activated when PINK1 phosphorylates it and phosphoubiquitin binds to it. Parkin produces a positive feedback loop by ubiquitinating proteins found on outer membranes. 4. Autophagy receptors like optineurin bind ubiquitin chains, which in turn bind LC3 on autophagosomal membranes. After consuming mitochondria, autophagosomes transport them to the lysosome for destruction. **b.** Cardiolipin-mediated mitophagy. 1. The phospholipid cardiolipin is found in the inner membrane of mitochondria under normal circumstances. 2. Cardiolipin relocates to the mitochondrial outer membrane in response to the blockage of the electron transport chain (e.g., by rotenone) or the administration of additional Parkinsonian toxins (e.g., 6-OHDA). 3. Cardiolipin and LC3 engage in an interaction that causes autophagosomes to engulf mitochondria.

Oxidative Stress

Oxidative stress occurs when steady-state levels of ROS surpass their catabolism or detoxification, which can affect mitophagy to varying degrees. Mitochondria can be harmed by oxidative stress conditions that encourage the production of ROS, such as hypoxia, dietary restriction, and ischemia/reperfusion (IR). Mitophagy is one of the early responses to excessive ROS that might minimize oxidative damage and ROS formation. Several lines of evidence have shown that ROS interacts with ubiquitin- and receptor-dependent mitophagy pathways. The effects of ROS generation within mitochondria using mitochondrial-targeted photosensitizer mtKillerRed (mtKR) are increased ROS levels in the mitochondrial (mt) matrix, the loss of MMP, and the activation of Parkin-dependent mitophagy (Wang et al. 2012; Ma et al. 2016). Overexpression of mtSOD2 and other antioxidant proteins reverses this effect. Peroxiredoxin-6 (PRDX6) regulates ROS homeostasis in the initial stage of PINK1-dependent mitophagy after being attracted to damaged mitochondria by ROS (Shefa et al. 2019).

Unsurprisingly, many age-related diseases, including Alzheimer's and Parkinson's, simultaneously display chronic oxidative stress, mitochondrial dysfunctions, and decreased mitophagy. Studying mitophagy in these illnesses and other diseases characterized by accelerated aging provided essential insights into the molecular interactions between mitophagy and oxidative stress throughout aging.

Mitophagy Defects and Oxidative Stress in Various Neurological Disorders

Parkinson's Disease (PD)

The substantia nigra's dopaminergic neurons are affected mainly by Parkinson's, the second most prevalent degenerative illness of the CNS (SN)(Clark et al. 2021). Further highlighting the crucial roles that oxidative stress and mitophagy play in the etiology of Parkinson's disease is the finding that the antioxidant activities of the NIX (gene) and AMBRA1 proteins can help delay cell death in Parkinson's disease.NIX can promote the clearance of damaged mitochondria, sparing dopaminergic neurons, when PINK1-mediated mitophagy is suppressed in Parkinson's disease dopaminergic neurons. The fact that phorbol 12-myristate 13-acetate (PMA) increases NIX expression and results in reactive oxygen species formation in the same cells makes it abundantly clear that activating mitophagy helps reduce oxidative stress. In a similar setting where PINK1-mediated mitophagy is defective due to PINK1 or PARK2 mutations, AMBRA1 can rescue mitophagy and decrease cell mortality in vitro caused by rotenone or 6-OHDA treatments. However, research on patients with early-stage Parkinson's disease revealed that oxidative stress develops before neuron loss occurs and supports the theory that ROS may trigger the degeneration of neurons (Koentjoro, Park, and Sue 2017; Strappazzon et al. 2015; Di Rita et al. 2018).

Alzheimer's Disease (AD)

Mitophagy has been observed to play a role in the neurodegeneration seen in Alzheimer's, according to numerous lines of evidence. ROS is one of the factors that contribute to the impairment of mitophagy, and oxidative stress markers that are seen to rise as one age, such as lipid oxidation, protein carbonylation, and mtDNA oxidation, are significantly pronounced in AD (Navarro and Boveris 2010). Additionally, several animal models of AD have shown an elevation in the levels of oxidative stress (Hauptmann et al. 2009; Rhein et al. 2009) and formation of Aβ in vitro and in vivo sufficiently triggered by mtROS (Leuner et al. 2012). The autophagic build-up of mitochondria in susceptible Alzheimer's neurons provided the initial indication of abnormal mitophagy in Alzheimer's patient brains, which was later verified by many groups (Moreira et al. 2007).

Amyotrophic Lateral Sclerosis (ALS)

The neurological disease ALS leads to paralysis and runs the risk of premature death mainly by affecting the motor neurons in the brain and spinal cord (Mancuso and Navarro 2017). In a yeast model, a mutation in TDP-43 resulted in a build-up of nuclear factor protein 2 (Nrf2), an oxidative species modulator, production of oxidative species, and mitochondrial dysfunction. Increased oxidative stress markers were seen in TDP-43-expressing cells. The motor neuron-like cell line (NSC34) that expressed wild-type or TDP-43 mutation in mice displayed oxidative damage, mitochondrial dysfunction, and inception of mitophagy apart from exhibiting a disease approaching ALS (Guareschi et al. 2012).

Thus, the information and findings presented show that under physiological or pathological circumstances, an imbalance in the equilibrium between mitophagy and dysfunction of mitochondria can initiate, fuel, or hasten the aging process. The factors driving this imbalance are still not apparent, i.e., whether ROS-induced mitochondrial damage starts the dysregulation of mitophagy, setting off a cycle that reduces the body's capacity to remove damaged mitochondria and causes more ROS damage, or if a change in mitophagy regulation is one of the preliminary events that lead to overproduction of ROS (Hong et al. 2012).

Accumulation of Aggregated Misfolded Proteins

In the nervous system, abnormal synaptic protein interactions and misfolding are studied extensively as important pathogenic events resulting in neurodegeneration in various neurological disorders. Existing evidence implies that these proteins can interact to damage mitochondrial and plasma membranes during protofibril translocation (Wang, Liu, and Lu 2019). An extensive study is being conducted to establish whether the misfolding (toxic conversion) and progressive accumulation of synaptic proteins in the nervous system could be a critical infective event leading to neurodegeneration in PD, AD, and other various neurological diseases (Brand and Nicholls 2011). Parkinson's disease is a common neurodegenerative disease characterized by dopaminergic neuron loss in mutant alpha-synuclein accumulation and substantia nigra (Cen et al. 2021).

Abnormal α-synuclein accumulation arises in neuronal synapses, axons, and cell bodies in Parkinson's disease and related conditions. Studies on the function of α -synuclein in Parkinson's disease have revealed a relationship between aberrant α -synuclein expression and mitochondrial dysfunction. α-synuclein accumulation can worsen oxidative stress and impair mitochondrial function. Several experimental studies have linked α-synuclein to oxidative stress and mitochondrial dysfunction. Furthermore, α-synuclein may be involved in neurotoxin-induced mitochondrial dysfunction (Risiglione et al. 2021).

Alzheimer's disease is the most prevalent neurodegenerative condition, with symptoms including cognitive impairment and inattention. It is distinguished by dysfunctional accumulation, damaged synapses, intracellular neurofibrillary tangles (NFTs), and disease-defining amyloid-β (Aβ) oligomers (Brand and Nicholls 2011; Lavie et al. 2018). According to new research, mitochondrial autophagy defects are strongly linked to AD development.

According to accumulating evidence, amyloid deposition is a typical pathogenic trait seen in many neurodegenerative illnesses, such as Alzheimer's. Through oxidative stress, mitochondrial dysfunction promotes amyloid production in AD. Furthermore, hyperphosphorylated Tau, another characteristic AD protein, appears to aggregate in patients with the disease which suggests that the protein aggregate deterioration could be a viable action plan for reducing CNS impairment and neuroprotection (Sharma 2019).

The accumulation of valosin-containing protein (VCP), an mtHtt-binding protein on the mitochondria, in the nuclei and neuronal cytosol is thought to cause neuronal death in Huntington's disease (HD) (Sharma et al. 2021).

Alterations in oxidative stress and mitochondrial function, on the other hand, may stimulate protein aggregation in AD, PD, and HD, implying that these proteins may play a dual role in neurodegenerative disorder pathology. Mitophagy can partially account for cellular homeostasis mechanisms; thus, adequate mitophagy must reduce abnormal protein aggregation and stimulate organelle removal (Wang, Liu, and Lu 2019).

Impairment of Degradative Processes, Including Mitophagy, and Ubiquitin Proteasome System

Degradation of protein is an essential function that, when impaired, can cause different types of diseases. Ubiquitin protease system (UPS) and mitophagy are the major protein degradation systems. UPS is the primary route for the degradation of short-lived proteins, providing a temporal control and specificity required for fine-tuningsteady-state levels of regulatory proteins (Ciechanover, Orian, and Schwartz 2000). During acute starvation, amino acid pools are maintained by UPS-mediated catabolism, which contributes significantly to the degradation of defective proteins (Ciechanover and Brundin 2003; Taylor, Hardy J Fau - Fischbeck, and Fischbeck 2002).

Many neurodegenerative diseases such as AD, PD, HD, and ALS are due to the dysregulation of autophagic and mitophagic functions.

Mutations in the PINK1 gene result in familial Parkinson's disease, whereas reports of dysregulation in mitochondrial motility and mitophagy have been shown for sporadic Parkinson's disease (Ryan et al. 2015; Deas, Wood, and Plun-Favreau 2011). In Huntington's disease, accumulation of dysfunctional mitochondria in the cytoplasm due to the defects in the autophagic cargo recognition has been reported (Martinez-Vicente et al. 2010) Additionally, the neurodegenerative disease 2A type, the mutations in the Mfn2 gene cause Charcot Marie Tooth (Choi BO et al., 2015), and type 3 Spinocerebellar ataxia (SCA) is caused by the mutation of ataxin3 (Durcan et al. 2010).

In diseases such as AD, ALS, A-T, Creutzfeldt-Jakob, HD, and PD, there have been reports regarding the accumulation of misfolded protein crystals in the affected regions of the brain (Huang and Figueiredo-Pereira 2010). In most cases, ubiquitin conjugates were the main component of the protein crystals; this suggests the failure of their degradation via UPS.

Kufor-Rakeb syndrome, previously known as PARK9, is caused by a primarily neuronal lysosomal ATPase, which a mutated ATP13A2 encodes. The mutation causes protein retention in the ER and proteasomal degradation, indicating that either loss of function in lysosomal protein degradation and/or overwhelming of the UPS is the main reason for neurodegeneration.

Therapeutic benefits regarding the intrinsic catabolic pathways are vitally important and can be achieved by further research on the relationship between UPS, mitophagy, and the relationship to human diseases.

Mitophagy and Neuroinflammation

The synergistic contribution of mitochondria and neuroinflammation to the degenerative process of neurodegeneration is now commonly accepted. Due to the pleiotropic nature of the genes, there is an interesting overlap between mitophagy and neuroinflammation (Zhao et al. 2018).

Neuroinflammation is a type of response mounted by glial cells such as astrocytes oligodendrocytes, and immune cells whenever any infection or damage occurs in the body. This response ultimately leads to cell death through a chronic and excessive production of inflammatory mediators. Apart from these cells, oligodendrocytes and pericytes are also involved in neuroinflammation (Dhapola et al. 2021). Mitochondrial DNA (mtDNA) has been linked increasingly to inflammation. Oxidatively damaged mtDNA causes inflammatory signaling in astrocytes, specifically concerning neuroinflammation. A plethora of pro-inflammatory compounds, including mtDNA, formyl peptides, adenosine triphosphate (ATP), mitochondrial transcription factor A (TFAM), cardiolipin, cytochrome C, and RNA, are found in mitochondria (Mathew et al. 2012).

A rising number of studies on ALS suggest that several genes involved in its pathogenesis encode mitophagy receptors, such as OPTN, p62, and TBK1(Martinez-Vicente 2017). While the role of p62 is controversial (Oakes, Davies, and Collins 2017), OPTN and TBK1 mutations interfere with the mitophagy process. Particularly in the case of TBK1 functional loss, the resulting impaired axonal transport of mitochondria would result in neuronal damage. Damage to neurons may cause the cells around them to respond innately, resulting in neuroinflammation. It has been shown that by controlling IFNα and IFNβ synthesis through phosphorylation, TBK1 participates in an innate immune response. Ligand binding of Toll-like receptors TLR3 and TLR4 triggers TBK1, resulting in the movement of glial cells to damaged regions for phagocytosis. Additionally, neurotoxic chemicals such as reactive oxygen species and pro-inflammatory cytokines are produced during this process. These chemicals may exacerbate neuronal injury, additional glial cell activation, and neuroinflammation (Oakes, Davies, and Collins 2017).

For the case of AD, one study (Fang et al. 2017) reported that AD post-mortem brain samples contained aggregates of damaged mitochondria. Another notable characteristic of AD is the presence of many activated microglia and astrocytes (Wilkins et al. 2015). Correlating these two, it has been observed that mitochondrial components can mimic pathogens and trigger inflammatory responses. Apart from this, amyloid precursor protein (APP) mRNA and protein levels were also observed to be elevated by mitochondrial lysates, as reported by the authors (Wilkins et al. 2015). The precursor peptide for amyloid beta (Aβ) is called amyloid precursor protein (APP), and the amyloid plaques found in AD patient brain samples are formed by aggregation of fibrils formed by Aβ. The interaction of Aβ and tau activates the glial cells, ultimately leading to an uncontrolled inflammatory response, as shown in Figure 2. These findings imply that mitochondrial lysates are neuroinflammation mediators and that, as a result, they may impact APP biology and AD as a whole.

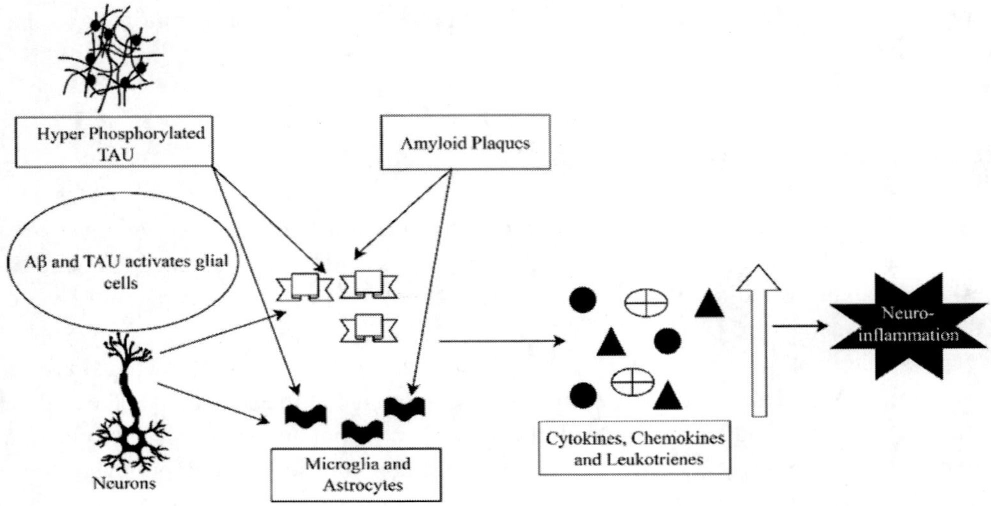

Figure 2. Neuroinflammation in Alzheimer's disease. The interaction between Aβ and tau activates glial cells. Microglia and astrocytes are the two primary cell types in charge of inflammatory reactions in the brain. As a result of the activation of glial cells, which produce various pro-inflammatory mediators, more inflammatory molecules are directed towards the site of injury, leading to neuroinflammation.

Mitophagy induction can reduce neuroinflammation. Damaged mitochondria emit DAMPs (damaged associated molecular patterns), which also entail a rise in the release of ROS and a decrease in ATP levels, among other things. Increased ROS levels brought on by the build-up of ROS-producing mitochondria due to mitophagy/autophagy blockage have previously been demonstrated to trigger NLRP3 (NLR family, pyrin domain containing 3 inflammasomes). Inflammation can also enhance ROS production, lower the mitochondrial membrane's potential, and cause mitochondrial pro-apoptotic proteins to leak into the cytoplasm of microglia. As a result, as cellular energy metabolism diminishes, levels of oxidative stress rise (Van Laar et al. 2011).

Talking about Obstructive Sleep Apnea (OSA), a recent study (Wu, X et al., 2021) revealed that chronic intermittent hypoxia (CIH), one of the primary causes of OSA, leads to neuroinflammation through dysregulation in the NLRP3-Parkin axis. mtROS has been implicated in hypoxia-related tissue injury, and the study demonstrated that through promoting Parkin-dependent mitophagy in CIH-induced neuroinflammation, NLRP3 deficiency served as a shielding promoter. Thus, potential treatment options for OSA-related neurocognitive impairment could be NLRP3 gene knockout or pharmacological blockade. A similar study found that Pinocembrin, a natural flavonoid, substantially improves the cognitive deterioration and neuroinflammation of intermittent hypoxia in IH mice. The findings specifically demonstrated that IH-induced cytotoxicity fails to damage microglial cells because of the protection by pinocembrin, which activates BNIP3-dependent mitophagy through the JNK-ERK signaling pathway. Pinocembrin-enhanced mitophagy was thus proposed as a plausible strategy to counter the effects of mtROS and NLRP3-mediated inflammation caused by IH-induced microglia in OSA patients (Gong et al. 2020).

All in all, neuroinflammation and mitophagy overlap considerably, making it an exciting field for further research. Neurodegeneration can lead to alterations that result in neuroinflammation and mitophagy defects, and there is a link between both such that changes in the functions of one can lead to significant alterations in the functions of the other (Wilkins et al. 2015).

Conclusion

In this chapter, we have discussed how mitophagy may play a role in the progression of neurological disorders. Some neurodegenerative diseases have been proposed to benefit from mitophagy regulation. Neurodegenerative disease phenotypes improve and are protected by mitophagy activation. Mitophagy is a double-edged sword, though, and can be damaging in excess and insufficient amounts. Mitochondrial dysfunction has a significant hand in many common neurological disorders as well. Mitochondrial autophagy is necessary to preserve the nervous system, but only recently has the function of the PINKI/Parkin pathway been elucidated, and some of the results are still up for debate. In the nervous system, most mitophagy research has focused on its molecular pathogenesis and prospective treatment targets. However, there is still discussion regarding the specific workings of the receptor-mediated autophagy pathway, discrepancies in research and environmental factors, the relative merits of targeting mitophagy for illness treatment, and the best ways to deal with these problems. Moreover, the evidence for mitophagy targeting in long-term chronic nervous system diseases is equivocal, and there is a dearth of thorough, rigorous studies in the field of short-term acute nervous system diseases. These issues warrant further investigation in the future, as mitophagy pathways can help in the discovery of potential therapeutic targets for neurological disorders (Wang, Liu, and Lu 2019; Zhang, Dai, and Li 2021).

References

Brand, M. D., and D. G. Nicholls. 2011. "Assessing mitochondrial dysfunction in cells." *Biochem J* 435 (2) :297-312.

Cai, Q., H. M. Zakaria, A. Simone, and Z. H. Sheng. 2012. "Spatial parkin translocation and degradation of damaged mitochondria via mitophagy in live cortical neurons." *Curr Biol* 22 (6) :545-52.

Cen, Xufeng, Manke Zhang, Mengxin Zhou, Lingzhi Ye, and Hongguang Xia. 2021. "Mitophagy Regulates Neurodegenerative Diseases." *Cells* 10 (8) :1876.

Chu, C. T., J. Ji, R. K. Dagda, J. F. Jiang, Y. Y. Tyurina, A. A. Kapralov, V. A. Tyurin, N. Yanamala, I. H. Shrivastava, D. Mohammadyani, K. Z. Q. Wang, J. Zhu, J. Klein-Seetharaman, K. Balasubramanian, A. A. Amoscato, G. Borisenko, Z. Huang, A. M. Gusdon, A. Cheikhi, E. K. Steer, R. Wang, C. Baty, S. Watkins, I. Bahar, H. Bayir, and V. E. Kagan. 2013. "Cardiolipin externalization to the outer mitochondrial membrane acts as an elimination signal for mitophagy in neuronal cells." *Nat Cell Biol* 15 (10) :1197-1205.

Ciechanover, Aaron, and Patrik Brundin. 2003. "The Ubiquitin Proteasome System in Neurodegenerative Diseases: Sometimes the Chicken, Sometimes the Egg." *Neuron* 40 (2) :427-446.

Ciechanover, Aaron, Amir Orian, and Alan L. Schwartz. 2000. "Ubiquitin-mediated proteolysis: biological regulation via destruction." *Bio Essays* 22 (5):442-451.

Clark, E. H., A. Vazquez de la Torre, T. Hoshikawa, and T. Briston. 2021. "Targeting mitophagy in Parkinson's disease." *J Biol Chem* 296 :100209.

Cummins, N., and J. Gotz. 2018. "Shedding light on mitophagy in neurons: what is the evidence for PINK1/Parkin mitophagy in vivo ?" *Cell Mol Life Sci* 75 (7) :1151-1162.

Deas, Emma, Nicholas W. Wood, and Hélène Plun-Favreau. 2011. "Mitophagy and Parkinson's disease: The PINK1–parkin link." *Biochimica et Biophysica Acta (BBA) - Molecular Cell Research* 1813 (4) :623-633.

Dhapola, R., S. S. Hota, P. Sarma, A. Bhattacharyya, B. Medhi, and D. H. Reddy. 2021. "Recent advances in molecular pathways and therapeutic implications targeting neuroinflammation for Alzheimer's disease." *Inflammopharmacology* 29 (6) :1669-1681.

Di Rita, A., P. D'Acunzo, L. Simula, S. Campello, F. Strappazzon, and F. Cecconi. 2018. "AMBRA1-Mediated Mitophagy Counteracts Oxidative Stress and Apoptosis Induced by Neurotoxicity in Human Neuroblastoma SH-SY5Y Cells." *Front Cell Neurosci* 12 : 92.

Durcan, Thomas M., Maria Kontogiannea, Thorhildur Thorarinsdottir, Lara Fallon, Aislinn J. Williams, Ana Djarmati, Tadeu Fantaneanu, Henry L. Paulson, and Edward A. Fon. 2010. "The Machado–Joseph disease-associated mutant form of ataxin-3 regulates parkin ubiquitination and stability." *Human Molecular Genetics* 20 (1) :141-154.

Fang, E. F., K. Palikaras, N. Sun, E. M. Fivenson, R. D. Spangler, J. S. Kerr, S. A. Cordonnier, Y. Hou, E. Dombi, H. Kassahun, N. Tavernarakis, J. Poulton, H. Nilsen, and V. A. Bohr. 2017. "In Vitro and In Vivo Detection of Mitophagy in Human Cells, C. Elegans, and Mice." *J Vis Exp* (129).

Gong, L. J., X. Y. Wang, W. Y. Gu, and X. Wu. 2020. "Pinocembrin ameliorates intermittent hypoxia-induced neuroinflammation through BNIP3-dependent mitophagy in a murine model of sleep apnea." *J Neuroinflammation* 17 (1) :337.

Grenier, K., G. L. McLelland, and E. A. Fon. 2013. "Parkin- and PINK1-Dependent Mitophagy in Neurons: Will the Real Pathway Please Stand Up ?" *Front Neurol* 4 :100.

Guareschi, S., E. Cova, C. Cereda, M. Ceroni, E. Donetti, D. A. Bosco, D. Trotti, and P. Pasinelli. 2012. "An over-oxidized form of superoxide dismutase found in sporadic amyotrophic lateral sclerosis with bulbar onset shares a toxic mechanism with mutant SOD1." *Proc Natl Acad Sci U S A* 109 (13) :5074-9.

Hauptmann, S., I. Scherping, S. Dröse, U. Brandt, K. L. Schulz, M. Jendrach, K. Leuner, A. Eckert, and W. E. Müller. 2009. "Mitochondrial dysfunction: An early event in Alzheimer pathology accumulates with age in AD transgenic mice." *Neurobiology of Aging* 30 (10) :1574-1586.

Hong, Kun, Yi Li, Weisong Duan, Yansu Guo, Hong Jiang, Wenju Li, and Chunyan Li. 2012. "Full-length TDP-43 and its C-terminal fragments activate mitophagy in NSC34 cell line." *Neuroscience Letters* 530 (2) :144-149.

Huang, Q., and M. E. Figueiredo-Pereira. 2010. "Ubiquitin/proteasome pathway impairment in neurodegeneration: therapeutic implications." *Apoptosis* 15 (11) :1292-311.

Koentjoro, B., J. S. Park, and C. M. Sue. 2017. "Nix restores mitophagy and mitochondrial function to protect against PINK1/Parkin-related Parkinson's disease." *Sci Rep* (2045-2322 (Electronic)).

Lavie, J., H. De Belvalet, S. Sonon, A. M. Ion, E. Dumon, S. Melser, D. Lacombe, J. W. Dupuy, C. Lalou, and G. Benard. 2018. "Ubiquitin-Dependent Degradation of Mitochondrial Proteins Regulates Energy Metabolism." *Cell Rep* 23 (10) :2852-2863.

Leuner, K., T. Schutt, C. Kurz, S. H. Eckert, C. Schiller, A. Occhipinti, S. Mai, M. Jendrach, G. P. Eckert, S. E. Kruse, R. D. Palmiter, U. Brandt, S. Drose, I. Wittig, M. Willem, C. Haass, A. S. Reichert, and W. E. Muller. 2012. "Mitochondrion-derived reactive oxygen species lead to enhanced amyloid beta formation." *Antioxid Redox Signal* 16 (12) :1421-33.

Ma, S., X. Zhang, L. Zheng, Z. Li, X. Zhao, W. Lai, H. Shen, J. Lv, G. Yang, Q. Wang, and J. Ji. 2016. "Peroxiredoxin 6 Is a Crucial Factor in the Initial Step of Mitochondrial Clearance and Is Upstream of the PINK1-Parkin Pathway." *Antioxid Redox Signal* 24 (9) :486-501.

Mancuso, Renzo, and Xavier Navarro. 2017. "Sigma-1 Receptor in Motoneuron Disease." In *Sigma Receptors: Their Role in Disease and as Therapeutic Targets*, edited by Sylvia B. Smith and Tsung-Ping Su, 235-254. Cham: Springer International Publishing.

Martinez-Vicente, M. 2017. "Neuronal Mitophagy in Neurodegenerative Diseases." *Front Mol Neurosci* 10 :64.

Martinez-Vicente, M., Z. Talloczy, E. Wong, G. Tang, H. Koga, S. Kaushik, R. de Vries, E. Arias, S. Harris, D. Sulzer, and A. M. Cuervo. 2010. "Cargo recognition failure is responsible for inefficient autophagy in Huntington's disease." *Nat Neurosci* 13 (5) :567-76.

Mathew, Alex, Tara A. Lindsley, Anna Sheridan, Devang L. Bhoiwala, Shazaan F. Hushmendy, Eric J. Yager, Elizabeth A. Ruggiero, and Dana R. Crawford. 2012. "Degraded Mitochondrial DNA is a Newly Identified Subtype of the Damage Associated Molecular Pattern (DAMP) Family and Possible Trigger of Neurodegeneration." *Journal of Alzheimer's Disease* 30 :617-627.

Moreira, P. I., S. L. Siedlak, X. Wang, M. S. Santos, C. R. Oliveira, M. Tabaton, A. Nunomura, L. I. Szweda, G. Aliev, M. A. Smith, X. Zhu, and G. Perry. 2007. "Increased autophagic degradation of mitochondria in Alzheimer's disease." *Autophagy* 3 (6) :614-5.

Navarro, A., and A. Boveris. 2010. "Brain mitochondrial dysfunction in aging, neurodegeneration, and Parkinson's disease." *Front Aging Neurosci* 2.

Oakes, J. A., M. C. Davies, and M. O. Collins. 2017. "TBK1 : A new player in ALS linking autophagy and neuroinflammation." *Mol Brain* 10 (1) :5.

Rhein, Virginie, Xiaomin Song, Andreas Wiesner, Lars M. Ittner, Ginette Baysang, Fides Meier, Laurence Ozmen, Horst Bluethmann, Stefan Dröse, Ulrich Brandt, Egemen Savaskan, Christian Czech, Jürgen Götz, and Anne Eckert. 2009. "Amyloid-beta and tau synergistically impair the oxidative phosphorylation system in triple transgenic Alzheimer's disease mice." *Proc Natl Acad Sci U S A* 106(47) (1091-6490 (Electronic) :20057-62.

Risiglione, P., F. Auid-Orcid Zinghirino, M. C. Di Rosa, A. Auid-Orcid Magrì, and A. Auid-Orcid Messina. 2021. "Alpha-Synuclein and Mitochondrial Dysfunction in Parkinson's Disease: The Emerging Role of VDAC. LID - 10.3390/biom11050718 [doi] LID - 718." *Biomolecules* 11(5) (2218-273X (Electronic)) :718.

Ryan, Brent J., Selim Hoek, Edward A. Fon, and Richard Wade-Martins. 2015. "Mitochondrial dysfunction and mitophagy in Parkinson's: from familial to sporadic disease." *Trends in Biochemical Sciences* 40 (4) :200-210.

Sharma, Aditi, Tapan Behl, Lalit Sharma, Lotfi Aelya, and Simona Bungau. 2021. "Mitochondrial Dysfunction in Huntington's Disease: Pathogenesis and Therapeutic Opportunities." *Current Drug Targets* 22 (14) :1637-1667.

Sharma, Vivek. 2019. "Alzheimer's Disease: A Consequence of Impaired Mitophagy?" *Asian Journal of Pharmaceutical and Clinical Research*: 75-80.

Shefa, U., N. Y. Jeong, I. O. Song, H. J. Chung, D. Kim, J. Jung, and Y. Huh. 2019. "Mitophagy links oxidative stress conditions and neurodegenerative diseases." *Neural Regen Res* 14 (5) :749-756.

Strappazzon, F., F. Nazio, M. Corrado, V. Cianfanelli, A. Romagnoli, G. M. Fimia, S. Campello, R. Nardacci, M. Piacentini, M. Campanella, and F. Cecconi. 2015. "AMBRA1 is able to induce mitophagy via LC3 binding, regardless of PARKIN and p62/SQSTM1." *Cell Death Differ* 22 (3) :419-32.

Taylor, J. P., Kenneth H. Hardy J Fau - Fischbeck, and K. H. Fischbeck. 2002. "Toxic proteins in neurodegenerative disease." *Science* 296(5575) (1095-9203 (Electronic)) :1991-5.

Van Laar, V. S., B. Arnold, S. J. Cassady, C. T. Chu, E. A. Burton, and S. B. Berman. 2011. "Bioenergetics of neurons inhibit the translocation response of Parkin following rapid mitochondrial depolarization." *Hum Mol Genet* 20 (5) :927-40.

Wang, Y., N. Liu, and B. Lu. 2019. "Mechanisms and roles of mitophagy in neurodegenerative diseases." *CNS Neurosci Ther* 25 (7) :859-875.

Wang, Y., Y. Nartiss, B. Steipe, G. A. McQuibban, and P. K. Kim. 2012. "ROS-induced mitochondrial depolarization initiates PARK2/PARKIN-dependent mitochondrial degradation by autophagy." *Autophagy* 8 (10) :1462-76.

Wilkins, H. M., S. M. Carl, S. G. Weber, S. A. Ramanujan, B. W. Festoff, D. A. Linseman, and R. H. Swerdlow. 2015. "Mitochondrial lysates induce inflammation and Alzheimer's disease-relevant changes in microglial and neuronal cells." *J Alzheimers Dis* 45 (1) :305-18.

Zhang, L., L. Dai, and D. Li. 2021. "Mitophagy in neurological disorders." *J Neuroinflammation* 18 (1) :297.

Zhao, Yan, Shaohui Huang, Jie Liu, Ximing Wu, Shuai Zhou, Ke Dai, and Yurong Kou. 2018. "Mitophagy Contributes to the Pathogenesis of Inflammatory Diseases." *Inflammation* 41 (5) :1590-1600.

Chapter 5

Mechanisms and Role of Mitophagy in Neurodegenerative Diseases

Rashmi Kumari[1,2]
Madhu Rani[1]
Siya Ram[2]
Moshahid Alam Rizvi[1]
and Anil Kumar[3,*]

[1]Genome Biology Laboratory, Department of Biosciences, Jamia Millia Islamia, New Delhi, India
[2]Indira Gandhi National Open University, Department of Life Sciences, New Delhi, India
[3]Directorate of Education, Rani Lakshmi Bai Central Agricultural University, Jhansi, Uttar Pradesh, India

Abstract

Selective autophagy is a long-term evolutionary phenomenon that eliminates abnormal intracellular elements and high protein accumulation. Strong mitophagy alters the number of mitochondria in some eukaryotic cells while preserving energy metabolism, such as in neurons. When mitophagy is interrupted, neurological diseases such as nerve degeneration, organelle dysfunction, and intracellular protein aggregation accumulation can result. Nevertheless, mitophagy hinders the accumulation of damaged mitochondria because of stress, injury, pathogenic protein, oligo-polymers, and the formation of protein mutations. Thus, mitophagy has neuroprotective effects in various neurological conditions, including stroke, hypoxic-ischemic brain injury, amyotrophic lateral sclerosis, Parkinson's disease, and Alzheimer's disease. It also protects against epilepsy and traumatic brain injury.

Furthermore, the multifaceted correlation between mitophagy and neurological disorders suggests that targeting mitophagy might be appropriate for the treatment and management of neurological disorders and acute brain injury. Conversely, this complex mitophagy process can be harmful and beneficial, so future practices should exploit its benefits. Here, we've covered various mitophagy signaling pathways and how it affects multiple neurological disorders. We have also emphasized the mitophagy inducers and briefly discussed the variations between mitophagy's beneficial and detrimental consequences.

---

[*] Corresponding Author's Email: anilkumar.mbge@gmail.com.

In: Mitophagy and Its Role in Health and Disease
Editors: Shalini Mani and Samiksha Wasnik
ISBN: 979-8-89113-470-6

Keywords: mitophagy, autophagy, neurological diseases, Alzheimer's disease, Parkinson's disease, Huntington's disease, traumatic brain injury

Introduction

Eukaryotic cells can control the function of mitochondria, which gives the organism crucial metabolic plasticity, important for various cellular functions.

Therefore, maintenance of mitochondrial function that depends on sufficient co-regulation of functions that control their turnover is known as mitochondrial biogenesis, which generates new mitochondria, whereas mitophagy maintains dysfunctional or damaged mitochondria.

An inadequate mitophagy process leads to the aggregation of impaired or stressed mitochondria, which decreases the tendency to produce adenosine triphosphate (ATP), which results in an upsurge in superoxide generation. Finally, itchanges the cellular cascade of intermediary metabolites having some pathological significance. Damaged or dysfunctional mitochondria are recognized as a hallmark of metabolic and neurodegenerative diseases, which are also associated with pathological developments (Chakravorty et al., 2019).

Mitophagy

Mitophagy, first revealed by Lewis and Lewis in cells, is selective autophagy. Autophagy is defined as the process by which a cell engulfs itself. Mitophagy is involved in maintaining damaged or dysfunctional mitochondria, which tend to gather subsequent impaired or stressed mitochondria. The latter pathway involves the transport of defective mitochondria for degradation into the lysosome (Li et al., 2022). In 1962, Ashford and Porter observed mitochondrial debris with the help of an electron microscope and reported that alterations in mitochondrial function might elicit autophagy. In 1998, the term mitophagy came into use.

Autophagy is categorized into chaperone-mediated autophagy, micro-autophagy, and macro-autophagy. These three kinds are divided based on the degraded structure size and being selective or non-selective based on the targeted particular cellular structure component. Notably, nonselective mitophagy is well-known as a primary mechanism in apoptosis. Recent studies have proposed that selective autophagy is very similar to non-selective macroautophagy, which has the only dissimilarity of having an extra phase targeting isolation membranes to cargo. However, it has been well recognized that, in macroautophagy pathways, various elements are frequently dispensable for selective autophagy. Thus, mitophagy is defined as a selective autophagy process of the organelles that occurs through the lysosomal pathway, which comprises isolation, sealing, and degradation. Not only mitochondria but also other subcellular structures such as the endoplasmic reticulum, Golgi, ribosomes, peroxisomes, lipid molecules, the midbody, and glycogen granules are also the most common targets of selective autophagy. Selective mitophagy is usually stimulated by aggregating damaged structures, such as accumulation of proteins, ER stress, or depolarization of mitochondria, which primarily contact particular autophagy signals and are then recognized by the specific autophagy receptor (Figure 1). In yeast, mitophagy was first described in 2004 (Kissova et al., 2004), where the mutated Uth1p presence on the outer membrane of mitochondria (OMM) was found to stop or inhibit autophagy during starvation. A related finding was conveyed later in cultured starved liver cells and observed that injured or impaired mitochondria are removed under contact with oxidative damage.

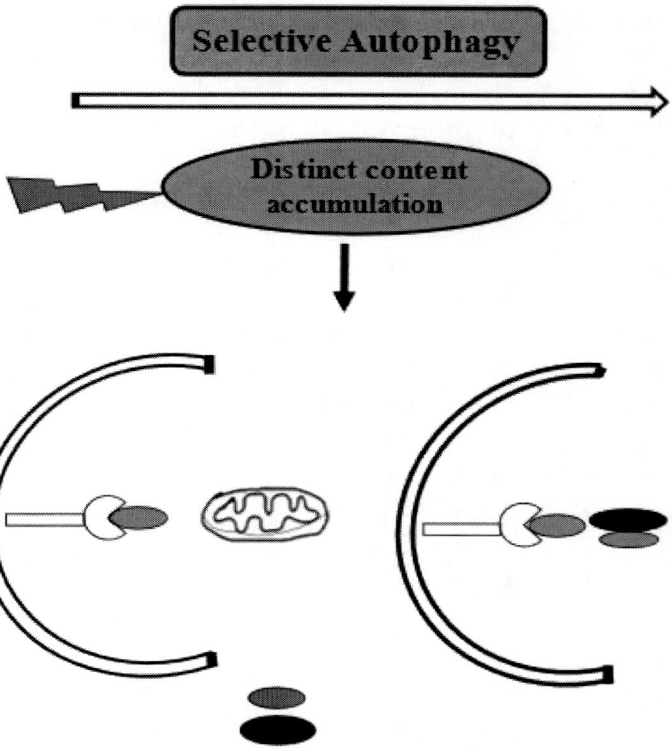

Figure 1. Depicts the selective mitophagy that is brought on by the assembly of numerous disrupted structures such as accumulations of proteins, ER stress, or depolarization of mitochondria that leads to interaction with particular signals of autophagy and then be recognized by the particular receptor of autophagy.

In general, the mitophagy is located on the inner side of the mitochondria in an autophagic vacuole known as mitophagosomes. Although mitophagy is categorized into three types, such as type 1 mitophagy, type 2 mitophagy, and type 3 mitophagy; in type 1, mitophagy occurs due to stimulation of deficient nutrients, whereas type 2 occurs due to stimulation of damage signals, and type 3 mitophagy, which is called micro-mitophagy, is associated with small mitochondria-derived vesicles. All types of mitophagy processes are different, as type 3 does not require fusion. On the other hand, both type 1 and type 2 require the fusion of a lysosome to form an autophagosome around mitochondria. Very recently, it has been investigated that mitophagy plays an essential function in normal growth and development by inspecting and quantifying it both in vitro and in vivo, which is observed in fluorescent transgenic mouse models such as mito-QC or mt-Keima (Sun et al., 2017).

Nonetheless, it is well known that every cell or tissue of an organism is a biological mechanism that functions and is controlled in regards to its modification of any energetic requirements. Some tissues, such as the kidneys, nervous systems, heart, the liver, and skeletal muscle, express great basal mitophagy activity. On the other hand, the spleen and thymus, for example, show a lesser level of mitophagy. In the mitophagy process, molecular and biochemical pathways are first emphasized in neurological disease, malignancy, cardiovascular disease, psychiatric disease, and aging. Balance, maintain synaptic plasticity, and remove stress or impaired mitochondria in the brain. Basal mitophagy plays an important role, although its unbalanced action is linked with age-associated neuronal impairment (Doblado et al., 2021).

Signaling Pathways in Mitophagy Mechanism

It is known that numerous receptor signaling pathways, including lipid- and protein-mediated pathways, are involved in mitochondrial mitophagy. The vital protein pathways often essential to the mitophagy mechanism include PINK1/Parkin, Nix/Bnip3, FUNDC1, p62/SQSTM1, and other related protein pathways. The three lipid-mediated pathways that also play a crucial part in the mitophagy mechanism are cardiolipids, fatty acids, and cholesterol (Zhang et al., 2021). Interestingly, a large number of neurodegenerative disease-associated proteins are also known to play a significant role in triggering mitophagy in these disease conditions (Table 1).

Table 1. Major neurodegenerative disease-associated proteins that play pivotal roles in mitophagy and their mechanisms (Wang et al., 2019)

Mitophagy regulatory proteins	Neurodegenerative diseases	Mechanisms
PINK1/Parkin	Alzheimer Disease	The dynamic equilibrium and mitophagy regulation is achieved by a process known as the interact
Tau		2. Modulate the expression of PSEN1, which is the major component of the catalytic center in the APP key enzyme γ-secretase
		1. Distress Parkin intracytoplasmic translocation to the mitochondrial membrane
Sirtuins		2. Regulate both Parkin and UCHL1 with the help of ANT-1-dependent ADP/ATP exchange
		1. SIRT1 plays an essential role in protecting neurons from the aggregation of Aβ -induced toxicity 2. Damage of mitochondria and microtubule function due to higher SIRT2 activation, which resulting in accumulation of Aβ and mitophagy
PINK1/Parkin	Parkinson's Disease	1. Various autophagy receptor proteins, for example; OPTN, p62, and NDP52, inducemitophagy thatinvolves the identification of ubiquitinate of the OMM proteins 2. Regulate the fission and fusion of mitochondria and also balance mitochondrial homeostasis;
α-Syn		3. Maintain the MMP by increasing the ubiquitination of Mfn
		1. A constituent of the Lewy bodies;
DJ-1		2. Upregulate oxidative stress and impair mitochondrial function;

Mitophagy regulatory proteins	Neurodegenerative diseases	Mechanisms
LRRK2		3. Triggering mPTP through interacting with the ANT or VDACand starting mitophagy affects the mitochondrial membrane potential, 1. Remove PD-associated aggregated p62 2. Modulate the elimination of endogenous ROS 3. Function in parallel with the PINK1/Parkin pathway 4. Regulate dimerization of α-Syn 1. Interact with the several regulators such as Drp1, OPA1 and Mfn of mitochondrial fission and fusion 2. Regulate the translocation of the CMA complex; 3. Induce the fission of mitochondria through DLP1, then stimulate the ULK1-dependent mitophagy 4. Initiate the JNK signal by interacting with the JIP3 and MKK4/7 5. Phosphorylate Bcl-2, regulate mitochondrial depolarization and autophagy; 6. Modulate the mitochondrial calcium uptake
PINK1/Parkin	Huntington's Disease	Maintain morphology of mitochondria, levels of ATP, and also neuronal health, together with the Mfn/MARF, as well as VDAC/Porin
GAPDH		Induction of amplified polyglutamine's leads to mitochondrial dysfunction
TG2		Cross-link with Htt, resulting in loss of mitochondrial membrane potential and amassing of unusual proteins in the brain
VCP		Causes mitophagy when combined with LC3, which is independent of the PINK1/Parkin pathway.
HAP1		Regulate neuronal autophagosome transport.
OPTN/TBK	Amyotrophic lateral sclerosis	1. Affect NF-κB activity and intracellular transport; 2. Maintain Parkin-dependent mitophagy homeostasis
SOD1		Modulate reverse transport of autophagosomes in axons in a PINK1/Parkin pathway-dependent manner.
VCP		Disrupt the mitophagy balance through the PINK1/Parkin pathway.

Protein Pathways

PINK1/Parkin-Mediated Mitophagy

The crucial roles of Parkin RBR E3 ubiquitin-protein ligase (PARKIN) and PTEN-induced kinase 1 (PINK1) in mitochondrial mitophagy, size, and motility have been documented (Quinn et al., 2020). These proteins have received the most research attention and are typically linked to neurodegenerative conditions like Parkinson's disease, Alzheimer's disease, and glaucoma (Table 1). Both PINK1 and Parkin are also engaged in various mitochondrial processes, including the dynamics of autophagic factors and their recruitment. Mitophagy that depends on ubiquitin is also governed by mitochondrial biogenesis. The PINK1 protein is encoded by the PARK6 gene and functions as a serine/threonine kinase. When PINK1 is directed to damage mitochondria, it phosphorylates itself, and when presenilins-associated rhomboid-like protein (PARL) cleaves it, the N-end rule route degrades it. However, when the mitochondrial membrane depolarizes, PINK1 accumulates on the outer membrane of the mitochondria and phosphorylates ubiquitin on the outer mitochondrial membrane (OMM) proteins at Ser65 (Figure 2) (Miller et al., 2019).

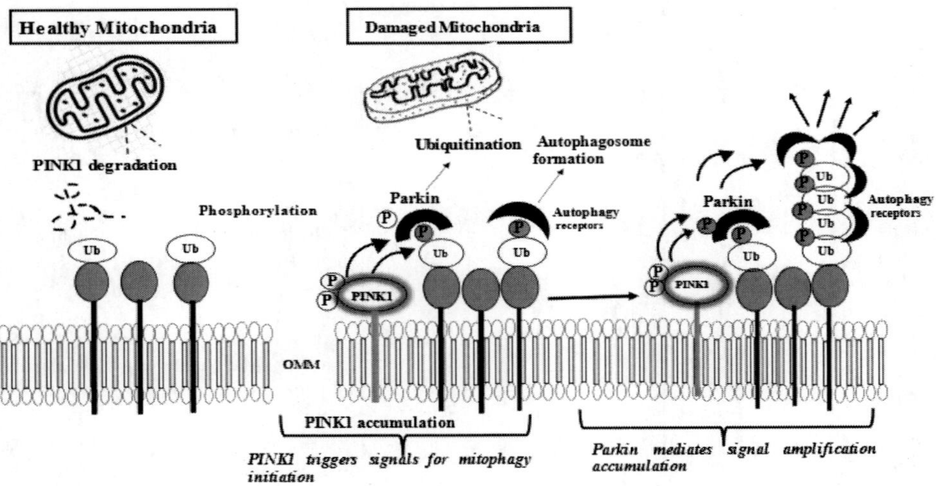

Figure 2. Mitophagy mediated by PINK1/Parkin: When PINK1 import is inhibited in damaged mitochondria that have lost their mitochondrial membrane potential, complete PINK1 builds up on the OMM. Dimeric PINK1 is precisely placed by the TOM complex, and auto-phosphorylation of PINK1 kinase activity prompts it to become active. When PINK1 phosphorylates ubiquitin, Parkin is drawn to mitochondria, and its E3 ligase activity is activated. The generation of autophagosomes is stimulated by the recruitment of autophagy receptors by phospho-ubiquitin. Parkin enhances this signaling by causing other mitochondrial proteins to become ubiquitinated.

On the other hand, the Parkin protein, an E3 ubiquitin ligase, is produced by the PARK2 gene. One of the finest regulating mechanisms in mitophagy for faulty and damaged mitochondria is the PINK1-dependent beginning of Parkin recruitment. Parkin utilizes the ubiquitin-proteasome system to ubiquitinate and destroy OMM proteins. Many autophagy receptors, including NDP52, p62, and OPTN, which are essential in the initialization phase of autophagy, may recognize ubiquitinated OMM proteins produced by Parkin during mitochondrial malfunction. Parkin appears to both support and increase the PINK1 impact but cannot do so, suggesting that PINK1 may play an irreversible function in mitophagy. While this happens, mitophagy begins without Parkin activation, and autophagy receptors are being recruited. To maintain mitochondrial homeostasis, which is done by regulating mitochondrial fusion and fission via ubiquitination of Mfn, both PINK1 and Parkin proteins are critical. The loss of PINK1 or Parkin causes Mfn to accumulate on OMM and impairs the mitophagy process (Mishra & Chan 2016; Pickrell & Youle 2015).

Additionally, the absence of Mfn lowers the membrane potential of mitochondria and inhibits Parkin recruitment, which is prevented by inhibiting the translocase of the OMM import system (TOM complex). Additionally, activating the TOM complex helps prevent mitophagy deficiency caused by Parkin mutations, indicating that TOM plays an essential mediator role in PINK1/Parkin-directed mitophagy. In contrast, the PINK1/Parkin pathway has been implicated in the mitophagy process in the Drosophila model; however, the mechanism of mitophagy in the human brain is unclear due to the complexity of the mammalian brain's structure and functions. Similarly, it has recently been proven that mitophagy can occur in various tissue types without the involvement of the PINK1 or Parkin proteins, or both (McWilliams et al., 2017; Szargel et al., 2016).

Nix/Bnip3-Mediated Mitophagy

The Bcl-2 family includes the apoptotic proteins Bnip3 and NIX, found at the mitochondria's outer membrane. Most Nix/Bnip3-mediated mitophagy similarities to Parkin/Pink1-mediated mitophagy have been demonstrated in recent research (Kumar et al., 2022). Both Bnip3 and NIX have been shown to feature an LC3-interacting region motif (LIR), which offers a superior LC3 and other LC3/GABARAP family members that take part in the start of autophagosome mechanisms have a relationship to the mitochondria. Initially, it was discovered that NIX performs as a mitophagy receptor, is crucial for erythropoiesis, and is also engaged in the removal of damaged mitochondria that display suppression of the synthesis of mature reticulocytes (Marinkovic et al., 2021). Recent studies have revealed that NIX/Bnip3 LIR domain 8 may interact with the autophagy-related gene protein 8 (Atg8), an autophagy protein that initiates the mitophagic process and generates a freshly created autophagic membrane.

Additionally, it has been discovered that NIX is overexpressed in PD patients' cells, which results in ATP generation in those cells. Additionally, hypoxia and HIF1 stabilization mediate NIX/Bnip3 mitophagy. It was subsequently suggested that using mitochondrial-mediated autophagy pathways could be beneficial for treating neurological illness. For instance, it has been demonstrated that NIX/Bnip3-facilitated mitophagy protects against cerebral ischemia injury (Wu et al., 2021).

FUNDC1-Mediated Mitophagy

One of the ways that mammalian cells undergo mitophagy is through FUNDC1, which is a domain 1 of the FUN14 protein. A transmembrane LIR (LC3-interacting region) motif may be found at the N-terminal location of FUNDC1, a well-known hypoxia-induced mitophagy receptor (Liu et al., 2022). Phosphoglycerate mutase family member 5 phosphatase (PGAM5) and ULK1 phosphorylated Ser17 and Ser13, respectively, during the hypoxic environment, increasing the FUNDC1 binding capacity with the LC3 motif. FUNDC1 interacts with Drp1, a mitochondrial fission protein, and OPA1, a mitochondrial fusion protein, and this connection is crucial for promoting hypoxia-induced mitophagy (Liu et al., 2021).

p62/SQSTM1-Mediated Mitophagy

Recent studies have identified p62/SQSTM1, a cytoprotein with a cytosolate LIR domain protein, as a selective autophagy receptor. It has been determined that the p62 LIR motif interacts with LC3 in many locations. Numerous studies have shown that p62 supports the PINK1/Parkin-mediated mitophagy pathway (Figure 3) (Liu et al., 2017).The PINK1/Parkin-mediated mitophagy pathway involves Parkin recruiting two proteins, ubiquitin and p62, that are necessary for initiating the autophagic engulfment of damaged mitochondria. This autophagic engulfment then leads to fusion with lysosomes, which culminates in the degradation of targeted mitochondria (Poon et al., 2021).

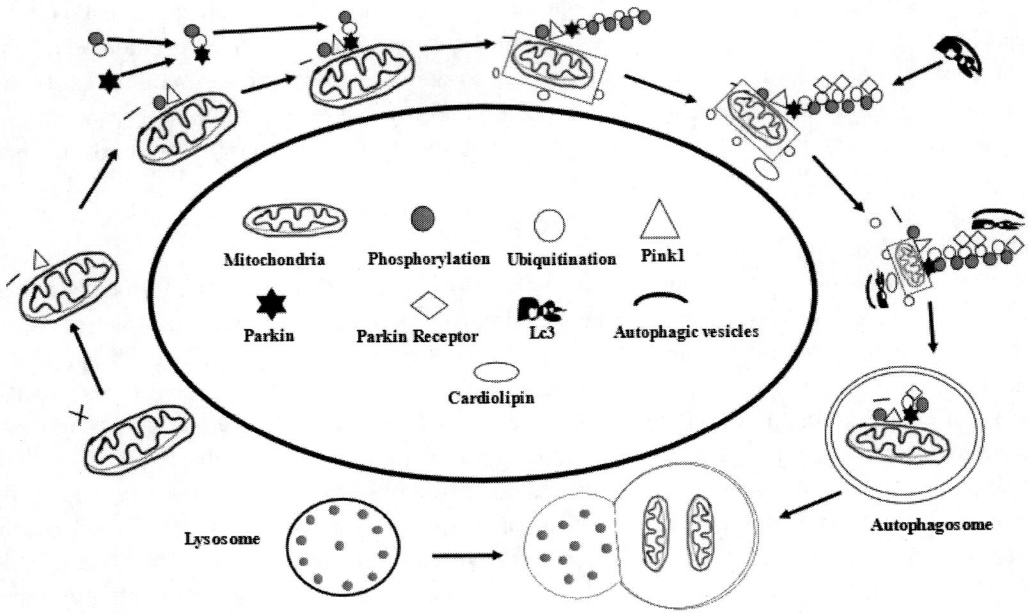

Figure 3. Anoverview of the mitophagy mechanism: PINK1 is phosphorylated on the outer membrane of mitochondria when they depolarize or gather misfolded membrane proteins. At the same time, PINK1 recruits phosphorylated Ub to the mitochondrial surface, where PINK1 phosphorylates it. Phosphorylated Ub can then connect to Parkin directly. In the whole cell, PINK1 can phosphorylate Ub and create polyubiquitin chains. The mitochondrial matrix is similarly ubiquitinated despite recruiting Parkin. Mitophagy receptors such as p62, NDP52, NBR1, and OPTN are attracted to mitochondria marked with ubiquitin. These receptors engage with the outer membrane situated autophagy protein LC3 through their LIR motifs, bind to the polyubiquitin chain through the ubiquitin-binding domain, and then initiate mitophagy. Additionally, the brain contains cardiolipids in the mitochondrial intimum, necessitating three translocases to expose the OMM to them and bind to LC3 to mediate mitophagy.

Other Proteins-Mediated Mitophagy

Furthermore, other more broadly distributed cytoplasmic receptor proteins, including NDP52, Optineurin, NBR1, and Optineurin, are also well-known for their functions in the mitophagy process. The protein ubiquitin-2 facilitates autophagy by binding polyubiquitinated substrates sent to a protease via its UBL domain. In several neurological disorders, the ubiquitin-selective partner protein Valosin-containing protein (VCP), which interacts with the lysosome by its composition, is also involved in the autophagic regulation of stress granule (SG) synthesis. The so-called OMM protein FKBP8, which binds to the FK506 gene, is one of the additional proteins that can recruit LC3A via the LIR motif to promote Parkin-independent mitophagy (Bhujabal et al., 2017).

Lipid Pathways

Cardiolipid-Mediated Pathway

Cardiolipids found in the inner membrane of mitochondria, contain a minimal amount of unsaturated fatty acids, and the neurological system is where cardiolipids mediate these pathways. Three translocases immediately open towards the OMM are needed to facilitate mitophagy, which aids in the direct binding with LC3. Recently, scientists have worked in vivo with the aid of anti-RNAi injections that cause PLS3, the enzyme in charge of cardiolipin translocation from the IMM to the OMM with PLS3-associated phosphorylation at amino acid threonine 21, to be knocked down. This decreased mitophagy activity was also seen in the controlled cortical impact model (He et al., 2007; Chao et al., 2019).

Cholesterol and Fatty Acids-Mediated Pathway

Synthetic steroids and lipids can support the stability of PINK1 in this pathway, and Parkin protein translocation when PINK1 is present, helping to regulate mitophagy. As a result, it has been discovered that fatty acids and cholesterol may affect mitophagy. On the other hand, cholesterol has a usually dual effect on PINK1-parkin-mediated mitophagy, causing damage to the lysosomes of the mitophagosomes and supporting an advanced oxidative-induced accumulation of OPTN aggregates that prevents its mitochondrial recruitment despite PINK1/parkin activation (Roca-Agujetas et al., 2021).

Mitochondrial Dynamics and Fusion

Dynamic balance is maintained in mitochondria through regular fission and fusion cycle events. Due to these modification events, changes occur in the shape of mitochondria and their function within the cell. Mitochondrial fusion permits matter exchange between intact and dysfunctional mitochondria, thus enabling damaged material turnover. Moreover, fusion has been found to stimulate the fidelity of oxidative phosphorylation, protein complementation, and repair of small extents of mitochondrial dysfunction (Zhang et al., 2016). Three mitochondrial enzymes responsible for hydrolyzing GTP are necessary for mitochondrial fusion. Mfn1 and Mfn2 are GTPase enzymes located on the outer membrane of mitochondria, promoting attachment between mitochondria during fusion (Cao et al., 2017). Optic atrophy (Opa) 1 (OPA1) has been identified as responsible for intimal fusion in the transmembrane region. These GTPases are involved in the fusion of mitochondria, which is closely associated with mitophagy. Various evidenceshave shown that mitochondrial fragmentation increases with BNIP3 when it interacts with OPA1.

In contrast, mitochondrial fission is supposed to separate highly impaired mitochondria from the healthy group. The fission of mitochondria produces two dissimilar populations of mitochondria, one with high mitochondrial membrane potential and the other with low mitochondrial membrane potential. The mitochondria, which possess high membrane potential, are then embedded into the mitochondrial network through fusion. On other hand, the mitochondria, which possess low membrane potential, are transported to autophagosomes and finally lead to degradation by mitophagy. Mitochondrial Dynamin-related protein 1 (Drp1) is a GTPase essential for mitochondria fission. Numerous Drp1 molecules are accompanied together to make an oligomer, which constricts to divide into two mitochondrion.

Several proteins found in the outer membrane of mitochondria serve as Drp1 receptors to facilitate the above process, such as mitochondrial fission factor (Mff), mitochondrial fission protein 1 (Fis1), mitochondrial dynamics proteins of 49 kDa (MiD49), and mitochondrial dynamics proteins of 51 kDa (MiD51). Drp1 can also interact with FUNDC1 to endorse mitophagy. In addition, PINK1 can also curb Drp1 activity. In brief, mitochondrial fission and selective fusion can distinguish functional mitochondria from dysfunctional ones; dysfunctional mitochondria are subsequently selected and degraded through mitophagy.

Inducers of Mitophagy

Apart from the mechanisms of mitophagy revealed in the above section, we have also briefly focused on mitophagy inducers such as reactive oxygen species (ROS) and hypoxia. ROS has a dual nature as cellular signaling and a lethal toxin, hence the ROS dynamic equilibrium is highly controlled. Various studies have identified that ROS act as a signal for activating the process of autophagy (Wang et al., 2018). Recent studies have shown that ROS induces mitophagy by prompting PINK1/Parkin axis in HeLa cells and activating the pathway of NIX/BNIP3L. In the muscle tissue, BNIP3L is upregulated, as are cells in vitro, when responding to hypoxia.

Similarly, it has also been revealed that hypoxia-induced BNIP3L-mediated mitophagy was also perceived in some non-muscular cell lines. Inflammation can activate mitophagy by increasing ROS production; on the other hand, upregulation of BNIP3L expression was observed in myocardial inflammatory rats. Therefore, we have undoubtedly found that these inducers may affect cell viability in various ways, such as autophagy, apoptosis, ferroptosis, mitophagy, and necrosis.

Mitophagy in Chronic Diseases of the Nervous System

Several studies have shown that neurodegeneration occurs due to the involvement of some pathogenic proteins. Genetic mutations also stimulate the self-accumulation of pathogenic proteins and can cause increased age-dependent generation. In contrast, mitochondrial dysfunction plays a crucial role in enhancing neurodegenerative disorders that can endorse the production and accumulation of pathogenic proteins through a decline in the generation of cellular ATP and high oxidative stress (Figure 4). Numerous studies have also identified that mutated proteins suppress the mitophagy mechanism. However, it has also been observed that the main reason for various neurodegenerative disorders is linked to pathogenic proteins found in the autophagy system and may lead to diminished autophagosome retrograde motility, inefficient cargo loading, and deficient cargo recognition.

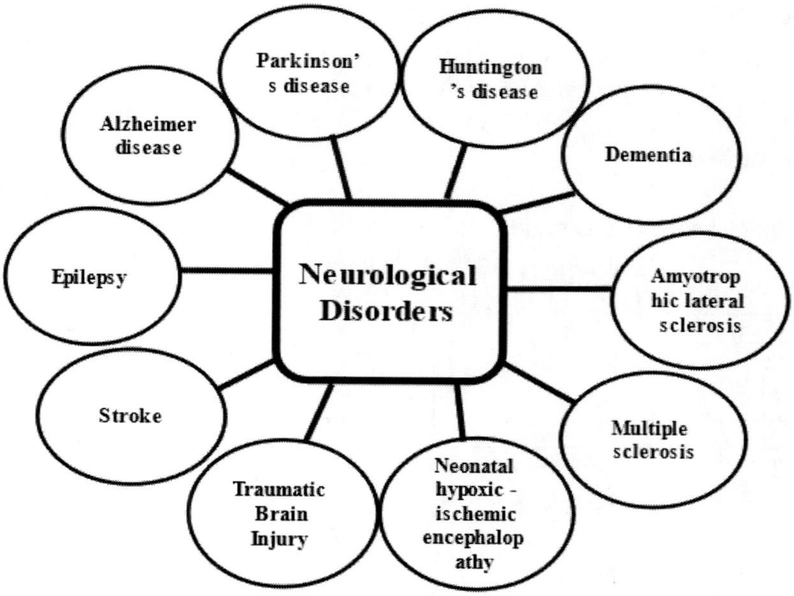

Figure 4. Various forms of human neurological disorders: PD, AD, HD, ALS, MS, TBI, stroke, epilepsy, dementia, and HIS.

Nonetheless, the aggregation of pathogenic proteins in various neurodegenerative diseases such as tau, α-synuclein, and Huntington's disease is a common and significant reason for malfunctioning autophagy (Table 2). The dissimilar pathological features of protein aggregation among affected and dysfunctional cytocells and the removal of proteins via mitophagy mechanism lead to the accumulation of damaged cytoblasts and polyubiquitinated proteins in the nucleus and cytoplasm of neurons. Dysfunctional or damaged mitochondrial accumulation stimulates mitophagy. Alterations in autophagic processes lead to neuronal degenerations that may be primarily or also reactive. For instance, in the pathogenesis of Alzheimer's disease (AD) implication, γ-secretase enzyme Presenilin1 (PS1) is involved in the cleavage of amyloid precursor protein that produces Abeta peptides, which enhances AD progression.

Furthermore, the genetic mutation of PSI represses autophagy by decreasing vATPase immunoreactivity on lysosomes, suppressing the activation of lysosomal protease, enhancing lysosomal pH, and delaying the turnover of autophagic protein in AD. It has been demonstrated that in neurons, using live cell imaging, 80% of autophagosome biogenesis occurs at the axon end, and autophagosomes are transported along the axons to dendrites and somatic cells. It has also been observed that accumulation of autophagosome membranes occurs in the endoplasmic reticulum (ER). In neurons, the motility of autophagosomes decreases due to the accumulation of pathogenic proteins. In contrast, pathogenic proteins binding at the mitochondrial membrane or surface of the autophagosome result in the autophagosome assemblage due to decreasing their retrograde transport. When the nascent autophagosome consumes the target cargo, it joins with lysosomes in the cytosol and decomposes the target cargo, such as mitochondria. The N-ethylmaleimide-sensitive factor attachment protein receptors (SNAREs), which are soluble membrane anchor proteins, directly or indirectly affect the fusion of both autophagosomes and lysosomes by facilitating the fusion of organelles or enhancing lysosomal biogenesis. Defective mitochondria and pathogenic proteins induce dysfunctional mitochondrial accumulation, including a 'vicious cycle reaction. However, the sequential order of dysfunctional mitochondria, defective mitophagy, and pathogenic proteins remains unclear. Thus, enhancing or restoring the mitophagy process might be a new treatment or therapy against multiple neurodegenerative diseases.

Table 2. Pharmacological regulation of autophagy/mitophagy in various neurological disorders adapted from Zhang et al., 2021

Neurological Disorder	Drug Use	Target	Model types	Function
Alzheimer disease	UA/UC	AMPK	iPSC neurons	Inhibit mitochondrial fission, increase ATP production ↑
Alzheimer disease	Rapamycin	mTOR	PDAPP mice	Inhibits mTOR activity ↑
Parkinson's disease	TFEB	LAMP1	Nigra neurons	Induces autophagosome and lysosome biogenesis
Huntington's disease	PINK1	Parkin	Drosophila	Inhibits accumulation of mitochondria ↑
Amyotrophic lateral sclerosis	Atg1	mTOR	Drosophila	Positively regulates the initiation of autophagy
Stroke	Melatonin	LC3, PINK1	SAH rat	Control dysfunctional mitochondria ↑
Stroke	BNIP3 Silencing	LC3	I/H mice	Inhibits higher levels of mitophagy
Neonatal hypoxic-ischemic encephalopathy	3-MA	LC3II,p62	HIBI rat	Inhibits autophagy flux
Neonatal hypoxic-ischemic encephalopathy	Glycine	p-AMPK	HI rat	Downregulates AMPK to alleviate autophagy flux. ↓
Neonatal hypoxic-ischemic encephalopathy	IPC	p-AMPK	pMCAO rat	Activates AMPK to induce autophagy
Neonatal hypoxic-ischemic encephalopathy	AG	miR-30d-5p	HI rat	Inhibits autophagy
Neonatal hypoxic-ischemic encephalopathy	AT	miR-30d-5p	HI rat	Enhance autophagy
Epilepsy	Rapamycin	mTORC	TSC mouse	Inhibits S6 phosphorylation ↑
Traumatic Brain Injury	Morin	LC3II/I, Beclin-1	TBI rat	Augments autophagy
Traumatic Brain Injury	NIX	LC3II/I,p62	TBI rat	Increases autophagy
Traumatic Brain Injury	17-AAG	LC3II/I, Beclin-1	TBI rat	Induces autophagy
Traumatic Brain Injury	Wnt3 α	β-proteasomal	TBI mouse	Activates Wnt/β-catenin pathway ↑

Alzheimer's Disease (AD)

Alzheimer's disease is one of the most well-known types of neurodegenerative diseases that are associated with aging. The gradual progression of Alzheimer's disease results in a loss of memory, severe cognitive decline, and neuronal cell death in the cerebral cortex and subcortical areas. About 60–80% of cases of dementia occur due to AD. AD is named after Dr. Alois Alzheimer, who identified some alterations in the brain tissue of dead women who were mentally deranged. Biologically, the brains of AD patients are pathologically characterized by the abnormal deposition of extracellular amyloid β (Aβ) peptides and the intracellular accumulation of tau-containing neurofibrillary tangles, which lead to endorse mitochondrial defects. A shred of evidence shows that in the medial front cortex and medial parietal cortex, the initial deposition of Aβ occurs in this region.

On the other hand, tau accumulates initially in the medial frontal lobe memory system, which disperses from the entorhinal cortex to the hippocampus and parahippocampal cortex and then spreads to other brain locations. Numerous pathological features have been conveyed regarding AD brains, such as oxidative stress, mitochondrial dysfunction, altered expression in mitochondrial DNA, and autophagic, lysosomal, and endosomal systems that stimulate the deposition of extracellular amyloid β peptides and phosphorylated tau protein. Recently, several studies have identified that defects in mitochondrial autophagy are the hallmarks of AD (Kerr et al., 2017). In the transgenic mouse model experiment, a large number of upsurge levels of mRNA expression of various proteins like BNIP3P, LC3, PINK1, PARK2, BENC1, and also p62, which are allied with mitochondrial autophagy, promoted the accumulation of the amyloid-β protein. In nematodes, mice, and humans, Aβ can surge both UPRmt and mitophagy conserved; nevertheless, the particular mechanism is still elusive. Induction of UPRmt and mitophagy are valuable in nematodes because they help delay disease development and improve life span. Likewise, disrupted-in-Schizophrenia-1 (DISC1), where mitophagy receptors have been found and observed, shows that accumulation of Aβ and DISCI downregulation in AD is mutually reinforced, forming a vicious circle. Moreover, it has been shown that DISCI protects synapses from the deposition of Aβ linked toxicity by promoting mitophagy (Burte et al., 2015). Correspondingly, in the case of AD-induced pluripotent stem cell (iPSC)-derived neurons, mitophagy moderates AD-associated tau hyper-phosphorylation and inhibits cognitive impairment, proposing that impaired elimination of damaged mitochondria is a crucial phenomenon in the development of AD and also showing a latent therapeutically mediated target for the treatment of AD (Fang et al., 2019).

Parkinson's Disease (PD)

Parkinson's disease is one of the most fatal neurodegenerative illnesses after AD. This disease has characteristic features of postural instability, static tremor, stiffness, and bradykinesia. PD mainly occurs due to the abnormal accumulation of α-synuclein in Lewy bodies and the early loss of dopaminergic neurons in the pars compacta of the substantia nigra region of the middle brain, cholinergic and monoaminergic brainstem neurons, and the olfactory system, affecting central synaptic pathology. Hence, the abnormal deposition of α-syn is the hallmark of PD, and numerous studies have recognized the connection between α-syn and mitochondrial dysfunction. Additionally, numerous studies have analyzed various characteristics that are primary contributors to PD pathogenesis, such as the depletion of mitochondrial DNA, the depletion of ATP, the release of caspase, and the overproduction of ROS.

Genetic mutations in several genes, like LRRK2, α-syn, Parkin, DJ-1, and PINK1, have been associated with familial PD. Mutations in parkin and PINK1 surging loss of function mechanisms, whereas mutations in α-syn and LRRK2 can cause autosomal dominant mutations. Henceforth, these genes are accompanied by mitochondrial function, and the proteins encoded by these genes play a mechanistic role in the association between Parkinson's disease pathogenesis and mitophagy. Studies on parkin- and PINK1 mutant fly models have displayed a clearer vision related to the phenotype of PD, such as dysfunctional mitochondria, loss of dopaminergic neurons, significant disabilities, and a reduced lifespan. Likewise, in the nigral dopaminergic neurons of MPTP mouse model of PD, overexpression of transcription factor EB (TFEB) improves the release of dopamine in the striatum. Also, it elicits the numerous other proteins expression pertinent to the dopaminergic phenotype, which concurrently drives a neurotrophic effect and leads to neuronal cell growth.

Dementia

Dementia is also a neurological disorder commonly occurring in older people and causes a severe deterioration in cognitive functions and social skills,such as loss of memory, speech, language, apraxia, and judgment (Duong et al., 2017). Dementia has various forms, with AD being the most common. On the other hand, vascular dementia, frontotemporal dementia (FTD), Lewy body dementia, and HIV-associated neurocognitive disorder generally progress in patients formerly diagnosed with dementia. Dementias are generally a well-known disease resulting from pathological deposition of misfolded proteins followed by progressive loss of synaptic function, while decreased vascular perfusion is also one of the significant eliciting causes. In the case of FTD, mutations are accompanied by the progression of this disease in many autophagy-related genes, such as p62. In addition, Valosin-containing protein (VCP) is involved in ubiquitin-dependent protein quality control from the perspective of membrane dynamics, with loss of function mutations found in FTD, resulting in decreased mitophagy.

Huntington's Disease (HD)

HD is the most commonly found autosomal dominant neurodegenerative disease in middle-aged people. HD occurs in people due to mutations in the huntingtin (Htt) gene found on chromosome 4. The expansion of the cytosine-adenine-guanine (CAG) trinucleotide repeat in exon 1 of Htt leads to the expansion of the amino-terminal region of polyglutamine (polyQ) of the mutant Htt (mHtt) protein. The main clinical features of HD are cognitive decline, motor dysfunction (Rothlind et al., 1993), and psychiatric disorders. Macromolecular aggregation forms when mutant Htt protein accumulates in cells, which leads to increased cytotoxicity. Numerous studies have reported that, at the molecular level, mutant Htt protein affects the normal function of nerve cells by several mechanisms, such as mitochondrial dysfunction, protein transport, apoptosis signaling, oxidative stress, and amino acid metabolism, that result in neurological alteration. Several shreds of evidence reveal that mutant Htt can impair mitochondria, which results in oxidative stress, energy metabolism disorders, and the release of proapoptotic factors. Numerous studies have also described that the involvement of mHtt is responsible for the pathogenesis of autophagic, endosomal, and lysosomal systems observed in the initial steps of the autophagic process. Not only can mHtt affect the transcription expression of several autophagy-associated genes (Metzger et al., 2013). Moreover, in the HD animal model, the pathway of PINK1/Parkin is unaffected; however, the limited fusion of mitochondria with autophagosomes results in destruction or damage to the mitophagy mechanism. When PINK1 is overexpressed, this destruction or damage is somewhat recovered, thus refining the integrity of mitochondria and protecting neural function (Khalil et al., 2015).

Amyotrophic Lateral Sclerosis (ALS)

It is a fatal neurological disorder in which progressive degeneration of motor neurons occurs in the brain and spinal cord. Progressive motor dysfunction is the pathological hallmark of ALS; however, sensory function is unaffected. Various studies have analyzed ALS's pathogenesis and reported the association of ALS with more than 30 gene mutations like VCP, OPTN, SOD1, C9ORF72, and TBK1. Mitophagy depends on OPTN, an autophagy receptor, and its kinase, TBK1. Damaged TBK1 and OPTN lead to destructive mitophagy, dysfunctional mitochondria, and protein aggregation. Mitochondrial homeostasis can be reconstructed by hindering OPTN or TBK1 mutations, pharmacological suppression or genetic knockdown of PINK1 or Parkin. In ALS models, the deposition of damaged and dysfunctional mitochondria has been recognized with the help of live-cell imaging. Changes in the TBK1/OPTN gene can considerably increase neuronal function and block the advancement of this disease (Moore et al., 2016). Over-expression of Atg1 leads to a subsequent increase in the life span of the ALS transgenic fruit fly model and controls disease progression by stimulating the mTOR target Atg1, which helps in the induction of autophagy (Marrone et al., 2018).

Multiple Sclerosis (MS)

Multiple sclerosis is also a neurological disorder of the central nervous systemlinked with an autoimmune response (Tafti et al., 2022) against the myelin sheath. In the central nervous system, auto-reactive T cells demyelinated white matter via targeting the myelin sheath, which led to a secondary loss of neuronal axons and neurodegeneration. The pathology of MS initiates with the development of a lesion with acute inflammation, which changes to a state of chronic inflammation followed by neurodegeneration. Mitophagy and mitochondria are crucial for immune signaling. ROS signals from mitochondria in MS can damage myelin and the blood-brain barrier, which impairs the disease. It has been found that DNA damage and oxidation of phospholipids occur during severe chronic inflammation with the disruption of neuronal axons. A piece of evidence has been refined to show that both autophagy and mitophagy mechanisms are impaired in MS humans and animal models. In the case of human subjects affected by MS, there is a diminished ratio of Lamp2 and LC3II/I, inhibiting autophagy. Studies also revealed that people affected by MS have an increased concentration of Atg5, which is found in serum and cerebral spinal fluid (CSF). This study also reported that the expression of Parkin was augmented in serum and CSF containing a higher lactate composition. Further research also observed that blood from MS subjects illustrates modified expression of various autophagy-associated genes such as PIK3R1, BCL2, GAA, ATG9AFAS, HTT, RAB24, FOXO1, ULK1, PIK3R1RGS19, and HGS.

Mitophagy in Acute Disease of the Nervous System

Mitophagy in acute neurological disorders includes multiple non-infectious diseases associated with the nervous system, such as stroke, ischemic and hypoxic brain injury, epilepsy, and traumatic brain injury. Studies on mitophagy in acute neurological diseases remain few in the context of chronic nervous system diseases. It remains elusive, whether mitophagy is useful or harmful. Recent research has observed that it may be due to various factors like variation between in vivo and in vitro research studies, differences in the periods of diseases, differences in apoptosis and mitophagy, and also different experimental methods and environments, as well as gender differences. In the case of gender difference studies, it has been shown that the dual endothelin-1/vascular signal peptide-activated receptor (DEspR) plays an essential role in the differentiation of the neuroepithelium and neural tube. DEspR+/-females showed good cognitive performance compared with wild-type females and showed no neuropathological changes. This well-designed study displayed gender-related cerebral neuronal vulnerability to autophagic dysregulation. In the sections below, we have highlighted some acute nervous system diseases and its association with the process of mitophagy. At the end of this section, a comprehensive table is also added, which highlights the pharmacological regulation of mitophagy in different neurological disorders (Table 2).

Neonatal Hypoxic-Ischemic Encephalopathy (HIE)

It is an acute neurological disorder where hypoxic-ischemic cerebral neuropathy is initiated by perinatal asphyxia. In severe conditions, HIE results in cerebral palsy, mental illness, epilepsy, and cognitive impairment. Brain tissue ischemia causes more significant amount of mitochondrial damage. Mitophagy is beneficial to remove this mitochondrial damage and ease brain injury. Bnip3L is an essential protein for initiatingmitophagy in cerebral tissue ischemia-reperfusion (Yuan et al., 2017). In models of hypoxic-ischemic brain injury in neonatal rats, hypoxic ischemia-induced more autophagy fluxes resulting in severe brain injury, such as augmented expression of LC3-II, decreased protein expression of P62/SQSTM1, and reduced cAMP-response element-binding protein (p-CREB). 3-methyladenine is an autophagy inhibitor that significantly attenuated the surge in LC3-II and the loss of P62/SQSTM1 and p-CREB, ameliorated neuronal death, improved the results of the MWM test, and ultimately improved the spatial learning of memory function in rats.

Stroke

A stroke is also called a "cerebrovascular accident," which occurs due to ischemic or hemorrhagic cerebrovascular disease that can create higher deprivation of oxygen and glucose levels and unrecoverable brain injury. Moreover, in ischemic cerebrovascular disease, major brain ischemias occur for a few minutes and cause irreparable damage to brain-to-brain cells. Melatonin decreases ROS generation and mitochondrial injury and also suppresses the activation of NLRP3-inflammatory entities by enhancing the expression of mitophagy-mediated proteins such as PINK1/Parkin in rat models of subarachnoid hemorrhage, thereby decreasing neuronal cell death, brain inflammation, and neuronal dysfunction after subarachnoid hemorrhage (Cao et al., 2017).

Epilepsy

It is a multifaceted, complicated neurological disorder well-known and defined by its chief symptoms, such as frequent and repetitive seizures that occur randomly. Epilepsy disorder is sub-categorized into two major types: generalized-onset seizures and partial or localized-onset seizures. Epilepsy is intensely triggered by an upsurge in the incidence of several co-occurring circumstances, such as sleep disorders, sadness, stress, cognitive impairment, anxiety, and sudden unusual deaths. It has been revealed that the decreased mitochondrial activity is a critical factor in epilepsy, indicating that abnormal mitophagy mechanisms cause epilepsy. Autophagy was identified by LC3B, an autophagy marker protein, and TOMM20, a mitochondrial marker present in the hippocampal tissue in patients with temporal lobe epilepsy (rTLE). In addition, numerous mutations in the various encoding genes, such as the Mtorc1 autophagy inhibitor, are also associated with epilepsy implications (Zeng et al., 2008).

Furthermore, several factors such as developmental disorders, metabolic abnormalities, and neurotoxic drugsare also primary triggersforepilepsy stimulation. For instance, some excitatory toxic substances such as glutamate act on postsynaptic endings, augmenting their depolarization and leading to uninterrupted inflow from synaptic gaps, mitochondrial damage, and inhibiting several mitophagy reactions. Additionally, it has been reported that epileptic seizures, or epilepsy are neuronal disorders caused by damaged or dysfunctional mitophagy. For example, loss of function of the HECT domain and ankyrin repeat consisting of the E3 ubiquitin ligase 1 (HACE1) enzyme, which ubiquitinates targets for degradation, has been linked to disrupted mitophagy flux and a decreased response to oxidative stress, which leads to spastic paraplegia and psychomotor retardation with or without seizures, resulting in epilepsy.

Traumatic Brain Injury (TBI)

TBI is a nondegenerative, traumatic structural injury where the brain becomes dysfunctional due to external mechanical forces on the brain. In children, the cerebrospinal fluid (CSF) is affected by traumatic brain injury; autophagy markers such as beclin and p62 are induced, as is reduced autophagy flux, suggesting that the upsurge in autophagy might be related to a reduction in autophagy flux (Au et al., 2017). Notably, CSF in brain cells does not directly affect autophagy. It has been observed that the use of a natural polyphenol named morin is helpful to augment autophagyand may decrease TNF- αand IL-6, inflammation markers, along with apoptosis, which are involved in enhancing memory after mild traumatic brain injury.

Conclusion

The molecular effects and prospective targets for managing and treating neurodegenerative diseases have received major attention in the current studies on the mitophagy process in the nervous system. Conflict regarding the precise mechanisms behind the receptor-mediated autophagy pathway is diverse across contexts of research and development. In light of this, a question has been raised as to whether much-targeted mitophagy processes for treating disease are valuable or detrimental and what the process will be to handle these problems. Further research is necessary to address these issues as mitophagy targeting has produced unsatisfactory results for long-term chronic disorders of the nervous system, and there is a lack of thorough, higher-level research and development as well as therapy for short-term acute diseases of the nervous system.

Future Perspective

The importance of nervous system-associated conditions and the mitophagy mechanism has been highlighted in this chapter, which may offer a fresh direction for numerous beneficial therapies to prevent and treat neurological disorders. A crucial characteristic for the identification of neurological disorders is abnormal mitophagy flux. In contrast, providing reasonable in vivo data and clinical studies for the novel therapeutic application involving mitophagy is critical. Additionally, researchers would want to recognize distinct biomarkers, beneficial flux, and specific measurements in identifying various mitophagy forms. To retain the properties of mitochondria, we also needed to design possible agonists and antagonists of mitophagy.

References

Au, A. K., Aneja, R. K., Bayır, H., Bell, M. J., Janesko-Feldman, K., Kochanek, P. M., & Clark, R. S. B. (2017). Autophagy Biomarkers Beclin 1 and p62 are Increased in Cerebrospinal Fluid after Traumatic Brain Injury. *Neurocritical care*. 26(3), 348–355.

Bhujabal, Z., Birgisdottir, Å. B., Sjøttem, E., Brenne, H. B., Øvervatn, A., Habisov, S., Kirkin, V., Lamark, T., Johansen, T. (2017). FKBP8 recruits LC3A to mediate Parkin-independent mitophagy. *EMBO reports,* 18(6), 947–961.

Burté, F., Carelli, V., Chinnery, P. F., & Yu-Wai-Man, P. (2015). Disturbed mitochondrial dynamics and neurodegenerative disorders. *Nature reviews. Neurology*. 11(1), 11–24.

Cao, S., Shrestha, S., Li, J., Yu, X., Chen, J., Yan, F., Ying, G., Gu, C., Wang, L., & Chen, G. (2017). Melatonin-mediated mitophagy protects against early brain injury after subarachnoid hemorrhage through inhibition of NLRP3 inflammasome activation. *Scientific reports*. 7(1), 2417.

Cao, Y. L., Meng, S., Chen, Y., Feng, J. X., Gu, D. D., Yu, B., Li, Y. J., Yang, J. Y., Liao, S., Chan, D. C., Gao, S. (2017). MFN1 structures reveal nucleotide-triggered dimerization critical for mitochondrial fusion. *Nature*. 372-376.

Cen, X., Zhang, M., Zhou, M., Ye, L., Xia, H. (2021). Mitophagy Regulates Neurodegenerative Diseases. *Cells*. 10(8):1876.

Chakravorty, A., Jetto, C. T., & Manjithaya, R. (2019). Dysfunctional Mitochondria and Mitophagy as Drivers of Alzheimer's Disease Pathogenesis. *Frontiers in aging neuroscience,* 11, 311.

Chao, H., Lin, C., Zuo, Q., Liu, Y., Xiao, M., Xu, X., Li, Z., Bao, Z., Chen, H., You, Y., Kochanek, P. M., Yin, H., Liu, N., Kagan, V. E., Bayır, H., & Ji, J. (2019). Cardiolipin-Dependent Mitophagy Guides Outcome after Traumatic Brain Injury. *The Journal of neuroscience: the official journal of the Society for Neuroscience*, 39(10), 1930–1943.

Doblado, L., Lueck, C., Rey, C., Samhan-Arias, A. K., Prieto, I., Stacchiotti, A., & Monsalve, M. (2021). Mitophagy in Human Diseases. *International journal of molecular sciences*. 22(8), 3903.

Duong, S., Patel, T., & Chang, F. (2017). Dementia: What pharmacists need to know. *Canadian pharmacists journal*: CPJ = Revue des pharmaciens du Canada: RPC, 150(2), 118–129.

Fang, E. F., Hou, Y., Palikaras, K., Adriaanse, B. A., Kerr, J. S., Yang, B., Lautrup, S., Hasan-Olive, M. M., Caponio, D., Dan, X., Rocktäschel, P., Croteau, D. L., Akbari, M., Greig, N. H., Fladby, T., Nilsen, H., Cader, M. Z., Mattson, M. P., Tavernarakis, N., & Bohr, V. A. (2019). Mitophagy inhibits amyloid-β and tau pathology and reverses cognitive deficits in models of Alzheimer's disease. *Nature neuroscience*. 22(3), 401–412.

He, Y., Liu, J., Grossman, D., Durrant, D., Sweatman, T., Lothstein, L., Epand, R. F., Epand, R. M., & Lee, R. M. (2007). Phosphorylation of mitochondrial phospholipid scramblase 3 by protein kinase C-delta induces its activation and facilitates mitochondrial targeting of tBid. *Journal of cellular biochemistry*, 101(5), 1210–1221.

Kerr, J. S., Adriaanse, B. A., Greig, N. H., Mattson, M. P., Cader, M. Z., Bohr, V. A., & Fang, E. F. (2017). Mitophagy and Alzheimer's Disease: Cellular and Molecular Mechanisms. *Trends in neurosciences*. 40(3), 151–166.

Khalil, B., El Fissi, N., Aouane, A., Cabirol-Pol, M. J., Rival, T., & Liévens, J. C. (2015). PINK1-induced mitophagy promotes neuroprotection in Huntington's disease. *Cell death & disease*. 6(1), e1617.

Kissová, I., Deffieu, M., Manon, S., & Camougrand, N. (2004). Uth1p is involved in the autophagic degradation of mitochondria. *The Journal of biological chemistry*. 279(37), 39068–39074.

Kumar, A. V., Mills, J., & Lapierre, L. R. (2022). Selective Autophagy Receptor p62/SQSTM1, a Pivotal Player in Stress and Aging. *Frontiers in cell and developmental biology*, 10, 793328.

Li, J., Yang, D., Li, Z., Zhao, M., Wang, D., Sun, Z., Wen, P., Dai, Y., Gou, F., Ji, Y., Zhao, D., & Yang, L. (2023). PINK1/Parkin-mediated mitophagy in neurodegenerative diseases. *Ageingresearch reviews*, 84, 101817.

Liu, H., Dai, C., Fan, Y., Guo, B., Ren, K., Sun, T., & Wang, W. (2017). From autophagy to mitophagy: the roles of P62 in neurodegenerative diseases. *Journal of bioenergetics and biomembranes*, 49(5), 413–422.

Liu, R., Xu, C., Zhang, W., Cao, Y.,Ye, J., Li, B., Shi, J., Weng, L., Liu, Y., Liu, L., Zheng, M. (2022). FUNDC1-mediated mitophagy and HIF1α activation drives pulmonary hypertension during hypoxia. *Cell Death Dis,* 13, 634.

Liu, L., Li, Y., & Chen, Q. (2021). The Emerging Role of FUNDC1-Mediated Mitophagy in Cardiovascular Diseases. *Frontiers in physiology*, 12, 807654.

Marrone, L., Poser, I., Casci, I., Japtok, J., Reinhardt, P., Janosch, A., Andree, C., Lee, H. O., Moebius, C., Koerner, E., Reinhardt, L., Cicardi, M. E., Hackmann, K., Klink, B., Poletti, A., Alberti, S., Bickle, M., Hermann, A., Pandey, U. B., Hyman, A. A., Sterneckert, J. L. (2018). Isogenic FUS-eGFP iPSC Reporter Lines Enable Quantification of FUS Stress Granule Pathology that Is Rescued by Drugs Inducing Autophagy. *Stem cell reports*. 10(2), 375–389.

Marinković, M., & Novak, I. (2021). A brief overview of BNIP3L/NIX receptor-mediated mitophagy. *FEBS open bio*, 11(12), 3230–3236.

McWilliams, T. G., Prescott, A. R., Montava-Garriga, L., Ball, G., Singh, F., Barini, E., Muqit, M. M. K., Brooks, S. P., & Ganley, I. G. (2018). Basal Mitophagy Occurs Independently of PINK1 in Mouse Tissues of High Metabolic Demand. *Cell metabolism,* 27(2), 439–449.e5.

Metzger, S., Walter, C., Riess, O., Roos, R. A. C., Nielsen, J. E., Craufurd, D., REGISTRY Investigators of the European Huntington's Disease Network; Nguyen, H. P. (2013). The V471A polymorphism in autophagy-related gene ATG7 modifies age at onset specifically in italian huntington disease patients. *PLoS One*. 8:e68951.

Miller, S., & Muqit, M. M. K. (2019). Therapeutic approaches to enhance PINK1/Parkin mediated mitophagy for the treatment of Parkinson's disease. *Neuroscience letters*, 705, 7–13.

Mishra, P., & Chan, D. C. (2016). Metabolic regulation of mitochondrial dynamics. *The Journal of cell biology*, 212(4), 379–387.

Moore, A. S., Holzbaur, E. L. (2016). Dynamic recruitment and activation of ALS-associated TBK1 with its target optineurin are required for efficient mitophagy. *Proc Natl Acad Sci USA*. 113:E3349-E3358.

Pickrell, A. M., & Youle, R. J. (2015). The roles of PINK1, parkin, and mitochondrial fidelity in Parkinson's disease. *Neuron,* 85(2), 257–273.

Poon, A., Saini, H., Sethi, S., O'Sullivan, G. A., Plun-Favreau, H., Wray, S., Dawson, L. A., & McCarthy, J. M. (2021). The role of SQSTM1 (p62) in mitochondrial function and clearance in human cortical neurons. *Stem cell reports*, 16(5), 1276–1289.

Rothlind, J. C., Bylsma, F. W., Peyser, C., Folstein, S. E., & Brandt, J. (1993). Cognitive and motor correlates of everyday functioning in early Huntington's disease. *The Journal of nervous and mental disease*. 181(3), 194–199.

Roca-Agujetas, V., Barbero-Camps, E., de Dios, C., Podlesniy, P., Abadin, X., Morales, A., Marí, M., Trullàs, R., & Colell, A. (2021). Cholesterol alters mitophagy by impairing optineurin recruitment and lysosomal clearance in Alzheimer's disease. *Molecular neurodegeneration*, 16(1), 15.

Szargel, R., Shani, V., Abd Elghani, F., Mekies, L. N., Liani, E., Rott, R., & Engelender, S. (2016). The PINK1, synphilin-1 and SIAH-1 complex constitutes a novel mitophagy pathway. *Human molecular genetics*, 25(16), 3476–3490.

Sun, N., Malide, D., Liu, J., Rovira, I. I., Combs, C. A., & Finkel, T. (2017). A fluorescence-based imaging method to measure in vitro and in vivo mitophagy using mt-Keima. *Nature protocols*, 12(8), 1576–1587.

Sung, H., Tandarich, L. C., Nguyen, K., & Hollenbeck, P. J. (2016). Compartmentalized Regulation of Parkin-Mediated Mitochondrial Quality Control in the Drosophila Nervous System In Vivo. *The Journal of neuroscience: the official journal of the Society for Neuroscience*. 36(28), 7375–7391.

Tafti, D., Ehsan, M., Xixis, K. L. (2022). Multiple Sclerosis. *In Stat Pearls*. StatPearls Publishing.

Wang, Y., Liu, N., Lu, B. (2019). Mechanisms and roles of mitophagy in neurodegenerative diseases. *CNS neuroscience & therapeutics*. 25(7), 859–875.

Wang, H., Yu, X., Su, C., Shi, Y., Zhao, L. (2018). Chitosan nanoparticles triggered the induction of ROS-mediated cytoprotective autophagy in cancer cells. *Artificial cells, nanomedicine, and biotechnology*. 46(sup1), 293–301.

Wu, X., Zheng, Y., Liu, M., Li, Y., Ma, S., Tang, W., Yan, W., Cao, M., Zheng, W., Jiang, L., Wu, J., Han, F., Qin, Z., Fang, L., Hu, W., Chen, Z., & Zhang, X. (2021). BNIP3L/NIX degradation leads to mitophagy deficiency in ischemic brains. *Autophagy*. 17(8), 1934–1946.

Yuan Y, Zheng Y, Zhang X, Chen Y, Wu X, Wu J, Shen Z, Jiang L, Wang L, Yang W, Luo J, Qin Z, Hu W, Chen Z. (2017). BNIP3L/NIX-mediated mitophagy protects against ischemic brain injury independent of PARK2. *Autophagy*. 2017; 13 (10):1754–66.

Zeng, L. H., Xu, L., Gutmann, D. H., Wong, M. (2008). Rapamycin prevents epilepsy in a mouse model of tuberous sclerosis complex. *Annals of neurology*. 63(4), 444–453.

Zhang, L., Dai, L., & Li, D. (2021). Mitophagy in neurological disorders. *Journal of neuroinflammation*. 18(1), 297.

Zhang, C., Shi, Z., Zhang, L., Zhou, Z., Zheng, X., Liu, G., Bu, G., Fraser, P. E., Xu, H., Zhang, Y. W. (2016). Appoptosin interacts with mitochondrial outer-membrane fusion proteins and regulates mitochondrial morphology. *J Cell Sci*. 129 (2016), 994-1002.

Chapter 6

Mitophagy in Carcinogenesis and Cancer Therapy

Madhu Rani[1]
Rashmi Kumari[1,2]
Moshahid Alam Rizvi[1]
Siya Ram[2]
and Anil Kumar[3,*]

[1]Genome Biology Laboratory, Department of Biosciences, Jamia Millia Islamia, New Delhi, India
[2]Indira Gandhi National Open University, Department of Life Sciences, New Delhi, India
[3]Directorate of Education, Rani Lakshmi Bai Central Agricultural University, Jhansi, Uttar Pradesh, India

Abstract

The role of crucial organelle, i.e., mitochondria, is well established in cell death, cytoplasmic protein degradation, and energy production. Mitophagy is an evolutionarily conserved cellular autophagic event that preferentially removes impaired mitochondria by autophagosome-mediated degradation to maintain cellular homeostasis. The sequestration of impaired mitochondria has been reported in deregulated mitophagy, a prominent cause of carcinogenesis and tumor development. The cell signaling receptors and adaptors that regulate the mitophagy process, such as PINK1, BNIP3, p62/SQSTM1, BNIP3L/NIX, and Parkin are defective in cancer. The role of mitophagy is also well established in sustaining the equilibrium between aerobic glycolysis and oxidative phosphorylation for the survival of cancer cells. ROS-mediated mitophagy competes for various aspects as per the stage of carcinogenesis. Mitophagy also facilitates plasticity in cancer stem cells via metabolic reconfiguration for the betterment of the tumor microenvironment. In this book chapter, we will describe the molecular and regulatory molecular mechanisms of mitophagy and also analyze the role of mitophagy in carcinogenesis and chemo-resistance. Further, we will discuss the effect of mitophagy on cancer therapy which proffers a promising strategy.

Keywords: mitochondria, mitophagy, cancer, drug resistance, cancer therapy

* Corresponding Author's Email: anilkumar.mbge@gmail.com.

In: Mitophagy and Its Role in Health and Disease
Editors: Shalini Mani and Samiksha Wasnik
ISBN: 979-8-89113-470-6

Introduction

Mitochondria are also called the "powerhouse of the cell," and are energy-producing cellular organelles responsible for aerobic respiration. Besides the core bioenergy function of mitochondria, it also provides basic stuff for apoptosis control, control for calcium homeostasis, transcriptional regulation, tumor anabolism, and tumor cell redox (Zhong et al., 2016). The host's immune system's effects on carcinogenesis, cancer progression, and therapy response likely depend on the tumor's mitochondrial metabolism. Therefore, the mitochondria might be exploited as an excellent target for novel cancer therapy (Klein K et al., 2020; 3. Missiroli S et al., 2020). Mitophagy is selective autophagy of mitochondria by removing dysfunctional mitochondria.

Impairment in mitophagy is linked with all defects in mitochondrial functions, oncogenesis, and cancer progression in various cancers (Poole LP and Macleod KF, 2021; Panigrahi DP et al., 2020). Hence, exploring regulatory molecular pathways of mitophagy that affect the cancer process is very important to study.

Mitochondria are well-versed charismatic structures that coordinate the cycle of mitochondrial division and fusion ("mitochondrial dynamics"), which is responsible for the morphological adaptation of mitochondria and involved in the modulation of cellular processes such as apoptosis, cell cycle, immunity of cells, and mitochondrial mass (Simula L et al., 2017). The impairment in mitochondrial dynamics causes damage in cells by improper amounts of ATP or high quantities of ROS and NOS, ultimately resulting in pathological conditions like cancer, neurodegenerative diseases, and autoimmune diseases. The association of human diseases is very well linked with alterations in the mechanical components of the mitochondrial core and impairment in mitochondrial dynamics (Chan DC, 2020; Rodrigues T and Ferraz LS, 2020).

This book chapter encompasses the mitochondrial dynamics and regulatory molecular mechanisms of mitophagy that affect carcinogenesis, metastasis, cancer therapy, and chemo-resistance. We have also discussed targeting mitophagy for developing cancer therapy.

Mitochondrial Dynamics

Mitochondrial dynamics is an event of mitochondrial fission and fusion, which play a crucial role in determining mitochondrial shape, quantity, and quality. This process is associated with functions of mitochondria such as cell migration, cell metabolism, and cell proliferation (Adebayo M et al., 2021; Maycotte P et al., 2017). Various types of proteins modulate mitochondrial dynamics.

Mitochondrial Fission and Fusion

Mitochondrial fission is an event where mitochondria split into two or more independent structures, forming novel mitochondria. In mitochondrial fusion, an amalgamation of various neighboring mitochondria occurs, where the content of partially dysfunctional mitochondria is mixed and complement mitochondria are formed (Srinivasan S et al., 2017). To avoid the loss of intra-mitochondrial content, these events need tightly controlled mechanisms for the production of novel, proper functional cell organelles. Mitochondria are split into smaller fragments; a crucial event to accelerate mitochondrial number in dividing cells is called mitochondrial fission, which also induces the discharge of mitochondria by mitophagy and also promotes the action of mitochondria by cytoskeletal network, and modulates calcium homeostasis and apoptosis (Srinivasan S et al., 2017). Mitochondrial fission (mitofission) is mainly controlled via Drp1 proteins. These proteins belong to the GTP-binding class proteins and might be hired from the cytoplasm to the mitochondrial membrane with help from mitochondrial receptor proteins like Fis1, MID49, MID51, and MFF. The Drp1 present on the membrane of mitochondria might produce a ring structure that intimately encircles the mitochondrion and facilitates damage in the membrane of mitochondria that needs hydrolysis of GTP. Mitofission is controlled by ubiquitination, phosphorylation, and sumoylation of Drp1 via affecting the stability of Drp1 and recruitment. CDK1/ Cyclin B catalyzed the phosphorylation of Drp1 on Ser585 and induced mitofission in mitotic cells, whereas phosphorylation at Ser637 of Drp1 can oppose the process of mitofission (Yapa NMB et al., 2021).

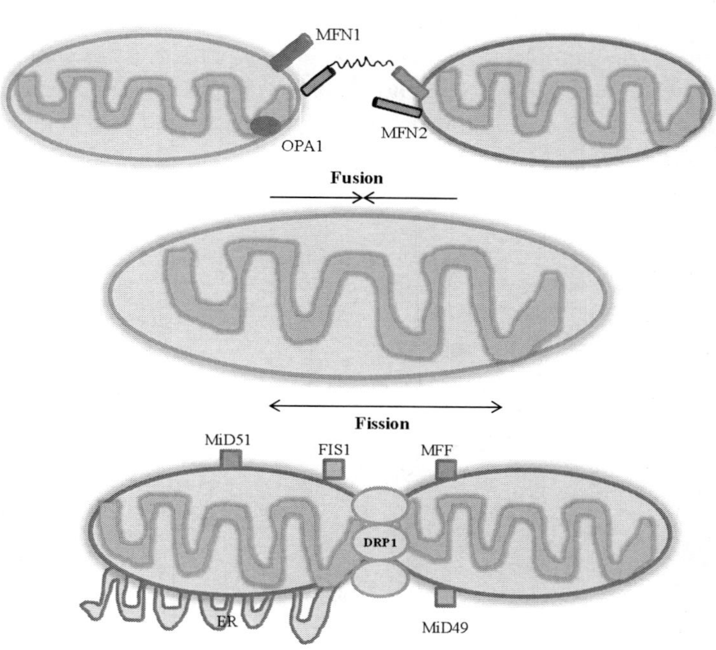

Figure 1. Model for mitochondrial fusion and fission: In mitochondrial fusion, two mitochondria are joined together, whereas, in fission, one mitochondrion separates into two. Mitochondrial fusion is controlled on OMM via mitofusins (MFN1 and MFN2) and the IMM through optic atrophy (OPA1). Mitochondrial fission starts if the endoplasmic reticulum (ER) is hired to the constriction site, as remarked by mtDNA. Further, numerous OMM-bound proteins like MFF, FIS1, MiD51, and MiD49 employed DRP1 on the mitochondrial surface and helped in ER-facilitated constriction.

The process of mitochondrial fusion (mitofusion) consists of outer mitochondrial membrane (OMM) fusion controlled by mitofusin1 (Mfn1) and mitofusin2 (Mfn2) and IMM (inner mitochondrial membrane) fusion facilitated by OPA1. At the molecular level, mitofusion is regulated by 3 dynamin family GTPases, i.e., Mitofusion (Mfn1), (Mfn2) in the OMM, and Opa1 (optic atrophy protein 1) in the IMM. Opa1 possesses two large isoforms (L-Opa1) and three minor isoforms (S-Opa1). The mitofusion of OMMs is well synchronized with a fusion of IMMs. The proteolytic division of IMM proteins and ubiquitination of OMM proteins control fusion proteins. If two mitochondria are located closely, then Mfn1 and Mfn2 begin to dimerize into homo-and heterotypic interactions that produce anti-parallel connections among mitochondria, permitting the fusion of OMM by blending the lipid bilayers (Lee H and Yoon Y, 2016). Moreover, an association of two combining mitochondria with ER is found to promote the event, as Mfn2 is also found located in ER, where it induces the interaction among mitochondria and ER and facilitates the transfer of Ca^{2+} in cell signaling. Mitofusion of IMMs is further confirmed via Opa1 along with Mfn1. To date, the accurate molecular mechanism for the fusion of IMM is not fully understood, but Opa1 is required. Mitofusion-linked proteins are modulated by post-translational modifications that influence their activity and abundance. The activities of Mfn1 and Mfn2 might be altered by unique phosphorylation, and the ubiquitination of proteins could result in their degradation (Gao S and Hu J, 2021) (Figure 1).

Molecular Mechanisms of Mitophagy Regulation

Mautophagy is a crucial type of autophagy that regulates mitochondrial turnover. Scott primarily used the term "mitophagy" in 1998 and also hypothesized the non-random nature of events in 2005 by Lemasters. The mitochondrion is targeted for mitophagy by two crucial known mechanisms: ubiquitin- and transmembrane receptor-mediated (Figure 2) (Choubey V et al., 2021; Onishi M et al., 2021).

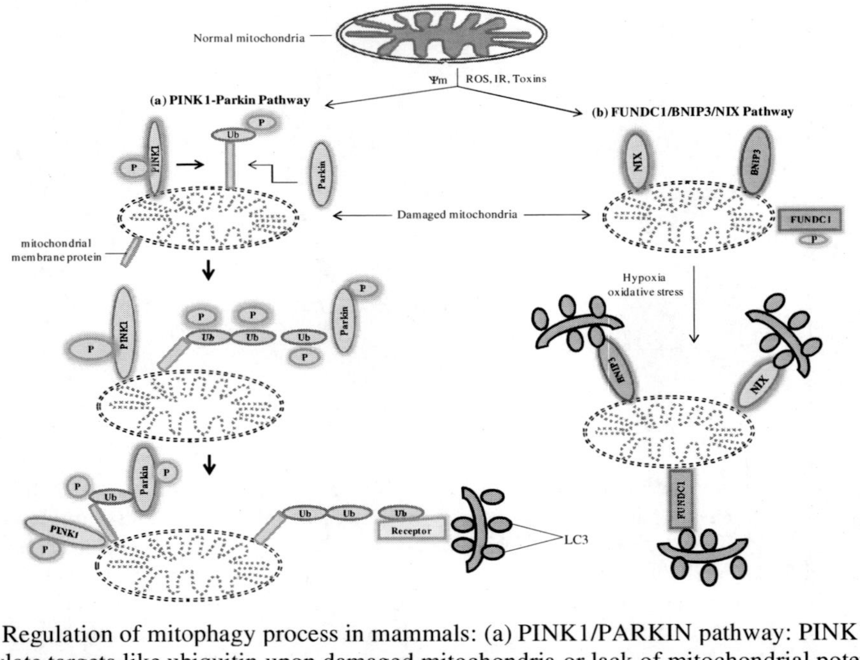

Figure 2. Regulation of mitophagy process in mammals: (a) PINK1/PARKIN pathway: PINK1 might phosphorylate targets like ubiquitin upon damaged mitochondria or lack of mitochondrial potential. Afterward, PINK1 recruits PARKIN which might potentiate the signal by ubiquitination of surface receptor proteins of mitochondria. These surface receptor proteins might identify ubiquitinated proteins that stimulate mitochondria to form autophagosomes, further degrading them. (b) FUNDC1/BNIP3/NIX pathway: Membrane surface receptor proteins recognize LC3 and facilitate complementation of phagophores, further addressing the mitochondrion for mitophagy. The interaction of OMM proteins like FUNDC1, BNIP3, or NIX with LC3-II by their motifs of cytoplasmic LIR occurred further aids in the stimulation of targeted removal of mitochondria. $\Delta\Psi_m$ (mitochondrial

potential), IR (Ischemia/Reperfusion), ROS (reactive oxygen species), poisons, and toxins formed naturally by organisms.

PINK1 (PTEN-induced putative kinase-1) facilitated mitophagy is a particular event in which mitochondria initially depolarize. Later, mitochondrial dysfunction is promoted by numerous factors like DNA damage of mitochondria or reactive oxygen species (ROS). The dysfunctional mitochondria might then hire autophagy receptors for autophagosome production on mitochondria and blend with lysosomes, hence suffering from the degradation process. Routine elimination of abnormal mitochondria by mitophagy is crucial to maintain the integrity of the cell and mitochondrial balance from the risk of damaged or aged mitochondria (Ivankovic D et al., 2016).

Molecular Mechanisms of Ubiquitin Type Mitophagy Pathway

The PINK1-Parkin Signaling Pathway in Mitophagy

It is an intensely studied classical pathway that causes mitophagy. In 2004, PINK1 was found to be a mitochondrial serine/threonine kinase in Parkinson's disease. The N-termini of PINK1 transversely transmit OMM to IMM, with the kinase domain situated near to C-termini projecting out into the cytosol under normal conditions (Onishi M et al., 2021; Choubey V et al., 2021).

A series of degradation events happen in PINK1 via matrix processing peptidases (MPP) in the matrix of mitochondria, presenilin-associated rhomboid-like (PARL) in IMM, and proteases in the cytoplasm, which results in decreased levels of PINK1 in normal cells. Therefore, PINK1 sequesters on OMM of damaged mitochondria via damaging inter-membrane transfer of the N-terminus domain to IMM under mitochondrial stressors such as membrane depolarization, mutagenic stress, mitochondrial complex dysfunction, and proteotoxicity. It has been revealed that PINK1 acts as a sensor probe for impaired mitochondria due to its capability to fast sequestration on OMM and activation upon mitochondrial stressors (Ivankovic D et al., 2016). The latest finding suggests that PINK-1 s (a short form of PINK1) senses the proteasomal abilities and rescues cells to overcome proteasome stress by suppressing the translation. Further on the OMM, homodimerization of PINK1 results in autophosphorylation, which further induces kinase activation and promotes the transfer of Parkin to the membrane of impaired mitochondria and employment of substrates ubiquitin to mitochondria. Parkin is activated by PINK1 by two mechanisms. Firstly, PINK1 phosphorylates ubiquitin at position S65 competes with an auto-inhibitory domain within Parkin, and sustains it in an active conformation. Second, it directly phosphorylated the Parkin at the position of S65 in Parkin's ubiquitin-such domain that promotes alteration in conformation, which is further responsible for activating Parkin activity. Parkin also plays a role in the ubiquitination of limited OMM proteins such as MIRO1, Mfn1, Mfn2, and VDAC. In the cytoplasm, ubiquitinated proteins like VDAC that recognize LC3 junction proteins like SQSTM1/P62, NBR1, NDP52, OPTN, and TAXIBP1 might interact with ubiquitinated proteins and facilitate mitophagy via recognizing and interacting with LC3 on the membrane of autophagy. Ambra1 is found to be the binding protein of Parkin during this process, and the crucial function of Ambra1 is to induce mitochondrial clearance and cellular autophagy by influencing the production of novel phagophores. Hence, the role of Ambra1 in autophagy and mitophagy is well-versed in literature (Onishi M et al., 2021; Choubey V. et al., 2021).

Molecular Mechanisms of the Receptor-Mediated Pathway

BNIP3 and NIX/BNLP3L Signaling Cascade of Mitophagy

BCL2 and adenovirus E1B 19-KDa interacting protein 3 (BNIP3) and Nip3-like protein X, also called BNIP3L (NIX), are known to be bifunctional mitochondrial proteins as they possess a BH3 domain that resembles BH3-only proapoptotic members of the Bcl-2 family members and hence can promote apoptosis. BNIP3 protein is mainly found at low levels in normal cells and tissues under physiological conditions. Numerous stress conditions, mainly hypoxia, can stimulate BNIP3-facilitated mitophagy because BNIP3 is transcriptionally stimulated by the hypoxia-inducible factor alpha 1 (HIFα-1) transcription factor in hypoxic. In particular conditions, FOXO3a and NF-κB have regulated the transcription of BNIP3 and NIX, which indicates their role in the signaling cascade besides hypoxia conditions. It has also been suggested that the BNIP3 and NIX produce homodimers and heterodimers and also separately connect with Mieap (a p53-regulated mitochondrial protein that is related to clearance of oxidized species) and cadherin6 (CDH6, protein regulator of mitophagy and DRP1-induced fission to assurance for mitochondrial homeostasis (Li H et al., 2022).

Further, Opa1 was found to bind with BNIP3 to stimulate fragmentation of mitochondria and phosphorylation of BNIP3 in its C-termini domain that disturbs the Opa1-BNIP3 binding, inducing mitophagy and decreasing apoptosis. It has also been revealed that the phosphorylation in the LIR domain of NIX and BNIP3 at serines 34 and 35 for NIX and serines 17 and 24 for BNIP3 results in the elevation of their affinity toward LC3 binding, implicating that the phosphorylation is the major molecular event which is modulating the pro-mitophagy or pro-death activity of these bifunctional membrane proteins. NIX performs the programmed elimination of mitochondria in reticulocytes and immature red blood cells. Hence, NIX acts as a modulated mitophagy receptor, and NIX-induced mitophagy also plays a chief role in controlling platelet mitochondria quality/activity (Panigrahi DP et al., 2020).

FUNDC1 Signaling Cascade in Mitophagy

FUNDC1 (Fun14 domain-containing protein 1), a protein situated in OMM, is found to be another new receptor that participates in mitophagy. FUNDC1-stimulated mitophagy participated in hypoxia-mediated mitophagy, is different from Pink Parkin cascade mitophagy since deletion of Parkin does not stop the degradation of mitochondria and also transition from LC3-I to II under hypoxia condition. Likewise, BNIP3 and NIX have a significant character in their conserved LC3 interaction region (LIR) that can bind with LC3, a biomarker protein of autophagosome in the cells of mammals. The binding between FUNDC1 and LC3B is significantly influenced by the phosphorylation of FUNDC1 (Hamacher-Brady A and Brady NR, 2016). Many studies show that FUNDC1 modulates hypoxia-stimulated selective mitophagy by changeable phosphorylation at crucial sites like Tyr18, Ser13, and Ser17. FUNDC1 binds with LC3B by its classical LIR-Y(18)xxL under normal conditions, whereas hypoxia-mediated dephosphorylation of FUNDC1 accelerates its binding with LC3 for selective mitophagy. A serine/threonine phosphatase named phosphoglycerate mutase family member 5 (PGAM5), situated at the mitochondria, can bind with and then dephosphorylate the LIR domain of FUNDC1 at the Ser13 position in response to FCCP treatment/hypoxia stimulation. Mitochondrial damage linked with FUNDC1 contributes to many pathophysiological conditions such as cardiovascular diseases, metabolic disorders, and intestinal ischemia/reperfusion injury (Lampert MA et al., 2019). In these diseases, FUNDC1 is protective because its induced mitophagy can improve the damage caused by intracellular stress like hypoxia and benefit overall outcomes. FUNDC1 protein is situated in MAMs (mitochondrial-associated endoplasmic reticulum membranes), and it plays an imperative role in communication among mitochondria and ER in the heart and maintains the normal functioning of the cardiac system (Zhang W, 2020).

Role of Mitophagy in Cancer

Mitochondria play crucial roles in cellular metabolism. Mitophagy sustains the homeostasis of mitochondria and normal cellular function, whereas abnormal mitophagy results in various pathophysiological conditions such as cardiovascular diseases, neurodegenerative diseases, muscle disease, and cancer. Hence, the involvement of mitophagy in tumor development and progression must be further explored. As per various findings, different degrees of mitophagy in numerous cancers such as liver, breast, colorectal, and lung established a close link between mitophagy and cancer. The alterations in complicated mitochondria function significantly influence cancer growth and progression (Vara-Perez M et al., 2019; Drake LE et al., 2017; Panigrahi DP et al., 2020). The role of mitophagy in cancer is depicted in Figure 3.

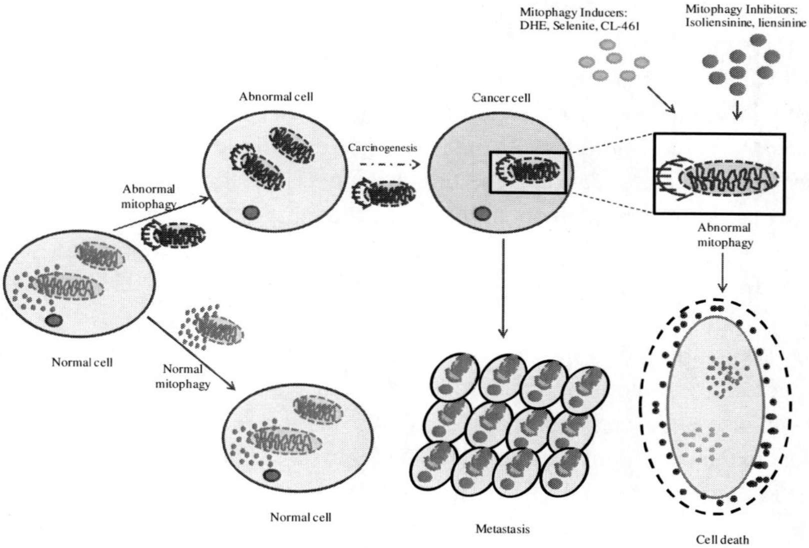

Figure 3. Function of mitophagy in cancer. Mitophagy occurs in different cell types. Mitophagy is generally accelerated or suppressed in cancer cells, i.e., distinct from normal cells. Mitophagy is also linked with incidences and metastasis of cancers. During the span of tumorigenesis, restricted mitophagy facilitates tumor development, whereas, in developed tumors, mitophagy might aid in the cellular proliferation of tumor cells. Mitophagy also plays a crucial role in the development of cancer metastasis. Inducers or inhibitors of mitophagy can observe anticancer effects via modulation of mitophagy. Dysfunctional mitochondria in cancer cells are quickly cleaned by mitophagy, which controls chemoresistance in cancer cells.

Mitophagy and Carcinogenesis

As reported in various studies, impaired mitophagy results in tumorigenesis and tumor development. Mitophagy plays differential roles in various stages of tumor development. Mitophagy maintains normal cellular metabolism in the early stage of tumor development and suppresses tumorigenesis, whereas, in the late stage, mitophagy improves the tolerance of cells and facilitates tumor development. The sequestration of dysfunctional mitochondria and suppression of mitophagy occur due to mutation, loss, or functional changes in various chief genes, which ultimately promotes carcinogenesis (Poole LP and Macleod KF, 2021; Panigrahi DP et al., 2020). The chief pathway of mitophagy is called the PINK1/Parkin pathway. The functional loss of the Parkin gene might suppress mitophagy and stimulate tumorigenesis in numerous cancer models. The repression of mitophagy further stimulates ROS, which affects the functional activities of cells, tissues, organs, and even entire systems. A previous study has shown that the silencing of Parkin in mice is more prone to sudden development of liver cancer than mock mice.

Furthermore, defective hepatocytes in the Parkin gene continue to proliferate via the up-surging of endogenous follistatin, exerting an anti-apoptotic effect that ultimately results in cancer progression. The Parkin protein, encoded by the PARK2 gene, is situated on human chromosome 6 (FRA6E, 6q26) and is very susceptible to gene mutation. Various researchers have reported that mutations in the Parkin gene are frequently revealed in numerous cancer types such as lung, colon, and glioma. The mutation types of the PARK2 gene are different in various cancer types. For instance, amplification in PARK2 is frequently found in uterine cancers and sarcomas, while hazardous mutation types such as deletion or loss of function of PARK2 are revealed in other cancer types. It has been suggested that up-regulation in the expression of Parkin might suppress tumor growth in breast cancer, colon cancer, glioblastoma, and liver cancer, whereas inhibition of expression of Parkin induces cellular proliferation and carcinogenesis in pancreatic cancer. The tumor suppressor function of the PARK2 gene has been proven by the rapid development of intestinal adenoma in neonatal mice and also, a variety of polyps elevated has been observed if PARK2-deleted mice were crossed with mice that display colorectal cancer adenomatous polyposis (Onishi M et al., 2021; Choubey V et al., 2021).

Under hypoxia, mitophagy has spontaneously been revealed as a crucial adaptation to cells that might decrease the number of mitochondria that results in limited production of ROS clusters but can also enhance the usage of available oxygen. BNIP3 and its ligand Nix are two chief receptors in the process of mitophagy under hypoxia conditions. As Parkin, Nix also plays a crucial role in carcinogenesis. HIF-1 stimulates the upsurge of BNIP3 and Nix expression in cancer tissues in response to hypoxia conditions (Li H et al., 2022). In hypoxic tumor cells, the level of Nix is highly elevated, and Nix-mediated mitophagy induces the survival of cells in glioblastoma. Nix knockout disturbs ROS clearance and conservation of cancer stem cells and suppresses cancer cells' survival. The oncogenic KRAS promotes mitophagy, i.e., induced by NIX, and also facilitates pancreatic tumorigenesis, while loss in function of NIX results in resume of mitochondrial function, which potently delays pancreatic tumor progression. Another study has revealed that under hypoxic conditions, an upsurge of NIX is stimulated by p53, which further induces apoptosis of cancer cells, but silencing the Nix gene elevates tumor growth. FUNDC1, a mitophagy-associated protein also involved in carcinogenesis. The level of FUNDC1 is highly expressed in early-stage cervical cancer patients compared with normal ones.

Furthermore, the knockout of FUNDC1 in cancer cells suppressed cell proliferation initiated apoptosis, and elevated the sensitivity of cancer cells toward chemo-radiotherapy such as cisplatin and ionizing radiation. It has also been observed that FUNDC1 gathers in most human HCCs. The silencing of FUNDC1 in hepatocytes induces the commencement and development of diethylnitrosamine (DEN)-promoted initial HCC; in contrast, if FUNDC1 is moved to hepatocytes, it suppresses liver carcinogenesis. The unique loss of FUNDC1 in hepatocytes results in the sequestration of damaged mitochondria and further induces a set of incidents like the awakening of inflammasome and JAK/STAT signaling. This implies that mitophagy can suppress the incidences and development of HCC by declining the awakening of inflammatory bodies (Zhang W, 2020; Lampert MA et al., 2019).

Mitophagy and Cancer Metastasis

Some cancer cells change the aerobic respiration of mitochondria into glycolysis to sustain the energy required during cancer progression. This process is known as the Warburg effect. A Warburg effect, the aberrant quality control of mitochondria, ROS production, regulation of the cell's redox state, and loss of apoptosis signals generally recognize cancer cells. Mitophagy can induce the survival of cancer cells as an adaptation to stress by removing damaged mitochondria. Parkin enhances the degree of oxidative metabolism and suppresses the Warburg effect downstream of the p53 tumor suppressor, which is likely to be completed by increasing the mitochondrial integrity (Poole LP and Macleod KF, 2021; Panigrahi DP et al., 2020). Therefore, as a p53-targeted gene, Parkin might display tumor-suppressive properties. Parkin suppresses cancer migration and invasion by directing the HIF-1α for its ubiquitination and degradation in breast cancer. With the introduction of the Parkin gene in breast cancer cells with a deleted Parkin gene, the advancement of breast cancer and metastatic rate potently decreased.

In contrast, Parkin also induces cancer metastasis, as per emerging evidence. According to pathological analysis, Parkin expression and cancer metastasis were elevated in melanoma compared with normal tissues. Moreover, it has also been suggested that the functional loss of Parkin suppressed the tumor production and metastasis of melanoma via inhibiting the ubiquitination of Mfn2. Parkin's differential roles in cancer metastasis might be due to various mechanisms of metastasis in different cancer types. BNIP3 is a pro-apoptotic protein that might suppress the fusion of dysfunctional mitochondria and increase mitophagy. Therefore, BNIP3 in cancer metastasis acts differently in various cancer types. The elevated expression of BNIP3 frequently occurs in many cancers like endometrial cancer, salivary gland adenoid cystic carcinoma, and DCIs of breast and cervical cancer are linked with the aggressiveness of cancer and poor prognosis. The functional loss of BNIP3 results in abnormal mitophagy, which enhances mitochondrial ROS levels linked with cancer metastasis in TNBC. The elevated expressions of BNIP3 and Nix are found in DCIS, whereas BNIP3 is not found in invasive carcinoma associated with cancer cell proliferation index and lymph node metastasis. The elevated level of BNIP3 facilitates excessive mitophagy that suppresses HCC metastasis. Rheb, a small GTPase protein, can connect with NIX and BNIP3 and efficiently act on the upstream domain of the mTOR pathway to stimulate cell growth. The suppression of BNIP3 via Rheb by the connection of Rheb and BNIP3 inhibits the property of the mTOR signal and decreases cell growth, which complements the role of BNIP3 in tumor repression. Therefore, similar to PINK1/PARKIN, BNIP3 is a cancer suppressor and exhibits oncogenic properties. The enhancement in aggressiveness and metastatic pancreatic cancer cells has been observed due to epigenetic silencing of BNIP3 and results in resistance to hypoxia-stimulated apoptosis in pancreatic cancer. In addition, elevated BNIP3 is responsible for the increase in cancer cell migration and invasion. The binary function of BNIP3, i.e., cancer progression and metastasis, might be because of different interactions with various determined factors via its BH3 (Bcl-2 homology 3) domains and also to the diversity of numerous cancers.

Mitophagy and Drug Resistance

Cancer resistance leads to complicated questions that fascinate various steps and factors. Both the effectiveness of cancer chemotherapy and the extent of drug resistance are affected by mitophagy. This is due to cancer cells quickly sweeping away dysfunctional mitochondria caused by mitophagy to facilitate their drug resistance. The findings align with the above facts; the toxicity of many chemotherapeutic drugs is at least partly caused by stimulation of dysfunctional mitochondria and oxidative stress (Bahar E et al., 2022). In the early treatment of multiple solid cancers, routine chemotherapeutic drugs like doxorubicin (Dox), cisplatin, Paclitaxel, and 5-fluorouracil (5-FU) might kill cancer cells. Thus, drug resistance might result in the treatment failure of chemotherapy in the course of autophagy or mitophagy. The generation of autophagy initiated by elevated galectin-1 results in drug resistance of epithelial ovarian cancer to cisplatin and a similar finding has been revealed in liver cancer (Bahar E et al., 2022; Kulikov AV et al., 2017; Wang Y et al., 2020).

Hypoxia-stimulated mitophagy is responsible for chemoresistance against cisplatin in a dependent style of HIF1-α and HIF2-α proteins (Wu H et al., 2015). ARIH1, an E3 ubiquitin ligase that is elevated in cancer cells, might initiate mitophagy in a PINK1-dependent and Parkin-defective manner that tackles chemotherapy-stimulated apoptosis in breast cancer and lung adenocarcinoma, ultimately leading to cancer cell resistance. Thus, appropriate anticancer results might be obtained by modulation of autophagy or mitophagy. The inhibition of autophagy via silencing of the Atg gene might drive the BNIP3-stimulated cancer apoptosis cascade post-treatment with cisplatin. Increased sensitivity of multidrug-resistant cancers towards betulinic acid analog B5G1, stimulated via the suppression of PINK1-Parkin-dependent mitophagy, leads to apoptosis in liver cancer. The stimulation of mitophagy might enhance in drug the sensitivity of cancer cells. The mitophagy-dependent necrosis and ferroptosis might be stimulated by the inhibition of the mitochondrial respiratory chain via BAY 87-2243 in melanoma cells. Besides this, mitophagy can induce the sensitivity of breast cancer cells toward liensinine by DNM1L-induced mitochondrial fusion. These findings indicate that mitophagy plays a role in drug resistance but requires further investigations in different cancer types. CSCs are generally resistant to chemotherapy. Fascinatingly, mitophagy also involves in CSCs-induced drug resistance. The cells of $CD44^{+}$/$ABCB1^{+}$/$ADAM17^{+}$ act with characters of CSCs and show chemoresistance by regulation of mitophagy. In colorectal cancer, BNIP3/NIX-induced mitophagy results in DOX resistance of CSCs and BNIP3L knockdown; both stop mitophagy and enhance chemo-sensitivity toward doxorubicin, which indicates the crucial role of mitophagy in chemoresistance and directing CSCs for cancer therapy (Praharaj PP et al., 2021).

Moreover, mitophagy regulates drug resistance and limits the effectiveness of radiotherapy. p53/BNIP3 induces mitophagy and mediates a crucial influence on the viability of cells of HNSCC cancer following radiotherapy by sustaining the mitochondrial integrity. In the radio-resistant cancer cells, BNIP3-dependent mitophagy depends on wild-type p53 status to restrict the glycolytic remodeling, reflecting the significant use of drugs targeting glycolysis as an alternative approach to curb radio-resistant cancers.

Mitophagy as a Target for Anticancer Therapeutics

Mitophagy induces cell survival via adaptation to stress, but it might result in apoptosis in the course of an enormous clearance of mitochondria. However, stimulators and inhibitors of mitophagy might be equally efficient in anticancer treatment. In order to enhance cancer therapy, there is immense attention on modifying mitophagy, although these approaches are considerably confusing. The purposeful result of mitophagy-promoted apoptosis or cell survival reliant on cancer therapy depends on numerous remedial treatments and cell types (Qiu YH et al., 2021).

Promotion of Mitophagy Increases Cancer Apoptosis and Chemotherapy Sensitivity

Ceramide is a fundamental particle of sphingolipid metabolism that modulates mitophagy (Dany M and Ogretmen B, 2015). It has been reported that CerS1/C18-ceramide selectively facilitates lethal mitophagy, which is B-cell lymphoma 2-associated X (BAX)/BCL2 antagonist/killer 1 (BAK1) and also caspase-independent. Ceramide promotes the formation of lapidated LC3 that interacts with ceramide on the mitochondrial membrane upon dynamin-related protein 1 (DRP1)-induced mitochondrial fission directing autophagolysosomes to mitochondria and resulting in lethal mitophagy in cancer cells (Sentelle RD et al., 2012; Dany M et al., 2016). In cancer cells, enormous mitochondrial clearance in a deficit of mitochondrial biogenesis or metabolic plasticity promotes cellular metabolic disorders and apoptosis. Furthermore, DRP1-induced fission is also required for mitophagy facilitated by BNIP3, a protective response driven by cardiac myocytes. The numerous influences of DRP1 among cancer cells and cardiac myocytes might aid in better anticancer drug formulation for understanding the underlying mechanism, i.e., remain unexplored. Immature hematopoietic cells expressed FLT3 (FMS-like tyrosine kinase 3), a membrane-bound receptor tyrosine kinase protein. Mutations in FLT3 have been found in 30% of patients with acute myelogenous leukemia (AML). The most frequently facilitating FLT3 mutation is an internal tandem duplication (ITD) in the juxtamembrane domain (FLT3-ITD). The inhibitors of FLT3-ITD, such as crenolanib, sorafenib, and quizartinib (AC220), exhibit effective therapy in preclinical models of AML while not working in clinical trials due to the development of chemoresistance (Stein EM and Tallman MS, 2016). LCL 461, a mitochondria-directed ceramide analog drug, effectively modifies crenolanib resistance by promoting lethal mitophagy in FLT3–ITD+ AML in vitro and in vivo (Dany M et al., 2016). The deadly pathways of mitophagy promotion via ceramide and ceramide analog drugs are possible targets in FLT3–ITD+ AML, and further investigations will be required to decipher if ceramide will be effective in other cancers. Figure 4 displays the promotion of mitophagy by various agents that trigger apoptosis of cancer cells.

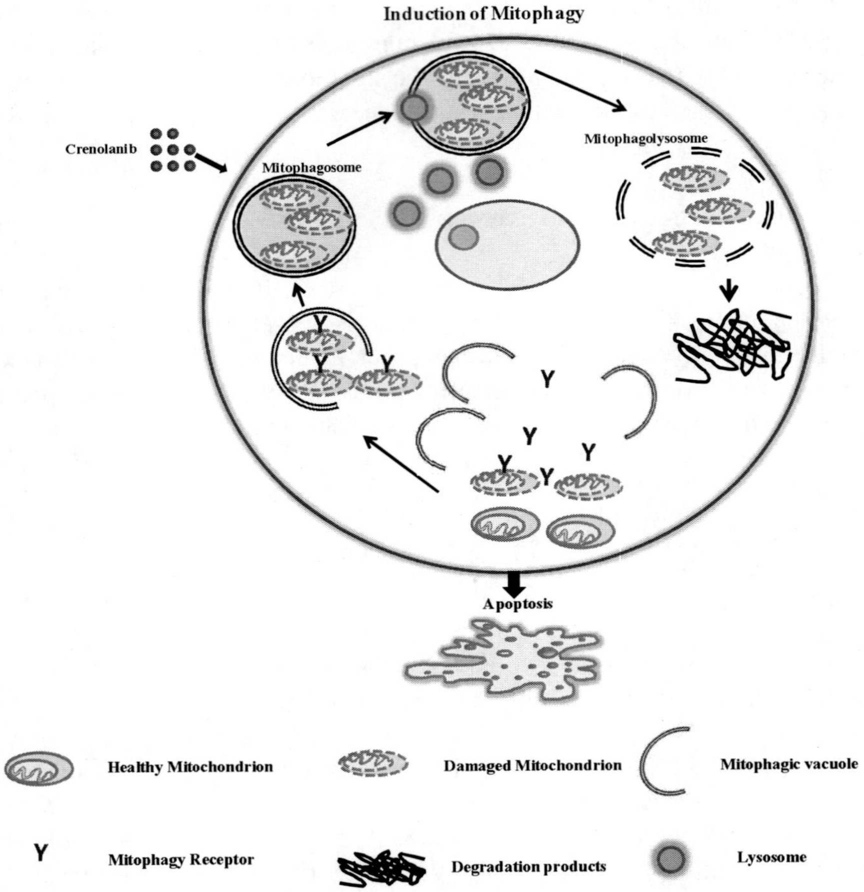

Figure 4. Induction of mitophagy.

Inhibition of Mitophagy Enhances Drug Sensitivity

Various studies have deciphered that mitophagy might be suppressed by genetically or pharmacologically directing numerous events of autophagy/mitophagy process like knock out of autophagic/mitophagic genes such as autophagy-related 5 (ATG5), BNIP3, BNIP3L, Parkin and PINK1), inhibiting the production of autophagosomes by phosphoinositide 3-kinase (PI3K) inhibitors like LY294002 and 3-methyladenine and suppressing the fusion of autophagosomes with lysosomes or the degradation capability of autophagolysosomes like bafilomycin A1, chloroquine, liensinine and leupeptin (Yan C et al., 2017; Radogna F et al., 2016). In Table 1, we have summarized effective agents that can be used as cancer therapeutics in different cancer types by influencing the mitophagy process.

Table 1. Mitophagy inhibitors and inducers in anticancer therapeutics

Name of Drug	Type of cancer	Effect of Drug	Mechanism of action
Mdivi-1	Lung cancer	Suppresses the Drp1 and pro-apoptotic effects	Increased the efficacy of chemotherapeutic drugs
Piperlongumine	Ovarian cancer	Stimulate Drp1 and facilitate equal fission and apoptosis	Enhances sensitivity of cancer cells toward a chemotherapeutic regimen
Lectin	Ovarian cancer	Stimulate AKT signaling cascade and de-phosphorylate Drp1	Provoke autophagy or necrosis of cancer cells
DHE	Lung cancer	Urge mitochondrial damage, apoptosis, and mitophagy	PINK1/Parkin-stimulated mitophagy enhanced the production of ROS, apoptosis promotion
Nanomicelle	NSCLC (Non-small cell lung carcinoma)	Promotes extreme autophagy/mitophagy and energy consumption	Upsurge the efficacy of chemotherapeutic drugs to overcome chemo-resistance and decrease drug toxicity
CerS1/C18 pyridinium ceramide	Hypopharyngeal cancer	Facilitate lethal mitophagy	Direct the autophagosomes to mitochondria
AT 101	Glioma	Stimulate mitochondrial damage and mitophagy	Mitophagic apoptosis
Sodium selenite	Glioma	Stimulate mitochondrial dysfunction and further mitophagy	Direct provoking of natural killer cells, oxidation of sulfhydryl groups leads to changes in their conformation
LCL-461	Acute myeloid leukemia	Urge deadly mitophagy	Debilitating crenolanib resistance via promoting deadly mitophagy in FLT3-ITD+AML *in vitro* and *in vivo*
Liensinine	Breast cancer	DNM1L-facilitated mitofission sensitizes cancer cells to chemotherapy	Enhances sensitivity of cancer cells toward chemotherapeutic regimen
Isoliensinine	Breast cancer	Promote cancer cell death by ROS and p38 MAPK/JNK stimulation	Enhances sensitivity of cancer cells toward chemotherapeutic regimen
Mito-CP	Colon cancer	Promote modifications in mitochondrial energy, prevent mTOR-induced cellular proliferation, and stimulate mitophagy.	Liberate ULK1 from mTOR-facilitated suppression, diminish mitochondrial membrane potential, and abate cellular proliferation in colorectal cancer.
Mito-Metformin Abrus agglutinin	Glioblastoma	Initiate ceramide production and stimulate ER stress and ROS to facilitate mitophagy.	Abate proliferation of cancer cells by mitophagy.
BAY 87-2243	Melanoma	Suppress mitochondrial complex I	Initiate mitophagy
B5G1	Liver cancer	Suppress PINK1-Parkin-dependent mitophagy	Activates PINK1 to recruit Parkin to mitochondria
Melatonin	Cervical cancer	Repress mitophagy and modulate cancer progression	Suppresses JNK (c-Jun N-terminal kinase) and further Parkin, provoke programmed cell death
Ketoconazole	HCC	Stimulates mitophagy	PINK1/Parkin-facilitated mitophagy, downregulation of COX-2, promotion of apoptosis
Sorafenib	Liver cancer, Renal cancer	Facilitate mitophagy	PINK1/Parkin-induced mitophagy, inhibition of ETC complexes II and III, promotion of apoptosis

Modified and adapted from Tang Y et al., 2022; Song C et al., 2022; Poole LP and Macleod KF, 2021; Panigrahi DP et al., 2020.

Classical Inhibitors of Mitophagy

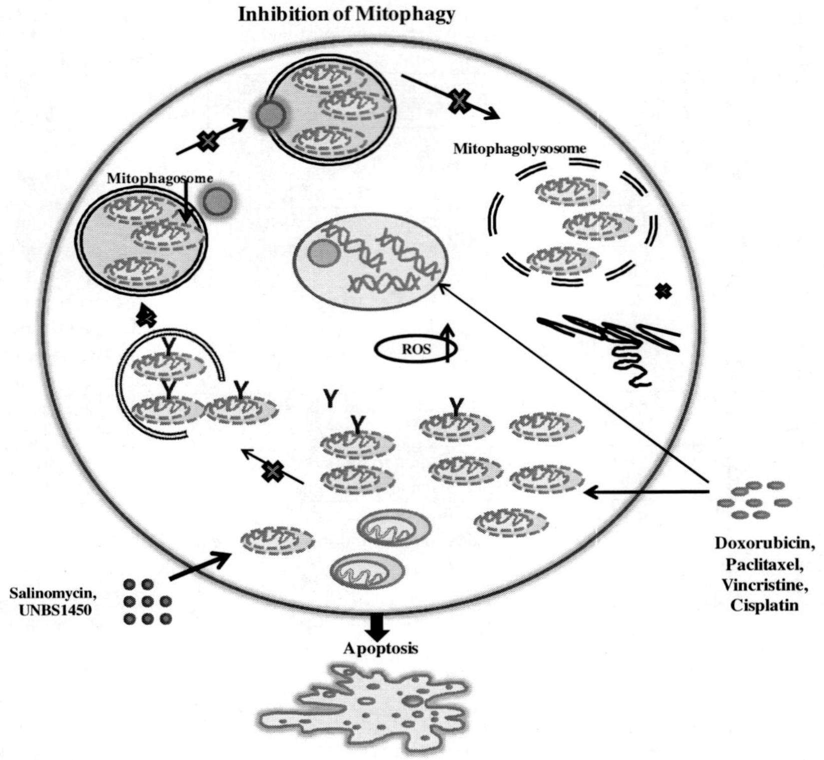

Figure 5. Inhibition of mitophagy.

Doxorubicin (DOX), a DNA-damaging agent, is found to be responsible for toxicity via promoting mitochondrial damage and increasing the superoxide formation (Luanpitpong S et al., 2012). The elevated expression of BNIP3L (regulator of mitophagy) has been observed after treatment of DOX in colorectal cancer stem cells. The knockout of the BNIP3L gene suppressed the mitophagy that further potently enhanced chemo-sensitivity toward DOX in colorectal cancer stem cells (Yan C et al., 2017). However, the molecular modulation mechanism of the BNIP3L gene in cancer stem cells remains unknown. Salinomycinis an antibacterial and coccidiostat ionophore remedy found to have new anticancer activity by hyperpolarizing mitochondria. In salinomycin-treated cells, mitochondrial dynamic fission and fusion mechanisms are responsible for mitochondrial mass changes and mitochondrial function sustenance. Mitophagy is a chief pro-survival mechanism that removes salinomycin-promoted damaged mitochondria. The defective mitochondria could deplete ATP, producing large amounts of detrimental ROS, leading to deprivation of cellular homeostasis. The co-administration of mitophagy and ATG5 knockout in cancer stem cells improved salinomycin's toxicity (Jangamreddy JR et al., 2013). Several significant drugs are now being utilized for anti-cancer treatments. The combination of the inhibition of mitophagy and a few significant anti-cancer drugs also enhances cytotoxicity. UNBS1450 is known for being a significant sodium channel antagonist in cancer treatment. Perfect mitophagy is the primary mechanism that causes malfunctioning in stimulating the apoptotic pathway that enhances resistance to UNBS1450 in the stromal neuroblastoma SH-N-AS cell line. Suppressing autophagy via small inhibitory RNAs directing ATG5, ATG7 (autophagy-related 7), and Beclin-1 sensitized SK-N-AS cells by resuming the apoptosis process (Radogna F et al., 2016). Figure 5 shows the inhibition of mitophagy by various agents.

Novel Inhibitors of Mitophagy

Liensinine is an isoquinoline alkaloid extracted from the seed embryo of *Nelumbonucifera* Gaertn. The mitophagy-suppressing influences of liensinine increase the chemo-sensitivity of breast cancer cells towards traditional chemotherapeutic drugs like cisplatin, doxorubicin, paclitaxel, and vincristine. Liensinine facilitated the gathering of mitophagosomes by suppressing the fusion of autophagosome-lysosome. However, the influence of liensinine is the same as that of bafilomycin A1, chloroquine and hydroxychloroquine; the liensinine-facilitated obstruction of autophagosome-lysosome fusion is dissimilar. The final agents alkalinized the pH of lysosomes to inhibit the fusion among autophagosomes and hamper the action of lysosomal hydrolases. However, upon response to treatment with liensinine, the pH of lysosomes is not changed, implicating that change in lysosomal pH is not required for suppression of fusion of autophagosome-lysosomes and maturation of cathepsin facilitated via liensinine. RAB7A (a small GTP-binding protein ras-related protein 7A) has played a significant role in lysosome biogenesis, the later endocytic cascade and the closing maturation of late autophagic vacuoles. Liensinine decreased the employment of RAB7A to lysosomes, suppressing the endocytic cascade-dependent cathepsin transfer to lysosomes and preventing the fusion of autophagosome-lysosome. Suppressing autophagy/mitophagy via liensinine sensitizes cancer cells to DOX-promoted apoptosis might be a potential molecular mechanism of chemosensitivity (Zhou J et al., 2015).

Chemoresistance to cisplatin inhibits the effectiveness of chemotherapy in cervical cancer cells. Melatonin (N-acetyl-5-methoxytryptamine) is a known antioxidant and an endogenous indoleamine that decreases hypoxia-ischemia damage and revises sleep. It can also tackle cancer progression and suppress mitophagy. Melatonin inhibits mitophagy by suppressing the JNK (c-Jun N-terminal kinase) and further by Parkin, and then it worsens cell death in cervical cancer cells (Chen L et al., 2018). A chemotherapeutic regimen like cisplatin that does not work well in removing cancer cells might be of inherent autophagy and mitophagy in liver cancer. Cisplatin also stimulates the dynamin-related protein 1(DRP1) to increase mitophagy. A lysosome inhibitor (bafilomycin) or an inhibitor of DRP1-facilitated mitophagy (MDiv-1) enhances the susceptibility of liver cancer cells towards cisplatin despite directly causing cell death.

Dihydroergotamine tartrate (DHE) is a migraine drug extracted from ergot alkaloids and promotes lung cancer apoptosis by mitophagy and apoptosis. DHE facilitated defective mitochondria that resulted in PINK1/PARKIN-dependent mitophagy stimulation linked with enhanced ROS formation and further cell death in lung cancer cells specifically (Chang SH et al., 2016). Ketoconazole is an oral antifungal agent that promotes PINK1/Parkin-facilitated mitophagy by inhibiting the COX2 expression, which further induces the cell death of liver cancer cells (Chen Y et al., 2019). Similarly, the multi-kinase inhibitor sorafenib induces cell death in liver and renal cancer cells by PINK1/Parkin-facilitated mitophagy (Zhang C et al., 2017). It has been revealed that sorafenib suppresses complexes II and III of the electron transport chain via establishing PINK1 on OMM and Parkin employment to dysfunctional mitochondria, thus starting mitophagy and Mito-CP and Mito-Metforminare two mitochondria-directed drugs that liberate ULK1 from mTOR-facilitated suppression, inhibit mitochondrial membrane potential and abate cell proliferation in colorectal cancer by mitophagy (Boyle K et al., 2018). Mitophagy-associated apoptosis is activated by promoting ceramide stress in various cancers (Sentelle RD et al., 2012). Hence, the overproduction of ceramide CerS1 causes mitophagy and caspase-independent apoptosis. Besides this, a novel BH-3 mimetic AT101 and sodium selenite prompted a large amount of mitophagic apoptosis in cells of glioma (Meyer N et al., 2018).

Conclusion

In order to maintain cellular homeostasis, mitophagy plays a crucial role in blocking the assembling of damaged mitochondria, which results in elevated production of ROS and damage to cellular functions. Evidence suggests that mitophagy participates in the modulation of carcinogenesis and cancer progression. Therefore, mitophagy regulation is likely a significant approach for cancer therapy. In cancer, mitophagy seems to display a dual role in tumor suppression and progression. On the one hand, mitophagy suppresses the cancer progression via restrictive the production of ROS. On the other side, mitophagy might induce cancer growth by adapting to cancer cells in changing microenvironments. In cancer cells, mitophagy in excess might result in mitophagy apoptosis, and at the same time, in various cancers, the suppression of mitophagy leads to the inhibition of cancer growth. However, the latest advancement in understanding the molecular insights of mitophagy in carcinogenesis is very complicated and varies from different cancer types and stages. Here, we shed light on the most crucial knowledge about the mechanisms of mitophagy and their function in cancer progression and therapy. However, there remains to be explored about the role of mitophagy in cancer. Further findings are needed to comprehend mitophagy's molecular mechanism and purpose better to develop effective/significant novel cancer therapy.

References

Adebayo M, Singh S, Singh AP, Dasgupta S. (2021). Mitochondrial fusion and fission: The fine-tune balance for cellular homeostasis. *FASEB J.* 35(6):e21620.

Bahar E, Han SY, Kim JY, Yoon H. (2022). Chemotherapy Resistance: Role of Mitochondrial and Autophagic Components. *Cancers (Basel).* 14(6):1462.

Boyle KA, Van Wickle J, Hill RB, Marchese A, Kalyanaraman B, Dwinell MB. (2018). Mitochondria-targeted drugs stimulate mitophagy and abrogate colon cancer cell proliferation. *J Biol Chem.* 293(38):14891-14904.

Chan DC. (2020). Mitochondrial Dynamics and Its Involvement in Disease. *Annu Rev Pathol.* 15:235-259.

Chang SH, Lee AY, Yu KN, Park J, Kim KP, Cho MH. (2016). Dihydroergotamine Tartrate Induces Lung Cancer Cell Death through Apoptosis and Mitophagy. *Chemotherapy.* 61(6):304-12.

Chen L, Liu L, Li Y, Gao J. (2018). Melatonin increases human cervical cancer HeLa cell apoptosis induced by cisplatin via inhibition of JNK/Parkin/mitophagy axis. *In Vitro Cell Dev Biol Anim.* 54(1):1-10.

Chen Y, Chen HN, Wang K, Zhang L, Huang Z, Liu J, Zhang Z, Luo M, Lei Y, Peng Y, Zhou ZG, Wei Y, Huang C. (2019). Ketoconazole exacerbates mitophagy to induce apoptosis by downregulating cyclooxygenase-2 in hepatocellular carcinoma. *J Hepatol.* 70(1):66-77.

Choubey V, Zeb A, Kaasik A. (2021). Molecular Mechanisms and Regulation of Mammalian Mitophagy. *Cells*. 11(1):38.

Dany M, Gencer S, Nganga R, Thomas RJ, Oleinik N, Baron KD, Szulc ZM, Ruvolo P, Kornblau S, Andreeff M, Ogretmen B. (2016). Targeting FLT3-ITD signaling mediates ceramide-dependent mitophagy and attenuates drug resistance in AML. *Blood*. 128(15):1944-1958.

Dany M, Ogretmen B. (2015). Ceramide induced mitophagy and tumor suppression. *Biochim Biophys Acta*. 1853(10 Pt B):2834-45.

Drake LE, Springer MZ, Poole LP, Kim CJ, Macleod KF. (2017). Expanding perspectives on the significance of mitophagy in cancer. *Semin Cancer Biol*. 47:110-124.

Gao S, Hu J. (2021). Mitochondrial Fusion: The Machineries In and Out. *Trends Cell Biol*. 31(1):62-74.

Hamacher-Brady A, Brady NR. (2016). Mitophagy programs: mechanisms and physiological implications of mitochondrial targeting by autophagy. *Cell Mol Life Sci*. 73(4):775-95.

Ivankovic D, Chau KY, Schapira AH, Gegg ME. (2016). Mitochondrial and lysosomal biogenesis are activated following PINK1/parkin-mediated mitophagy. *J Neurochem*. 136(2):388-402.

Jangamreddy JR, Ghavami S, Grabarek J, Kratz G, Wiechec E, Fredriksson BA, Rao Pariti RK, Cieślar-Pobuda A, Panigrahi S, Łos MJ. (2013). Salinomycin induces activation of autophagy, mitophagy and affects mitochondrial polarity: differences between primary and cancer cells. *Biochim Biophys Acta*. 1833(9):2057-69.

Klein K, He K, Younes AI, Barsoumian HB, Chen D, Ozgen T, Mosaffa S, Patel RR, Gu M, Novaes J, Narayanan A, Cortez MA, Welsh JW. (2020). Role of Mitochondria in Cancer Immune Evasion and Potential Therapeutic Approaches. *Front Immunol*. 11:573326.

Kulikov AV, Luchkina EA, Gogvadze V, Zhivotovsky B. (2017). Mitophagy: Link to cancer development and therapy. *Biochem Biophys Res Commun*. 482(3):432-439.

Lampert MA, Orogo AM, Najor RH, Hammerling BC, Leon LJ, Wang BJ, Kim T, Sussman MA, Gustafsson ÅB. (2019). BNIP3L/NIX and FUNDC1-mediated mitophagy is required for mitochondrial network remodeling during cardiac progenitor cell differentiation. *Autophagy*. 15(7):1182-1198.

Lee H, Yoon Y. (2016). Mitochondrial fission and fusion. *Biochem Soc Trans*. 44(6):1725-1735.

Li H, Zhang C, Zhang Q, Jia J, Wang X. (2022). BNIP3 enhances pancreatic cancer cell migration and proliferation via modulating autophagy under hypoxia. *Heliyon*. 8(10): e11190.

Luanpitpong S, Chanvorachote P, Nimmannit U, Leonard SS, Stehlik C, Wang L, Rojanasakul Y. (2012). Mitochondrial superoxide mediates doxorubicin-induced keratinocyte apoptosis through oxidative modification of ERK and Bcl-2 ubiquitination. *Biochem Pharmacol*. 83(12):1643-54.

Maycotte P, Marín-Hernández A, Goyri-Aguirre M, Anaya-Ruiz M, Reyes-Leyva J, Cortés-Hernández P. (2017). Mitochondrial dynamics and cancer. *Tumour Biol*. 39(5):1010428317698391.

Meyer N, Zielke S, Michaelis JB, Linder B, Warnsmann V, Rakel S, Osiewacz HD, Fulda S, Mittelbronn M, Münch C, Behrends C, Kögel D. (2018). AT 101 induces early mitochondrial dysfunction and HMOX1 (heme oxygenase 1) to trigger mitophagic cell death in glioma cells. *Autophagy*. 14(10):1693-1709.

Missiroli S, Perrone M, Genovese I, Pinton P, Giorgi C. (2020). Cancer metabolism and mitochondria: Finding novel mechanisms to fight tumours. *EBioMedicine*. 59:102943.

Onishi M, Yamano K, Sato M, Matsuda N, Okamoto K. (2021). Molecular mechanisms and physiological functions of mitophagy. *EMBO J*. 40(3): e104705.

Panigrahi DP, Praharaj PP, Bhol CS, Mahapatra KK, Patra S, Behera BP, Mishra SR, Bhutia SK. (2020). The emerging, multifaceted role of mitophagy in cancer and cancer therapeutics. *Semin Cancer Biol*. 66:45-58.

Poole LP, Macleod KF. (2021). Mitophagy in tumorigenesis and metastasis. *Cell Mol Life Sci*. 78(8):3817-3851.

Praharaj PP, Panigrahi DP, Bhol CS, Patra S, Mishra SR, Mahapatra KK, Behera BP, Singh A, Patil S, Bhutia SK. (2021). Mitochondrial rewiring through mitophagy and mitochondrial biogenesis in cancer stem cells: A potential target for anti-CSC cancer therapy. *Cancer Lett*. 498:217-228.

Qiu YH, Zhang TS, Wang XW, Wang MY, Zhao WX, Zhou HM, Zhang CH, Cai ML, Chen XF, Zhao WL, Shao RG. (2021). Mitochondria autophagy: a potential target for cancer therapy. *J Drug Target*. 29(6):576-591.

Radogna F, Cerella C, Gaigneaux A, Christov C, Dicato M, Diederich M. (2016). Cell type-dependent ROS and mitophagy response leads to apoptosis or necroptosis in neuroblastoma. *Oncogene*. 35(29):3839-53.

Rodrigues T, Ferraz LS. (2020). Therapeutic potential of targeting mitochondrial dynamics in cancer. *Biochem Pharmacol*. 182:114282.

Sentelle RD, Senkal CE, Jiang W, Ponnusamy S, Gencer S, Selvam SP, Ramshesh VK, Peterson YK, Lemasters JJ, Szulc ZM, Bielawski J, Ogretmen B. (2012). Ceramide targets autophagosomes to mitochondria and induces lethal mitophagy. *Nat Chem Biol*. 8(10):831-8.

Simula L, Nazio F, Campello S. (2017). The mitochondrial dynamics in cancer and immune-surveillance. *Semin Cancer Biol*. 47:29-42.

Song C, Pan S, Zhang J, Li N, Geng Q. (2022). Mitophagy: A novel perspective for insighting into cancer and cancer treatment. *Cell Prolif*. 55(12):e13327.

Srinivasan S, Guha M, Kashina A, Avadhani NG. (2017). Mitochondrial dysfunction and mitochondrial dynamics-The cancer connection. *Biochim Biophys Acta Bioenerg*. 1858(8):602-614.

Stein EM, Tallman MS. (2016). Emerging therapeutic drugs for AML. *Blood*. 127(1):71-8.

Tang Y, Wang L, Qin J, Lu Y, Shen HM, Chen HB. (2022). Targeting mitophagy to promote apoptosis is a potential therapeutic strategy for cancer. *Autophagy*. 22:1-3.

Vara-Perez M, Felipe-Abrio B, Agostinis P. (2019). Mitophagy in Cancer: A Tale of Adaptation. *Cells*. 8(5):493.

Wang Y, Liu HH, Cao YT, Zhang LL, Huang F, Yi C. (2020). The Role of Mitochondrial Dynamics and Mitophagy in Carcinogenesis, Metastasis and Therapy. *Front Cell Dev Biol*. 8:413.

Yan C, Luo L, Guo CY, Goto S, Urata Y, Shao JH, Li TS. (2017). Doxorubicin-induced mitophagy contributes to drug resistance in cancer stem cells from HCT8 human colorectal cancer cells. *Cancer Lett*. 388:34-42.

Yapa NMB, Lisnyak V, Reljic B, Ryan MT. (2021). Mitochondrial dynamics in health and disease. *FEBS Lett*. 595(8):1184-1204.

Zhang C, Liu Z, Bunker E, Ramirez A, Lee S, Peng Y, Tan AC, Eckhardt SG, Chapnick DA, Liu X. (2017). Sorafenib targets the mitochondrial electron transport chain complexes and ATP synthase to activate the PINK1-Parkin pathway and modulate cellular drug response. *J Biol Chem*. 292(36):15105-15120.

Zhang W. (2020). The mitophagy receptor FUN14 domain-containing 1 (FUNDC1): A promising biomarker and potential therapeutic target of human diseases. *Genes Dis*. 8(5):640-654.

Zhou J, Li G, Zheng Y, Shen HM, Hu X, Ming QL, Huang C, Li P, Gao N. (2015). A novel autophagy/mitophagy inhibitor liensinine sensitizes breast cancer cells to chemotherapy through DNM1L-mediated mitochondrial fission. *Autophagy*. 11(8):1259-79.

Zong WX, Rabinowitz JD, White E. (2016). Mitochondria and Cancer. *Mol Cell*. 61(5):667-676.

Chapter 7

Mitophagy in Aging Muscle: Current Understanding and Future Perspectives

Kshitij B. Mankar[1]
Pranali Chandurkar[1]
Nitu L. Wankhede[1]
Shubhada V. Mangrulkar[1]
Manjit B. Chauhan[1]
Brijesh G. Taksande[1]
Aman B. Upaganlawar[2]
Milind J. Umekar[1]
and Mayur B. Kale[1,*]
[1]Smt. Kishoritai Bhoyar College of Pharmacy, Kamptee, Nagpur, Maharashtra, India
[2]SNJB's Shriman Sureshdada Jain College of Pharmacy, Neminagar, Nashik, Maharashtra, India

Abstract

Numerous human illnesses are connected to age. In recent years, a strong correlation has been shown between lifespan and mitochondrial homeostasis and age-related diseases. Mitochondrial homeostasis entails the production, biogenesis, and elimination of damaged mitochondria by mitophagy. Two stages are necessary to remove damaged mitochondria by autophagy: inducing general autophagy and prepping damaged mitochondria for selective autophagic identification. Recent developments in mitophagy research show that the mitophagy receptors Nix and Bnip3 or the Pink1-Parkin signaling pathway induce mitochondrial priming. We summarize what we now know about the mechanics of mitophagy in this review. We also discuss the pathophysiological roles that mitophagy plays and the available assays for monitoring mitophagy. Various mitochondrial and extra-mitochondrial factors, including morphology, oxidative stress, and DNA damage, regulate mitophagy. For many years, it has been believed that age-related illnesses are determined mainly by DNA damage and ineffective DNA repair. Although it is generally known that deficiencies in DNA damage identification, repair, and mitophagy play a significant role

* Corresponding Author's Email: mayur.kale28@gmail.com.

In: Mitophagy and Its Role in Health and Disease
Editors: Shalini Mani and Samiksha Wasnik
ISBN: 979-8-89113-470-6

in age-associated illnesses, the interplay between these is poorly understood. Age-related ailments like cancer, diabetes, inflammatory diseases, neurodegenerative diseases, and many more are made worse by gathering defective mitochondria caused by the decreased effectiveness of mitophagy. Mitophagy is therefore being addressed for treatments of illnesses linked to aging.

Keywords: aging, mitophagy, DNA damage, mitochondria, autophagy

Introduction

As we age, autophagy and its subtype mitochondrial autophagy, also known as mitophagy, can cause an imbalance in proteostasis and an increase in mitochondrial malfunction, linked to a decline in physiological integrity. The term "autophagy" refers to the method cells use to digest and recycle their parts. In low-nutrient environments, autophagy can refer to the elimination of specific superfluous or damaged organelles and the general, cell-wide degradation of organelles or misfolded proteins. According to a recent study, upregulating autophagy may postpone the onset and diminish the symptoms of age-related phenotypes. Aging and illnesses associated with it are related to autophagy (Rubinsztein et al., 2011; Madeo et al., 2010). As we age, autophagy and its subtype mitochondrial autophagy, also known as mitophagy, which can lead to an imbalance in proteostasis and an increase in mitochondrial malfunction, are linked to a loss of physiological integrity. The process of breaking down and recycling cellular components within the cell is known as autophagy. Autophagy in nutrient-poor environments can relate to the removal of specific defective or superfluous organelles and general cell-wide breakdown of organelles or misfolded proteins. According to a recent study, upregulating autophagy may prevent or postpone the emergence of age-related phenotypes and diminish their symptoms. Autophagy is linked to aging and age-related disorders (Rubinsztein et al., 2011), (Madeo et al., 2010). Since then, mitophagy has been linked to a plethora of disorders, such as Parkinson's disease, Huntington's disease, and Alzheimer's disease, as well as normal physiological aging (Ryan et al., 2015; Khalil et al., 2015; Palikaras and Tavernarakis, 2012; Sun et al., 2016).

A double-membrane structure termed the autophagosome is created during the genetically preprogrammed and evolutionarily conserved catabolic process of macroautophagy, which breaks down cellular proteins and damaged or extra organelles (Mizushima, 2007; Nakatogawa et al., 2009; Yang and Klionsky, 2010). When autophagosomes combine with lysosomes to form autolysosomes, the contents of the package are obliterated. Yeast has discovered around 30 autophagy-related genes (Atg), the majority of which have mammalian homologs that take part in autophagy or a process related to autophagy, even if the precise mechanism of autophagosome production is still unknown (Klionsky et al., 2003; Nakatogawaxet al., 2009). The Atg9-Atg2-Atg18 complex, Beclin 1-VPS34 class III phosphoinositide 3-kinase (PI3K) complex, Atg5-Atg12-Atg16, and the Atg8/LC3 conjugation systems are a few examples of multi-molecule complexes that regulate the formation of autophagosomes. The crucial roles of these Atg complexes in regulating autophagy have been discussed in great detail (Ravikumar et al., 2010; Ding et al., 2011; Mizushima and Komatsu, 2011; Upaganlawar et al., 2021).

Mitochondrial Fission and Fusion: A Dynamic Role in Aging and a Potential Target for Age-Related Diseases

The Basics of Mitochondrial Structure

Because mitochondria are dynamically connected, they can exchange proteins, solutes, metabolites, solutes, membranes, and electrochemical gradients. Each mitochondrion comprises two phospholipid bilayers: the outer mitochondrial membrane (OMM), which separates the intermembrane space from the cytosol, and the inner mitochondrial membrane (IMM), which forms cristae into the matrix (Tilokani et al., 2018). In contrast to the IMM, the OMM is mainly impermeable to ions and larger molecules. In the IMM, an electrochemical gradient is created that enables the production of ATP through oxidative phosphorylation, heat through proton leakage, and ROS through oxygen reduction (Vafai and Mootha, 2012; Wankhede et al., 2021; Marchi et al., 2014; Soto Heredero et al., 2017) In addition to communicating with one another, mitochondria can link up and exchange materials with other cellular organelles, like the endoplasmic reticulum (ER) and lysosomes (Marchi et al., 2014; Soto Heredero et al., 2017; Tiwari et al., 2021).

Mitochondrial Fusion and Fission Proteins and Processes

Fission is splitting a single mitochondrion into two, whereas fusion is uniting two nearby mitochondria (Sebastián and Zorzano, 2018). The inactivation of one of these two processes results in the unopposed operation of the other, and the resulting imbalance determines the shape of the mitochondria (Scorrano, 2013). The molecular mechanisms of these two processes have already been well-studied (Tilokani et al., 2018). Thus, we will only discuss the conserved proteins (Table 1) and the regulating systems.

Table 1. Orthologs of fusion and fission

	Mammalian	*C. elegans*	*D. melanogaster*
Fusion proteins	MFN1 MFN2 OPA1	FZO-1 EAT-3	FZO Marf Opa1
Fission proteins	DRP1 FIS1 MFF MID49 MID49 DNM2	DRP-1 FIS-1;FIS-2	Drp1 Fis Dmel

Fusion

Separate sequential events involving OMM-located mitofusin (MFN) 1, MFN2, and the IMM-located optic atrophy 1 (OPA1) coordinate fusion in mammalian cells (Malka et al., 2005; Song et al., 2009). The mitofusins are conserved catalytic GTP-binding domains at their N-termini and C-terminal transmembrane domains that attach them to the OMM. They are dynamin-like GTPases (Wai and Langer, 2016). They fuse when their two hydrophobic heptad repeats interact with neighboring mitochondria (Koshiba et al., 2004). The hydrolysis of GTP, which results in a conformational change that brings the opposing membranes together, initiates the fusing of the OMMs (Cao et al., 2017; Qi et al., 2016; Kale et al., 2022). MFN1 and MFN2 are nearly identical in 80% of ways (Santel et al., 2003). MFN2 is also found in the ER, where it regulates ER tethering to the mitochondria (Basso et al., 2018; de Brito and Scorrano, 2008; Naon et al., 2016), which, as mentioned later, aids in the fission process' mitochondrial constriction (Cohen et al., 2018).

OPA1, a dynamin-like GTPase tethered to the IMM via its N-terminal transmembrane domain, fuses inner mitochondrial membranes (Delettre et al., 2000). Due to alternative splicing, OPA1's long forms (L-OPA1) can be transformed into short forms via proteolysis (S-OPA1). OMA1 and YME1L, two IM peptidases, are in charge of this cleavage (Anand et al., 2014).

Cardiolipin, a lipid exclusive to mitochondria, is also necessary for IMM fusion (Ban et al., 2017, Tilokani et al., 2018). The two IMMs, which merge during OPA1-dependent GTP hydrolysis, are tethered to the membrane via their association with L-OPA1 (Liesa et al., 2009). It was postulated that S-OPA1 activates OPA1-CL contact and fusion (DeVry et al., 2009; Rujiviphat et al., 2009; Umare et al., 2021). Transcriptional and post-transcriptional mechanisms regulate the synthesis of mitofusins, whereas ubiquitylation and phosphorylation control their degradation. OPA1 is managed both post-transcriptionally and post-translationally (Wai and Langer, 2016). Proteolytic processing, in particular, is crucial for controlling mitochondrial dynamics (Dietz et al., 2019). A deficiency or loss of fusion proteins results in mitochondrial fragmentation (Ichishita et al., 2008; Kanazawa et al., 2008).

Fission

The replication of mitochondrial DNA in the matrix indicates the location where the endoplasmic reticulum (ER) should be recruited to begin the process of mitochondrial fission (Friedman et al., 2011; Lewis et al., 2016). OMM constriction starts at the mitochondria-ER contact sites before DRP1 oligomerization. Many mitochondrial-bound proteins, such as FIS1, MFF, MiD49, and MiD51, help to attract DRP1 to the mitochondria. Nevertheless, several of these proteins have overlapping functions (Loson et al., 2013). FIS1 was recently discovered to be involved in the lysosomal labeling of mitochondrial fission sites (Wong et al., 2018), suggesting that inter-organelle communication is a critical component of this process and a topic of future research. Although DRP1 recruitment to the OMM and subsequent fission induction do not require FIS1, excessive FIS1 expression in the absence of DRP1 can result in mitochondrial fragmentation (Liesa et al., 2019).

According to recent studies, FIS1 promotes fission while inhibiting fusion, which increases mitochondrial fragmentation (Yu et al., 2019). He also found that human FIS1 inhibits the fusion machinery by blocking the GTPases MFN1, MFN2, and OPA1 activities. This research highlights the intimate connection between the fission and fusion processes and suggests that one process may have additional methods to inhibit the other. While the IMM division mechanism is still unclear, the OMM constriction mechanism is well established (Tilokani et al., 2018). IMM constriction is Ca2+-dependent and occurs at mitochondria-ER contact sites, according to recent research (Chakrabarti et al., 2018; Cho et al., 2017), and S-OPA1 may play a role in untethering the OMM from the IMM during fission (Chakrabarti et al., 2018). DRP1-deficient mitochondria in yeast, worms, and mammals develop a hypertubular network with a highly coupled and extended network (Labrousse et al., 1999; Weir et al., 2017). Peroxisomal membranes are bound by DRP1 and its receptors, which cause membrane fission (Schrader et al., 2016; Waterham et al., 2007; Marde et al., 2022). Hence, investigations that interfere with DRP1 may affect cellular functions other than mitochondrial fission. DRP1 can be activated or inactivated depending on the phosphorylated residue, and DRP1 can also be further controlled by ubiquitylation and SUMOylation (Galvan et al., 2019; Jhun et al., 2018; van der Bliek et al., 2013; Marde et al., 2021). Various factors, including exercise, can influence DRP1 phosphorylation (Moore et al., 2019) and the 24-hour clock (Schmitt et al., 2018), although the mechanisms linking these activities to phosphorylation are unknown.

Mitophagy

In the evolutionarily conserved process of autophagy, lysosomes are employed to degrade cytoplasmic substrates. Mitophagy is the term for the process of selectively removing mitochondria. It is connected to physiological and pathological processes and entails eliminating damaged mitochondria to preserve mitochondrial quality and turnover (Gong et al., 2015). Mitophagy contributes to human life in several ways, including the removal of paternal mitochondria during fertilization, metabolic switching of cardiomyocytes after birth, erythrocyte differentiation during maturation, fibroblast to adipocyte differentiation, generation of iPS cells, and differentiation of retinal ganglion cells during embryogenesis. Defective mitophagy and the buildup of damaged mitochondria are associated with aging, metabolic diseases, cancer, senescence, inflammation, and genomic instability (Fanget al., 2016).

In addition, mitochondrial dynamics, biogenesis, unbalanced calcium levels, mtDNA damage, oxidative stress, and general mitochondrial health affect mitophagy. The damaged or unnecessary mitochondria must primed to form phagophores and swallowed by autophagosomes. Two processes that regulate mitochondrial homeostasis are mitochondrial biogenesis and mitophagy. Mitochondrial synthesis and breakdown coordination controls mitochondrial content, quality, and bioenergetics. AMP-activated protein kinase (AMPK), CaMK, PKD-NFB, and the Parkin-interacting substrate (PARIS) pathway are among the common pathways that regulate mitophagy and biogenesis (Palikaraset al., 2014). In injured mitochondria, mitochondrial membrane potential (MMP) changes lead to depolarization. By controlling the fission/fusion mechanism, MMP also influences the shape of mitochondria. While MFN1, MFN2, and OPA1 are the primary regulators of mitochondrial fusion, DRP1 is the primary regulator of mitochondrial fission. Mitophagy requires mitochondrial fission, which also helps distinguish between unhealthy and faulty mitochondria (Kubliet al., 2012). Given the significance of mitophagy in human diseases, it is essential to comprehend its molecular mechanism to develop therapeutic interventions that target it. To help phagophores find defective mitochondria, mitophagy receptors recognize and tag depolarized, fragmented, and damaged mitochondria.

Given the significance of mitophagy in human diseases, it is essential to comprehend its molecular mechanism to develop therapeutic interventions that target it. To help phagophores find defective mitochondria, mitophagy receptors recognize and tag depolarized, fragmented, and damaged mitochondria. Less conserved than the proteins of the autophagic machinery are proteins related to mitophagy. Mitophagy receptors that contain the ubiquitin-binding domain attract mitochondria to a phagophore in ubiquitin-mediated mitophagy after they have been ubiquitinated by PINK1/Parkin (UBD). Mitophagy receptors, which are OMM proteins with

LC3 interacting regions (LIRs) that bind to LC3 to create phagophores, include BNIP3, NIX, and FUN14 domain-containing protein 1 (FUNDC1) (Palikaras et al., 2018) (Figure 1).

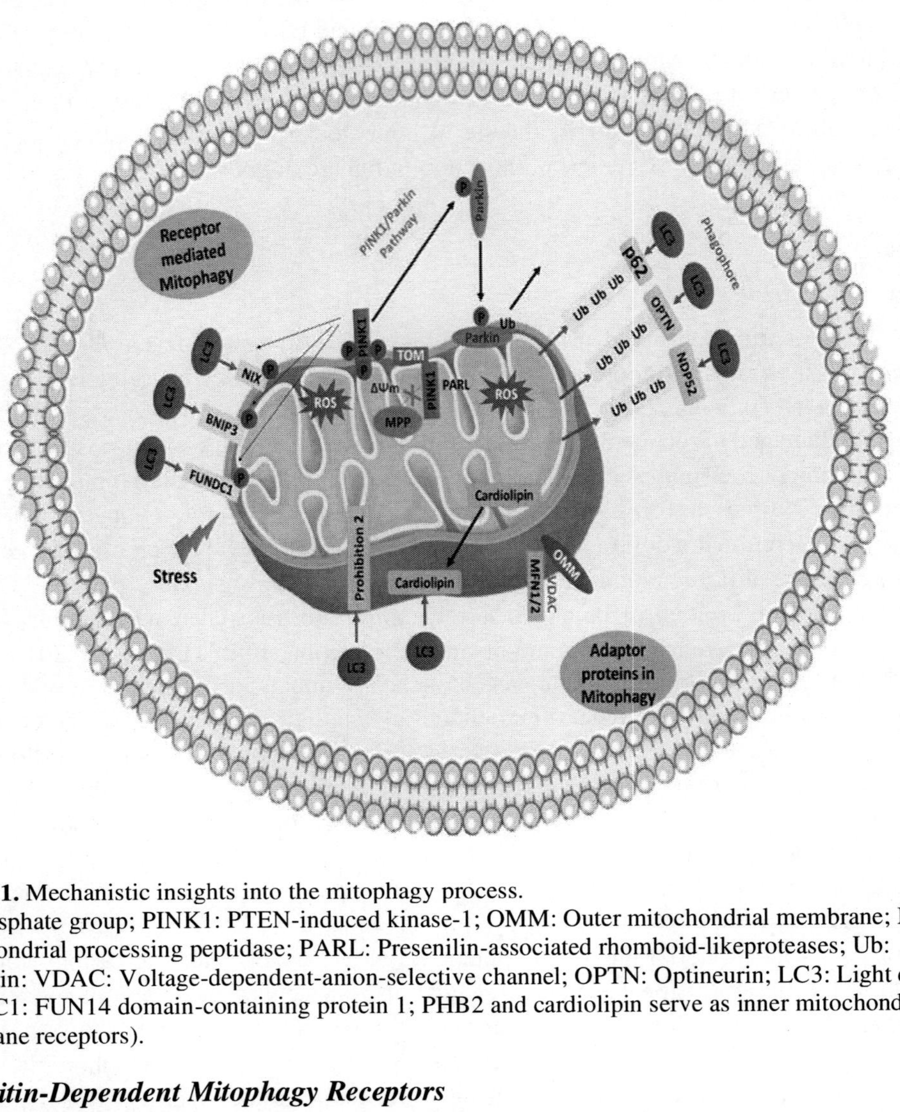

Figure 1. Mechanistic insights into the mitophagy process.
(P: Phosphate group; PINK1: PTEN-induced kinase-1; OMM: Outer mitochondrial membrane; MPP: Mitochondrial processing peptidase; PARL: Presenilin-associated rhomboid-likeproteases; Ub: Ubiquitin: VDAC: Voltage-dependent-anion-selective channel; OPTN: Optineurin; LC3: Light chain 3; FUNDC1: FUN14 domain-containing protein 1; PHB2 and cardiolipin serve as inner mitochondrial membrane receptors).

Ubiquitin-Dependent Mitophagy Receptors

The best-studied mitophagy route is ubiquitin-dependent mitophagy. E3 ligases are used to ubiquitinate damaged/inactive mitochondria. Parkin, p62, MARCH 5, VPS13D, MUL1, ARIH1, and SIAH1 are among the ubiquitin ligases linked to the ubiquitination of faulty mitochondria (Yang et al., 2019).

E3 ubiquitin ligases, on the other hand, are primarily unable to recognize the targeted mitochondria, necessitating mediators such as PINK1. PINK1 ubiquitinates OMM proteins by labeling depolarized mitochondria and providing a platform for E3 ligases to do so. USP15, USP30, and USP35 deubiquitination enzymes that fight pro-mitophagy forces and uphold homeostasis further regulate polyubiquitin chains on OMM proteins (Gersch, et al., 2017). Other proteins that influence ubiquitin-dependent mitophagy include mitochondrial RhoGTase and AMBRA1 (MIRO). Mitophagy receptors that use ubiquitin-dependent processes use ubiquitin-binding domain (UBD) E3 ligases and UBD-motif proteins. Mitophagy receptors containing UBD motifs, such as p62, NBR1, NDP52, and optineurin, contain UBD and LIR. They can identify ubiquitinated mitochondria and help localize them to phagophores that contain LC3 for the production of mitophagosomes and autophagic flux. Some UBD-motif mitophagy receptors, including RABGEF1, lack LIR. After ubiquitination, RABGEF1 recognizes mitochondria and collaborates with ATG9A to encourage the development of phagophores and autophagosomes (Yamano et al., 2018).

Ubiquitin-Independent Mitophagy Receptors

Mitophagy receptors that depend on LIR-motif proteins or lipids rather than ubiquitin do not have UBDs. LIR and other mitophagy receptors, in contrast to PINK1-Parkin, are consistently expressed on the OMM. Transcription or post-translational changes can regulate the function of the mitophagy receptor. This review examined some LIR-motif-containing proteins (BNIP3/Nix, FUNDC1, BCL3L13, and FKBP8) implicated in ubiquitin-independent mitophagy.

Lipid-Mediated Mitophagy

In mitochondria, lipids are calledcardiolipins and ceramides that aid in mitophagy preparation. When there is mitochondrial stress, ceramide is either generated by mitochondrial ceramide synthase 1 or confined to the OMM (CerS1). OMM ceramide binds to LC3 and initiates mitophagy. Usually in the inner mitochondrial membrane (IMM), cardiolipin is also moved from the IMM to the OMM following stress. This relocalized cardiolipin brought on by stress interacts with LC3 and promotes the formation of phagophores. Specific mitochondria are eliminated via a combination of ubiquitin-dependent, ubiquitin-independent, and lipid-mediated mitophagy. More investigation is necessary to understand these complex circuits' spatiotemporal distribution completely.

Mitophagy Inducers

As mentioned, mitophagy is a crucial physiological process supporting several cellular processes, including preserving mitochondrial integrity. Different tissues exhibit varying degrees of mitophagy. A few examples include the tissue of the heart, skeletal muscles, nervous system, hepatocellular, and renal systems. Mitophagy is more prevalent in the thymus and spleen (Sun et al., 2015). External or internal stress can influence the level of mitophagy in a specific tissue or organ. More research is needed to understand how particular pressures activate particular mitophagy pathways. Oxidative stress, dietary restriction, and mitochondrial heteroplasmy are a few of the stress-related factors that harm or damage the mitochondrial DNA. This protein triggers the mitophagic process. In this part, we will go through some of these mitophagy-provoking details (Figure 2).

Nutrient-Regulated Mitophagy

The degeneration of the mitochondria has been linked to prolonged food starvation. A cell's response to a deficiency in several metabolites, such as glucose, oxygen, amino acids, vitamins, and minerals, is known as mitophagy. Mitophagy is primarily impacted by hypoxia or oxygen deprivation through the upregulation and stabilization of HIF-1. HIF-1 produces pro-survival effects by lowering oxygen consumption. HIF-1 controls mitophagy through the BNIP3,

FUNDC1, and PINK1/Parkin pathways (Yang et al., 2019). Minerals and vitamins are essential for preserving cellular balance. The lack of vitamins B1, B2, B3, B6, B9, C, or D impairs mtDNA maintenance and mitophagy. For instance, lack of vitamin D inhibits complex I and raises mitophagy. Several studies have connected vitamin B3 (nicotinamide) insufficiency to the regulation of mitophagy (Fang et al., 2017). The mitophagy mechanism, which includes dct-1, pdr-1, and PINK-1, is directly activated by nicotinamide supplementation to cause mitophagy in C. elegans (Fang et al., 2019). Iron, selenium, magnesium, manganese, calcium, and zinc are trace minerals stored in mitochondria. Controlling the function of mitochondrial enzymes requires certain minerals. A lack of iron reduces the activity of the enzymes in the electron transport chain (ETC), altering the potential of the mitochondrial membrane and resulting in Parkin, PINK1, and BNIP3-mediated mitophagy. Since selenium and manganese are cofactors for glutathione peroxidase and MnSOD, manganese controls oxidative stress-induced mitophagy. Indirectly, the accumulation of calcium and zinc can influence the opening of the mitochondrial permeability transition pore and mitophagy (Yang et al., 2019).

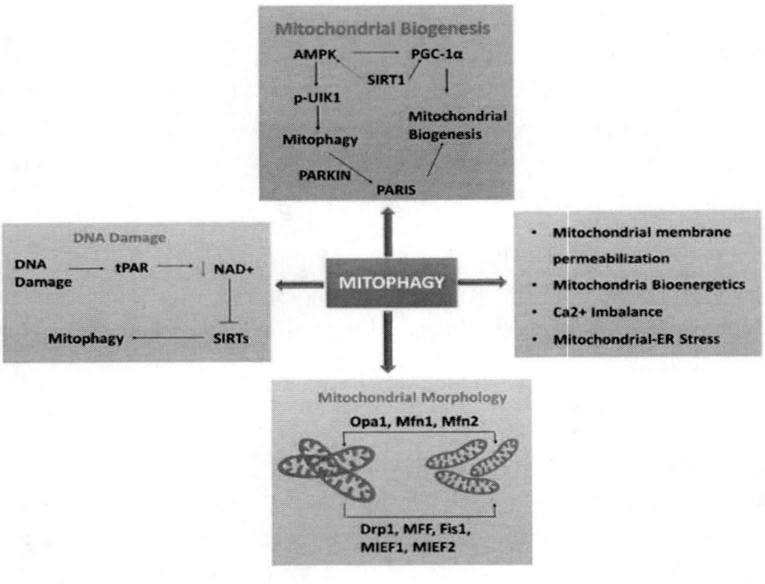

Figure 2. Factors in the mitochondria that control mitophagy. NAD+: Nicotinamide adenine dinucleotide; AMPK: AMP-activated protein kinase; PARIS: Parkin-interacting substrate; Opa 1: Optic atropy 1; MFF: Mitochondrial fission factor; Mfn1/2: Mitofusion 1/2; Drp1: Dynamin-related protein 1; MIEF1/2: Mitochondrial elongation factor 1/2.

Oxidative Stress

Both mitophagy and autophagy play a role in promoting and preventing survival. A significant element in this process is oxidative stress. In contrast to severe oxidative stress, which causes cell death, moderate oxidative stress promotes autophagy and mitophagy. Many cellular signaling mechanisms involve ROS. Using transcriptional or post-transcriptional processes, oxidative stress—caused by an increase in cellular or mitochondrial ROS/RNS—activates mitophagy. Increased oxidative stress influences transcriptional regulation, activating factors such as HIF-1 and Forkhead box O3a (FOXO3a), which boosts transcription of the mitophagy proteins Bnip3 and Nix (Higgins et al., 2014). The same applies to oxidative stress, which activates nuclear factor (erythroid-derived 2)-like 2 (Nrf2) and raises the transcription of p62.

Several proteins undergo post-translational ROS/RNS species modifications that either promote or hinder mitophagic flow. DJ-1's mitochondrial translocation is aided by reactive oxygen species (ROS), which do this by oxidizing the cysteine residue at position 106 (Cys106). While inside mitochondria, DJ-1 interacts with Parkin to regulate the PINK1/Parkin mitophagy pathway and initiates mitophagy caused by oxidative stress. On the other hand, oxidative stress is caused by the buildup of damaged mitochondria due to mitophagy suppression. Different proteins can undergo S-nitrosylation that is changed by NO, H202, and 4-hydroxynonenal produced due to lipid peroxidation. S-nitrosylation of IKK and JNK1 can cause mTOR inhibition and Beclin release (Lee et al., 2012). S-nitrosylation of Drp-1, on the other hand, encourages mitochondrial fragmentation and subsequently mitophagy. However, evidence indicates that Parkin can be oxidized and/or S-nitrosylated by ROS, which lowers Parkin's E3 ligase activity and, as a result, mitophagy (Yao, 2004).

Mitochondrial DNA Damage

Since the mitochondria produce most free radicals in cells, mtDNA is especially vulnerable to ROS damage. Thisemphasizes the importance and requirement of maintaining the integrity of the mitochondrial genome with the aid of the DNA repair machinery. As damage to mitochondria and its DNA is directly related to its functional decline and worsening oxidative phosphorylation capacity, organs that depend on mitochondrial energy are likely the most affected. Many stressors have the potential to damage mtDNA. The respiratory complex is inhibited, and membrane potential is lost in mitochondria, despite having a specialized DNA repair system that can recognize and repair some types of damaged DNA. Key components of the autophagy system, including NIX, E3 ligase, and LC3, identify and ubiquitinate severely damaged and malfunctioning mitochondria, which an autophagosome then chooses for apoptosis. The precise mechanism by which the mtDNA repair system fails to function and induces mitophagy is currently unknown (Bess et al., 2013). NIX or BNIP3, on the other hand, has been established in prior research to play a significant role in mitophagy beginning (Novak et al., 2010; Ding et al., 2010; Matsuda et al., 2015). Furthermore, previous research has shown that blocking mitophagy after DNA damage might result in a buildup of defective mitochondria, which causes apoptosis (Cotan et al., 2011). It is uncertain if it has a direct function in mitophagy induction (Czarny et al., 2015). According to a recent study, the p62 complex interacts with the ubiquitinated membrane of damaged mitochondria and activates LC3II and other autophagy-related machinery. However, it is still unclear how specifically p62 contributes to mitophagy induction (Czarny et al., 2015). The data in the section before strongly supports the idea that mtDNA damage contributes to mitochondrial dysfunction. Furthermore, as people age, the machinery intended to repair DNA malfunctions successfully causes the accumulation of mtDNA bases with damage (Trifunovic et al., 2006). More research is needed to fully comprehend mitophagy's direct and specific roles in mitochondrial quality control and turnover and how particular types of mtDNA damage trigger it. Even though some studies have identified mitophagy as one of the contributing mechanisms for the efficient turnover of damaged mitochondria and mtDNA, more studies are still needed.

Mitophagy Pathways

Mitophagy in Mammals

Damaged and redundant mitochondria are selectively destroyed in nutrient-rich environments to preserve mitochondrial homeostasis. It is one of the most well-researched clearing routes. The PINK1/Parkin protein is found in damaged mitochondria in mammalian cells (Pick Youle, 2015). Mitophagy is PINK1-Parkin-dependent. When the potential of the mitochondrial membrane falls below a certain level, a cascade of events occurs. The body's stability is compromised by mitochondrial dysfunction. Ubiquitin is phosphorylated here by the mitochondria's ubiquitin kinase (PTEN)-induced kinase 1 (PINK1) membrane, which induces Parkin, an E3 ubiquitin ligase, to be recruited. When Parkin is activated by PINK1 phosphorylation polyubiquitinate mitochondrial proteins, they join with the ubiquitin-binding domains of the autophagy receptors to create the autophagosome.

The mitochondria are then degraded due to the autophagosome and lysosome joining forces (Lazarou et al., 2015). Alternatively, in the absence of Parkin, PINK1 can directly recruit autophagy receptors, resulting in low levels of mitophagy (Lazarou et al., 2015). Parkin substrates can be deubiquitinated to prevent it from inducing mitophagy. In human cell lines, USP8 (Ubiquitin Specific Peptidase 8) has been shown to influence mitophagy (Durcan et al., 2014), and lower levels of USP30 and USP15 can increase mitophagy and correct mitochondrial defects in fly models.

Parkin (Kitada et al., 1998) and PINK1 (Valente et al., 2004) mutations cause Parkinson's disease to be autosomal recessive. PINK1 mice have mitochondrial dysfunction at 34 months of age, including reduced mitochondrial respiration in the striatum. By 24 months of age, only control mice had this issue, demonstrating that PINK1 animals are susceptible to damage from aging (Gautier et al., 2008). Parkin animals have modest metabolic abnormalities and decreased levels of proteins involved in oxidative stress responses. However, they do not have significant mitochondrial malfunction or neuronal death, suggesting the existence of Parkin-independent mitochondrial clearance pathway(s) (Perez and Palmiteret al., 2005). Polk mutant mice, which have increased mtDNA damage, are bred with Parkinson's disease-associated traits such as mitochondrial dysfunction and loss of dopaminergic neurons in Parkin mice (Pickrell et al., 2015).

According to reports, Parkin is not required for AMBRA1, Bcl2-L-13, or other mitophagy proteins to function. The Nix/BNIP3L pathway, FUNDC1, MUL1, and the Activating Molecule in Beclin1-Regulated Autophagy (AMBRA1) pathway are crucial for embryogenesis and the development of the nervous system (Fimia et al., 2007). AMBRA, independent of P62/SQSTM1, may directly interact with LC3 and hence induce the mitophagy (Strappazzon et al., 2015). Similarly, the BH domains of Bcl2-L-13 can also damage mitochondria on their own by attaching to LC3, and thus in the absence of Parkin, Bcl2-L-13 continues to function and triggers mitophagy (Murakawa et al., 2015). The mitochondrial outer-membrane protein FUNDC1 was initially discovered to be a crucial component of hypoxia-induced mitophagy (Liu et al., 2012). By attracting LC3 through its (21 LC3-binding motifs) and by controlling mitochondrial dynamics through interactions with fission and fusion proteins (DNM1L/DRP1 and OPA1, respectively), FUNDC1 regulates mitophagy (Liu et al., 2012; Chen et al., 2016). During mitophagy, MUL1 (mitochondrial ubiquitin ligase 1), an E3 protein ligase, is located on the outer membrane of the mitochondria (Yun et al., 2014). In PINK1 or parkin-deficient Drosophila, MUL1 can reduce disease symptoms through ubiquitin-dependent Mitofusin degradation (Yun et al., 2014). In zygotes, MUL1 can also cleave mitochondria generated from sperm, but only in the presence of PINK1 (Rojansky et al., 2016).

Nix/BNIP3L can also encourage mitophagy without PINK1/Parkin (Song et al., 2016). Ambra1, Nix, and FUNDC1 have been linked to Parkin-independent mitophagy pathways, but they also function in traditional Parkin-dependent mitophagy (Van Humbeeck et al., 2011; Ding et al., 2010; Wu et al., 2014). Because of this, mammalian cells have various mitophagy pathways that can replace those lost. In addition to removing damaged mitochondria, mammalian mitophagy mechanisms have been developed to destroy mitochondria during developmental processes, including fertilization and blood cell maturation. When reticulocytes mature into red blood cells, the Nix/BNIP3L-dependent mitophagy pathway entirely clears the mitochondria (Sandoval et al., 2008). Nix, a member of the Bcl2 family, induces mitophagy by decreasing the potential of the mitochondrial membrane.

Mitophagy in C. Elegans

Decades of research in cross-species investigations have revealed that many cellular processes involved in cerebral function and aging are evolutionarily conserved. The lifespan and health of C. elegans depend on mitophagy (Palikaras et al., 2015). The human mitophagy protein Nix/BNIP3L is detected in C. elegans mitochondria, and DCT-1 may be an orthologue. DCT-1, a protein that works downstream of the PINK-1/Parkin pathway, is upregulated by both the oxidative stress response transcription factor SKN-1 (ortholog of mammalian NRF2) and the FOXO transcription factor DAF-16 (ortholog of mammalian FOXO3) (Palikaras et al., 2015). Like mammals, C. elegan's sperm-derived mitochondria are likewise eliminated after conception. According to Song et al. (2016) and Zhou et al. (2016), CPS-6, a mitochondrial endonuclease G, is required to degrade the paternal mitochondrial genome in C. elegans. Further investigation into the mechanisms of autophagy and mitophagy in C. elegans will hasten both mechanistic and translational investigations in biological aging and neurodegeneration.

Mitophagy and Aging

Sarcopenia indicates a reduction in muscle strength and bulk with aging. Mitochondria have been associated with the onset of sarcopenia associated with aging because of their defective traits and increasing activity. Apoptosis, protein degradation, and skeletal muscle atrophy are all factors that contribute to skeletal muscle atrophy and ultimately to muscular dysfunction (Carter et al., 2018; Kang et al., 2017; Sandri et al., 2004). Skeletal muscle atrophy is a result of the accumulation of reactive oxygen species (ROS) in aged muscle FOXO pathways are activated. Moreover, modifications in mitochondrial DNA are related to ROS formation. As a result, the electron transport chain's (ETC) proteins are changed (Hepple et al., 2014; Hood et al., 2001). When the ETC is not working, ATP production is reduced, oxidative phosphorylation decreases, and ROS levels rise significantly (Hood et al., 2019; Romanello et al., 2016). Problems will also arise if the mitochondrial production of ROS grows. Pink1 and Parkin mediate the steps in importing proteins and the circumstances leading to Mitophagy.

The literature has competing theories regarding the connection between muscle autophagy and aging. Some sources claim that autophagy is slowed in muscles. People gain damaged mitochondria as they age (Hood et al., 2019; Taiet al., 2017). This appears to be supported by elevated levels of the normally degraded p62 and LC3-II proteins in the presence of autophagy.

After administering medications like those mentioned above to animals or cells, mitochondria are separated, and the extracts are utilized to examine the organelle-specific localization of mitophagy markers. Our techniques revealed that LC3-II mitophagy flux levels are more significant in aged than young muscle. This hypothesis was supported by increased p62 mitophagy flux and location in mitochondria and increased expression of mitophagy receptors in the mitochondria of aged muscle. Along with increased Parkin ubiquitination with age, there has also been an increase in Parkin (Carter et al., 2017; Chen et al., 2018; Leduce et al., 2019). Older muscle has increased levels of mitophagy, and there is evidence of decreasing mitochondrial content in this tissue. However, mitochondrial malfunction still poses a challenge. Aging suggests that enhanced mitophagy, while made to preserve a high-quality mitochondrial pool, is insufficient. As a result, despite being destroyed by mitophagy, skeletal muscle mitochondria nonetheless display an increase in ROS, ultimately contributing to the sarcopenia that comes with aging because the rise in mitochondrial dysfunction cannot be reversed (Chenet al., 2018).

The Regulation of Mitophagy in Muscles

Muscle mitochondrial health depends on organellar turnover, a byproduct of the equilibrium between creating new, healthy organelles through biogenesis and recycling harmed or malfunctioning mitochondria through mitophagy. Mitophagy encourages the survival of a tightly connected mitochondrial reticulum, which efficiently produces ATP and generates sufficient ROS to initiate processes that protect homeostasis. Mitophagy is the process by which segmental mitochondrial dysfunction, indicated by varying degrees of respiratory impairment, excessive ROS production, or membrane potential loss, is digested at the lysosomes. This turnover keeps the muscle's metabolic capacity constant when combined with a change in mitochondrial biogenesis (Hoodet al., 2019; Kim et al., 2017). The general health of the mitochondrial pool will be jeopardized if these mechanisms are out of balance, a characteristic found in both aging and muscle disease (Hoodet al., 2019; Kim et al., 2017; Ziaaldini et al.,2018; Marzetti et al., 2013).

Mitochondrial Dynamics in Muscle-Implications for Mitophagy

A reticular network connects the mitochondria of skeletal muscle, enhancing the tissue's metabolic potential (Ogata et al., 1985; Glancy et al., 2015, Glancy et al., 2017). The opposing processes of fusion and fission maintain this network. The pro-fission proteins dynamic-related protein 1 (Drp1), mitochondrial fission protein 1 (Mfp1), as well as the pro-fusion proteins mitofusin 1/2 (Mfn1/2) and the optic atrophy protein 1 (Opa1), as well as more recently (Fis1), regulate mitochondrial dynamics (Romanello et al., 2021). When a network node malfunctions, mitochondrial fission sequesters it, preparing the organelle segment for its eventual degradation through the mitophagy pathway (Iqbal et al., 2015; Twig et al., 2008). The significance of these fusion- and fission-regulating proteins has been shown using genetic models. According to Romanello et al. (2010), Dulac et al. (2010), Favaro et al. (2019), and Zhamg et al. (2019), the pro-fission proteins Fis1 and Drp1 knockdown promote hyperfused mitochondria with impaired function and disturb mitophagy (Favaro et al., 2019; Zhamg et al., 2019). These results demonstrate that the supply of damaged mitochondria from the mitophagy route requires fission to sustain muscle organellar health. Mfn2 and Opa1 deletion result in mitochondrial fragmentation, excessive ROS production, and muscle atrophy in terms of fusion (Sebastián et al., 2016; Sebastián et al., 2012). Our results suggest that fusion and fission must be balanced for a healthy mitochondrial pool, which affects muscle health as we age.

Mitochondrial Dynamics in Aged Muscles

Although there have been some inconsistencies, numerous studies have demonstrated that fission proteins are more prevalent in elderly muscles than fusion proteins (Koltai et al., 2012; Sebastián et al., 2016; Yeo et al., 2012; O'Leary et al., 2016; Joseph et al., 2012; Iqbal et al., 2013), (Konopka et al., 2016; Leduc-Gaudet et al., 2015). Old muscle has elongated mitochondria, which the researchers attribute to inherent mitochondrial defects that promote fusion as a coping mechanism (Leduc-Gaudet et al., 2015). The aging model that was applied, species differences, or the level of age assessed could all account for discrepancies in the outcomes. The discovery of fragmented mitochondria in older muscles is not surprising. Nevertheless, this organellar pool may be primed for mitophagy because mitochondrial failure is evident in old muscle, and fission occurs before mitophagy.

Mitophagy in Aging Skeletal Muscle

Mitophagy Targeting in Aged Muscles

Given that senescent muscle exhibits mitochondrial breakage and malfunction, mitophagy signaling would be enhanced to eliminate these faulty organelles. Numerous studies (Carter et al., 2018; Sebastián et al., 2016; Chen et al., 2018; Zampieri et al., 2015; Ko et al., 2016) have shown that it increased the expression of the mitophagy proteins Bnip3, Nix, Pink1, and Parkin. However, other studies have found inconsistent results and have discovered either no change or decreased levels of Bnip3 and Nix in mice (Yeo et al., 2019; Distefano et al., 2017; Triolo et al., 2021)the muscle that has aged. Although this has not been studied in the context of aging muscle, the phosphorylation status of these proteins would signal whether or not these mitophagy mechanisms are functional. Our results and others showed that the levels of Parkin protein (Leduc et al., 2019) and mitochondrial Parkin (Chen et al., 2018) were lower in 18-month-old mice. PINK1, Parkin, and ubiquitin levels are higher in older rodents (24-month-old mice and 35-month-old rats) (Yeo et al., 2019; O'Leary et al., 2013), suggesting an age-specific PINK1-Parkin mitophagic response.

Mitophagy Flux in Aged Muscles

Evidence of enhanced LC3-II and p62 (Carter et al., 2018; O'Leary et al., 2013) in these isolated organelles, our team has discovered increased autophagosomal tethering to mitochondria in aged muscle, which supports improved targeting. These results might point to either an increase in the rate of organellar targeting and degradation or a decrease in the effectiveness of mitophagic breakdown. Colchicine-induced increases in mitophagy flow were observed in aged mice (Chen et al., 2018) and rats (Carter et al., 2018) muscle, which is consistent with deficits in mitochondrial biogenesis (Hood et al., 2019; Kim et al., 2017; Tryon et al., 2014). The presence of damaged organelles shows that the removal rate of damaged mitochondria may not be high enough to eliminate all dysfunctional cellular components, including mitochondria, despite higher flux suggesting that damaged mitochondria are being removed faster. In addition to the decreased metabolic capacity, this remaining organellar pool will cause oxidative muscle damage, which could cause further dysfunction. Hence, resuming mitophagy in aging muscles may slow the onset of age-related sarcopenia. Aging is associated with both a decline in physical activity and neuromuscular denervation. Since denervation and muscular disuse have independent effects on mitophagy (Huang et al., 2019; Shen et al., 2020; O'Leary et al., 2013; Vainshtein et al., 2015; Tamura et al., 2015; Yang et al., 2020; Graham et al., 2018; Furuya et al., 2014). In young muscles, acute endurance exercise also promotes mitophagy, which reduces the mitochondrial pool and creates healthier tissue (Hood et al., 2019). However, there is evidence of diminished acute exercise-induced mitophagy in aging muscles (Chen et al., 2018). In addition, denervation-induced autophagic upregulation (O'Leary et al., 2013) and chronic contractile activity-induced mitophagy flux decrease (Carter et al., 2018) are hampered in elderly muscle. As a result, it is becoming clear that aging influences the mitophagy phenotype of muscle, and a deficiency in adequate stimulus-induced mitophagy alterations may restrict advantageous adaptations in maintaining a healthy organelle population.

Conclusion

Genomic instability and mitochondrial dysfunction have been linked to diabetes, neurodegeneration, cancer, and the pathophysiology of normal aging. Understanding the molecular biology of controlling mitophagy and mitochondrial activity in response to DNA damage will identify therapeutic targets in addition to mechanistic relationships. Mitophagy problems have been linked to defects in DNA repair mechanisms. Additionally, nothing is known about how deficiencies in DNA repair affect different mitochondrial quality control systems. Undoubtedly, the state of the mitochondria affects how quickly people age. Understanding the function of mitophagy in nuclear DNA repair may be a key research direction for a combination therapy aimed at extending life. Mitophagy plays various physiological and pathological roles in various tissues and exhibits spatiotemporal dispersion. It is critical to comprehend how mitophagy affects stem cell renewal and functionality. The inhibition of autophagy and/or mitophagy has also been well-established in recent years as a factor in the pathophysiology of several illnesses. Therapeutic measures that target mitophagy may have a significant effect on human health. Future research must provide targeted gene treatments and medications for mitophagy induction.

References

Anand, R., Wai, T., Baker, M. J., Kladt, N., Schauss, A. C., Rugarli, E. and Langer, T. (2014). The i-AAA protease YME1L and OMA1 cleave OPA1 to balance mitochondrial fusion and fission. *Journal of Cell Biology*, *204*(6), pp.919-929.

Babbar, M., Basu, S., Yang, B., Croteau, D. L. and Bohr, V. A. (2020). Mitophagy and DNA damage signaling in human aging. *Mechanisms of ageing and development*, *186*, p.111207.

Ban, T., Ishihara, T., Kohno, H., Saita, S., Ichimura, A., Maenaka, K., Oka, T., Mihara, K. and Ishihara, N. (2017). Molecular basis of selective mitochondrial fusion by heterotypic action between OPA1 and cardiolipin. *Nature cell biology*, *19*(7), pp.856-863.

Basso, V., Marchesan, E., Peggion, C., Chakraborty, J., von Stockum, S., Giacomello, M., Ottolini, D., Debattisti, V., Caicci, F., Tasca, E. and Pegoraro, V. (2018). Regulation of ER-mitochondria contacts by Parkin via Mfn2. *Pharmacological Research*, *138*, pp.43-56.

Bess, A. S., Ryde, I. T., Hinton, D. E. and Meyer, J. N. (2013). UVC-induced mitochondrial degradation via autophagy correlates with mtDNA damage removal in primary human fibroblasts. *Journal of biochemical and molecular toxicology*, *27*(1), pp.28-41.

Bingol, B., Tea, J. S., Phu, L., Reichelt, M., Bakalarski, C. E., Song, Q., Foreman, O., Kirkpatrick, D. S. and Sheng, M. (2014). The mitochondrial deubiquitinase USP30 opposes parkin-mediated mitophagy. *Nature*, *510*(7505), pp.370-375.

Carter, H. N., Kim, Y., Erlich, A. T., Zarrin-khat, D. and Hood, D. A. (2018). Autophagy and mitophagy flux in young and aged skeletal muscle following chronic contractile activity. *The Journal of Physiology*, *596*(16), pp.3567-3584.

Chakrabarti, R., Ji, W. K., Stan, R. V., de Juan Sanz, J., Ryan, T. A. and Higgs, H. N. (2018). INF2-mediated actin polymerization at the ER stimulates mitochondrial calcium uptake, inner membrane constriction, and division. *Journal of Cell Biology*, *217*(1), pp.251-268.

Chen, C. C. W., Erlich, A. T. and Hood, D. A. (2018). Role of Parkin and endurance training on mitochondrial turnover in skeletal muscle. *Skeletal muscle*, *8*(1), pp.1-14.

Chen, C. C. W., Erlich, A. T., Crilly, M. J. and Hood, D. A. (2018). Parkin is required for exercise-induced mitophagy in muscle: impact of aging. *American Journal of Physiology-Endocrinology and Metabolism*, *315*(3), pp. E404-E415.

Chen, M., Chen, Z., Wang, Y., Tan, Z., Zhu, C., Li, Y., Han, Z., Chen, L., Gao, R., Liu, L. and Chen, Q. (2016). Mitophagy receptor FUNDC1 regulates mitochondrial dynamics and mitophagy. *Autophagy*, *12*(4), pp.689-702.

Chen, Y. and Dorn, G. W. (2013). PINK1-phosphorylated mitofusin 2 is a Parkin receptor for culling damaged mitochondria. *Science*, *340*(6131), pp.471-475.

Cho, B., Cho, H. M., Jo, Y., Kim, H. D., Song, M., Moon, C., Kim, H., Kim, K., Sesaki, H., Rhyu, I. J. and Kim, H. (2017). Constriction of the mitochondrial inner compartment is a priming event for mitochondrial division. *Nature communications*, *8*(1), pp.1-17.

Clark, I. E., Dodson, M. W., Jiang, C., Cao, J. H., Huh, J. R., Seol, J. H., Yoo, S. J., Hay, B. A. and Guo, M. (2006). Drosophila pink1 is required for mitochondrial function and interacts genetically with parkin. *Nature*, *441*(7097), pp.1162-1166.

Cornelissen, T., Haddad, D., Wauters, F., Van Humbeeck, C., Mandemakers, W., Koentjoro, B., Sue, C., Gevaert, K., De Strooper, B., Verstreken, P. and Vandenberghe, W. (2014). The deubiquitinase USP15 antagonizes Parkin-mediated mitochondrial ubiquitination and mitophagy. *Human molecular genetics*, *23*(19), pp.5227-5242.

Cotán, D., Cordero, M. D., Garrido-Maraver, J., Oropesa-Avila, M., Rodríguez-Hernández, Á., Izquierdo, L. G., Mata, M. D. L., Miguel, M. D., Lorite, J. B., Infante, E. R. and Jackson, S. (2011). Secondary coenzyme Q10 deficiency triggers mitochondria degradation by mitophagy in MELAS fibroblasts. *The FASEB Journal*, *25*(8), pp.2669-2687.

Czarny, P., Pawlowska, E., Bialkowska-Warzecha, J., Kaarniranta, K. and Blasiak, J. (2015). Autophagy in DNA damage response. *International journal of molecular sciences*, *16*(2), pp.2641-2662.

De Brito, O. M. and Scorrano, L. (2008). Mitofusin 2 tethers endoplasmic reticulum to mitochondria. *Nature*, *456*(7222), pp.605-610.

Deng, H., Dodson, M. W., Huang, H. and Guo, M. (2008). The Parkinson's disease genes pink1 and parkin promote mitochondrial fission and/or inhibit fusion in Drosophila. *Proceedings of the National Academy of Sciences*, *105*(38), pp.14503-14508.

Deval, C., Calonne, J., Coudy-Gandilhon, C., Vazeille, E., Bechet, D., Polge, C., Taillandier, D., Attaix, D. and Combaret, L. (2020). Mitophagy and mitochondria biogenesis are differentially induced in rat skeletal muscles during immobilization and/or remobilization. *International journal of molecular sciences*, *21*(10), p.3691.

DeVay, R. M., Dominguez-Ramirez, L., Lackner, L. L., Hoppins, S., Stahlberg, H. and Nunnari, J. (2009). Coassembly of Mgm1 isoforms requires cardiolipin and mediates mitochondrial inner membrane fusion. *Journal of Cell Biology*, *186*(6), pp.793-803.

Dietz, J. V., Bohovych, I., Viana, M. P. and Khalimonchuk, O. (2019). Proteolytic regulation of mitochondrial dynamics. *Mitochondrion*, *49*, pp.289-304.

Ding, D., Sallis, J. F., Kerr, J., Lee, S. and Rosenberg, D. E. (2011). Neighborhood environment and physical activity among youth: a review. *American journal of preventive medicine*, *41*(4), pp.442-455.

Ding, W. X. and Yin, X. M. (2012). Mitophagy: mechanisms, pathophysiological roles, and analysis. *Biological chemistry*, *393*(7), pp.547-564.

Ding, W. X., Ni, H. M., Li, M., Liao, Y., Chen, X., Stolz, D. B., Dorn, G. W 2nd., Yin, X. M. (2010). Nix is critical to two distinct phases of mitophagy, reactive oxygen speciesmediated autophagy induction and Parkin-ubiquitin-p62-mediated mitochondrial priming. *J Biol Chem*. 285(36): p. 27879-90.

Distefano, G., Standley, R. A., Dubé, J. J., Carnero, E. A., Ritov, V. B., Stefanovic-Racic, M., Toledo, F. G., Piva, S. R., Goodpaster, B. H. and Coen, P. M. (2017). Chronological age does not influence ex-vivo mitochondrial respiration and quality control in skeletal muscle. *Journals of Gerontology Series A: Biomedical Sciences and Medical Sciences*, *72*(4), pp.535-542.

Dulac, M., Leduc-Gaudet, J. P., Reynaud, O., Ayoub, M. B., Guérin, A., Finkelchtein, M., Hussain, S. N. and Gouspillou, G. (2020). Drp1 knockdown induces severe muscle atrophy and remodelling, mitochondrial dysfunction, autophagy impairment and denervation. *The Journal of physiology*, *598*(17), pp.3691-3710.

Durcan, T. M., Tang, M. Y., Pérusse, J. R., Dashti, E. A., Aguileta, M. A., McLelland, G. L., Gros, P., Shaler, T. A., Faubert, D., Coulombe, B. and Fon, E. A. (2014). USP 8 regulates mitophagy by removing K 6-linked ubiquitin conjugates from parkin. *The EMBO journal*, *33*(21), pp.2473-2491.

Erlich, A. T. and Hood, D. A. (2019). Mitophagy regulation in skeletal muscle: Effect of endurance exercise and age. *Journal of Science in Sport and Exercise*, *1*(3), pp.228-236.

Esteban-Martínez, L., Sierra-Filardi, E., McGreal, R. S., Salazar-Roa, M., Mariño, G., Seco, E., Durand, S., Enot, D., Graña, O., Malumbres, M. and Cvekl, A. (2017). Programmed mitophagy is essential for the glycolytic switch during cell differentiation. *The EMBO journal*, *36*(12), pp.1688-1706.

Fang, E. F. and Bohr, V. A. (2017). NAD+: The convergence of DNA repair and mitophagy. *Autophagy*, *13*(2), pp.442-443.

Fang, E. F., Hou, Y., Palikaras, K., Adriaanse, B. A., Kerr, J. S., Yang, B., Lautrup, S., Hasan-Olive, M. M., Caponio, D., Dan, X. and Rocktäschel, P. (2019). Mitophagy inhibits amyloid-β and tau pathology and reverses cognitive deficits in models of Alzheimer's disease. *Nature neuroscience*, *22*(3), pp.401-412.

Fang, E. F., Scheibye-Knudsen, M., Chua, K. F., Mattson, M. P., Croteau, D. L. and Bohr, V. A., (2016). Nuclear DNA damage signalling to mitochondria in ageing. *Nature reviews Molecular cell biology*, *17*(5), pp.308-321.

Favaro, G., Romanello, V., Varanita, T., Andrea Desbats, M., Morbidoni, V., Tezze, C., Albiero, M., Canato, M., Gherardi, G., De Stefani, D. and Mammucari, C. (2019). DRP1-mediated mitochondrial shape controls calcium homeostasis and muscle mass. *Nature communications*, *10*(1), pp.1-17.

Fivenson, E. M., Lautrup, S., Sun, N., Scheibye-Knudsen, M., Stevnsner, T., Nilsen, H., Bohr, V. A. and Fang, E. F. (20170. Mitophagy in neurodegeneration and aging. *Neurochemistry international*, *109*, pp.202-209.

Friedman, J. R., Lackner, L. L., West, M., DiBenedetto, J. R., Nunnari, J. and Voeltz, G. K. (2011). ER tubules mark sites of mitochondrial division. *Science*, *334*(6054), pp.358-362.

Furuya, N., Ikeda, S. I., Sato, S., Soma, S., Ezaki, J., Trejo, J. A. O., Takeda-Ezaki, M., Fujimura, T., Arikawa-Hirasawa, E., Tada, N. and Komatsu, M. (2014). PARK2/Parkin-mediated mitochondrial clearance contributes to proteasome activation during slow-twitch muscle atrophy via NFE2L1 nuclear translocation. *Autophagy*, *10*(4), pp.631-641.

Galvan, D. L., Long, J., Green, N., Chang, B. H., Lin, J. S., Schumacker, P., Truong, L. D., Overbeek, P. and Danesh, F. R. (2019). Drp1S600 phosphorylation regulates mitochondrial fission and progression of nephropathy in diabetic mice. *The Journal of clinical investigation*, *129*(7), pp.2807-2823.

Gautier, C. A., Kitada, T. and Shen, J. (2008). Loss of PINK1 causes mitochondrial functional defects and increased sensitivity to oxidative stress. *Proceedings of the National Academy of Sciences*, *105*(32), pp.11364-11369.

Gersch, M., Gladkova, C., Schubert, A. F., Michel, M. A., Maslen, S. and Komander, D. (2017). Mechanism and regulation of the Lys6-selective deubiquitinase USP30. *Nature structural & molecular biology*, *24*(11), pp.920-930.

Glancy, B., Hartnell, L. M., Malide, D., Yu, Z. X., Combs, C. A., Connelly, P. S., Subramaniam, S. and Balaban, R. S. (2015). Mitochondrial reticulum for cellular energy distribution in muscle. *Nature*, *523*(7562), pp.617-620.

Glancy, B., Hartnell, L. M., Combs, C. A., Fenmou, A., Sun, J., Murphy, E., Subramaniam, S., Balaban, R. S. (2017). Power Grid Protection of the Muscle Mitochondrial Reticulum. *Cell Rep*. 19, 487–496.

Gong, G., Song, M., Csordas, G., Kelly, D. P., Matkovich, S. J. and Dorn, G. W. (2015). Parkin-mediated mitophagy directs perinatal cardiac metabolic maturation in mice. *Science*, *350*(6265), p.aad2459.

Graham, Z. A., Harlow, L., Bauman, W. A. and Cardozo, C. P. (2018). Alterations in mitochondrial fission, fusion, and mitophagic protein expression in the gastrocnemius of mice after a sciatic nerve transection. *Muscle & nerve*, *58*(4), pp.592-599.

Greene, J. C., Whitworth, A. J., Kuo, I., Andrews, L. A., Feany, M. B. and Pallanck, L. J. (2003). Mitochondrial pathology and apoptotic muscle degeneration in Drosophila parkin mutants. *Proceedings of the National Academy of Sciences*, *100*(7), pp.4078-4083.

Hara, T., Nakamura, K., Matsui, M., Yamamoto, A., Nakahara, Y., Suzuki-Migishima, R., Yokoyama, M., Mishima, K., Saito, I., Okano, H. and Mizushima, N. (2006). Suppression of basal autophagy in neural cells causes neurodegenerative disease in mice. *Nature*, *441*(7095), pp.885-889.

Hepple, R. (2014). Mitochondrial involvement and impact in aging skeletal muscle. *Frontiers in aging neuroscience*, *6*, p.211.

Higgins, G. C. and Coughlan, M. T. (2014). Mitochondrial dysfunction and mitophagy: the beginning and end to diabetic nephropathy? *British journal of pharmacology*, *171*(8), pp.1917-1942.

Hood, D. A. (2001). Invited Review: contractile activity-induced mitochondrial biogenesis in skeletal muscle. *Journal of applied physiology*, *90*(3), pp.1137-1157.

Hood, D. A., Memme, J. M., Oliveira, A. N. and Triolo, M. (2019). Maintenance of skeletal muscle mitochondria in health, exercise, and aging. *Annual Review of Physiology*, *81*, pp.19-41.

Huang, Z., Fang, Q., Ma, W., Zhang, Q., Qiu, J., Gu, X., Yang, H. and Sun, H. (2019). Skeletal muscle atrophy was alleviated by salidroside through suppressing oxidative stress and inflammation during denervation. *Frontiers in pharmacology*, p.997.

Huang, Z., Zhong, L., Zhu, J., Xu, H., Ma, W., Zhang, L., Shen, Y., Law, B. Y. K., Ding, F., Gu, X. and Sun, H. (2020). Inhibition of IL-6/JAK/STAT3 pathway rescues denervation-induced skeletal muscle atrophy. *Annals of Translational Medicine*, *8*(24).

Ichishita, R., Tanaka, K., Sugiura, Y., Sayano, T., Mihara, K. and Oka, T. (2008). An RNAi screen for mitochondrial proteins required to maintain the morphology of the organelle in Caenorhabditis elegans. *Journal of biochemistry*, *143*(4), pp.449-454.

Iqbal, S., Ostojic, O., Singh, K., Joseph, A. M. and Hood, D. A. (2013). Expression of mitochondrial fission and fusion regulatory proteins in skeletal muscle during chronic use and disuse. *Muscle & nerve*, *48*(6), pp.963-970.

Iqbal, S.; Hood, D. A. (2015). The Role of Mitochondrial Fusion and Fission in Skeletal Muscle Function and Dysfunction. *Front. Biosci*.20, 157–172.

Janssen, I., Heymsfield, S. B. and Ross, R. (2002). Low relative skeletal muscle mass (sarcopenia) in older persons is associated with functional impairment and physical disability. *Journal of the American Geriatrics Society*, *50*(5), pp.889-896.

Jhun, B. S., O-Uchi, J., Adaniya, S. M., Cypress, M. W. and Yoon, Y. (2018). Adrenergic regulation of Drp1-driven mitochondrial fission in cardiac physio-pathology. *Antioxidants*, *7*(12), p.195.

Joseph, A. M., Adhihetty, P. J., Buford, T. W., Wohlgemuth, S. E., Lees, H. A., Nguyen, L. M. D., Aranda, J. M., Sandesara, B. D., Pahor, M., Manini, T. M. and Marzetti, E. (2012). The impact of aging on mitochondrial function and biogenesis pathways in skeletal muscle of sedentary high-and low-functioning elderly individuals. *Aging cell*, *11*(5), pp.801-809.

Kale, M. B., Bajaj, K., Umare, M., Wankhede, N. L., Taksande, B. G., Umekar, M. J. and Upaganlawar, A. (2022). Exercise and Nutraceuticals: Eminent Approach for Diabetic Neuropathy. *Current Molecular Pharmacology*, *15*(1), pp.108-128.

Kanazawa, T., Zappaterra, M. D., Hasegawa, A., Wright, A. P., Newman-Smith, E. D., Buttle, K. F., McDonald, K., Mannella, C. A. and van der Bliek, A. M. (2008). The C. elegans Opa1 homologue EAT-3 is essential for resistance to free radicals. *PLoS genetics*, *4*(2), p.e1000022.

Kang, S. H., Lee, H. A., Kim, M., Lee, E., Sohn, U. D., Kim, I. (2017). Forkhead box O_3 plays a role in skeletal muscle atrophy through expression of E3 ubiquitin ligases MuRF-1 and atrogin-1 in Cushing's syndrome. *Am J Physiol Endocrinol Metabol*.312(6):E495– 507.

Kang, C., Yeo, D. and Ji, L. L. (2016). Muscle immobilization activates mitophagy and disrupts mitochondrial dynamics in mice. *Acta physiologica*, *218*(3), pp.188-197.

Kim, Y., Triolo, M. and Hood, D. A. (2017). Impact of aging and exercise on mitochondrial quality control in skeletal muscle. *Oxidative Medicine and Cellular Longevity*, 2017.

Kitada, T., Asakawa, S., Hattori, N., Matsumine, H., Yamamura, Y., Minoshima, S., Yokochi, M., Mizuno, Y. and Shimizu, N. (1998). Mutations in the parkin gene cause autosomal recessive juvenile parkinsonism. *nature*, *392*(6676), pp.605-608.

Klionsky, D. J., Codogno, P., Cuervo, A. M., Deretic, V., Elazar, Z., Fueyo-Margareto, J., Gewirtz, D. A., Kroemer, G., Levine, B., Mizushima, N. and Rubinsztein, D. C. (2010). A comprehensive glossary of autophagy-related molecules and processes. *Autophagy*, *6*(4), pp.438-448.

Klionsky, D. J., Cregg, J. M., Dunn, W. A., Emr, S. D., Sakai, Y., Sandoval, I. V., Sibirny, A., Subramani, S., Thumm, M., Veenhuis, M. and Ohsumi, Y. (2003). A unified nomenclature for yeast autophagy-related genes. *Developmental cell*, *5*(4), pp.539-545.

Ko, F., Abadir, P., Marx, R., Westbrook, R., Cooke, C., Yang, H. and Walston, J. (2016). Impaired mitochondrial degradation by autophagy in the skeletal muscle of the aged female interleukin 10 null mouse. *Experimental gerontology*, *73*, pp.23-27.

Koltai, E., Hart, N., Taylor, A. W., Goto, S., Ngo, J. K., Davies, K. J. and Radak, Z. (2012). Age-associated declines in mitochondrial biogenesis and protein quality control factors are minimized by exercise training. *American Journal of Physiology-Regulatory, Integrative and Comparative Physiology*, *303*(2), pp. R127-R134.

Komatsu, M., Waguri, S., Chiba, T., Murata, S., Iwata, J., Tanida, I., Ueno, T., Koike, M., Uchiyama, Y., Kominami, E., Tanaka, K.(2006). Loss of autophagy in the central nervous system causes neurodegeneration in mice. *Nature*, 441 (7095), 880e884.

Konopka, A. R., Suer, M. K., Wolff, C. A. and Harber, M. P. (2014). Markers of human skeletal muscle mitochondrial biogenesis and quality control: effects of age and aerobic exercise training. *Journals of Gerontology Series A: Biomedical Sciences and Medical Sciences*, *69*(4), pp.371-378.

Koshiba, T., Detmer, S. A., Kaiser, J. T., Chen, H., McCaffery, J. M. and Chan, D. C. (2004). Structural basis of mitochondrial tethering by mitofusin complexes. *Science*, *305*(5685), pp.858-862.

Kubli, D. A. and Gustafsson, Å. B. (2012). Mitochondria and mitophagy: the yin and yang of cell death control. *Circulation research*, *111*(9), pp.1208-1221.

Labrousse, A. M., Zappaterra, M. D., Rube, D. A. and van der Bliek, A. M. (1999). C. elegans dynamin-related protein DRP-1 controls severing of the mitochondrial outer membrane. *Molecular cell*, *4*(5), pp.815-826.

Lazarou, M., Sliter, D. A., Kane, L. A., Sarraf, S. A., Wang, C., Burman, J. L., Sideris, D. P., Fogel, A. I., Youle, R. J. (2015). The ubiquitin kinase PINK1 recruits autophagy receptors to induce mitophagy. *Nature*, 524 (7565), 309e314.

Leduc-Gaudet, J. P., Picard, M., Pelletier, F. S. J., Sgarioto, N., Auger, M. J., Vallée, J., Robitaille, R., St-Pierre, D. H. and Gouspillou, G. (2015). Mitochondrial morphology is altered in atrophied skeletal muscle of aged mice. *Oncotarget*, *6*(20), p.17923.

Leduc-Gaudet, J. P., Reynaud, O., Hussain, S. N. and Gouspillou, G. (2019). Parkin overexpression protects from ageing-related loss of muscle mass and strength. *The Journal of Physiology*, *597*(7), pp.1975-1991.

Lee, J., Giordano, S. and Zhang, J. (2012). Autophagy, mitochondria and oxidative stress: cross-talk and redox signalling. *Biochemical Journal*, *441*(2), pp.523-540.

Leermakers, P. A., Kneppers, A. E., Schols, A. M., Kelders, M. C., de Theije, C. C., Verdijk, L. B., van Loon, L. J., Langen, R. C. and Gosker, H. R. (2019). Skeletal muscle unloading results in increased mitophagy and decreased mitochondrial biogenesis regulation. *Muscle & nerve*, *60*(6), pp.769-778.

Lemasters, J. J. (2005). Selective mitochondrial autophagy, or mitophagy, as a targeted defense against oxidative stress, mitochondrial dysfunction, and aging. *Rejuvenation research*, *8*(1), pp.3-5.

Lewis, S. C., Uchiyama, L. F. and Nunnari, J. (2016). ER-mitochondria contacts couple mtDNA synthesis with mitochondrial division in human cells. *Science*, *353*(6296), p.aaf5549.

Liesa, M. and Shirihai, O. S. (2013). Mitochondrial dynamics in the regulation of nutrient utilization and energy expenditure. *Cell metabolism*, *17*(4), pp.491-506.

Liu, L., Feng, D., Chen, G., Chen, M., Zheng, Q., Song, P., Ma, Q., Zhu, C., Wang, R., Qi, W. and Huang, L. (2012). Mitochondrial outer-membrane protein FUNDC1 mediates hypoxia-induced mitophagy in mammalian cells. *Nature cell biology*, *14*(2), pp.177-185.

López-Otín, C., Blasco, M. A., Partridge, L., Serrano, M. and Kroemer, G. (2013). The hallmarks of aging. *Cell*, *153*(6), pp.1194-1217.

Losón, O. C., Song, Z., Chen, H. and Chan, D. C. (2013). Fis1, Mff, MiD49, and MiD51 mediate Drp1 recruitment in mitochondrial fission. *Molecular biology of the cell*, *24*(5), pp.659-667.

Madeo, F., Tavernarakis, N. and Kroemer, G. (2010). Can autophagy promote longevity?. *Nature cell biology*, *12*(9), pp.842-846.

Malka, F., Guillery, O., Cifuentes-Diaz, C., Guillou, E., Belenguer, P., Lombès, A. and Rojo, M. (2005). Separate fusion of outer and inner mitochondrial membranes. *EMBO reports*, *6*(9), pp.853-859.

Manczak, M. and Reddy, P. H. (2012). Abnormal interaction between the mitochondrial fission protein Drp1 and hyperphosphorylated tau in Alzheimer's disease neurons: implications for mitochondrial dysfunction and neuronal damage. *Human molecular genetics*, *21*(11), pp.2538-2547.

Marde, V. S., Atkare, U. A., Gawali, S. V., Tiwari, P. L., Badole, S. P., Wankhede, N. L., Taksande, B. G., Upaganlawar, A. B., Umekar, M. J. and Kale, M. B. (2022). Alzheimer's disease and sleep disorders: Insights into the possible disease connections and the potential therapeutic targets. *Asian Journal of Psychiatry*, *68*, p.102961.

Marde, V. S., Tiwari, P. L., Wankhede, N. L., Taksande, B. G., Upaganlawar, A. B., Umekar, M. J. and Kale, M. B. (2021). Neurodegenerative disorders associated with genes of mitochondria. *Future Journal of Pharmaceutical Sciences*, *7*(1), pp.1-8.

Maria Fimia, G., Stoykova, A., Romagnoli, A., Giunta, L., Di Bartolomeo, S., Nardacci, R., Corazzari, M., Fuoco, C., Ucar, A., Schwartz, P. and Gruss, P. (2007). Ambra1 regulates autophagy and development of the nervous system. *Nature*, *447*(7148), pp.1121-1125.

Marzetti, E., Calvani, R., Cesari, M., Buford, T. W., Lorenzi, M., Behnke, B. J. and Leeuwenburgh, C. (2013). Mitochondrial dysfunction and sarcopenia of aging: from signaling pathways to clinical trials. *The international journal of biochemistry & cell biology*, *45*(10), pp.2288-2301.

Masiero, E., Agatea, L., Mammucari, C., Blaauw, B., Loro, E., Komatsu, M., Metzger, D., Reggiani, C., Schiaffino, S. and Sandri, M. (2009). Autophagy is required to maintain muscle mass. *Cell metabolism*, *10*(6), pp.507-515.

Matsuda, S., Nakanishi, A., Minami, A., Wada, Y. and Kitagishi, Y. (2015). Functions and characteristics of PINK1 and Parkin in cancer. *Front Biosci (Landmark Ed)*, *20*, pp.491-501.

McLelland, G. L., Soubannier, V., Chen, C. X., McBride, H. M. and Fon, E. (2014). Parkin and PINK 1 function in a vesicular trafficking pathway regulating mitochondrial quality control. *The EMBO journal*, *33*(4), pp.282-295.

Meng, F., Yao, D., Shi, Y., Kabakoff, J., Wu, W., Reicher, J., Ma, Y., Moosmann, B., Masliah, E., Lipton, S. A. and Gu, Z. (2011). Oxidation of the cysteine-rich regions of parkin perturbs its E3 ligase activity and contributes to protein aggregation. *Molecular neurodegeneration*, *6*(1), pp.1-15.

Mizushima, N. (2007). Autophagy: process and function. *Genes & development*, *21*(22), pp.2861-2873.

Moore, T. M., Zhou, Z., Cohn, W., Norheim, F., Lin, A. J., Kalajian, N., Strumwasser, A. R., Cory, K., Whitney, K., Ho, T. and Ho, T. (2019). The impact of exercise on mitochondrial dynamics and the role of

Drp1 in exercise performance and training adaptations in skeletal muscle. *Molecular metabolism*, *21*, pp.51-67.

Murakawa, T., Yamaguchi, O., Hashimoto, A., Hikoso, S., Takeda, T., Oka, T., Yasui, H., Ueda, H., Akazawa, Y., Nakayama, H. and Taneike, M. (2015). Bcl-2-like protein 13 is a mammalian Atg32 homologue that mediates mitophagy and mitochondrial fragmentation. *Nature communications*, *6*(1), pp.1-14.

Nakatogawa, H., Suzuki, K., Kamada, Y. and Ohsumi, Y. (2009). Dynamics and diversity in autophagy mechanisms: lessons from yeast. *Nature reviews Molecular cell biology*, *10*(7), pp.458-467.

Naon, D., Zaninello, M., Giacomello, M., Varanita, T., Grespi, F., Lakshminaranayan, S., Serafini, A., Semenzato, M., Herkenne, S., Hernández-Alvarez, M. I. and Zorzano, A. (2016). Critical reappraisal confirms that Mitofusin 2 is an endoplasmic reticulum–mitochondria tether. *Proceedings of the National Academy of Sciences*, *113*(40), pp.11249-11254.

Nemazanyy, I., Blaauw, B., Paolini, C., Caillaud, C., Protasi, F., Mueller, A., Proikas-Cezanne, T., Russell, R. C., Guan, K. L., Nishino, I. and Sandri, M. (2013). Defects of Vps15 in skeletal muscles lead to autophagic vacuolar myopathy and lysosomal disease. *EMBO molecular medicine*, *5*(6), pp.870-890.

Novak, I., Kirkin, V., McEwan, D. G., Zhang, J., Wild, P., Rozenknop, A., Rogov, V., Löhr, F., Popovic, D., Occhipinti, A. and Reichert, A. S. (2010). Nix is a selective autophagy receptor for mitochondrial clearance. *EMBO reports*, *11*(1), pp.45-51.

Ogata, T. and Yamasaki, Y. (1985). Scanning electron-microscopic studies on the three-dimensional structure of mitochondria in the mammalian red, white and intermediate muscle fibers. *Cell and tissue research*, *241*(2), pp.251-256.

O'Leary, M. F., Vainshtein, A., Iqbal, S., Ostojic, O. and Hood, D. A. (2013). Adaptive plasticity of autophagic proteins to denervation in aging skeletal muscle. *American Journal of Physiology-Cell Physiology*, *304*(5), pp.C422-C430.

Palacino, J. J., Sagi, D., Goldberg, M. S., Krauss, S., Motz, C., Wacker, M., Klose, J. and Shen, J. (2004). Mitochondrial dysfunction and oxidative damage in parkin-deficient mice. *Journal of Biological Chemistry*, *279*(18), pp.18614-18622.

Palikaras, K. and Tavernarakis, N. (2014). Mitochondrial homeostasis: the interplay between mitophagy and mitochondrial biogenesis. *Experimental gerontology*, *56*, pp.182-188.

Palikaras, K., Lionaki, E. and Tavernarakis, N. (2015). Coordination of mitophagy and mitochondrial biogenesis during ageing in C. elegans. *Nature*, *521*(7553), pp.525-528.

Palikaras, K., Lionaki, E. and Tavernarakis, N. (2018). Mechanisms of mitophagy in cellular homeostasis, physiology and pathology. *Nature cell biology*, *20*(9), pp.1013-1022.

Park, D., Jeong, H., Lee, M. N., Koh, A., Kwon, O., Yang, Y. R., Noh, J., Suh, P. G., Park, H. and Ryu, S. H. (2016). Resveratrol induces autophagy by directly inhibiting mTOR through ATP competition. *Scientific reports*, *6*(1), pp.1-11.

Park, J., Lee, S. B., Lee, S., Kim, Y., Song, S., Kim, S., Bae, E., Kim, J., Shong, M., Kim, J. M. and Chung, J. (2006). Mitochondrial dysfunction in Drosophila PINK1 mutants is complemented by parkin. *Nature*, *441*(7097), pp.1157-1161.

Perez, F. A. and Palmiter, R. D. (2005). Parkin-deficient mice are not a robust model of parkinsonism. *Proceedings of the National Academy of Sciences*, *102*(6), pp.2174-2179.

Pickrell, A. M. and Youle, R. J. (2015). The roles of PINK1, parkin, and mitochondrial fidelity in Parkinson's disease. *Neuron*, *85*(2), pp.257-273.

Poole, A. C., Thomas, R. E., Andrews, L. A., McBride, H. M., Whitworth, A. J. and Pallanck, L. J. (2008). The PINK1/Parkin pathway regulates mitochondrial morphology. *Proceedings of the National Academy of Sciences*, *105*(5), pp.1638-1643.

Qi, Y., Yan, L., Yu, C., Guo, X., Zhou, X., Hu, X., Huang, X., Rao, Z., Lou, Z. and Hu, J. (2016). Structures of human mitofusin 1 provide insight into mitochondrial tethering. *Journal of Cell Biology*, *215*(5), pp.621-629.

Ravikumar, B., Sarkar, S., Davies, J. E., Futter, M., Garcia-Arencibia, M., Green-Thompson, Z. W., Jimenez-Sanchez, M., Korolchuk, V. I., Lichtenberg, M., Luo, S. and Massey, D. C. (2010). Regulation of mammalian autophagy in physiology and pathophysiology. *Physiological reviews*, *90*(4), pp.1383-1435.

Rojansky, R., Cha, M. Y. and Chan, D. C. (2016). Elimination of paternal mitochondria in mouse embryos occurs through autophagic degradation dependent on PARKIN and MUL1. *elife*, *5*, p.e17896.

Romanello, V. and Sandri, M. (2016). Mitochondria quality control and muscle mass maintenance. *Frontiers in physiology*, *6*, p.422.

Romanello, V. and Sandri, M. (2021). The connection between the dynamic remodeling of the mitochondrial network and the regulation of muscle mass. *Cellular and Molecular Life Sciences*, *78*(4), pp.1305-1328.

Romanello, V., Guadagnin, E., Gomes, L., Roder, I., Sandri, C., Petersen, Y., Milan, G., Masiero, E., Del Piccolo, P., Foretz, M. and Scorrano, L. (2010). Mitochondrial fission and remodelling contributes to muscle atrophy. *The EMBO journal*, *29*(10), pp.1774-1785.

Rubinsztein, D. C., Mariño, G. and Kroemer, G. (2011). Autophagy and aging. *Cell*, *146*(5), pp.682-695.

Rujiviphat, J., Meglei, G., Rubinstein, J. L. and McQuibban, G. A. (2009). Phospholipid association is essential for dynamin-related protein Mgm1 to function in mitochondrial membrane fusion. *Journal of Biological Chemistry*, *284*(42), pp.28682-28686.

Sandri, M., Sandri, C., Gilbert, A., Skurk, C., Calabria, E., Picard, A., Walsh, K., Schiaffino, S., Lecker, S. H. and Goldberg, A. L. (2004). Foxo transcription factors induce the atrophy-related ubiquitin ligase atrogin-1 and cause skeletal muscle atrophy. *Cell*, *117*(3), pp.399-412.

Santel, A., Frank, S., Gaume, B., Herrler, M., Youle, R. J. and Fuller, M. T. (2003). Mitofusin-1 protein is a generally expressed mediator of mitochondrial fusion in mammalian cells. *Journal of cell science*, *116*(13), pp.2763-2774.

Scheibye-Knudsen, M., Fang, E. F., Croteau, D. L., Wilson III, D. M. and Bohr, V. A. (2015). Protecting the mitochondrial powerhouse. *Trends in cell biology*, *25*(3), pp.158-170.

Schmitt, K., Grimm, A., Dallmann, R., Oettinghaus, B., Restelli, L. M., Witzig, M., Ishihara, N., Mihara, K., Ripperger, J. A., Albrecht, U. and Frank, S. (2018). Circadian control of DRP1 activity regulates mitochondrial dynamics and bioenergetics. *Cell metabolism*, *27*(3), pp.657-666.

Schrader, M., Costello, J. L., Godinho, L. F., Azadi, A. S. and Islinger, M. (2016). Proliferation and fission of peroxisomes—An update. *Biochimica Et Biophysica Acta (BBA)-Molecular Cell Research*, *1863*(5), pp.971-983.

Scorrano, L. (2013). Keeping mitochondria in shape: a matter of life and death. *European journal of clinical investigation*, *43*(8), pp.886-893.

Sebastián, D. and Zorzano, A. (2018). Mitochondrial dynamics and metabolic homeostasis. *Current Opinion in Physiology*, *3*, pp.34-40.

Sebastián, D., Hernández-Alvarez, M. I., Segalés, J., Sorianello, E., Muñoz, J. P., Sala, D., Waget, A., Liesa, M., Paz, J. C., Gopalacharyulu, P. and Orešič, M. (2012). Mitofusin 2 (Mfn2) links mitochondrial and endoplasmic reticulum function with insulin signaling and is essential for normal glucose homeostasis. *Proceedings of the National Academy of Sciences*, *109*(14), pp.5523-5528.

Sebastián, D., Sorianello, E., Segalés, J., Irazoki, A., Ruiz-Bonilla, V., Sala, D., Planet, E., Berenguer-Llergo, A., Muñoz, J. P., Sánchez-Feutrie, M. and Plana, N. (2016). Mfn2 deficiency links age-related sarcopenia and impaired autophagy to activation of an adaptive mitophagy pathway. *The EMBO journal*, *35*(15), pp.1677-1693.

Shen, Y., Zhang, Q., Huang, Z., Zhu, J., Qiu, J., Ma, W., Yang, X., Ding, F. and Sun, H. (2020). Isoquercitrin delays denervated soleus muscle atrophy by inhibiting oxidative stress and inflammation. *Frontiers in physiology*, p.988.

Smirnova, E., Griparic, L., Shurland, D. L. and Van Der Bliek, A. M. (2001). Dynamin-related protein Drp1 is required for mitochondrial division in mammalian cells. *Molecular biology of the cell*, *12*(8), pp.2245-2256.

Song, W. H., Yi, Y. J., Sutovsky, M., Meyers, S. and Sutovsky, P. (2016). Autophagy and ubiquitin–proteasome system contribute to sperm mitophagy after mammalian fertilization. *Proceedings of the National Academy of Sciences*, *113*(36), pp.E5261-E5270.

Stouth, D. W., Ng, S. Y., Webb, E. K., Manta, A., Moll, Z. and Ljubicic, V. (2020). CARM1 regulates AMPK signaling in skeletal muscle. *Iscience*, *23*(11), p.101755.

Strappazzon, F., Nazio, F., Corrado, M., Cianfanelli, V., Romagnoli, A., Fimia, G. M., Campello, S., Nardacci, R., Piacentini, M., Campanella, M. and Cecconi, F. (2015). AMBRA1 is able to induce mitophagy via LC3 binding, regardless of PARKIN and p62/SQSTM1. *Cell Death & Differentiation*, *22*(3), pp.419-432.

Sun, N., Yun, J., Liu, J., Malide, D., Liu, C., Rovira, I. I., Holmström, K. M., Fergusson, M. M., Yoo, Y. H., Combs, C. A. and Finkel, T. (2015). Measuring in vivo mitophagy. *Molecular cell*, *60*(4), pp.685-696.

Tai, H., Wang, Z., Gong, H., Han, X., Zhou, J., Wang, X., Wei, X., Ding, Y., Huang, N., Qin, J. and Zhang, J. (2017). Autophagy impairment with lysosomal and mitochondrial dysfunction is an important characteristic of oxidative stress-induced senescence. *Autophagy*, *13*(1), pp.99-113.

Taksande, B. G., Sharma, O., Aglawe, M. M., Kale, M. B., Gawande, D. Y., Umekar, M. J. and Kotagale, N. R. (2017). Acute orexigenic effect of agmatine involves interaction between central α2-adrenergic and GABAergic receptors. *Biomedicine & Pharmacotherapy*, *93*, pp.939-947.

Tamura, Y., Kitaoka, Y., Matsunaga, Y., Hoshino, D. and Hatta, H. (2015). Daily heat stress treatment rescues denervation-activated mitochondrial clearance and atrophy in skeletal muscle. *The Journal of physiology*, *593*(12), pp.2707-2720.

Tezze, C., Romanello, V., Desbats, M. A., Fadini, G. P., Albiero, M., Favaro, G., Ciciliot, S., Soriano, M. E., Morbidoni, V., Cerqua, C. and Loefler, S. (2017). Age-associated loss of OPA1 in muscle impacts muscle mass, metabolic homeostasis, systemic inflammation, and epithelial senescence. *Cell metabolism*, *25*(6), pp.1374-1389.

Tilokani, L., Nagashima, S., Paupe, V. and Prudent, J. (2018). Mitochondrial dynamics: overview of molecular mechanisms. *Essays in biochemistry*, *62*(3), pp.341-360.

Tiwari, P., Wankhede, N., Badole, S., Umare, M., Taksande, B., Upaganlawar, A., Umekar, M. and Kale, M. (2021). Mitochondrial Dysfunction in Ageing: Involvement of Oxidative Stress and Role of Melatonin. *Bull. Env. Pharmacol. Life Sci*, *10*, pp.156-172.

Trifunovic, A. (2006). Mitochondrial DNA and ageing. *Biochimica et Biophysica Acta (BBA)-Bioenergetics*, *1757*(5-6), pp.611-617.

Triolo, M. and Hood, D. A. (2021). Manifestations of age on autophagy, mitophagy and lysosomes in skeletal muscle. *Cells*, *10*(5), p.1054.

Tryon, L. D., Vainshtein, A., Memme, J. M., Crilly, M. J. and Hood, D. A. (2014). Recent advances in mitochondrial turnover during chronic muscle disuse. *Integrative Medicine Research*, *3*(4), pp.161-171.

Twig, G., Elorza, A., Molina, A. J., Mohamed, H., Wikstrom, J. D., Walzer, G., Stiles, L., Haigh, S. E., Katz, S., Las, G. and Alroy, J. (2008). Fission and selective fusion govern mitochondrial segregation and elimination by autophagy. *The EMBO journal*, *27*(2), pp.433-446.

Umare, M. D., Wankhede, N. L., Bajaj, K. K., Trivedi, R. V., Taksande, B. G., Umekar, M. J., Mahore, J. G. and Kale, M. B. (2021). Interweaving of Reactive Oxygen Species and Major Neurological and Psychiatric Disorders. In *Annales PharmaceutiquesFrançaises*. Elsevier Masson.

Upaganlawar, A. B., Wankhede, N. L., Kale, M. B., Umare, M. D., Sehgal, A., Singh, S., Bhatia, S., Al-Harrasi, A., Najda, A., Nurzyńska-Wierdak, R. and Bungau, S. (2021). Interweaving epilepsy and neurodegeneration: Vitamin E as a treatment approach. *Biomedicine & Pharmacotherapy*, *143*, p.112146.

Vafai, S. B. and Mootha, V. K. (2012). Mitochondrial disorders as windows into an ancient organelle. *Nature*, *491*(7424), pp.374-383.

Vainshtein, A., Desjardins, E., Armani, A., Sandri, M. and Hood, D. A. (2015). PGC-1α modulates denervation-induced mitophagy in skeletal muscle. *Skeletal muscle*, *5*(1), pp.1-17.

Valente, E. M., Abou-Sleiman, P. M., Caputo, V., Muqit, M. M., Harvey, K., Gispert, S., Ali, Z., Del Turco, D., Bentivoglio, A. R., Healy, D. G. and Albanese, A. (2004). Hereditary early-onset Parkinson's disease caused by mutations in PINK1. *science*, *304*(5674), pp.1158-1160.

Van der Bliek, A. M., Shen, Q. and Kawajiri, S. (2013). Mechanisms of mitochondrial fission and fusion. *Cold Spring Harbor perspectives in biology*, *5*(6), p.a011072.

Van Humbeeck, C., Cornelissen, T., Hofkens, H., Mandemakers, W., Gevaert, K., De Strooper, B. and Vandenberghe, W. (2011). Parkin interacts with Ambra1 to induce mitophagy. *Journal of Neuroscience*, *31*(28), pp.10249-10261.

Wai, T. and Langer, T. (2016). Mitochondrial dynamics and metabolic regulation. *Trends in Endocrinology & Metabolism*, *27*(2), pp.105-117.

Wallace, D. C. (2013). A mitochondrial bioenergetic etiology of disease. *The Journal of clinical investigation*, *123*(4), pp.1405-1412.

Wan, Q., Zhang, L., Huang, Z., Zhang, H., Gu, J., Xu, H., Yang, X., Shen, Y., Law, B. Y. K., Zhu, J. and Sun, H. (2020). Aspirin alleviates denervation-induced muscle atrophy via regulating the Sirt1/PGC-1α axis and STAT3 signaling. *Annals of Translational Medicine*, *8*(22).

Wankhede, N. L., Samrit, S. R., Umare, M. D., Bajaj, K. K., Trivedi, R. V., Umekar, M. J. and Kale, M. B. (2021). Biomarkers of cancer: A Comprehensive Review.

Wei, H., Liu, L. and Chen, Q. (2015). Selective removal of mitochondria via mitophagy: distinct pathways for different mitochondrial stresses. *Biochimica et Biophysica Acta (BBA)-Molecular Cell Research, 1853*(10), pp.2784-2790.

Wong, Y. C., Ysselstein, D. and Krainc, D. (2018). Mitochondria–lysosome contacts regulate mitochondrial fission via RAB7 GTP hydrolysis. *Nature, 554*(7692), pp.382-386.

Wu, W., Tian, W., Hu, Z., Chen, G., Huang, L., Li, W., Zhang, X., Xue, P., Zhou, C., Liu, L. and Zhu, Y. (2014). ULK 1 translocates to mitochondria and phosphorylates FUNDC 1 to regulate mitophagy. *EMBO reports, 15*(5), pp.566-575.

Yamano, K., Wang, C., Sarraf, S. A., Münch, C., Kikuchi, R., Noda, N. N., Hizukuri, Y., Kanemaki, M. T., Harper, W., Tanaka, K. and Matsuda, N. (2018). Endosomal Rab cycles regulate Parkin-mediated mitophagy. *Elife, 7*, p.e31326.

Yang, X., Xue, P., Chen, H., Yuan, M., Kang, Y., Duscher, D., Machens, H. G. and Chen, Z. (2020). Denervation drives skeletal muscle atrophy and induces mitochondrial dysfunction, mitophagy and apoptosis via miR-142a-5p/MFN1 axis. *Theranostics, 10*(3), p.1415.

Yang, X., Zhang, R., Nakahira, K. and Gu, Z. (2019). Mitochondrial DNA mutation, diseases, and nutrient-regulated Mitophagy. *Annual review of nutrition, 39*, pp.201-226.

Yang, Y., Ouyang, Y., Yang, L., Beal, M. F., McQuibban, A., Vogel, H. and Lu, B. (2008). Pink1 regulates mitochondrial dynamics through interaction with the fission/fusion machinery. *Proceedings of the National Academy of Sciences, 105*(19), pp.7070-7075.

Yang, Z. and Klionsky, D. J. (2010). Eaten alive: a history of macroautophagy. *Nature cell biology, 12*(9), pp.814-822.

Yao, D., Gu, Z., Nakamura, T., Shi, Z. Q., Ma, Y., Gaston, B., Palmer, L. A., Rockenstein, E. M., Zhang, Z., Masliah, E. and Uehara, T. (2004). Nitrosative stress linked to sporadic Parkinson's disease: S-nitrosylation of parkin regulates its E3 ubiquitin ligase activity. *Proceedings of the National Academy of Sciences, 101*(29), pp.10810-10814.

Yeo, D., Kang, C., Gomez-Cabrera, M. C., Vina, J. and Ji, L. L. (2019). Intensified mitophagy in skeletal muscle with aging is downregulated by PGC-1alpha overexpression in vivo. *Free Radical Biology and Medicine, 130*, pp.361-368.

Yoo, S. Z., No, M. H., Heo, J. W., Park, D. H., Kang, J. H., Kim, S. H. and Kwak, H. B. (2018). Role of exercise in age-related sarcopenia. *Journal of exercise rehabilitation, 14*(4), p.551.

Youle, R. J. and Narendra, D. P. (2011). Mechanisms of mitophagy. *Nature reviews Molecular cell biology, 12*(1), pp.9-14.

Yu, R., Jin, S. B., Lendahl, U., Nistér, M. and Zhao, J. (2019). Human Fis1 regulates mitochondrial dynamics through inhibition of the fusion machinery. *The EMBO journal, 38*(8), p.e99748.

Zampieri, S., Pietrangelo, L., Loefler, S., Fruhmann, H., Vogelauer, M., Burggraf, S., Pond, A., Grim-Stieger, M., Cvecka, J., Sedliak, M. and Tirpáková, V. (2015). Lifelong physical exercise delays age-associated skeletal muscle decline. *Journals of Gerontology Series A: Biomedical Sciences and Medical Sciences, 70*(2), pp.163-173.

Zhang, Z., Sliter, D. A., Bleck, C. K. and Ding, S. (2019). Fis1 deficiencies differentially affect mitochondrial quality in skeletal muscle. *Mitochondrion, 49*, pp.217-226.

Zhou, J., Tan, S. H., Nicolas, V., Bauvy, C., Yang, N. D., Zhang, J., Xue, Y., Codogno, P. and Shen, H. M. (2013). Activation of lysosomal function in the course of autophagy via mTORC1 suppression and autophagosome-lysosome fusion. *Cell research, 23*(4), pp.508-523.

Zhou, Q., Li, H., Li, H., Nakagawa, A., Lin, J. L., Lee, E. S., Harry, B. L., Skeen-Gaar, R. R., Suehiro, Y., William, D. and Mitani, S. (2016). Mitochondrial endonuclease G mediates breakdown of paternal mitochondria upon fertilization. *Science, 353*(6297), pp.394-399.

Ziaaldini, M. M., Marzetti, E., Picca, A. and Murlasits, Z. (2017). Biochemical pathways of sarcopenia and their modulation by physical exercise: a narrative review. *Frontiers in Medicine, 4*, p.167.

Chapter 8

Mitophagy and Aging

Divya Jindal
Vinayak Agarwal
and Manisha Singh*
Department of Biotechnology, Jaypee Institute of Information Technology, Noida, Uttar Pradesh, India

Abstract

The cellular constituents are transported into lysosomes for breakdown and recycling via the primary protein turnover mechanism known as mitophagy. Under stressful circumstances, this intracellular process can preserve cellular homeostasis, and its dysregulation may result in physiological changes. Age-related declines in autophagic activity are considered to be a factor in the buildup of harmed macromolecules and organelles. Interestingly, the failure of these mitophagy pathways has also been linked to worsening age-related ailments, including cancer and neurological problems. Similarly, it has been suggested that maintaining adequate mitophagy functions extends longevity in other creatures. In this review, we address current studies that demonstrate the molecular, signaling, and mitochondrial bases of aging. The chapter also covered the effects of mitophagy on many organs and disorders linked to aging, emphasizing how pertinent this process is to the symptoms of aging. Understanding how autophagy affects aging thus opens up new opportunities for identifying biochemical and pharmacological targets, i.e., synthetic and natural compounds to modulate mitophagy during aging and the creation of cutting-edge anti-aging therapeutic strategies.

Keywords: mitophagy, mitochondrial dysfunction, aging, oxidative stress, mitochondrial dynamics, parkin-dependent mitophagy

Introduction

The principal energy source for maintaining neural homeostasis and its associated processes is provided by mitochondria, which produces adenosine triphosphate (ATP) by oxidative phosphorylation (OXPHOS) (Sheng and Cai 2012b). In addition, mitochondria are involved in

* Corresponding Author's Email: manishasingh1295@gmail.com.

In: Mitophagy and Its Role in Health and Disease
Editors: Shalini Mani and Samiksha Wasnik
ISBN: 979-8-89113-470-6

producing presynaptic neurotransmitters in synapses and are necessary for developing critical centers and heme in neurons (Sheng and Cai 2012a). Numerous studies have shown that aging patients have substantial mitochondrial abnormalities, and malfunctioning mitochondria are an early and obvious sign of aging (Amorim et al., 2022).

Mitophagy is selective autophagy in which cells target mitochondria with defects or damage that tend to clump together in response to trauma or stress.

Mitophagy is necessary for cellular growth and differentiation, removing damaged mitochondria and preserving a constant rate of mitochondrial turnover (Ding and Yin 2012). This mechanism is essential to mitochondrial dynamics and ensures that mitochondria are integrated into eukaryotic cells. Damaged mitochondria may not be adequately cleared, leading to increased oxidative stress and cell death (Zorov, Juhaszova, and Sollott 2014). Many proteins strictly control mitophagy because they are responsible for eliminating defective mitochondria and their related difficulties (Youle and Narendra, 2011). These include Parkin, which yeast lacks, and PTEN-induced putative kinase 1 (PINK1), known to be a crucial mitophagy regulator in metazoans. This suggests that the mitophagy regulation varies depending on the species. One of the most prevalent methods for removing harmed and defective mitochondria from various cells is parkin-dependent mitophagy. Recently, it has been discovered that poor mitophagy and the accumulation of damaged mitochondria are the causes of various neurological diseases, including Alzheimer's disease (AD). In these situations, activating mitophagy may be a possible therapeutic strategy. The published literature reports that a broad range of medications/methods have been investigated and found to activate mitophagy through various mechanisms.

The idea that mitochondrial dysfunction contributes significantly to biological aging is widely accepted. Decades ago, it was proposed that mitochondria have a central role in aging, with one theory suggesting that the excessive creation of reactive oxygen species (ROS) damages biomolecules over time, leading to aging (Amorim et al., 2022). Since mitochondria are the primary source of ROS in cells, this theory evolved into the mitochondrial theory of aging (Nita and Grzybowski 2016). The hypothesis states that as mitochondria age, ROS-induced damage accumulates and causes malfunction, leading to a decline in cell function and ultimately aging and death. Animal studies, such as using "mtDNA mutator mice" with high rates of mitochondrial DNA mutations, have provided scientific evidence to support this idea (Guo et al., 2013, Shabalina et al., 2017). However, several recent studies have also produced data that refute this idea. For instance, Caenorhabditis elegans' longevity was unaffected by superoxide dismutase gene deletion (Van Raamsdonk and Hekimi 2009). Indeed, it appears that mitochondria play a highly intricate function in aging (Amorim et al., 2022).

Epidemiology of Aging

Richard Doll, a renowned epidemiologist, claimed that aging was not a single, consistent phenomenon twenty years ago. Doll argued that it was preferable to view each of the independent disorders of aging as distinct and thus as having distinct potential treatments, given the weight of the data at the time (Richmond, 2005). Since then, gerontology has made significant strides, demonstrating that many age-related disorders have shared origins and may be treated to reverse or prevent them. Since then, there has been a significant movement in the scientific consensus from viewing aging as an unstoppable, entropic process to one mediated by clear-cut, crucially adjustable biochemical pathways (da Costa et al., 2016). Initial research on nematodes showed that longevity may be genetically manipulated. The numerous genetic factors affecting human longevity are just starting to be teased out by later research. The cells of elderly adults have aggregated proteins, dysfunctional mitochondria, epigenetic alterations, and short telomeres. Senescence and the accompanying decline in active stem cell populations may be a consequence, reducing the capacity for tissue regeneration and speeding up the aging process (Christensen, Johnson, and Vaupel 2006).

Significant ethical and economic reasons exist to address and prevent the pathologies associated with aging, given that aging is the primary cause of many disabling diseases that render people weak and deprive individuals of their autonomy and dignity. With a wide range of human aging, numerous companies and government-funded projects have emerged to address human aging as a problem in and of itself rather than concentrating on aging-related ailments separately (Franceschi et al., 2018). Increased mitochondrial damage is a characteristic of aging and age-related neurodegeneration, highlighting that these neuronal cells are particularly susceptible to age-dependent mitophagy failure. The nuclear genome expresses multiple proteins that control mitophagy (Grimm and Eckert 2017).

Drugs that have already reached the market can be repurposed for various benefits. Developing a drug, from initial development clinical trials to market-ready goods, can frequently cost billions and take around 15 years (Hughes et al., 2011). By reusing existing medications, much of the exploratory effort required to identify the mechanisms of action, formulation, manufacturing process, and pharmacokinetics is avoided. Market-tested medications are significant because they already have undergone clinical trials and have shown to be safe and effective for the original intended use. As a result, the time needed to develop repurposed medications can be significantly decreased (3–12 years). Medication-centered and disease-centered strategies for drug repurposing are the two main methods. Briefly stated physicochemical features (such as structural similarities to other medications) or biological effects are the foundation of drug-centered repurposing. It is also possible to leverage indirect qualities, such as shared side effects between medications prescribed for various disorders. More recently, high-throughput screenings of potential interaction partners have been made possible by computer modeling of drug-protein interactions. Disease-centered strategies indicate possible drug repurposing by drawing on similarities between various disorders. In general, treatments used to treat one disease may also be effective for another if the molecular etiologies of the two diseases are comparable (Mallikarjun and Swift 2016).

Molecules or Signal Transduction Pathways in Aging

Molecules in Aging

MicroRNAs (miRNAs)

Endogenous noncoding RNAs (eukaryotic; 20–25 nucleotides; microRNAs) play regulatory roles in eukaryotes. It has recently been demonstrated that miRNAs substantially influence aging (Ying, Chang, and Lin, 2008). Du et al. found that miR-17 regulates two targets, ADCY5, and IRS1, to enhance MKP and FoxO3 and reduce mTOR and JNK. Transgenic mice now live longer (Du et al., 2014). Furthermore, it was found in this study that ADCY5 or IRS1 can promote autophagy and ward off aging and apoptosis in cells. MiR-83 has been shown by

Dzakah et al. to affect Caenorhabditis elegans longevity (Liu et al., 2019). Their research revealed that miR-83 expression decreased with aging and that C. elegans lived longer when miR-83 was deleted. MiR-83's life-extension impact was made possible by the increased transcription factors expression din-1 and daf-16 (Przybysz et al., 2009). Lyu et al., claim that the regulation of the transforming growth factor (TGF) signaling promotes senescence through the miR-29-induced deletion of H4K20me3. They discovered that miR-29 accelerated cell senescence suppressed the expression of H4K20me3 in senescent murine fibroblast cells and aided in the loss of suv4-20h2 (Lyu et al., 2018). TGF-, on the other hand, accelerated cellular aging by promoting the loss of H4K20me3 brought on by miR-29. According to Fan et al., MiR-1292 is connected to the cellular senescence of human adipose-derived mesenchymal stem cells (hADSCs). They found that miR-1292 may have been a target in FZD4 downregulation, which led to an overexpression of miR-1292 and aided in the aging and osteogenic differentiation of hADSC (Fan et al., 2018). It was discovered that this activity included the Wnt/-catenin signaling pathway. MiR-335-3p, which is neuron-enriched, is highly associated with aging and age-related neurological disorders, according to mounting data.

Telomere

The cell division cycle is regulated, and the integrity of the genome is safeguarded by the nucleoprotein structures known as telomeres, which are situated at the ends of chromosomes and are composed of the DNA sequence of the telomere and its protein (Srinivas, Rachakonda, and Kumar 2020). Many studies have shown reductions in telomere damage and its activity, two of the main factors in aging and age-related damage leading to cellular senescence. The relationship between unfavorable social circumstances and DNA damage and increased telomere shortening is the most well-established cause. Using standardized surveys created by Epel et al., 58 healthy premenopausal women were selected from this group and asked to assess their stress levels from the previous month. The biological mother of a child with a chronic disease was included in the trial, while the control group consisted of women with at least one healthy biological child. To gauge the effects of stress, average telomere length and telomerase activity were evaluated. The findings demonstrated that perhaps the experimental group experienced significantly more stress than the control group. Compared to the control group, female members of the experimental group similarly showed lower telomerase activity and shorter telomere length. These discoveries provide light on how stress operates at the cellular level, influencing cell aging and potentially triggering the emergence of age-related disorders earlier than expected (Epel et al., 2004).

p16, p53/p21

Cell cycle arrest is the foundation of aging, and studies have shown that an increase in certain proteins, such as p16 or p53/p21, can lead to cell aging. While cellular aging is a complex process that involves various gene expressions and morphological changes, cycle stagnation has been identified as a key factor. In mouse embryonic fibroblasts, the overexpression of miR-20a was found to enhance p16 and induce cell senescence by increasing INK4a/ARF transcriptional activity (Overhoff et al., 2014). P53 has been identified as both a catalyst for cellular aging and an anti-aging agent. Its anti-aging effects are linked to its role in controlling the mTOR pathway, which is associated with autophagy (Mrakovcic and Fröhlich 2018). By inhibiting mTOR activity, P53 can activate autophagy and slow aging. Meanwhile, p53 was found to cause cell cycle arrest and aging through its downstream p53/p21/CDK2 signaling pathway. Studies have shown that azithromycin can accelerate aging in VSMCs by increasing the expression of p53/p21/p16 and activating the mTOR signaling pathway. Inhibiting mTOR activity can induce autophagy and reduce the expression of these proteins, thus slowing down the aging of VSMCs (Sung et al., 2017).

Signaling Pathway in Aging

Mammalian Target of Rapamycin (mTOR) Pathway

Nutrition and growth factors stimulate mTOR, which inhibits autophagy and boosts protein synthesis. Over time, mTOR may promote cellular stress brought on by events such as protein aggregate formation and oxidative damage. A buildup of damage and a cell function decline could ensue from this, subsequently promoting the emergence of age-related disorders. The standard path mTOR takes is the PI3K/Akt/mTOR signaling pathway. According to Tan et al., (Tan et al., 2016), PI3K/Akt/mTOR is crucial for VSMC replication and aging, and this role may be related to the control of oxidative stress and telomere function. They transfected mTOR siRNA and scrambled siRNA into human VSMCs. Furthermore, stem cell depletion brought on by mTOR activation decreased tissue healing and exacerbated tissue failure. Additionally, experimental investigations have demonstrated that delaying the aging of several biological models, such as yeast, worms, fruit flies, and mice, by blocking the mTOR signaling system using gene deletion, rapamycin, or dietary restriction (Lamming et al., 2013).

Nuclear Factor of Activated B-Cell (NF-κB) Signaling Pathway

The transcription of genes related to oxidative stress, immunity, inflammation, cell proliferation, apoptosis, and aging is regulated by NF-KB, activation of the transcription factor protein family. According to studies, the REL protein family, of which NF-B is a member, also includes the proteins p50, p52, REL, REL-A, and REL-B (DiDonato, Mercurio, and Karin 2012). A 300-amino acid-long, highly conserved REL homologous domain is also a part of NF-B. Inflammatory factors linked to senescence trigger the NF-KB signaling pathway (Zhao et al., 2015). The immunological response of a cell is aided by activated NF-KB when it enters the nucleus and attaches to DNA. Numerous senile degenerative disorders have been linked through research to the aging signaling system controlled by NF-KB (O'Neill and Kaltschmidt 1997). Alzheimer's disease (AD) patients' brain tissues had increased NF-KB activity, according to a postmortem investigation. In addition, neurons and glial cells near degenerating neurons and senile plaques showed immunological activity of p65. The buildup of beta-amyloid (A) is connected to NF-KB activation. Studies have shown that the production of NO products associated with oxidative stress and amyloid-beta deposition can activate NF-KB in cultured neurons. According to autopsy investigations, persons with Parkinson's disease had 70 times more NF-KB-positive dopaminergic neurons in their brains than healthy individuals, indicating that the pathogenic mechanism underlying Parkinson's disease may be related to NF-KB activation (Hunot et al., 1997).

Nuclear Factor-E2-Related Factor 2 (Nrf2) Signaling Pathway

Nrf2 significantly influences the antioxidant activity in cells. After being transported to the nucleus in response to oxidative stress, Nrf2 binds with the antioxidant response element (ARE) and controls the expression of several antioxidant proteins and detoxifying enzymes, ultimately aiding in endogenous defense (Zhang, Davies, and Forman 2015). According to Suh et al., (Suh et al., 2004) Rat liver cells' total and nuclear Nrf2 protein levels drastically decrease as age increases. One of the major factors contributing to ovarian aging is the increasing death of ovarian cells, which is caused by an imbalance between oxidation and antioxidants as people age. According to research, Nrf2 gene deletion can make mouse ovaries more sensitive to toxins and age their ovaries more quickly (Meng et al., 2015). According to Chen et al., (Chen et al., 2019) increasing Nrf2 expression can reduce DNA damage and oxidative stress while also blocking the p53-p21-p16-rb signaling pathway, which slows cell aging. Age-related skeletal muscle mitochondrial dysfunction and muscle atrophy are increasingly promoted by Nrf2 deficiency, which can regulate mitochondrial biogenesis and kinetics to maintain muscle mass and function (Huang et al., 2019, Kitaoka et al., 2019). The study also discovered that Nrf2 activation could prevent oxidative stress and inflammatory reactions linked to aging and postpone the onset of aging and age-related disorders (Sahin et al., 2019). Aging mice given D-galactose (D-gal) had better learning and memory after Nrf2 activation (Ho et al., 2019).

Wnt/β-Catenin Signaling Pathway

A highly conserved signaling pathway throughout evolution, the Wnt/-catenin pathway serves various biological purposes. According to studies, mesenchymal stem cells may undergo senescence-related changes if this pathway is activated, which has been linked to an important regulatory function in cell aging (Suzuki, Minamide, and Iwata 2018). Studies have shown a connection between the activation of this system and enhanced p53 protein production and DNA damage response, which may be one of the main reasons underlying stem cell senescence (Cui et al., 2019). The importance of the p53/p21 pathway and the DNA oxidative damage response in Wnt/-catenin signal pathway-induced aging of hematopoietic stem/progenitor cells has been demonstrated (Wang et al., 2010). Aging of the skin is the most notable external indicator of human aging, and tissues from old skin have exceptionally high concentrations of WNT/-catenin signal pathway components (Zhang, Wang, and Tan 2011). It was shown that the WNT/-catenin signal pathway could be inhibited to treat skeletal muscle regeneration injuries caused by aging. In a mouse model of aging, it was discovered that the WNT/-catenin signal pathway was enhanced (Zhang et al., 2012).

Mitochondrial Basis of Aging

Somatic Mitochondrial DNA Mutations

Mitochondria are unique because they contain their genome, a circular molecule called mtDNA that has 37 genes and produces various types of RNA (DiMauro and Schon, 2003). Point mutations and deletions are the most common types of mutations that occur in mtDNA as people age, usually due to errors in replication or repair. Many studies suggest that the accumulation of mtDNA mutations contributes to mitochondrial dysfunction during aging, but the precise mechanism is still unclear. Evidence shows that different types of mtDNA mutations increase with age in various tissues of humans and animals, including the brain, heart, skeletal muscles, and liver (Linnane et al., 1989, Trounce, Byrne, and Marzuki 1989, Müller-Höcker

1990, 1989, Münscher, Müller-Höcker, and Kadenbach 1993, Cottrell et al., 2001). The accumulation of mtDNA deletions has been linked to various muscle abnormalities that lead to sarcopenia, and the common deletion known as mtDNA 4977 bp deletion has been shown to increase with age in some areas of the human brain (Bua et al., 2006, Wanagat et al., 2001).

A study that used a susceptible mutation detection method contradicted the idea that mtDNA mutations accelerate aging by revealing that elderly wild-type mice had a significantly lower mtDNA mutation frequency in their brains and hearts than in previous reports. In contrast, mtDNA mutator heterozygous mice had a much higher mtDNA mutation load, but their lifespan remained normal (Vermulst et al., 2007). However, this study did not account for significant mtDNA deletions, which accumulate in aging human tissues and could not control for species-specific variations that might affect the relationship between mtDNA mutations and aging (Kraytsberg et al., 2006, Khrapko and Vijg, 2007). Further studies are necessary to determine whether the lower mtDNA mutation frequency observed in elderly mice has any functional significance for human aging and to clarify species-specific differences in the accumulation of somatic mtDNA mutations during aging.

ROS and Oxidative Stress

Reactive oxygen species, which include superoxide anions (O2•), hydrogen peroxide (H_2O_2), and hydroxyl (•OH) radicals, are significant consequences of oxidative metabolism. While in the Fenton reaction, H_2O_2 is transformed into highly reactive •OH radicals without reduced transition metals. Single electron reduction of molecular O_2 results in O_2• anions, which are easily transformed into H_2O_2 by superoxide dismutase (SODs) (Turrens 2003, Brand 2010). ROS dynamically oxidizes proteins, lipids, and nucleic acids, causing damage to macromolecular structures (Fraga et al. 1990, Niki 2009, Ugarte, Petropoulos, and Friguet 2010). Following the free radical hypothesis of aging, oxidative damage to cellular macromolecules is brought on by an increase in ROS generation brought on by an increase in mitochondrial dysfunction as people age. Then begins the vicious cycle of molecular damage accumulation and ROS generation (Harman 1972, 1956). An imbalance between excessive ROS production and the ability of cells to defend against antioxidants results in oxidative stress (Turrens, 2003).

Numerous studies have supported the role of ROS in aging by showing that ROS levels and oxidative damage rise with age, ROS production rises when mitochondrial function is inhibited, mitochondrial dysfunction rises with age, and several age-related pathologies are linked to elevated ROS production and oxidative stress (Fraga et al., 1990, Ugarte, Petropoulos, and Friguet 2010, Sohal and Sohal 1991, Asensi et al., 1999, Balaban, Nemoto, and Finkel 2005, Cho, Hur, and Walker 2011). The question of whether elevated ROS production is a cause or effect of aging is still up for debate, even though these data point to a strong correlation between ROS creation and oxidative damage throughout aging. In the past decade, numerous studies have discovered that increasing oxidative stress does not shorten longevity in various model animals and that a moderate rise in ROS levels in response to stress acts as a survival signal that lengthens a lifetime. For instance, it has been demonstrated that *Caenorhabditis elegans* and yeast live longer when more ROS are produced (Doonan et al., 2008, Mesquita et al., 2010). Reduced superoxide dismutase 2 (Sod2) upregulated oxidative DNA damage however did not speed up aging in heterozygous Sod2+/mutant mice. Mice lacking the antioxidant enzymes Sod2 and glutathione peroxidase-1 (Gpx-1) showed higher oxidative damage but not a shorter lifespan (Zhang et al., 2009).

Contrary to popular belief, mtDNA is not more susceptible to oxidative damage than nuclear DNA, despite being closer to the site where mitochondrial ROS are formed (Lim et al., 2005). Furthermore, mtDNA mutations that do not cause oxidative stress accelerate aging in mice. Together, these findings challenge the validity of the free radical hypothesis for aging by indicating that increased ROS generation or oxidative stress is not the primary or fundamental cause of aging but rather one of its effects. Future research is required to ascertain if bioenergetic shortage is the main source of mitochondrial dysfunction in many age-related diseases or whether increased ROS generation and oxidative stress are the underlying causes.

The continual fission and fusion processes, along with mitophagy that define the morphology of mitochondria and control their quantity, size, and function, make them highly dynamic structures.

Mitochondrial Dynamics

As shown in Figure 1 for mitochondria to be viable and to respond to changes in the bioenergetic status of the cell, mitochondrial dynamics must be present. Mitochondrial fission is essential for mitotic mitochondrial partition to daughter cells, mitochondrial distribution to subcellular sites, and mitophagy (Trotta and Chipuk 2017). Unchecked fission causes mtDNA depletion, mitochondrial fragmentation, loss of OXPHOS activity, and the creation of ROS, all of which are linked to metabolic inefficiency or disease. For the maintenance of mitochondrial membrane potential, ATP generation, and maximum respiratory capacity, mitochondria must fuse. Unopposed fusion creates a network of hypo-perfused mitochondria that produce more ATP, produce fewer ROS, are more resistant to metabolic shocks, and can fend off autophagy and death. Variations in metabolic demand and nutritional availability dictate how quickly mitochondrial fission and fusion occur, leading to fragmented or hypertubular mitochondria (Geto et al., 2020). While mitochondrial fusion is a two-step process involving the actions of mitofusins for the inner and outer membranes, respectively (i.e., optic atrophy 1 protein and MFN1 and MFN2), mitochondrial fission is facilitated by the action of dynamin-related protein (DRP1), which is drawn to mitochondria by a variety of receptors (e.g., Fission 1 protein (FIS1), mitochondrial fission factor (MFF), mitochondria (Song et al., 2009). It has been demonstrated that in mice, the lack of mitochondrial fission and fusion proteins results in embryonic mortality (Chen, Chomyn, and Chan, 2005). Over the past ten years, several studies have demonstrated the modulation of mitochondrial function and metabolism, mtDNA integrity, calcium homeostasis, autophagy, mitophagy, and apoptosis (Büeler 2010, Zhang et al., 2022, Osellame, Blacker, and Duchen 2012). Therefore, alterations in mitochondrial dynamics are closely associated with aging and several age-related disorders (Zhang et al., 2021). For instance, it has been shown that Mfn2 expression in skeletal muscle gradually decreases with age, which encourages mitochondrial malfunction and accounts for age-related changes in metabolic balance and sarcopenia (Sebastián et al., 2016). Obesity and type 2 diabetes mellitus in mice and humans have been associated with decreased expression of Mfn1 and/or Mfn2 in skeletal muscle (Dai and Jiang 2019). Additionally, it has been demonstrated that in mice, liver specific Mfn2 deletion leads to glucose intolerance and increased gluconeogenesis, which impairs insulin signaling and glucose homeostasis *in vivo* (Sebastián et al., 2012).

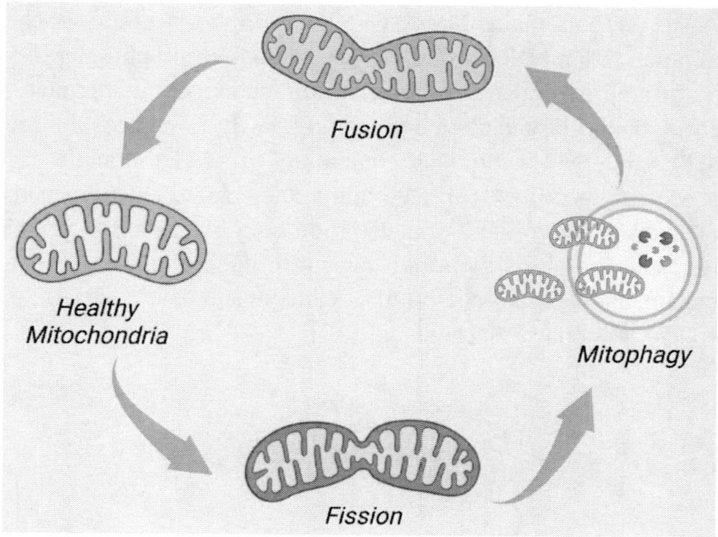

Figure 1. Showing the dynamic nature of mitochondria and their significance in neuronal plasticity.

Diseases affecting the heart and neurological system and muscle atrophy have been linked to variations in how mitochondria break apart and merge. Reduced Mfn1 and/or Mfn2 expression in rodents has been linked to atherosclerosis, restenosis, ventricular hypertrophy, and cardiomyopathy (Bonora et al., 2019, Ong et al., 2017). The absence of Mfn1 and Mfn2 in the heart has also been linked to myocardial infarction caused by acute ischemia/reperfusion injury (Hall et al., 2016). Lower levels of Opa1 have been associated with cardiomyopathy and heart failure in mice, whereas higher levels of Opa1 have been found to protect the heart and brain from ischemia injury (Le Page et al., 2016). Studies on animal models of Huntington's disease have shown that preventing Drp1-mediated mitochondrial fragmentation can help protect against long-term cardiac failure. Mitofusins can have various roles in response to acute and chronic cardiac damage (Le Page et al., 2016).

Mitophagy Process

Parkin Dependent

The PARK6 gene encodes the serine/threonine kinase PINK1, while the PARK2 gene encodes the E3 ubiquitin ligase Parkin. Initially discovered in autosomal recessive familial Parkinson's disease cases, mutations in the PINK1 and Parkin genes have been associated with abnormalities in mitochondria (Hattori et al., 1998, Lücking et al., 1998, Valente et al., 2002). This can be the underlying cause of several neurodegenerative diseases, including the death of dopaminergic neurons in Parkinson's disease (Valente et al., 2004, Exner et al., 2007, Morais et al., 2009). The PINK1/Parkin-dependent mitophagy pathway is critical for removing damaged mitochondria and ubiquitin-dependent mitophagy. PINK1 comprises several domains, including a mitochondrial targeting sequence (MTS), transmembrane domain, serine/threonine kinase domain, and C-terminal autoregulatory domain. PINK1 functions as a mitochondrial damage sensor, even under normal conditions, and is transported to healthy mitochondria via its MTS domain. The MTS directs PINK1 to the mitochondria, detecting mitochondrial depolarization or ROS buildup and initiating mitophagy. PINK1 is transported to mitochondria through several translocases in mitochondrial membranes under stress-free conditions. The mitochondrial membrane potential heavily influences the movement of MTS through the IMM, and the N-terminal containing the MTS domain is cut off by the mitochondrial processing peptidase (MPP) found in the matrix. The transmembrane domain of PINK1 is then cleaved by Presenilin-associated rhomboid-like protease, producing a shortened protein. This truncated protein is then translocated to the cytoplasm for degradation via the proteasomal mechanism (Meissner et al., 2015, Yamano and Youle, 2013). Figure 2 illustrates the parkin-dependent mitophagy mechanism.

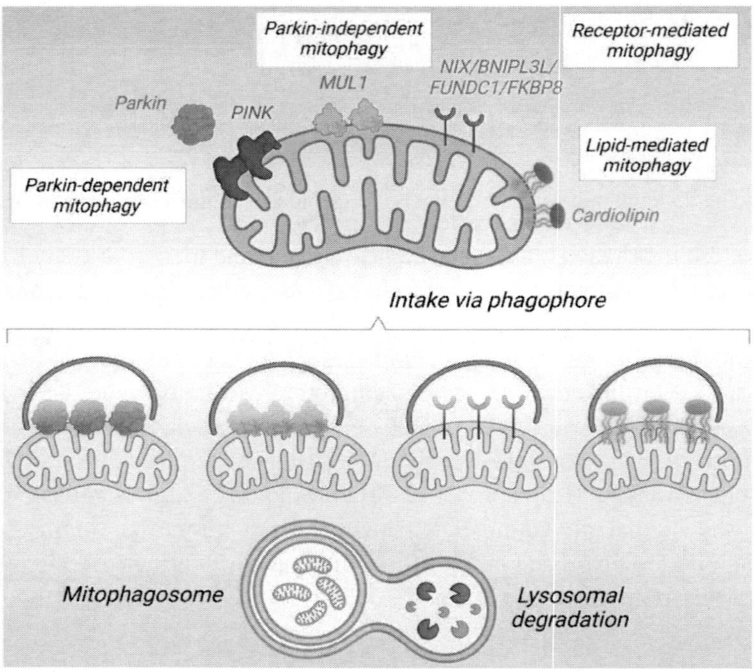

Figure 2. Showing different mitophagy pathways with a significant emphasis on the Parkin-dependent pathway.

Parkin Independent

Parkin-independent mitophagy refers to the mechanisms that allow mitophagy to occur even in the absence of Parkin. The main ways that it is mediated are through ubiquitin ligases, receptors, and lipids. The mechanism of parkin-independent mitophagy is illustrated in Figure 2.

Ubiquitin Ligase Mediated Mitophagy

Recent advances in research suggest that mitophagy is also regulated by receptors other than PINK1 and Parkin. In the study by Lazarou et al., (Lazarou et al., 2015), it was discovered that Parkin proteins are not necessary for PINK1 to draw OPTN and NDP52 to the injured mitochondria. Furthermore, it was proposed that Parkin catalyzes the PINK1-generated ubiquitin phosphorylation signals, which enable quick and reliable induction of mitophagy, rather than being necessary for mitophagy itself (Harper, Ordureau, and Heo 2018). To control the removal of damaged mitochondria, ubiquitin E3 ligases other than Parkin may also activate PINK1/Parkin-independent pathways (Fu et al., 2013, Yun et al., 2014).

Receptor-Mediated Mitophagy

Different mitochondrial receptors, including the FUN14 domain-containing protein 1 (FUNDC1), NIX/BNIP3L, Bcl-2/adenovirus E1B 19 kDa interacting protein 3 (BNIP3), and FK506 Binding Protein 8 (FKBP8), can also encourage parkin-independent mitophagy (Bhujabal et al., 2017, Kanki and Klionsky 2010, Okamoto, Kondo-Okamoto, and Ohsumi 2009). Skin fibroblasts from PD patients with mutations in PARK6 or PARK2 showed corrected mitophagy after NIX/BNIP3L overexpression. This demonstrates that PINK1/Parkin proteins are not involved in the mitophagy pathways mediated by NIX/BNIP3L (Koentjoro, Park, and Sue 2017). Further research has shown that hypoxia-inducible factor (HIF) and forkhead box O3 (FOXO3) regulate the transcription of NIX/BNIP3L and BNIP3L, suggesting that these proteins have a role in inducing hypoxia-induced mitophagy (Mammucari et al., 2007). Under neuronal stress, it has also been discovered that NIX/BNIP3L and BNIP3 are upregulated (Schweers et al., 2007). These proteins have LIR domains on their N-termini, facilitating their interaction with LC3 or GABARAP (Hanna et al., 2012, Rogov et al., 2017). Both proteins may bind to OMM via their C-termini. The specific mitophagy mechanism is mediated by NIX/BNIP3L and BNIP3, which are phosphorylated at Ser17 and Ser24 residues to increase their binding to LC3, which is yet unknown. Uncommonly expressed and localized in OMM, Bcl2-like 13 (BCL2L13) is a new mitophagy receptor that functions as yeast's Atg32 (Zhu et al., 2013).

Lipid-Mediated Mitophagy

Substantial evidence supports the idea that some lipids, like cardiolipin, can bind to LC3 and act as an elimination signal to encourage the recruitment of defective mitochondria for mitophagy in a PINK1/Parkin-independent way. A mitochondrial phospholipid called cardiolipin makes up around 20% of the lipids in IMM. Cardiolipin found in the IMM of healthy mitochondria, mediates the stabilization of cristae to improve the formation and operation of oxidative phosphorylation (OXPHOS) complexes. Cardiolipin responds to mitochondrial membrane depolarization similarly to Parkin, but differently. These proteins are activated in response to varying levels of mitochondrial membrane depolarization (Kagan et al., 2016). Cardiolipin serves as a mitophagy receptor when it is externalized from IMM to OMM. It attracts LC3 and the autophagosome machinery because it has the LC3 binding domain, facilitating mitophagy (Chu et al., 2013). Cardiolipin and LC3 interact primarily through the charged head group and negatively charged basic residues of LC3 and cardiolipin, respectively. As demonstrated in SH-SY5Y cells, phospholipid scramblase-3 in mitochondria controls the externalization of cardiolipin. However, it does so through mitochondrial nucleoside diphosphate kinase D. As shown in mouse embryonic fibroblasts; it is a hexameric complex between IMM and OMM (Kagan et al., 2016). Nothing else is necessary for the interaction between cardiolipin and LC3. Additionally, it has been proposed that the movement of cardiolipin from the IMM to the OMM is a marker of mitochondrial malfunction and facilitates mitophagy removal of these organelles (Chu, Bayır, and Kagan 2014). Cardiolipin synthase's role in mitophagy was further highlighted by the finding that RNAi suppression of the enzyme inhibited the recruitment of mitochondria to autophagosomes by translocating cardiolipin to the OMM (Kagan et al., 2016).

Role of Mitophagy in Aging: Regulating Autophagy in Different Organ Systems

Studies have shown that different organs, and perhaps even regions within an organ, can change at different rates as the human body ages. Since the detrimental effects of aging on each of our organs vary from person to person, we should focus more on calculating our "biological age," which records the impacts of aging on each of our organ systems, even though our body and its numerous organ systems are of identical age chronologically. However, there are several critical structural and functional changes that the primary organ system goes through, which we will briefly go over in this section.

Heart

The circulatory system revolves around the heart, and our hearts alter in structure as we age. Cardiomyocytes and sinoatrial pacemaker cells are two organ systems that experience cell loss as they age. This loss coincides with the deposited extracellular matrix, leading to fibrosis and a reduction in the strength and flexibility of the heart walls (Lu et al., 2011). Ventricular and arterial hypertrophy is brought on by this shift, which also causes adaptive remodeling of the heart anatomy. Together, these remodeling alterations significantly alter the heart's healthy electrical and muscular activities, raising the dangers of heart failure and arrhythmia. The vascular periphery also exhibits this extracellular matrix remodeling, which results in arterial stiffening and endothelial dysfunction and raises the risk of hypertension, ischemia, stroke, and heart disease. In addition to structural changes in the heart, aging affects the immune system in the organ, and the heart experiences chronic inflammation, similar to many other organs. It has been proposed that CD4 T lymphocytes and macrophages are the primary agents causing inflammation in the cardiac tissue (Lafuse, Wozniak, and Rajaram, 2020, Chen et al., 2018).

Brain

The brain and central nervous system are two essential organ systems related to aging. Although the precise reason for this decline in brain capacity is unknown, aging is connected. The most common explanation for this result involves the loss of neuronal cells or decreased neuronal volume. In addition, alterations in neuronal structure, such as synaptic pruning, dendritic sprouting, and myelin sheath degradation, have also been noted. These changes have been shown to impact cognition. At various ages, different parts of the brain shrink at varying rates. The cross-sectional whole brain, temporal lobe, and hippocampus volumes shrank with age, with an increased atrophy rate after age 70, according to longitudinal research (Peters 2006). In addition, several cross-sectional studies were examined, and the authors concluded that the prefrontal cortex was the part of the brain that aged the most. Aging causes cognitive impairment, degenerative disorders, and a decrease in brain capacity. Age-related changes to the blood-brain barrier and decreased cerebral blood flow are linked to white matter lesions, correlated with cognitive decline (Mokhber et al., 2021). Stroke and Alzheimer's disease are also linked to hypertension and changed small and large cerebro-vasculature (Faraco and Iadecola 2013). According to immunological theory, the inflammatory phenotype in the elderly brain increases (Chee and Solito 2021). Microglia are one important cell group thought to be involved in inflammation of the brain. In the research on aged mice, the innate immune cell known as microglia was found to be primed for greater activation and production of inflammatory markers. According to the hypothesis that elevated brain inflammation causes cognitive decline, cytokines like type 1 IFN have been found to have a deleterious influence on brain function, but IL-4 has been demonstrated to have a positive impact on cognitive performance in mice (Sartori et al., 2012). According to human research, peripheral blood CD28+ CD4 T cells in Alzheimer's patients grow more than those in healthy controls, causing low-grade inflammation. As a result, cognitive impairment may be linked to an elevated pro-inflammatory phenotype in the central nervous system, which might worsen any cognitive decline brought on by age-related structural changes (Jóźwik et al., 2012).

Musculoskeletal

The musculoskeletal system is a different system that is frequently connected to age. As we age, the muscles that control our mobility, strength, fragility, and the underlying skeletal system deteriorate structurally. Bones are intricate structures made up of various minerals, an organic matrix, a vasculature, and cells that change over a person's lifespan. As people age, their risk of fracture increases due to several factors, including increased brittleness brought on by changes in protein matrix and bone mineral density, structural weakening brought on by morphological changes, and an imbalance between bone formation, resorption, and repair. The skeletal system and muscles collaborate to enable motion and movement (Wohlgemuth et al., 2010).

Gastrointestinal

The gastrointestinal system is another important system that interacts heavily with the world outside the body. Its health and aging physiology are inextricably tied to the microbiome it contains, making it the most challenging system to investigate of those so far stated. There is general agreement that as people age, their microbiomes change, with older people living in their communities showing more subtle alterations and older people living in nursing homes showing more pronounced ones. However, as people age, their gut microbiomes become less diverse, and pathobionts proliferate more in the gastrointestinal tract. The gut environment, and the entire organism, are both significantly impacted by the changes in the microbiome. Indeed, a wide range of gastrointestinal dysfunction and disorders are linked to aging, including internal bleeding, loss of intestinal barrier integrity, decreased gut motility, colitis, ulcers, and malignancies. The ability of intestinal dysbiosis to cause inflammatory reactions in an organism is noteworthy for this review. One study showed that transferring the microbiome of old mice to young mice led to local and systemic inflammatory reactions in the recipients, indicating that gut dysbiosis might fuel chronic inflammation and result in an inflammation phenotype. However, studies on the human gut must consider whether their findings are inextricably tied to age, shifting microbiota, or both. While many disorders mentioned above are linked to aging, they are not directly related. Additionally, several other investigations found contradictory findings. For instance, some animal studies claimed that myenteric neurons decline with aging, while others claimed no appreciable alterations. It has not yet been established whether this is also seen in people. This is particularly true considering that studying aging in the gut is challenging since human aging is influenced by the complex combination of genetic, environmental, socioeconomic, and infectious variables (Soenen et al., 2016).

Lungs

The pathology of aging is also linked to the lungs. The lungs serve a crucial role as a physical and immunological barrier to the environment since they are one of the organ systems constantly exposed to environmental stimuli and allergens (Guttenberg, Vose, and Tighe 2021). The respiratory muscles' strength is impacted by age-associated muscular degeneration, which causes breathing problems and a weaker cough. During infection and sickness, it is especially harmful when the airways cannot efficiently evacuate air, fluids, and particles. This is worsened by the diminished capacity of the cilia to effectively transport mucus and foreign particles up and out of the airways. Structural changes in the average alveolus size and elasticity impact air exchange efficiency and susceptibility to alveolar damage during infections. Since the lungs are an organ exposed to the outside world, this also introduces us to another crucial consideration: how aging affects the immune system. To prevent long-term harm from the allergen/infectious agent or the body's immunological response, the immune system must mount a delicately balanced response to the challenge after exposure to environmental allergens or infections (Chaplin 2010).

Synthetic and Natural Compounds to Modulate Mitophagy during Aging

Selective stimulation of defective organelles may harm the physiology of cells and tissues because a wide range of human illnesses are associated with abnormal mitochondrial aggregation and poor energy metabolism. Mitophagy control of mitochondrial homeostasis provides optimism for developing new therapeutic modalities to handle various age-related ailments. To control mitochondrial elimination, increase cellular viability, and improve life expectancy as humans age, a range of synthetic and/or natural substances have been proposed. By changing mitophagy and restoring mitochondrial function, sirtuin-activating compounds (STACs), NAD^+ precursors (such as nicotinamide mononucleotide and nicotinamide riboside), and resveratrol, among others, have been proven to lessen the disorders connected with aging

(Feige et al., 2008, Yoshino et al., 2011, Mouchiroud et al., 2013). Additionally, NAD supplementation maintains mitochondrial metabolism, guards against several neurodegenerative symptoms, and extends lifespan in progeria-prone worms and mice. Although it has been claimed that resveratrol and other sirtuin activators positively affect cellular metabolism and age-related decline, several studies have questioned the substances' longevity-extending abilities. Antibiotics significantly negatively impact mitochondrial homeostasis, according to the postulated endosymbiotic theory. According to recent research, antibiotics like actinonin and doxycycline affect how energy is metabolized and how mammalian cells undergo mitophagy (Richter et al., 2016). To maintain mitochondrial homeostasis and ward off disorders that shorten life expectancy, it may be advantageous to use antibiotics in moderation (Sun et al., 2015, Xing et al., 2017). Spermidine and urolithin A are naturally occurring substances that retain mitochondrial function and lengthen lifespan by inducing mitophagy. It has been shown that spermidine supplementation encourages lifetime extension in various model organisms, including yeast, worms, flies, and mice (Ryu et al., 2016). In spermidine-mediated longevity, autophagy plays a crucial role. Notably, spermidine ingestion promotes autophagy, whereas genetic or pharmacological suppression of the autophagic pathway eliminates its positive effects on animals' physiology as they age (Tong and Hill 2017). Additionally, giving spermidine to aged animals improves cardiovascular diseases and age-related cardiac changes such as arterial stiffness, diastolic dysfunction, arterial hypertrophy, and cardiac activity (Eisenberg et al., 2016). It is interesting to note that spermidine's cardio-protective effects are mediated through mitophagy activation and the restoration of the energy balance of cardiomyocytes. Urolithin A is an additional chemical that induces mitophagy that is first in class. When consumed, the pomegranate seed metabolite urolithin A, which is generated from ellagitannins, triggers mitochondrial elimination (Al-Harbi et al., 2021). The significant indicators of urolithin A's potential therapeutic effectiveness are the general changes in mitochondrial metabolism that lead to higher exercise capacity, elevated muscle function, and lifespan extension in both nematodes and rats. It is interesting to note that long-term urolithin A administration results in both SKN-1-dependent lifetime extension and improved mitochondrial biogenesis, which suggests that urolithin A may coordinate mitophagy and mitochondrial biogenesis to maintain energy balance (Palikaras, Lionaki, and Tavernarakis 2015). Tomatadine, a naturally occurring compound in unripe tomatoes, increases mitochondrial clearance in addition to urolithin A, lengthening lifespan and enhancing muscle function in nematodes and mice. Treatment with tomatidine mediates mitohormesis, resulting in increased ROS levels and activation of SKN-1/NRF2. Thus, coordinated mitophagy and mitochondrial biogenesis help maintain tissue homeostasis and human survival (Fang et al., 2017).

Conclusion

Since mitochondria play a crucial role in both cellular life and death, it follows that when cells act under stress, mitochondrial homeostasis needs to be carefully monitored and adjusted. Multiple systems regulate mitochondrial quantity and quality management, originating from the fundamental mechanism of mitophagy. According to research, there is much communication across the various mitophagy pathways, suggesting they can work together to meet environmental difficulties. However, additional research is required to determine the precise mechanisms connecting the many routes in the intricate network of mitophagy management. One of the critical elements of aging and age-related illness is mitochondrial dysfunction. According to mounting evidence, mitophagy—which involves eliminating damaged mitochondria—plays a crucial role in reversing age-related clinical diseases. Thus, boosting mitophagy while chronically stimulating mitochondrial turnover is a viable strategy to postpone age-related disorders and to lengthen health and lifespan.

References

Al-Harbi, S. A., A. O. Abdulrahman, M. A. Zamzami, and M. I. Khan. (2021). "Urolithins: The Gut Based Polyphenol Metabolites of Ellagitannins in Cancer Prevention, a Review." *Front Nutr* 8:647582.

Amorim, J. A., G. Coppotelli, A. P. Rolo, C. M. Palmeira, J. M. Ross, and D. A. Sinclair. (2022). "Mitochondrial and metabolic dysfunction in ageing and age-related diseases." *Nat Rev Endocrinol* 18 (4):243-258.

Asensi, M., J. Sastre, F. V. Pallardo, A. Lloret, M. Lehner, J. Garcia-de-la Asuncion, and J. Viña. (1999). "Ratio of reduced to oxidized glutathione as indicator of oxidative stress status and DNA damage." *Methods Enzymol* 299:267-76. doi: 10.1016/s0076-6879(99)99026-2.

Balaban, R. S., S. Nemoto, and T. Finkel. (2005). "Mitochondria, oxidants, and aging." *Cell* 120 (4):483-95.

Bhujabal, Zambarlal, Åsa B. Birgisdottir, Eva Sjøttem, Hanne B. Brenne, Aud Øvervatn, Sabrina Habisov, Vladimir Kirkin, Trond Lamark, and Terje Johansen. (2017). "FKBP8 recruits LC3A to mediate Parkin-independent mitophagy." *EMBO reports* 18 (6):947-961.

Bonora, M., M. R. Wieckowski, D. A. Sinclair, G. Kroemer, P. Pinton, and L. Galluzzi. (2019). "Targeting mitochondria for cardiovascular disorders: therapeutic potential and obstacles." *Nat Rev Cardiol* 16 (1):33-55.

Brand, M. D. (2010). "The sites and topology of mitochondrial superoxide production." *Exp Gerontol* 45 (7-8):466-72.

Bua, E., J. Johnson, A. Herbst, B. Delong, D. McKenzie, S. Salamat, and J. M. Aiken. (2006). "Mitochondrial DNA-deletion mutations accumulate intracellularly to detrimental levels in aged human skeletal muscle fibers." *Am J Hum Genet* 79 (3):469-80.

Büeler, H. 2010. "Mitochondrial dynamics, cell death and the pathogenesis of Parkinson's disease." *Apoptosis* 15 (11):1336-53.

Chaplin, D. D. (2010). "Overview of the immune response." *J Allergy Clin Immunol* 125 (2 Suppl 2):S3-23.

Chee, S. E. J., and E. Solito. (2021). "The Impact of Ageing on the CNS Immune Response in Alzheimer's Disease." *Front Immunol* 12:738511.

Chen, Hsiuchen, Anne Chomyn, and David C. Chan. (2005). "Disruption of Fusion Results in Mitochondrial Heterogeneity and Dysfunction." *Journal of Biological Chemistry* 280 (28):26185-26192.

Chen, L., H. Deng, H. Cui, J. Fang, Z. Zuo, J. Deng, Y. Li, X. Wang, and L. Zhao. (2018). "Inflammatory responses and inflammation-associated diseases in organs." *Oncotarget* 9 (6):7204-7218.

Chen, Lulu, Renlei Yang, Wanxin Qiao, Wei Zhang, Jie Chen, Li Mao, David Goltzman, and Dengshun Miao. (2019). "1,25-Dihydroxyvitamin D exerts an antiaging role by activation of Nrf2-antioxidant signaling and inactivation of p16/p53-senescence signaling." *Aging Cell* 18 (3):e12951.

Cho, J., J. H. Hur, and D. W. Walker. (2011). "The role of mitochondria in Drosophila aging." *Exp Gerontol* 46 (5):331-4.

Christensen, K., T. E. Johnson, and J. W. Vaupel. (2006). "The quest for genetic determinants of human longevity: challenges and insights." *Nat Rev Genet* 7 (6):436-48.

Chu, Charleen T., Hülya Bayır, and Valerian E. Kagan. (2014). "LC3 binds externalized cardiolipin on injured mitochondria to signal mitophagy in neurons." *Autophagy* 10 (2):376-378.

Chu, Charleen T., Jing Ji, Ruben K. Dagda, Jian Fei Jiang, Yulia Y. Tyurina, Alexandr A. Kapralov, Vladimir A. Tyurin, Naveena Yanamala, Indira H. Shrivastava, Dariush Mohammadyani, Kent Zhi Qiang Wang, Jianhui Zhu, Judith Klein-Seetharaman, Krishnakumar Balasubramanian, Andrew A. Amoscato, Grigory Borisenko, Zhentai Huang, Aaron M. Gusdon, Amin Cheikhi, Erin K. Steer, Ruth Wang, Catherine Baty,

Simon Watkins, Ivet Bahar, Hülya Bayır, and Valerian E. Kagan. (2013). "Cardiolipin externalization to the outer mitochondrial membrane acts as an elimination signal for mitophagy in neuronal cells." *Nature Cell Biology* 15 (10):1197-1205.

Cottrell, D. A., E. L. Blakely, M. A. Johnson, P. G. Ince, G. M. Borthwick, and D. M. Turnbull. (2001). "Cytochrome c oxidase deficient cells accumulate in the hippocampus and choroid plexus with age." *Neurobiol Aging* 22 (2):265-72.

Cui, Hui, Duozhuang Tang, George B. Garside, Ting Zeng, Yiting Wang, Zhendong Tao, Liu Zhang, and Si Tao. (2019). "Wnt Signaling Mediates the Aging-Induced Differentiation Impairment of Intestinal Stem Cells." *Stem Cell Reviews and Reports* 15 (3):448-455.

da Costa, J. P., R. Vitorino, G. M. Silva, C. Vogel, A. C. Duarte, and T. Rocha-Santos. (2016). "A synopsis on aging-Theories, mechanisms and future prospects." *Ageing Res Rev* 29:90-112.

Dai, W., and L. Jiang. (2019). "Dysregulated Mitochondrial Dynamics and Metabolism in Obesity, Diabetes, and Cancer." *Front Endocrinol (Lausanne)* 10:570.

DiDonato, Joseph A., Frank Mercurio, and Michael Karin. (2012). "NF-κB and the link between inflammation and cancer." *Immunological Reviews* 246 (1):379-400.

DiMauro, S., and E. A. Schon. (2003). "Mitochondrial respiratory-chain diseases." *N Engl J Med* 348 (26):2656-68.

Ding, Wen-Xing, and Xiao-Ming Yin. (2012). "Mitophagy: mechanisms, pathophysiological roles, and analysis." *Biological chemistry* 393 (7):547-564.

Doonan, R., J. J. McElwee, F. Matthijssens, G. A. Walker, K. Houthoofd, P. Back, A. Matscheski, J. R. Vanfleteren, and D. Gems. (2008). "Against the oxidative damage theory of aging: superoxide dismutases protect against oxidative stress but have little or no effect on life span in Caenorhabditis elegans." *Genes Dev* 22 (23):3236-41.

Du, W. W., W. Yang, L. Fang, J. Xuan, H. Li, A. Khorshidi, S. Gupta, X. Li, and B. B. Yang. (2014). "miR-17 extends mouse lifespan by inhibiting senescence signaling mediated by MKP7." *Cell Death Dis* 5 (7):e1355.

Eisenberg, T., M. Abdellatif, S. Schroeder, U. Primessnig, S. Stekovic, T. Pendl, A. Harger, J. Schipke, A. Zimmermann, A. Schmidt, M. Tong, C. Ruckenstuhl, C. Dammbrueck, A. S. Gross, V. Herbst, C. Magnes, G. Trausinger, S. Narath, A. Meinitzer, Z. Hu, A. Kirsch, K. Eller, D. Carmona-Gutierrez, S. Büttner, F. Pietrocola, O. Knittelfelder, E. Schrepfer, P. Rockenfeller, C. Simonini, A. Rahn, M. Horsch, K. Moreth, J. Beckers, H. Fuchs, V. Gailus-Durner, F. Neff, D. Janik, B. Rathkolb, J. Rozman, M. H. de Angelis, T. Moustafa, G. Haemmerle, M. Mayr, P. Willeit, M. von Frieling-Salewsky, B. Pieske, L. Scorrano, T. Pieber, R. Pechlaner, J. Willeit, S. J. Sigrist, W. A. Linke, C. Mühlfeld, J. Sadoshima, J. Dengjel, S. Kiechl, G. Kroemer, S. Sedej, and F. Madeo. (2016). "Cardioprotection and lifespan extension by the natural polyamine spermidine." *Nat Med* 22 (12):1428-1438.

Epel, E. S., E. H. Blackburn, J. Lin, F. S. Dhabhar, N. E. Adler, J. D. Morrow, and R. M. Cawthon. (2004). "Accelerated telomere shortening in response to life stress." *Proc Natl Acad Sci U S A* 101 (49):17312-5.

Exner, N., B. Treske, D. Paquet, K. Holmström, C. Schiesling, S. Gispert, I. Carballo-Carbajal, D. Berg, H. H. Hoepken, T. Gasser, R. Krüger, K. F. Winklhofer, F. Vogel, A. S. Reichert, G. Auburger, P. J. Kahle, B. Schmid, and C. Haass. 2007. "Loss-of-function of human PINK1 results in mitochondrial pathology and can be rescued by parkin." *J Neurosci* 27 (45):12413-8.

Fan, J., X. An, Y. Yang, H. Xu, L. Fan, L. Deng, T. Li, X. Weng, J. Zhang, and R. Chunhua Zhao. 2018. "MiR-1292 Targets FZD4 to Regulate Senescence and Osteogenic Differentiation of Stem Cells in TE/SJ/Mesenchymal Tissue System via the Wnt/β-catenin Pathway." *Aging Dis* 9 (6):1103-1121.

Fang, E. F., T. B. Waltz, H. Kassahun, Q. Lu, J. S. Kerr, M. Morevati, E. M. Fivenson, B. N. Wollman, K. Marosi, M. A. Wilson, W. B. Iser, D. M. Eckley, Y. Zhang, E. Lehrmann, I. G. Goldberg, M. Scheibye-Knudsen, M. P. Mattson, H. Nilsen, V. A. Bohr, and K. G. Becker. 2017. "Tomatidine enhances lifespan and healthspan in C. elegans through mitophagy induction via the SKN-1/Nrf2 pathway." *Sci Rep* 7:46208.

Faraco, G., and C. Iadecola. 2013. "Hypertension: a harbinger of stroke and dementia." *Hypertension* 62 (5):810-7.

Feige, Jérôme N., Marie Lagouge, Carles Canto, Axelle Strehle, Sander M. Houten, Jill C. Milne, Philip D. Lambert, Chikage Mataki, Peter J. Elliott, and Johan Auwerx. (2008). "Specific SIRT1 Activation Mimics

Low Energy Levels and Protects against Diet-Induced Metabolic Disorders by Enhancing Fat Oxidation." *Cell Metabolism* 8 (5):347-358.

Fraga, C. G., M. K. Shigenaga, J. W. Park, P. Degan, and B. N. Ames. (1990). "Oxidative damage to DNA during aging: 8-hydroxy-2'-deoxyguanosine in rat organ DNA and urine." *Proc Natl Acad Sci U S A* 87 (12):4533-7.

Franceschi, C., P. Garagnani, C. Morsiani, M. Conte, A. Santoro, A. Grignolio, D. Monti, M. Capri, and S. Salvioli. 2018. "The Continuum of Aging and Age-Related Diseases: Common Mechanisms but Different Rates." *Front Med (Lausanne)* 5:61.

Fu, M., P. St-Pierre, J. Shankar, P. T. Wang, B. Joshi, and I. R. Nabi. 2013. "Regulation of mitophagy by the Gp78 E3 ubiquitin ligase." *Mol Biol Cell* 24 (8):1153-62. doi: 10.1091/mbc.E12-08-0607.

Geto, Z., M. D. Molla, F. Challa, Y. Belay, and T. Getahun. 2020. "Mitochondrial Dynamic Dysfunction as a Main Triggering Factor for Inflammation Associated Chronic Non-Communicable Diseases." *J Inflamm Res* 13:97-107.

Grimm, A., and A. Eckert. 2017. "Brain aging and neurodegeneration: from a mitochondrial point of view." *J Neurochem* 143 (4):418-431.

Guo, C., L. Sun, X. Chen, and D. Zhang. 2013. "Oxidative stress, mitochondrial damage and neurodegenerative diseases." *Neural Regen Res* 8 (21):2003-14.

Guttenberg, M. A., A. T. Vose, and R. M. Tighe. 2021. "Role of Innate Immune System in Environmental Lung Diseases." *Curr Allergy Asthma Rep* 21 (5):34. -0.

Hall, A. R., N. Burke, R. K. Dongworth, S. B. Kalkhoran, A. Dyson, J. M. Vicencio, G. W. Dorn, II, D. M. Yellon, and D. J. Hausenloy. 2016. "Hearts deficient in both Mfn1 and Mfn2 are protected against acute myocardial infarction." *Cell Death Dis* 7 (5):e2238.

Hanna, Rita A., Melissa N. Quinsay, Amabel M. Orogo, Kayla Giang, Shivaji Rikka, and Åsa B. Gustafsson. 2012. "Microtubule-associated Protein 1 Light Chain 3 (LC3) Interacts with Bnip3 Protein to Selectively Remove Endoplasmic Reticulum and Mitochondria via Autophagy. *Journal of Biological Chemistry* 287 (23):19094-19104.

Harman, D. 1956. "Aging: a theory based on free radical and radiation chemistry." *J Gerontol* 11 (3):298-300.

Harman, D. 1972. "The biologic clock: the mitochondria?" *J Am Geriatr Soc* 20 (4):145-7.

Harper, J. W., A. Ordureau, and J. M. Heo. 2018. "Building and decoding ubiquitin chains for mitophagy." *Nat Rev Mol Cell Biol* 19 (2):93-108.

Hattori, N., T. Kitada, H. Matsumine, S. Asakawa, Y. Yamamura, H. Yoshino, T. Kobayashi, M. Yokochi, M. Wang, A. Yoritaka, T. Kondo, S. Kuzuhara, S. Nakamura, N. Shimizu, and Y. Mizuno. 1998. "Molecular genetic analysis of a novel Parkin gene in Japanese families with autosomal recessive juvenile parkinsonism: evidence for variable homozygous deletions in the Parkin gene in affected individuals." *Ann Neurol* 44 (6):935-41.

Ho, Shang-Tse, Yueh-Tung Hsieh, Sheng-Yao Wang, and Ming-Ju Chen. 2019. "Improving effect of a probiotic mixture on memory and learning abilities in d-galactose-treated aging mice." *Journal of Dairy Science* 102 (3):1901-1909.

Huang, Dong-Dong, Sheng-Dong Fan, Xi-Yi Chen, Xia-Lin Yan, Xian-Zhong Zhang, Bing-Wei Ma, Ding-Ye Yu, Wen-Yu Xiao, Cheng-Le Zhuang, and Zhen Yu. 2019. "Nrf2 deficiency exacerbates frailty and sarcopenia by impairing skeletal muscle mitochondrial biogenesis and dynamics in an age-dependent manner." *Experimental Gerontology* 119:61-73.

Hughes, J. P., S. Rees, S. B. Kalindjian, and K. L. Philpott. 2011. "Principles of early drug discovery." *Br J Pharmacol* 162 (6):1239-49.

Hunot, Stephane, Bernard Brugg, Damien Ricard, Patrick P. Michel, Marie-Paule Muriel, Merle Ruberg, Baptiste A. Faucheux, Yves Agid, and Etienne C. Hirsch. 1997. "Nuclear translocation of NF-kB is increased in dopaminergic neurons of patients with Parkinson disease." *Proceedings of the National Academy of Sciences* 94 (14):7531-7536.

Joshi, A. U., A. E. Ebert, B. Haileselassie, and D. Mochly-Rosen. 2019. "Drp1/Fis1-mediated mitochondrial fragmentation leads to lysosomal dysfunction in cardiac models of Huntington's disease." *J Mol Cell Cardiol* 127:125-133.

Jóźwik, A., J. Landowski, L. Bidzan, T. Fülop, E. Bryl, and J. M. Witkowski. 2012. "Beta-amyloid peptides enhance the proliferative response of activated CD4CD28 lymphocytes from Alzheimer disease patients and from healthy elderly." *PLoS One* 7 (3):e33276.

Kagan, V. E., J. Jiang, Z. Huang, Y. Y. Tyurina, C. Desbourdes, C. Cottet-Rousselle, H. H. Dar, M. Verma, V. A. Tyurin, A. A. Kapralov, A. Cheikhi, G. Mao, D. Stolz, C. M. St. Croix, S. Watkins, Z. Shen, Y. Li, M. L. Greenberg, M. Tokarska-Schlattner, M. Boissan, M. L. Lacombe, R. M. Epand, C. T. Chu, R. K. Mallampalli, H. Bayır, and U. Schlattner. 2016. "NDPK-D (NM23-H4)-mediated externalization of cardiolipin enables elimination of depolarized mitochondria by mitophagy." *Cell Death & Differentiation* 23 (7):1140-1151.

Kanki, Tomotake, and Daniel J. Klionsky. 2010. "The molecular mechanism of mitochondria autophagy in yeast." *Molecular Microbiology* 75 (4):795-800.

Khrapko, K., and J. Vijg. 2007. "Mitochondrial DNA mutations and aging: a case closed?" *Nat Genet* 39 (4):445-6.

Kitaoka, Yu, Yuki Tamura, Kenya Takahashi, Kohei Takeda, Tohru Takemasa, and Hideo Hatta. 2019. "Effects of Nrf2 deficiency on mitochondrial oxidative stress in aged skeletal muscle." *Physiological Reports* 7 (3):e13998.

Koentjoro, Brianada, Jin-Sung Park, and Carolyn M. Sue. 2017. "Nix restores mitophagy and mitochondrial function to protect against PINK1/Parkin-related Parkinson's disease." *Scientific Reports* 7 (1):44373.

Kraytsberg, Y., E. Kudryavtseva, A. C. McKee, C. Geula, N. W. Kowall, and K. Khrapko. 2006. "Mitochondrial DNA deletions are abundant and cause functional impairment in aged human substantia nigra neurons." *Nat Genet* 38 (5):518-20.

Lafuse, W. P., D. J. Wozniak, and M. V. S. Rajaram. 2020. "Role of Cardiac Macrophages on Cardiac Inflammation, Fibrosis and Tissue Repair." *Cells* 10 (1).

Lamming, Dudley W., Lan Ye, David M. Sabatini, and Joseph A. Baur. 2013. "Rapalogs and mTOR inhibitors as anti-aging therapeutics." *The Journal of Clinical Investigation* 123 (3):980-989.

Lazarou, M., D. A. Sliter, L. A. Kane, S. A. Sarraf, C. Wang, J. L. Burman, D. P. Sideris, A. I. Fogel, and R. J. Youle. 2015. "The ubiquitin kinase PINK1 recruits autophagy receptors to induce mitophagy." *Nature* 524 (7565):309-314.

Le Page, S., M. Niro, J. Fauconnier, L. Cellier, S. Tamareille, A. Gharib, A. Chevrollier, L. Loufrani, C. Grenier, R. Kamel, E. Sarzi, A. Lacampagne, M. Ovize, D. Henrion, P. Reynier, G. Lenaers, D. Mirebeau-Prunier, and F. Prunier. 2016. "Increase in Cardiac Ischemia-Reperfusion Injuries in Opa1+/- Mouse Model." *PLoS One* 11 (10):e0164066.

Lim, K. S., K. Jeyaseelan, M. Whiteman, A. Jenner, and B. Halliwell. 2005. "Oxidative damage in mitochondrial DNA is not extensive." *Ann N Y Acad Sci* 1042:210-20.

Linnane, A. W., S. Marzuki, T. Ozawa, and M. Tanaka. 1989. "Mitochondrial DNA mutations as an important contributor to ageing and degenerative diseases." *Lancet* 1 (8639):642-5.

Liu, Y., W. Weng, R. Gao, and Y. Liu. 2019. "New Insights for Cellular and Molecular Mechanisms of Aging and Aging-Related Diseases: Herbal Medicine as Potential Therapeutic Approach." *Oxid Med Cell Longev* 2019:4598167.

Lu, P., K. Takai, V. M. Weaver, and Z. Werb. 2011. "Extracellular matrix degradation and remodeling in development and disease." *Cold Spring Harb Perspect Biol* 3 (12).

Lücking, C. B., N. Abbas, A. Dürr, V. Bonifati, A. M. Bonnet, T. de Broucker, G. De Michele, N. W. Wood, Y. Agid, and A. Brice. 1998. "Homozygous deletions in parkin gene in European and North African families with autosomal recessive juvenile parkinsonism. The European Consortium on Genetic Susceptibility in Parkinson's Disease and the French Parkinson's Disease Genetics Study Group." *Lancet* 352 (9137):1355-6.

Lyu, G., Y. Guan, C. Zhang, L. Zong, L. Sun, X. Huang, L. Huang, L. Zhang, X. L. Tian, Z. Zhou, and W. Tao. 2018. "TGF-β signaling alters H4K20me3 status via miR-29 and contributes to cellular senescence and cardiac aging." *Nat Commun* 9 (1):2560.

Mallikarjun, Venkatesh, and Joe Swift. 2016. "Therapeutic Manipulation of Ageing: Repurposing Old Dogs and Discovering New Tricks." *EBioMedicine* 14.

Mammucari, Cristina, Giulia Milan, Vanina Romanello, Eva Masiero, Ruediger Rudolf, Paola Del Piccolo, Steven J. Burden, Raffaella Di Lisi, Claudia Sandri, Jinghui Zhao, Alfred L. Goldberg, Stefano Schiaffino, and Marco Sandri. 2007. "FoxO3 Controls Autophagy in Skeletal Muscle In Vivo." *Cell Metabolism* 6 (6):458-471.

Meissner, C., H. Lorenz, B. Hehn, and M. K. Lemberg. 2015. "Intramembrane protease PARL defines a negative regulator of PINK1- and PARK2/Parkin-dependent mitophagy." *Autophagy* 11 (9):1484-98.

Meng, Xiaoyin, Hongguang Chen, Guolin Wang, Yonghao Yu, and Keliang Xie. 2015. "Hydrogen-rich saline attenuates chemotherapy-induced ovarian injury via regulation of oxidative stress." *Exp Ther Med* 10 (6):2277-2282.

Mesquita, A., M. Weinberger, A. Silva, B. Sampaio-Marques, B. Almeida, C. Leão, V. Costa, F. Rodrigues, W. C. Burhans, and P. Ludovico. 2010. "Caloric restriction or catalase inactivation extends yeast chronological lifespan by inducing H2O2 and superoxide dismutase activity." *Proc Natl Acad Sci U S A* 107 (34):15123-8.

Mokhber, N., A. Shariatzadeh, A. Avan, H. Saber, G. S. Babaei, G. Chaimowitz, and M. R. Azarpazhooh. 2021. "Cerebral blood flow changes during aging process and in cognitive disorders: A review." *Neuroradiol J* 34 (4):300-307.

Morais, V. A., P. Verstreken, A. Roethig, J. Smet, A. Snellinx, M. Vanbrabant, D. Haddad, C. Frezza, W. Mandemakers, D. Vogt-Weisenhorn, R. Van Coster, W. Wurst, L. Scorrano, and B. De Strooper. 2009. "Parkinson's disease mutations in PINK1 result in decreased Complex I activity and deficient synaptic function." *EMBO Mol Med* 1 (2):99-111.

Mouchiroud, Laurent, Riekelt H Houtkooper, Norman Moullan, Elena Katsyuba, Dongryeol Ryu, Carles Cantó, Adrienne Mottis, Young-Suk Jo, Mohan Viswanathan, Kristina Schoonjans, Leonard Guarente, and Johan Auwerx. 2013. "The NAD+/Sirtuin Pathway Modulates Longevity through Activation of Mitochondrial UPR and FOXO Signaling." *Cell* 154 (2):430-441.

Mrakovcic, M., and L. F. Fröhlich. 2018. "p53-Mediated Molecular Control of Autophagy in Tumor Cells." *Biomolecules* 8 (2).

Müller-Höcker, J. 1989. "Cytochrome-c-oxidase deficient cardiomyocytes in the human heart--an age-related phenomenon. A histochemical ultracytochemical study." *Am J Pathol* 134 (5):1167-73.

Müller-Höcker, J. 1990. "Cytochrome c oxidase deficient fibres in the limb muscle and diaphragm of man without muscular disease: an age-related alteration." *J Neurol Sci* 100 (1-2):14-21.

Münscher, C., J. Müller-Höcker, and B. Kadenbach. 1993. "Human aging is associated with various point mutations in tRNA genes of mitochondrial DNA." *Biol Chem Hoppe Seyler* 374 (12):1099-104.

Niki, E. 2009. "Lipid peroxidation: physiological levels and dual biological effects." *Free Radic Biol Med* 47 (5):469-84.

Nita, M., and A. Grzybowski. 2016. "The Role of the Reactive Oxygen Species and Oxidative Stress in the Pathomechanism of the Age-Related Ocular Diseases and Other Pathologies of the Anterior and Posterior Eye Segments in Adults." *Oxid Med Cell Longev* 2016:3164734.

O'Neill, L. A. J., and C. Kaltschmidt. 1997. "NF-kB: a crucial transcription factor for glial and neuronal cell function." *Trends in Neurosciences* 20 (6):252-258. doi: 10.1016/S0166-2236(96)01035-1.

Okamoto, Koji, Noriko Kondo-Okamoto, and Yoshinori Ohsumi. 2009. "Mitochondria-Anchored Receptor Atg32 Mediates Degradation of Mitochondria via Selective Autophagy." *Developmental Cell* 17 (1):87-97.

Ong, S. B., S. B. Kalkhoran, S. Hernández-Reséndiz, P. Samangouei, S. G. Ong, and D. J. Hausenloy. 2017. "Mitochondrial-Shaping Proteins in Cardiac Health and Disease - the Long and the Short of It!" *Cardiovasc Drugs Ther* 31 (1):87-107.

Osellame, Laura D., Thomas S. Blacker, and Michael R. Duchen. 2012. "Cellular and molecular mechanisms of mitochondrial function." *Best Practice & Research Clinical Endocrinology & Metabolism* 26 (6):711-723.

Overhoff, M. G., J. C. Garbe, J. Koh, M. R. Stampfer, D. H. Beach, and C. L. Bishop. 2014. "Cellular senescence mediated by p16INK4A-coupled miRNA pathways." *Nucleic Acids Res* 42 (3):1606-18.

Palikaras, K., E. Lionaki, and N. Tavernarakis. 2015. "Balancing mitochondrial biogenesis and mitophagy to maintain energy metabolism homeostasis." *Cell Death Differ* 22 (9):1399-401.

Peters, R. 2006. "Ageing and the brain." *Postgrad Med J* 82 (964):84-8.

Przybysz, A. J., K. P. Choe, L. J. Roberts, and K. Strange. 2009. "Increased age reduces DAF-16 and SKN-1 signaling and the hormetic response of Caenorhabditis elegans to the xenobiotic juglone." *Mech Ageing Dev* 130 (6):357-69.

Richmond C 2005. "Sir Richard Doll." *BMJ*. 30;331(7511):295. PMCID: PMC1181278.

Richter, Benjamin, Danielle A. Sliter, Lina Herhaus, Alexandra Stolz, Chunxin Wang, Petra Beli, Gabriele Zaffagnini, Philipp Wild, Sascha Martens, Sebastian A. Wagner, Richard J. Youle, and Ivan Dikic. 2016. "Phosphorylation of OPTN by TBK1 enhances its binding to Ub chains and promotes selective autophagy of damaged mitochondria." *Proceedings of the National Academy of Sciences* 113 (15):4039-4044.

Rogov, Vladimir V., Hironori Suzuki, Mija Marinković, Verena Lang, Ryuichi Kato, Masato Kawasaki, Maja Buljubašić, Matilda Šprung, Natalia Rogova, Soichi Wakatsuki, Anne Hamacher-Brady, Volker Dötsch, Ivan Dikic, Nathan R. Brady, and Ivana Novak. 2017. "Phosphorylation of the mitochondrial autophagy receptor Nix enhances its interaction with LC3 proteins." *Scientific Reports* 7 (1):1131.

Ryu, D., L. Mouchiroud, P. A. Andreux, E. Katsyuba, N. Moullan, A. A. Nicolet-Dit-Félix, E. G. Williams, P. Jha, G. Lo Sasso, D. Huzard, P. Aebischer, C. Sandi, C. Rinsch, and J. Auwerx. 2016. "Urolithin A induces mitophagy and prolongs lifespan in C. elegans and increases muscle function in rodents." *Nat Med* 22 (8):879-88.

Sahin, Kazim, Cemal Orhan, Mehmet Tuzcu, Hakki Tastan, Birdal Bilir, Nurhan Sahin, Deniz Aslar Oner, and Omer Kucuk. 2019. "Tomato Powder Modulates NF-κB, mTOR, and Nrf2 Pathways during Aging in Healthy Rats." *Journal of Aging Research* 2019:1643243.

Sartori, A. C., D. E. Vance, L. Z. Slater, and M. Crowe. 2012. "The impact of inflammation on cognitive function in older adults: implications for healthcare practice and research." *J Neurosci Nurs* 44 (4):206-17.

Schweers, R. L., J. Zhang, M. S. Randall, M. R. Loyd, W. Li, F. C. Dorsey, M. Kundu, J. T. Opferman, J. L. Cleveland, J. L. Miller, and P. A. Ney. 2007. "NIX is required for programmed mitochondrial clearance during reticulocyte maturation." *Proc Natl Acad Sci U S A* 104 (49):19500-5.

Sebastián, D., M. I. Hernández-Alvarez, J. Segalés, E. Sorianello, J. P. Muñoz, D. Sala, A. Waget, M. Liesa, J. C. Paz, P. Gopalacharyulu, M. Orešič, S. Pich, R. Burcelin, M. Palacín, and A. Zorzano. 2012. "Mitofusin 2 (Mfn2) links mitochondrial and endoplasmic reticulum function with insulin signaling and is essential for normal glucose homeostasis." *Proc Natl Acad Sci U S A* 109 (14):5523-8.

Sebastián, D., E. Sorianello, J. Segalés, A. Irazoki, V. Ruiz-Bonilla, D. Sala, E. Planet, A. Berenguer-Llergo, J. P. Muñoz, M. Sánchez-Feutrie, N. Plana, M. I. Hernández-Álvarez, A. L. Serrano, M. Palacín, and A. Zorzano. 2016. "Mfn2 deficiency links age-related sarcopenia and impaired autophagy to activation of an adaptive mitophagy pathway." *Embo j* 35 (15):1677-93.

Shabalina, I. G., M. Y. Vyssokikh, N. Gibanova, R. I. Csikasz, D. Edgar, A. Hallden-Waldemarson, Z. Rozhdestvenskaya, L. E. Bakeeva, V. B. Vays, A. V. Pustovidko, M. V. Skulachev, B. Cannon, V. P. Skulachev, and J. Nedergaard. 2017. "Improved health-span and lifespan in mtDNA mutator mice treated with the mitochondrially targeted antioxidant SkQ1." *Aging (Albany NY)* 9 (2):315-339.

Sheng, Z. H., and Q. Cai. 2012a. "Mitochondrial transport in neurons: impact on synaptic homeostasis and neurodegeneration." *Nat Rev Neurosci* 13 (2):77-93.

Sheng, Zu-Hang, and Qian Cai. 2012b. "Mitochondrial transport in neurons: impact on synaptic homeostasis and neurodegeneration." *Nature reviews. Neuroscience* 13 (2):77-93.

Soenen, S., C. K. Rayner, K. L. Jones, and M. Horowitz. 2016. "The ageing gastrointestinal tract." *Curr Opin Clin Nutr Metab Care* 19 (1):12-8.

Sohal, R. S., and B. H. Sohal. 1991. "Hydrogen peroxide release by mitochondria increases during aging." *Mech Ageing Dev* 57 (2):187-202.

Song, Z., M. Ghochani, J. M. McCaffery, T. G. Frey, and D. C. Chan. 2009. "Mitofusins and OPA1 mediate sequential steps in mitochondrial membrane fusion." *Mol Biol Cell* 20 (15):3525-32.

Srinivas, N., S. Rachakonda, and R. Kumar. 2020. "Telomeres and Telomere Length: A General Overview." *Cancers (Basel)* 12 (3).

Suh, Jung H., Swapna V. Shenvi, Brian M. Dixon, Honglei Liu, Anil K. Jaiswal, Rui-Ming Liu, and Tory M. Hagen. 2004. "Decline in transcriptional activity of Nrf2 causes age-related loss of glutathione synthesis, which is reversible with lipoic acid." *Proceedings of the National Academy of Sciences* 101 (10):3381-3386.

Sun, Nuo, Jeanho Yun, Jie Liu, Daniela Malide, Chengyu Liu, Ilsa I Rovira, Kira M Holmström, Maria M Fergusson, Young Hyun Yoo, Christian A Combs, and Toren Finkel. 2015. "Measuring *In Vivo* Mitophagy." *Molecular Cell* 60 (4):685-696.

Sung, Jin, Kyung Lee, Jae-Ryong Kim, and Hyoung Choi. 2017. "Interaction between mTOR pathway inhibition and autophagy induction attenuates adriamycin-induced vascular smooth muscle cell senescence through decreased expressions of p53/p21/p16." *Experimental Gerontology* 109.

Suzuki, Akiko, Ryohei Minamide, and Junichi Iwata. 2018. "WNT/β-catenin signaling plays a crucial role in myoblast fusion through regulation of nephrin expression during development." *Development* 145 (23).

Tan, Pan, Yan-Jiao Wang, Shuang Li, Yi Wang, Je-Yu He, Yi-Yin Chen, Hui-Qian Deng, Wu Huang, Jun-Kun Zhan, and You-Shuo Liu. 2016. "The PI3K/Akt/mTOR pathway regulates the replicative senescence of human VSMCs." *Molecular and Cellular Biochemistry* 422 (1):1-10.

Tong, D., and J. A. Hill. 2017. "Spermidine Promotes Cardioprotective Autophagy." *Circ Res* 120 (8):1229-1231.

Trotta, A. P., and J. E. Chipuk. 2017. "Mitochondrial dynamics as regulators of cancer biology." *Cell Mol Life Sci* 74 (11):1999-2017.

Trounce, I., E. Byrne, and S. Marzuki. 1989. "Decline in skeletal muscle mitochondrial respiratory chain function: possible factor in ageing." *Lancet* 1 (8639):637-9.

Turrens, J. F. 2003. "Mitochondrial formation of reactive oxygen species." *J Physiol* 552 (Pt 2):335-44.

Ugarte, N., I. Petropoulos, and B. Friguet. 2010. "Oxidized mitochondrial protein degradation and repair in aging and oxidative stress." *Antioxid Redox Signal* 13 (4):539-49.

Valente, E. M., P. M. Abou-Sleiman, V. Caputo, M. M. Muqit, K. Harvey, S. Gispert, Z. Ali, D. Del Turco, A. R. Bentivoglio, D. G. Healy, A. Albanese, R. Nussbaum, R. González-Maldonado, T. Deller, S. Salvi, P. Cortelli, W. P. Gilks, D. S. Latchman, R. J. Harvey, B. Dallapiccola, G. Auburger, and N. W. Wood. 2004. "Hereditary early-onset Parkinson's disease caused by mutations in PINK1." *Science* 304 (5674):1158-60.

Valente, E. M., F. Brancati, V. Caputo, E. A. Graham, M. B. Davis, A. Ferraris, M. M. Breteler, T. Gasser, V. Bonifati, A. R. Bentivoglio, G. De Michele, A. Dürr, P. Cortelli, A. Filla, G. Meco, B. A. Oostra, A. Brice, A. Albanese, B. Dallapiccola, and N. W. Wood. 2002. "PARK6 is a common cause of familial parkinsonism." *Neurol Sci* 23 Suppl 2:S117-8.

Van Raamsdonk, J. M., and S. Hekimi. 2009. "Deletion of the mitochondrial superoxide dismutase sod-2 extends lifespan in Caenorhabditis elegans." *PLoS Genet* 5 (2):e1000361.

Vermulst, M., J. H. Bielas, G. C. Kujoth, W. C. Ladiges, P. S. Rabinovitch, T. A. Prolla, and L. A. Loeb. 2007. "Mitochondrial point mutations do not limit the natural lifespan of mice." *Nat Genet* 39 (4):540-3.

Wanagat, J., Z. Cao, P. Pathare, and J. M. Aiken. 2001. "Mitochondrial DNA deletion mutations colocalize with segmental electron transport system abnormalities, muscle fiber atrophy, fiber splitting, and oxidative damage in sarcopenia." *Faseb j* 15 (2):322-32.

Wang, Yong, Lingbo Liu, Senthil K. Pazhanisamy, Hongliang Li, Aimin Meng, and Daohong Zhou. 2010. "Total body irradiation causes residual bone marrow injury by induction of persistent oxidative stress in murine hematopoietic stem cells." *Free Radical Biology and Medicine* 48 (2):348-356.

Wohlgemuth, S. E., A. Y. Seo, E. Marzetti, H. A. Lees, and C. Leeuwenburgh. 2010. "Skeletal muscle autophagy and apoptosis during aging: effects of calorie restriction and life-long exercise." *Exp Gerontol* 45 (2):138-48.

Xing, Yue-Xian, Liu Yang, Hong-Yu Kuang, Xin-Yuan Gao, and Hao-Ling Liu. 2017. "Function of obestatin in the digestive system." *Nutrition* 34:21-28.

Yamano, K., and R. J. Youle. 2013. "PINK1 is degraded through the N-end rule pathway." *Autophagy* 9 (11):1758-69.

Ying, S. Y., D. C. Chang, and S. L. Lin. 2008. "The microRNA (miRNA): overview of the RNA genes that modulate gene function." *Mol Biotechnol* 38 (3):257-68.

Yoshino, Jun, Kathryn F Mills, Myeong Jin Yoon, and Shin-ichiro Imai. 2011. "Nicotinamide Mononucleotide, a Key NAD+ Intermediate, Treats the Pathophysiology of Diet- and Age-Induced Diabetes in Mice." *Cell Metabolism* 14 (4):528-536.

Youle, R. J., and D. P. Narendra. 2011. "Mechanisms of mitophagy." *Nat Rev Mol Cell Biol* 12 (1):9-14.

Yun, J., R. Puri, H. Yang, M. A. Lizzio, C. Wu, Z. H. Sheng, and M. Guo. 2014. "MUL1 acts in parallel to the PINK1/parkin pathway in regulating mitofusin and compensates for loss of PINK1/parkin." *Elife* 3:e01958.

Zhang, Cuiping, Peng Chen, Yang Fei, Bo Liu, Kui Ma, Xiaobing Fu, Zhili Zhao, Tongzhu Sun, and Zhiyong Sheng. 2012. "Wnt/β-catenin signaling is critical for dedifferentiation of aged epidermal cells in vivo and in vitro." *Aging Cell* 11 (1):14-23.

Zhang, D., F. Wang, P. Li, and Y. Gao. 2022. "Mitochondrial Ca(2+) Homeostasis: Emerging Roles and Clinical Significance in Cardiac Remodeling." *Int J Mol Sci* 23 (6).

Zhang, Da-yong, Hai-jie Wang, and Yu-zhen Tan. 2011. "Wnt/β-Catenin Signaling Induces the Aging of Mesenchymal Stem Cells through the DNA Damage Response and the p53/p21 Pathway." *PLOS ONE* 6 (6):e21397.

Zhang, Hongqiao, Kelvin J. A. Davies, and Henry Jay Forman. 2015. "Oxidative stress response and Nrf2 signaling in aging." *Free Radical Biology and Medicine* 88:314-336.

Zhang, J., Z. He, J. Fedorova, C. Logan, L. Bates, K. Davitt, V. Le, J. Murphy, M. Li, M. Wang, E. G. Lakatta, D. Ren, and J. Li. 2021. "Alterations in mitochondrial dynamics with age-related Sirtuin1/Sirtuin3 deficiency impair cardiomyocyte contractility." *Aging Cell* 20 (7):e13419.

Zhang, Y., Y. Ikeno, W. Qi, A. Chaudhuri, Y. Li, A. Bokov, S. R. Thorpe, J. W. Baynes, C. Epstein, A. Richardson, and H. Van Remmen. 2009. "Mice deficient in both Mn superoxide dismutase and glutathione peroxidase-1 have increased oxidative damage and a greater incidence of pathology but no reduction in longevity." *J Gerontol A Biol Sci Med Sci* 64 (12):1212-20.

Zhao, Jing, Xuesen Li, Sara McGowan, Laura J. Niedernhofer, and Paul D. Robbins. 2015. "NF-κB Activation with Aging: Characterization and Therapeutic Inhibition." In *NF-kappa B: Methods and Protocols*, edited by Michael J. May, 543-557. New York, NY: Springer New York.

Zhu, Yanyan, Stefan Massen, Marco Terenzio, Verena Lang, Silu Chen-Lindner, Roland Eils, Ivana Novak, Ivan Dikic, Anne Hamacher-Brady, and Nathan R. Brady. 2013. "Modulation of Serines 17 and 24 in the LC3-interacting Region of Bnip3 Determines Pro-survival Mitophagy versus Apoptosis. *Journal of Biological Chemistry* 288 (2):1099-1113.

Zorov, Dmitry B., Magdalena Juhaszova, and Steven J. Sollott. 2014. "Mitochondrial reactive oxygen species (ROS) and ROS-induced ROS release." *Physiological reviews* 94 (3):909-950.

Chapter 9

Mitophagy and Age-Associated Diseases

Aanchal Verma
and Shalini Mani*
Department of Biotechnology, Jaypee Institute of Information Technology, Noida, Uttar Pradesh, India

Abstract

Aging is a natural phenomenon that has been associated with different biological changes that occur inside the organism's body, such as telomere shortening, mitochondrial dysfunction, ROS production, changes in signal pathways, and gene expression. Mitochondria plays an important role in ATP production and regulates pathways leading to cell death. Some experimental results have shown that mitochondrial mutations lead to accelerated aging and age-related phenotypes, proving defective mitochondrion's important role in aging and age-related diseases. To prevent mutation and maintain its correct organization, mitochondria have evolved a vital process called mitophagy which deletes mutated, fragmented, or dysfunctional mitochondria from the cell and maintains its quality via different signaling pathways. In this chapter, we will discuss the mitochondrial theory of aging, mitochondrial dynamics, the role of mitophagy in age-related diseases, and how we can target it to treat them.

Keywords: mitophagy, aging, mitochondria, ROS, mitochondrial dynamics

Introduction

Life is an inevitable transition into old age. Numerous theories attempt to explain why people age. D. Harman defined aging as the gradual accumulation of changes that are connected to or accountable for the ever-increasing susceptibility to disease and mortality that comes with aging (Harman, D., 1981). Various processes such as nuclear DNA damage, telomere shortening, loss of epigenetic structure, increased DNA methylation, altered post-translational modifications, and damage by reactive oxygen species (ROS) oxidizes lipids, proteins, and DNA and cause mutation (Mkrtchyan, Garik V et al., 2020; Redza-Dutordoir, Maureen, et al., 2016; Johnson, Adiv A et al., 2012; Granville, D J et al., 2001). These modifications lead to the

* Corresponding Author's Email: Shalini.mani@jiit.ac.in.

In: Mitophagy and Its Role in Health and Disease
Editors: Shalini Mani and Samiksha Wasnik
ISBN: 979-8-89113-470-6

malfunction of cellular organelles, notably mitochondria, which deteriorates cells and tissues. The mitochondria are the engine of the cell, producing energy through oxidative phosphorylation (OXPHOS), β- oxidation, which controls fatty acid metabolism, and the TCA cycle, which produces intermediate metabolites needed by all organelles of the cell for healthy growth and survival.

The discovery that mitochondria release several pro-apoptotic proteins, such as cytochrome c, SMAC/DIABLO, ENDOG, OMI/HTR, and AIF, which irrevocably lead to cellular death, demonstrates the critical function that mitochondria play in controlling the pathways that lead to cell death. (Du, C et al., 2000; Green, Douglas R, and Guido Kroemer., 2004; Wang, Chunxin, and Richard J Youle., 2009; Bua, Entela et al., 2006; Cortopassi, G A, and N Arnhei., 1990). Studies have shown that mutations in mitochondrial DNA increase with age and contribute majorly to aging processes and diseases associated with aging (Trifunovic, Aleksandra, et al., 2004; Chen, Hsiuchen, and David C Chan., 2009). In an experiment by Trifunovic et al., the molecular mice (Knock in) that express defective mitochondrial proofreading polymerase show accelerated age-related phenotypes and reduced lifespan, which explains the significance of mitochondria in aging (Cell Ashrafi, G., & Schwarz, T. L., 2013). The continual fission and fusion that occurs in mitochondria help to prevent the build-up of damaged mitochondria, making it a highly organized and dynamic organelle (Palikaras, K., & Tavernarakis, N., 2012). Mitochondria have evolved the process of mitophagy to maintain its quality and reduce mtDNA accumulation (Khalil B et al., 2015). Mitophagy is the primary type of autophagy specifically used for mitochondria. This reduces and over-mitophagy causes various diseases and age-related issues. Recent research has identified the role of mitophagy in neurodegenerative disorder (Ye, X., Sun, X., Starovoytov, V., & Cai, Q., 2015), Huntington's (Kitada, T et al., 1998), and Alzheimer's disease (Valente, Enza Maria, et al., 2004).

Mitochondrial Theory of Aging

The life cycle of multicellular organisms is like a hyperbolic curve starting from embryogenesis, followed by the growth and reproductive maturation phase, characterized by the organ systems' maximum efficiency. With increasing age, the deceleration or death phase initiates, causing a reduction in the organ system's working capacity, causing various physiochemical changes in the body that lead to aging and death of the individual. Much research has been done to understand the mechanism and cause of aging; despite recent advances in biological science and technology, the reason why and how we age is still unrevealed. Numerous theories have been developed to explain the aging process, including programmed longevity, wear and tear theory, the immune theory, the endocrine theory, the rate of living theory, the cross-linking theory, and the free radical theory. (Jin, Kunlin., 2010). Among them, Dr. Denham Harman's free radical theory is the most palatable, they proposed that naturally produced free radicals and superoxides react with cellular constituents and damage the cell, which eventually leads to inhibition of cell function and cellular senescence (Harman, D., 1956). In 1972 Dr. D. Harman further revised his theory of aging by describing the probable contribution of mitochondria to the generation of aging-causing free radicals. He also states that organisms with high energy requirement have shorter life span because of high oxygen transport followed by increased oxidative changes in the cell causing a rise in the generation of dangerous reactive oxygen species, or free radicals, which oxidize DNA, proteins, and lipids and promote cellular senescence (Harman, D., 1972; Bokov et al., 2004). In 1980 Miquel *et al. postulated the "mitochondrial free radical theory of aging" in which he explained mitochondria as inducers of free radicals and mitochondrial DNA as the primary target that leads to age-related pathologies (*Miquel, J et al., 1980).

The discovery of free radical theory gave a new and different direction to researchers in the field of aging. Various experimental results have been found to support this theory of aging. Scientists have found that old animals have higher oxidative DNA, proteins, and lipids than young ones (Stadtman, E R., 1992; Sohal, R S et al., 1993). The presence of antioxidants alone is insufficient; a rather specific pool of antioxidants affects longevity, which is explained in an experiment with transgenic drosophila. It was found that Drosophila with increased expression of both catalase and Cu/Zn-superoxide dismutase showed an increased life span, whereas Drosophila overexpressing both catalase and Mn-superoxide dismutase showed resistance to oxidative stress but did not increase the life span indicating that not only the level of antioxidants affect longevity instead the specific pool of antioxidant is also essential (Bayne, Anne-Cécile V et al., 2005; Orr, W C, and R S Sohal., 1994).

In the past three decades, much research has been done in support and the aging theory of free radicals that causes changes in understanding the significance of ROS in cellular physiology. Unlike previous characterizations of ROS as being toxic, it was found that they are also crucial for proper cell signaling and other metabolic activities. After this discovery, scientists modified the theory to "the redox stress theory of aging," which declares that ROS are also required by our body but in a limited amount if they exceed the generally required concentration, then they will start harming the cellular physiology that lead to various age-related pathologies (Sohal, Rajindar S, and William C Orr, 2012).

In 2013 Viña, Jose et al. further modified and postulated "the cell signaling disruption theory of aging," which describes that ROS in a small amount can be tolerated by the cell by inhibiting their effects with the help of cellular antioxidants, but the high concentration ROS will be harmful and disrupt the various signaling pathways that ultimately alter the cellular physiology (Viña, Jose, et al., 2013). Another modification of the free radical theory is the "Adaptive Homeostasis Theory of Ageing," which explains that young individuals can activate adaptive homeostasis in response to any cellular damage or increased oxidative stress conditions. However, this capability is reduced in old individuals, leading to increased oxidative and cellular stress-causing age-related conditions. With the recent advancement in science and technologies, different aspects of free radical theory are revealing, but still, it appears both viable and robust.

Mitochondrial DNA Damage

Other than the nucleus, mitochondria are the only cell organelles with genetic material and machinery for synthesizing protein and lipids. It is an essential cell organelle that performs various functions that are important for the maintenance and proper functioning of the cell such as generating energy, regulating immune responses, redox signaling, calcium homeostasis, and stem cell programming Angelika S et al., 2018; Nikoletopoulou, Vassiliki, et al., 2013; Kamer, Kimberli J, et al., 2015). The inner mitochondrial membrane holds the mitochondrial respiratory chain, which consists of the five components: complex I, II, III, IV, and V working together to form adenosine triphosphate (ATP) (Chen, Jin-Qiang, et al., 2005). In the mitochondrial respiratory chain, two distinct processes co-occur: the first is the transfer of an electron through the complexes to molecule oxygen to generate water. (Sas, Katalin et al., 2007) Moreover, complexes I, III, and IV pump protons through the inner mitochondrial membrane; the return of these protons to the inner mitochondrial membrane produces ATP. (Porter, R K, and M D Brand., 1995). Under normal circumstances, complex I AND III convert 1-5% oxygen to ROS (Selivanov, Vitaly A et al., 2011). These ROS can damage macromolecules of mitochondria like DNA, lipids, and proteins (Van Houten, Bennett et al., 2005) and if left untreated, will eventually affect complexes I and III, which will cause an excess of oxygen to be reduced to superoxide. (Castro, María del R et al., 2012). Because mitochondria are where ROS are generated, mitochondrial DNA is highly susceptible to the damage caused by it compared to nuclear DNA. Increased oxidative stress and decreased production of enzymes and proteins necessary for the regular operation of the respiratory chain caused by mutation and modification in the mitochondrial DNA resulted in cell death. (Van Houten, Bennett et al., 2005).

ROS and superoxides react with mtDNA at two different sites: the first is the sugar-phosphate backbone causing strand fragmentation, and the second is the base that results in a damaged, dysfunctional base (Richter, C., 1992). The DNA repair system of mitochondria is significantly less effective than the nucleus and is not able to repair the harm done by ROS, but mitochondria try to eliminate the mutated mtDNA by the process of mitophagy (Richter, C., 1992).

The inner mitochondrial membrane is permeable to neutral molecules only, therefore the exogenous antioxidant cannot pass through it. Peroxidation of lipids occurs at the site of ROS production, which will increase the membrane's permeability to protons that will affect cellular signaling (Stewart, Victoria C, and Simon J R Heales., 2003). Calcium is an essential signaling molecule in muscle contraction, releasing neurotransmitters and other secretions, cell division, gene expression, and cell death (Fu, Ji-Dong, and Huang-Tian Yang., 2006). Increased permeability of the mitochondrial membrane to protons alters calcium homeostasis, affecting the oxidation of specific thiol groups of proteins (Toescu, Emil C., 2005). An excessive influx of calcium affects the mitochondrial potential, which increases the production of superoxides and the formation of mitochondrial transition pores that will eventually cause rupture of the outer mitochondrial membrane and osmotic edema (Joshi, Gururaj, et al., 2005). Increased calcium uptake triggers cell death pathways and neural degeneration (Ross, William N., 2012).

Mitochondrial Dynamics

Mmitochondrial size, shape, and location are critical factors in controlling mitochondrial activity. The various morphological states of mitochondria represent many physiological states (Nunnari, Jodi, and Anu Suomalainen., 2012). By keeping a healthy balance between fusion and fission processes, mitochondria may control their morphology (Yoon, Yisang et al. 2010). The efficient operation of the mitochondria and the appropriate response to cellular requirements depend on the harmony between these two processes.

Mitochondrial Fission

One mitochondrion splits into two daughter mitochondria, referred to as mitochondrial fission. It performs a key role in maintaining mitochondrial quality control by distinguishing between healthy and faulty mitochondria. The mitochondrial fission process eliminates the damaged mitochondria. Dynamin-related protein-1 (Drp1) and fission protein 1 (Fis1) are the two proteins that play crucial roles in controlling mitochondrial fission (Westermann, Benedikt., 2008).

Drp1 is a cytosolic protein that plays a significant part in apoptosis, calcium signaling, and mitochondrial fission. Yeast and Caenorhabditis elegans were the first organisms to describe their function in mitochondrial fission, respectively (Bleazard, W et al., 1999). Structurally, the N-terminal GTPase domain, the middle domain, the variable domain or B insert, and the GED domain in the C-terminus are the four domains that makeup DRP1 (Fröhlich, Chris, et al., 2013). Although mitochondrial fission factor (MFF) and mitochondrial dynamics proteins 49 and 51 (MiD49 and MiD51) act as receptors for DRP1 in mammals (Palmer, Catherine S et al., 2011), which aids in its recruitment since DRP1 is unable to bind OMM directly. Assembly of DRP1 is done with the help of its middle domain followed by DRP1 oligomerization into an X-shaped dimer on the mitochondrial outer membrane in GTPase dependent manner (Mears, Jason A et al., 2011; Fröhlich, Sajjan, et al., 2013, Chris et al., 2013; Koirala,). DRP 1 forms a ring-like multimeric spiral structure around the mitochondria followed by GTPase activation causing GTP hydrolysis that leads to the narrowing of the mitochondrial membrane which acts as a constriction site for mitochondrial scission (Fröhlich, Chris, et al., 2013; Ingerman, Elena, et al., 2005; Mears, Jason A et al., 2011). Dynamin 2 (Dnm2) is another GTPase protein recruited to the mitochondrial fission-causing DRP1-induced constriction location (Lee, Jason E et al., 2016). Drp1 also plays a crucial part in eliminating defective mitochondria via mitophagy. It helps in the oligomerization of BAX (a proapoptotic BCL-2 family member) at the mitochondrial constriction site (Lee, Yang-ja et al., 2004).

The outer membrane of the mitochondria contains an even distribution of the membrane protein known as human fission protein 1 (Fis1). It attaches to MOM by exposing its N terminal domain to the cytosol and C terminal transmembrane domain, leaving its c terminal tail in the intermembrane gap (Yoon, Yisang et al., 2003). It helps attach Dnm1 to the MOM surface of yeast but does not have any role in mammalian DRP1 recruitment. Therefore, Fis1 knockdown will not impact Drp1's mitochondrial localization (Koirala, Sajjan, et al., 2010; Lee, Yang-ja, et al., 2004). Experimental studies revealed that fis1 is an obstacle in mitochondrial fission as its abundance was found to amplify mitochondrial fragmentation (Stojanovski Diana et al., 2004).

A recent discovery in mitochondrial dynamics is ER involvement in mitochondrial fission. It was found that ER tubules make contact with mitochondria and wrap around them to decrease the diameter of mitochondria, facilitating the DRP1 ring formation (Friedman, Jonathan R et al., 2011). Before the assembly of DRP1 at the mitochondria-ER interface, it is found that two actin-nucleating proteins, the mitochondrial-anchored formin-binding Spire1C and ER-bound inverted-formin 2 (INF2), are crucial for the constriction of the mitochondria. Experimental studies have found that mitochondrial DNA division requires the endoplasmic reticulum-mitochondrial interface. High-resolution microscopy and live cell imaging techniques have found mtDNA replication at these constriction sites to reveal mtDNA division as one of the first steps of the mitochondrial fission process (Lewis, Samantha C et al., 2016).

Recently, a mechanism for the inner mitochondrial membrane was postulated, according to which the region of the ER-mitochondrial interface experiences calcium-dependent IMM constriction before the recruitment of DRP1 (Chakrabarti, Rajarshi, et al., 2018; Cho, Bongki, et al., 2017). According to Cho, Bongki et al., calcium influx in the mitochondria cause a reduction in the mitochondrial membrane potential, which causes OPA1 processing to disrupt the mitochondrial connect site and cristae organizing system (MICOS), which in turn causes IMM constriction (Cho, Bongki et al., 2017).

Mitochondrial Fusion

This process is known as mitochondrial fusion when two mitochondria fuse to produce a single mitochondrion. Two mitochondria approach one another and form a ring-like structure around the contact point, which is followed by the fusion of the outer mitochondrial membrane with the aid of the MOM protein, mitofusins Mfn1, and Mfn2, and finally, the fusion of the inner mitochondrial membrane with the aid of the IM localized protein OPA1 (Malka, Florence, et al., 2005; Brandt, Tobias, et al., 2016). Mfn1 and Mfn2 have three crucial domains in their structural makeup: an N-terminal GTPase, two hydrophobic heptads, HR1 and HR2 at the C-terminus, and two transmembrane domains in between the HR1 and HR2. These transmembrane areas were found to be responsible for the U shape of mitofusin, due to which both N and C terminals face cytosol (Chan, David C., 2006; Rojo, Manuel, et al., 2002). These two mitofusins can promote mitochondrial fusion by dimerizing to generate three distinct functional complexes: Three types of oligomers exist for Mfn: Mfn1 homotypic, Mfn2 homotypic, and Mfn1-Mfn2 heterotypic. (Chen, Hsiuchen et al., 2003). Dimerization of mitofusins is followed by GTP hydrolysis causing conformational changes leading to the fusion of MOM (Choi, Seok-Yong et al., 2006). Mfn1 shows ubiquitous expression, and its overexpression causes an increased level of mitochondrial fusion compared to Mfn2 overexpression (Ishihara, Naotada et al., 2004). Mfn2 interacts with Bax and MITOL to play a significant role in apoptosis and mitochondrial-ER association, respectively (Karbowski, Mariusz, et al., 2002; de Brito, Olga Martins et al., 2008). Ubiquitination of mitofusins leading to their degradation plays a crucial regulatory role in mitofusins activity (Cohen, Mickael M J, et al., 2008).

OPA1 is another dynamin-related GTPase protein that plays a significant role in the fusion of the inner membrane of mitochondria after the fusion of OMM. It consists of five domains: N terminal mitochondrial import signal (MIS), GTPase domain, GED domain, proteolytic cleavage site, and a variable domain which helps in its interaction with mitochondrial inner membrane protein, cardiolipin (Delettre, C et al., 2000; Meglei, Gabriela, et al., 2000). OPA1 is essential in maintaining Cristea morphology, apoptosis, calcium homeostasis, mtDNA level, mitophagy, and cytochrome c release. OPA1 C terminus and mitofusins loop created by u-shaped topology interact to coordinate OMM and IMM fusion (Cipolat, Sara et al., 2004). Although in various physiological processes, this interaction was not found that advocates the contribution of some other proteins in regulating OPA1-mediated fusion of mitochondria. According to experimental research, the cleavage of OPA1 into its two isoforms, long OPA1 and short OPA1, which aid in OPA1 coordination with mitofusins, is a crucial step in OPA1-mediated fusion (Song, Zhiyin, et al., 2007). OPA1 long isoform is located at the inner membrane and faces the inner membrane space, while the short isoform is located within the inner membrane space (Olichon, Aurélien et al., 2002). OPA1 has three proteolytic cleavage sites (S1, S2, and S3) that produce several OPA1 isoforms that impact the regulation of mitochondrial dynamics. Proteolytic cleavage of OPA1 is regulated by different factors like MPP, apoptosis, ATP level, and mtDNA stability (Ishihara, Naotada, et al., 2006; Baricault, Laurent, et al., 2007; Chan, David C., 2012).

In an experimental investigation, it was discovered that a short OPA1 isoform in mammals promotes the interaction between cardiolipin, an inner mitochondrial membrane lipid, and OPA1, which leads to IMM fusion (DeVay, Rachel M et al., 2009; Rujiviphat, Jarungjit, et al., 2009). This study demonstrates that minimal fusion machinery can assist fusion when cardiolipin and OPA1 are present on either membrane side.

Mitophagy

An advanced form of autophagy known as mitophagy preserves the health of mitochondria by removing unhealthy mitochondria from the cell. Mitophagy receptors recognize damaged, fragmented, and mutant mitochondria and send them to phagophores for destruction. Mitochondrial dynamics, oxidative stress, energetics, and homeostasis regulate it. Reduced or increased expression of mitophagy leads to various health problems and age-related pathologies, which will become dangerous for the individual.

For a complete understanding of mitophagy, it is essential to learn its mechanism to know how it maintains the quality of mitochondria. Mitophagy follows different pathways for dysfunctional mitochondria presentation to phagophores for their degradation, broadly divided into two pathways: ubiquitin-independent and ubiquitin-dependent. Ubiquitin-dependent pathways include PINK1/ Parkin mediated mitophagy, in which mitochondria are transported to phagophores with the help of the ubiquitin-binding domain, while the ubiquitin-independent mitophagy pathways involve different membrane proteins with LC3 binding regions that bind to LC3 domains to produce phagophores include BNIP3, NIX, FUNDC1, and others.

PINK1 and Parkin Regulated Mitophagy

A serine/threonine protein kinase called PINK1 is encoded by the PARK6 gene that localizes to mitochondria and is responsible for the protection of cells from stress triggered by mitochondrial malfunction. In contrast, a multiprotein E3 ubiquitin ligase complex that facilitates the targeting of substrate proteins for proteasomal destruction contains the protein Parkin as one of its components. Autosomal recessive early-onset Parkinson's disease is caused by mutations in Pink1 and Parkin (Valente, Enza Maria, et al., 2004; Kitada, T et al., 1998). The pink1/parkin pathway has been associated with modulating mitochondrial dynamic machinery by increasing mitochondrial fission or inhibiting fusion (Deng, Hansong, et al., 2008).

PINK1 is imported into the inner mitochondrial membrane under steady-state conditions, where it undergoes degradation by the rhomboid intramembrane protease PARL (presenilin-associated, rhomboid-like), which releases it from the outer mitochondrial membrane into the cytoplasm, where it is cleaved and degraded by the N-end rule pathway. According to Pickrell,

Alicia M., and Richard J. Youle (2014), the stabilization of PINK1 in the outer mitochondrial membrane results from reducing the membrane potential of mitochondria caused by any form of mitochondrial damage. The accumulation of PINK1 causes parkin to be recruited from the cytoplasm to damage mitochondria in its autoinhibited conformation (Chaugule, Viduth K et al., 2011). Parkin gets triggered when activated PINK1 phosphorylates ser65 in ubiquitin. Phosphorylated ubiquitin acts as an allosteric activator and binds to the parkin causing alteration in conformation that causes the release of the ubi domain from the parkin core, allowing phosphorylation of parkin ubi domain by PINK1 (Kane, Lesley A et al., 2014; Kazlauskaite, Agne, et al., 2014; Wauer, Tobias, et al., 2015; Kazlauskaite, Agne, et al., 2015; Sauvé, Véronique, et al., 2015). Phosphorylation of parkin causes the formation of multiple poly-ub chains on damaged mitochondria where they draw autophagy receptors: NBR1, NDP52, OPTN, p62/ SQSTM1, TAX1BP1, TOLLIP and TANK binding kinase 1 (TBK1) after which LC3-positive phagophores are formed, which destroy mitochondria by way of the lysosome (Yamano, Koji, et al., 2016; Lazarou, Michael, et al., 2015).

Ubiquitin Independent Mitophagy

Even though PINK1 and parkin play a significant role in mitophagy, other receptor-mediated mitophagy pathways also exist. Some of them are listed below.

BNIP3 and NIX-Dependent Mitophagy

Proapoptotic BCL2 family members BNIP3 and NIX are found in mitochondria and play a role in stress-induced cell death. They both have a crucial role in mitophagy (Zhang, J., and P. A. Ney., 2009) and are hypoxia-inducible proteins (Sowter, H. M. et al., 2001). After activation, BNIP3 and NIX enter the outer mitochondrial membrane, where they cause the MPTP to open, the membrane potential to drop, depolarization and recruitment of LC3/GABARAP for autophagosome formation (Zhang, J, and P A Ney., 2009; Zhang, Ji et al., 2012). LC3/GABARAP proteins and other signals, such as ubiquitination of mitochondrial proteins, mediate the incorporation of mitochondria into autophagosomes (Novak, Ivana, et al., 2010).

FUNDC-1 Mediated Mitophagy

FUNDC-1 is a protein present outside the mitochondria that facilitates mitophagy by decreasing the level of essential mitochondrial proteins and by communicating with the LC3 autophagy receptor via its LC3-interaction region (LIR) located at its cytosolic N- terminal (Liu, Lei, et al., 2012). Under ordinary circumstances, protein kinases SRC proto-oncogene, non-receptor tyrosine kinase (SRC), and casein kinase 2 (CK2) constitutively phosphorylate FUNDC1, which blocks the connection between its LIR and LC3 receptor. When PGAM5 dephosphorylates FUNDC1 under conditions of hypoxia or loss of mitochondrial membrane potential, a Ser/Thr phosphatase activates mitophagy facilitated by FUNDC1 (Chen, Guo, et al., 2014; Lim, Yunki, et al., 2019). BCL2L1 acts like a mitophagy regulator, which remains bound to PGAM5 under normal conditions and inhibits its activity. Hypoxia causes the degradation of BCL2L1 along with other mitochondrial proteins and activates PGAM5 (Wu, Hao et al., 2014).

The connection between the mitochondrial fission factor DNM1L and the mitochondrial inner membrane fusion protein OPA1 has an impact on FUNDC1-mediated mitophagy as well.

Under normal conditions, OPA1 interacts with FUNDC1; stressed conditions reduced the interaction between these two and recruited DNM1L for mitochondrial fission followed by mitophagy (Chen, Ming, et al., 2016).

Status of Mitophagy in Aging

It is abundantly evident that mitophagy is crucial for maintaining the quality control of mitochondria, which is necessary for their appropriate function and the proper functioning of the cell. Mitochondrial defects are the hallmark of aging and age-associated pathologies. Many cardiovascular, neurodegenerative, muscular, and skeletal disorders have been registered due to mitochondria defects. However, the exact cause and mechanism behind the diseases are still unclear, but the increased number of altered or defective mitochondria reflects the presence of dysregulated mitophagy which can be due to defects in mitophagy machinery or proteins involved in mitophagy. Though scientists are not sure, in many diseases, various proteins involved in mitophagy are found to be the potential target for their treatment. The status and function of mitophagy in the pathophysiology of a few age-related diseases are described below.

Mitophagy and Neurodegenerative Diseases

When nerve cells in the brain or peripheral nervous system begin to lose their functionality over time and eventually die, this is known as a neurodegenerative illness. The prevalence, incidence, and death rate of neurodegenerative diseases in India is increasing daily. Mitochondrial dysfunction is found to be associated with various neurodegenerative diseases, such as Parkinson's disease (PD), Alzheimer's (AD), Huntington's disease (HD), Friedreich's ataxia (FRDA), and amyotrophic lateral sclerosis (ALS).

Alzheimer's Disease

Alzheimer's disease (AD) is a widespread neurological disorder associated with age. Some of its hallmarks are significant neuronal loss, dysfunctional synapses, mitochondrial damage, neuroinflammation, and an accumulation of amyloid plaques and neurofibrillary tangles. (NFTs) (Kocahan and Dogan, 2017; Magalingam et al., 2018). The two main hallmarks of AD are beta protein buildup in extracellular senile plaques, and tau protein hyperphosphorylation creates intracellular neurofibrillary tangles, which cause defective synapses and neuroinflammation, followed by neuronal loss and clinical symptoms (De Strooper, Bart, and Eric Karran., 2016; Doblado, Laura, et al., 2021). Many genetic, molecular, and cellular events are linked with AD, but the two critical events that are seen in the early onset of the disease are synapse dysfunction and mitochondrial dysfunction related (Selkoe, Dennis J., 2002; Reddy, P Hemachandra, et al., 2008).

Various mitochondrial abnormalities associated with AD were identified such as recently, abnormalities in the mitochondria have been discovered, including mutations in the mitochondrial DNA, decreased enzyme activity, aberrant gene expression, increased fragmentation, and decreased fusing. (Reddy P. H., 2009). Aβ aggregates reduce the activity of enzymes involved in mitochondrial respiration leading to a reduction in energy production which impairs the metabolic activity of the brain. They also influence pathways related to oxidative stress (Tobore, Tobore Onojighofia., 2019). Aβ aggregates also impair the calcium homeostasis of the cell and increase the concentration of calcium ions that disrupt the membrane potential of mitochondria, ultimately leading to decreased ATP production and increased oxidative stress.

Additionally, by promoting fission and causing neurological dysfunction, the Aβ precursor protein (AβPP) and Aβ aggregates upset the equilibrium between fusion and fission mitochondrial dynamics (Flannery, Padraig J et al., 2019). Mitochondrial defects cause increased oxidative stress, leading to high lipid peroxidation and forming γ-secretase modifying protein called 4-hydroxynonenal. Increased activity of γ-secretase leads to a more significant accumulation of Aβ (Esposito, Luke, et al., 2006; Chen, Liuji, et al., 2015).

Postpartum reports of AD brain reveal that p-tau and Aβ interact with the Drp1 protein responsible for the fission of mitochondria, which is increased with the progression of the disease and is found to be responsible for mitochondrial fragmentation, synaptic damage, and defective mitochondrial dynamics (Manczak, Maria, et al., 2011; Manczak, Maria, et al., 2012). Stress-induced mitophagy is induced by advanced glycation end products (AGEs). In AD patients, AGE and their receptors are upregulated, which favors the expression of the fission proteins DRP1 and OPA1 over fusion proteins. AGEs impair neural cells by cross-linking with elevated kinases and GTPases, increasing the production of ROS and neural inflammation (Li, X-H et al., 2013; Stock, Amanda J et al., 2016). It has been found that Aβ also causes a decrease in the production of parkin, leading to PINK1 accumulation, causing impaired mitophagy. Mutation in presenilin 1 and apolipoprotein e4 (ApoE4) gene are also associated with AD pathology leading to lysosomal dysregulation, which results in decreased mitophagy (Nuriel, Tal, et al., 2017).

In sum, total research has revealed that decreasing the accumulation of dysregulated mitochondria by increasing mitophagy can be the therapeutic target for treating AD. Decreasing the production of DRP1 and other fission-promoting proteins is essential for treating AD pathologies.

Parkinson's Disease

The second most prevalent neurological ailment, Parkinson's disease (PD), is characterized by bradykinesia, tremors, rigidity, postural instability, and other symptoms, including impaired smell perception, constipation, and depression. Loss of substantia nigra (SN) dopaminergic neurons is the disease's primary identifying feature. Accumulation of hyperphosphorylated tau (p-tau) and amyloid deposition, as well as dysregulated autophagy and lysosomal dysfunction, are typical pathogenic alterations in PD (Wills, Jonathan, et al., 2010; Tanji, Kunikazu, et al., 2011). It is the first neurological disorder in which the presence and significance of dysregulated mitochondria are described. PD patients' studies have found increased production of ROS, altered activity of ETC complexes, and impaired mitophagy (Schapira, A H et al., 1989, Onyango, Isaac G. et al., 2017).

Leucine-rich repeat kinase2 [LRRK2], alpha-synuclein [SNCA], glucocerebrosidase [GBA], and autosomal recessive genes (DJ-1, PRKN, PINK1, and ATP13A2) have all been identified as having mutations in genetic studies of PD patients which are closely associated with mitophagy and mitophagy machinery (Lopes da Fonseca et al., 2014; Klein, Christine, et al., 2012). Loss of function mutation in PARK6 and PARK2 genes is associated with familial

PD, leading to defective mitophagy. A study of mutant drosophila flies shows dopaminergic degeneration, reduced lifespan, and locomotive defects (Clark, Ira E et al., 2006; Chu, Charleen T., 2019). Mutations in the gene LRRK2 lead to impaired mitochondrial trafficking, PINK1/Parkin-dependent mitophagy, and impaired trafficking of fission proteins (Malpartida, Ana Belen, et al., 2021). Homozygous mutations in gene GBA lead to the development of lysosomal storage disease called Gaucher's disease along with dysfunctional mitophagy due to impairment in lysosomal degradation (Moors, Tim E et al., 2019; Malpartida, Ana Belen, et al., 2021).

Defects in mitophagy cause faulty mitochondria to build up inside the cell, which in turn causes an increase in ROS generation and a decrease in ATP synthesis. Dopamine neurons are metabolically highly active and require high energy; therefore, dysregulated mitophagy negatively affects their activity (Haddad, Dominik, and Ken Nakamura., 2015). Experimental studies have found that α-synuclein and tau accumulation causes autophagic-lysosomal degradation and impacts mitochondrial and lysosomal functions (Moors Tim et al., 2016; Wang, Yipeng et al., 2009). Accumulated α-synuclein interacts with various outer mitochondrial components such as α-synuclein TOM20, VDAC, and F1Fo-ATP synthase that causes the mitochondrial transition pore (MPTP) to open and inhibits the protein import in mitochondria resulting in mitochondrial dysfunction (Malpartida, Ana Belen, et al., 2021).

Amyotrophic Lateral Sclerosis (ALS)

A fatal neurodegenerative condition called amyotrophic lateral sclerosis (ALS) causes the motor neurons in the brain and spinal cord to deteriorate gradually, which causes several pathologies, including neuromuscular denervation, atrophy of the voluntary skeletal muscles, and ultimately paralysis and death. Infectious disease, chronic neuroinflammation, trauma or injury, and exposure to toxins are some possible sporadic ALS causes; however, aging is the critical risk factor for both familial and sporadic ALS. Neurons are high energy demanding cells of our body as well as they are highly polarized and susceptible cells therefore, they require proper ionic and energy maintenance, but due to mutations mitochondria of ALS patients are not working correctly and are not able to fulfill the energy requirements and ionic homeostasis in the cell that leads to neuron degeneration along with the high production of ROS. Approximately 20 genes, including SOD1, TAR DNA-binding protein 43 (TDP-43), OPTN, fused in sarcoma (FUS), p62, chromosome 9 open reading frame 72 (C9ORF72), TBK1, SQSTM1, C9ORF72, VCP, CHMP2B, VAPB, ALS2, and DCTN1, have mutations that have been linked to ALS so far (Abel, Olubunmi, et al., 2012). These genes are directly or indirectly involved in mitophagy, leading to changes in the process, causing the accumulation of dysfunctional mitochondria that affects the cells negatively (Taylor J Paul et al., 2016). The protein translated from these genes interacts with the mitochondria, affecting its permeability, ion channels, MPTP, and mtDNA. Mutation in TDP-43 causes interaction of this protein with mtDNA leading to the formation of dysfunctional complex 1 of ETC, affecting ATP production of mitochondria (Smith, Emma F et al., 2019). Mutation in SOD1 and Fus gene causes aggregation of mutant protein in IMS, imbalanced calcium homeostasis, disrupted mitochondrial structure decreased ATP generation, impaired mitochondrial transport, and disturbance in mitophagy (Smith, Emma F et al., 2019).

Expulsing dysregulated mitochondria from the cell and maintaining the cell's good health depends heavily on mitochondrial dynamics. ALS patients have shown dysregulated mitophagy and mitochondrial dynamics due to various mutations. TDP-43 gene protein regulated the expression of parkin, and the mutation in TDP-43 leads to decreased parkin availability to mitochondria and decreased the parkin-dependent mitophagy pathway. An experimental study of parkin knock-out mice resulted in an increased survival rate in parkin deficient mice, from which the scientists have concluded that the short-term activity of parkin is healthy and neuroprotective, but in the long run, it causes dysfunctionality in the mitochondrial dynamics (Lagier-Tourenne, Clotilde, et al., 2012; Chen, Yanbo, et al., 2016). p62/SQSTM1 and OPTN are two essential ubiquitin-binding proteins that regulate dysregulated mitochondria clearance from the cell via the PINK1/Parkin-dependent mitophagy pathway. Mutation in OPTN, P62, and TBK1 genes causes a loss of function that disrupts the process of mitophagy (Cirulli Elizabeth T et al., 2015).

As mentioned before, all the mutation that occurs in LAS directly or indirectly alters the proteins involved in the process of mitophagy resulting in the accumulation of dysregulated mitochondria in the cell. For the efficient supply of energy and proper functioning of neurons, correct mitophagy is essential.

Huntington's Disease

Huntington's disease is an autosomal neurodegenerative disorder characterized by preferential atrophy in the striatum as a result of the loss of GABAergic medium spiny neurons in the caudate nucleus and putamen, as well as abnormal expansion of the CAG triplet encoding polyglutamine (polyQ) tract at the N-terminus of the huntingtin (HTT) protein (Ross, Christopher A, and Sarah J Tabrizi., 2011; Rosas, H D et al., 2003). The ability to stabilize other proteins around it, such as protein complexes involved in transcriptional regulation, axonal trafficking of vesicles, autophagosomes, and mitochondria, nuclear export/import, and ubiquitin (UB)--mediated proteolysis, is provided by HTT, a large protein with multiple protein-protein interactions. Wild-type HTT forms a complex with huntingtin-associated protein1(HAP1), which regulates the fusion of the autophagosome with the lysosome, but the mHTT forms a complex with HAP1 that lead to the accumulation of autophagosomes and dysregulated mitochondria in the cells (Li, X J et al., 1995; Wang, David B et al., 2014). Interaction of mHTT with mitochondria leads to decreased production of ATP as the postpartum reports of HD patients state that their ETC complexes were not working correctly, and impaired calcium homeostasis was also found in the reports, which reflects the presence of dysregulated mitochondria in the brain cells of HD patients (Šonský, I et al., 2021). mHTT leads to the direct interaction between Fis1 and Drp1 protein, leading to the excessive fission of mitochondria (Qi, Xin, et al., 2013). In an experiment with an R6/2 mouse model, inhibiting the direct structure of mitochondria found in cardiac tissues is improved due to the interaction between Fis1 and Drp1. (Joshi, A U et al., 2019). mHTT strongly interacts with the fission protein DRP1 and Fis1 proteins, leading to an increase in their expression and fission of mitochondria resulting in impaired mitophagy. Postpartum reports of the brain of HD patients reveal the decreased expression of fusion proteins OPA1, TOMM40, MFN1, MFN2, and MFN2 and increased expression of the proteins Drp1 and Fis1 (Shirendeb, Ulziibat, et al., 2011). Glyceraldehyde 3-phosphate dehydrogenase (GAPDH) is a crucial component of the glycolytic cascade and aids in the autophagic engulfment of damaged mitochondria; however, in HD patients, GAPDH interacts improperly with the polyglutamine tracts, impairing mitophagy (Hwang, Sunhee, et al., 2015).

Cardiovascular Diseases

Cardiovascular diseases include coronary heart disease, cerebrovascular illness, peripheral arterial disease, rheumatic heart disease, and congenital heart disease. They are a category of conditions affecting the heart and blood arteries. Acute occurrences like heart attacks and strokes are brought on by blockages that stop blood from flowing to the heart or brain. In India, cardiovascular illnesses are the leading killers. Age-standardized CVD death rate in India is 272 per 100,000 people, significantly higher than the global average of 235, according to the survey "Global Burden of Disease" (Prabhakaran, Dorairaj, et al. 2016). The prevalence of cardiovascular disease is highest in India. The number of CVD-related deaths in India has climbed from 2.26 million in 1990 to 4.77 million in 2020. (Murray, C J, and A D Lopez., 2020).

Another significant indicator of aging is cardiac aging. Since the heart is one of the most energy-demanding organs and contains defective mitochondria, it has been discovered that reduced mitochondrial functioning with age is linked to cardiac aging (Jang, Ji Yong, et al., 2018). By lowering NAD+ levels, decreasing the efficiency of the respiratory chain, mutating the mtDNA, leaking electrons, and dysregulating mitochondrial synthesis, dysfunctional mitochondria impair the bioenergetics of the cell. In these circumstances, mitophagy is crucial for clearing the dysregulated mitochondria and preserving the cardiac cells' homeostasis. The following section discusses the role of mitochondria in certain significant cardiovascular disorders.

Ischemia Strokes

Ischemia strokes account for 87% of all stroke cases (Benjamin, Emelia J et al., 2019) and are caused by brain tissue damage due to sudden rupturing of blood vessels or blockage of blood flow in the brain. Blockage of blood flow further produces depletion of Oxygen and ATP production in the cells that lead to reduced activity of ATP-dependent ion channels, causing intracellular hyperosmolarity, cytotoxic edema, interfering with the glutamate receptor's ability to function, which eventually results in DNA damage, structural protein damage, or even cell death (Deb, Prabal, et al., 2010; Nishizawa, Y., 2001). Much evidence has been found that supports the role of dysfunctional mitochondria and impaired mitophagy in the occurrence of ischemia strokes (Ham, P Benson 3rd, et al., 2017; Sun, Ming-Shuo, et al., 2018). I/R causes mitochondrial cristae to disappear, and lowers the membrane potential, and opens the mPTP (mitochondrial permeability transition pore) that leads to the excessive activation or inhibition of mitophagy in the brain cells (Chouchani, Edward T et al., 2014).

The study with perfused rabbit hearts reveals the presence of lysosomal alteration after 40 min of ischemia and an increase in the number of autophagosomes upon reperfusion, showing that both conditions cause mitophagy to occur (Decker, R S, and K Wildenthal., 1980). Increased mitochondrial fragmentation and fission of brain cells have been found in patients having ischemia strokes mediated by DRP1 and Mfn1/2 (Barsoum, Mark J et al., 2006; Martorell-Riera, Alejandro, et al., 2014) along with the activation of mitophagy via multiple pathways. During the ischemic phase, in ATP-depleting conditions, mitophagy is activated via the adenosine monophosphatase-activated protein kinase (AMPK) pathway that leads to the activation of ULK1 followed by FUNDC1 mitophagy receptor activation (Kundu, Mondira, et al., 2008; Kim, Joungmok, et al., 2011). Lan et al. (2018) have found PINK1 accumulation on the surface of the outer mitochondrial membrane suggesting that Pink1/parkin-dependent mitophagy is also activated during ischemia and reperfusion conditions in brain cells. The role of mitophagy in ischemic strokes is still controversial. Some studies have found that mitophagy plays a neuroprotective role by activating mitochondrial fission and removing the dysfunctional mitochondria from the cell. It has been found that in the case of middle cerebral artery occlusion (MCAO), mitophagy gets activated within an hour and removes the dysfunctional mitochondria via the DRP1-dependent pathway, while in the reperfusion phase also after MCAO, mitophagy plays an essential role in the inhibition of the post-stroke apoptosis (Zuo, Wei, et al., 2014; Zhang, Xiangnan, et al., 2013). Mitochondrial fission is essential for removing dysregulated mitochondria, but upregulation of it harms the cells. I/R injury causes excessive production of PINK1, Parkin, DRP1, and BNIP3, leading to the over-activation of mitophagy, causing dysregulated mitochondrial dynamics.

Heart Failure

The last phase of many cardiac illnesses, heart failure, is distinguished by having the most significant death and morbidity rates. Breathlessness, ankle swelling, weariness, increased jugular venous pressure, pulmonary crackles, and peripheral edema are some indications and symptoms caused by structural or functional cardiac problems (Ponikowski, Piotr, et al., 2016). The current treatment methods used for HT are not able to cure it. It is essential to know about the factors responsible for proper treatment options. Mitophagy starts decreasing with age, and the onset of the disease-causing accumulation of dysregulated mitochondria in the cell leads to excessive production of ROS and peroxide levels in the cell, which are the primary pathophysiological conditions of the disease (Shires, Sarah E, and Åsa B Gustafsson, 2015). Mitochondria of failing hearts show impaired morphology, disorganized, and reduced cristae density which have also been found in parkin-deficient mice, along with reduced mitophagy due to dysfunctional mitochondria (Kubli, Dieter A et al., 2013; Bugger, Heiko et al., 2010). Reduced levels of mitophagy receptors such as Beclin1, LC3-II, Parkin, and PINK1 have been found in patients of HF, suggesting reduced levels of mitophagy (Morales, Pablo E et al., 2020). Parkin-dependent mitophagy declines with age, causing an increased accumulation of dysregulated mitochondria in the cell and making the heart more susceptible to heart failure in stressful conditions. Under hypoxic conditions and in vitro models of chronic HF, BNIP3-mediated mitophagy gets activated (Thompson, John W et al., 2015). DRP1-mediated mitochondrial fission regulates parkin recruitment and gives a protective response to cardiomyocytes (Lee, Youngil, et al., 2011). In patients with heart failure, decreased levels of DRP1 and MFN2 have led to the accumulation of sizeable structured parking in the heart, which could not interact with the LC3-mediated autophagosomes, leading to impaired mitophagy. DRP1 knockout mice have shown pressure-induced cardiac dysfunction followed by I/R injury suggesting the critical role of DRP1 in regulating mitochondrial dynamics (Thai, Phung N et al., 2019).

Atherosclerosis

One of the leading causes of death in industrialized countries is atherosclerosis which causes a considerable narrowing of the arteries due to plaque build-up inside the arterial lumen and is characterized by lipid buildup, increased vascular smooth muscle cell (VSMC) proliferation, matrix turnover, calcification, and inflammation (Wu, Meng-Yu et al., 2017). Studies with a mouse model and human specimen reveal the decreased level of mitophagy based on detecting mitophagy markers p62 and LC3-II isolated from plaques. Lower expression of LC3-II was lower in unstable plaques compared to stable plaques (Sergin, Ismail, et al., 2016; Morciano, Giampaolo, et al., 2020). Impaired mitophagy, which has been linked to the disease, causes a buildup of malfunctioning mitochondria, which increases ROS generation and causes LDL to be oxidized, which is involved in the apoptosis induction pathway of macrophages and the calcium-dependent mitochondrial pathway (Vindis, Cécile, et al., 2005). Oxidized LDL disrupts mitochondrial structure, increases the production of ROS, and induces mitochondrial dysfunction (Kattoor, Ajoe, John, et al., 2017). A study of thrombi harvested from animal models and humans reveals the reduced expression of the Mfn2 mitophagy receptor, and it is experimentally proved that overexpression of Mfn2 can reduce atherosclerosis lesions in rabbits and balloon-injured arteries of rats (Chen, Kuang-Hueih, et al., 2004). In aortic ring test in ex vivo and apolipoprotein E knockout mice, DRP1 inhibition reduced the level of VSMC proliferation, endothelial dysfunction, and atherosclerosis (Lim, Soyeon et al., 2015; Wang, Qilong et al., 2017). Collectively, these findings imply that mitophagy can be the potential target of atherosclerosis.

Diabetic Cardiomyopathy

One of the most prominent organ damage manifestations in diabetes complications is diabetic cardiomyopathy (DCM). Among people with diabetes, DCM is a significant source of morbidity and mortality. It is a non-ischemic, non-hypersensitive cardiomyopathy brought on by metabolic diseases such as diabetes. Although the exact etiology of DCM diseases is unknown, mitochondrial malfunction, oxidative stress, inflammation, and cardiomyocyte death are potential contributors (Wu, Shengnan et al., 2019). Cardiomyocytes of animal models of type 1&2 diabetes and atrial tissues of human diabetic subjects show dysregulated mitochondria and increased ROS production, causing cellular oxidative damage.

Fatty acids are the primary source of mitochondrial ATP synthesis in cardiac tissues through acetyl-CoA supply to the TCA cycle and NADH/FADH2 supply to the respiratory chain. A characteristic of DCM is cardiac steatosis, which indicates that the mitochondrion does not correctly metabolize fatty acids in DCM (McGavock, Jonathan M et al., 2007; Rijzewijk, Luuk J et al., 2008). Antioxidants that can reduce ROS in the cardiomyocytes can restore cardiac function and mitochondrial morphology in diabetic mice, suggesting the ROS's involvement in heart dysfunction and mitochondrial morphology (Shen, Xia et al., 2006; Liang, Qigong et al., 2002). Changes in mitochondrial morphology occur co-ordinately with changes in flux, and acute hyperglycaemic conditions trigger the activation of the mitochondrial fission protein DLP1, creating tiny mitochondria. Small mitochondrial formation elevates the level of ROS, followed by increased mitochondrial fragmentation and cell death (Yu, Tianzheng et al., 2006; Galloway, Chad A. et al., 2012; Makino, A. et al., 2010). In the mouse model of chronic hyperglycemia, there is a decreased level of the mitochondrial fusion protein OPA1 and an increased level of the fission protein DLP1, which together boost the cell's generation of ROS (Makino A et al., 2010).

Diabetes 1 and 2 have variable levels of mitophagy. However, it is yet unknown if mitophagy prevents or causes harm during DCM. In some previous studies, it was found that insulin inhibits mitophagy. However, due to its deficiency in type 1 diabetes, mitophagy would expect to be activated (Riehle, Christian, et al., 2013), but researchers have found that mitophagy is decreasing in diabetes 1 suggesting that mitophagy is inhibited at a later stage of DCM (Xie, Zhonglin, et al., 2011; Wang, Shanjie, et al., 2018). Low levels of PINK1 and Parkin mitophagy proteins have also been found in DCM1, along with a high level of GTPase

RAB9, suggesting the occurrence of atypical mitophagy (Xu, Xianmin et al., 2013). the study of mice having DCM2 reveals that the activation of mitophagy in the cardiac cell would be harmful (Russo, Sarah Brice, et al., 2012). Recent research has revealed that the Pink1/Parkin pathway is unnecessary for AMPK-mediated mitophagy by mitochondrial fission to engulf the mitophagosome (Seabright, Alex P et al., 2020).

Targeting Mitophagy in Aging

It is obvious now that mitophagy plays a vital role in pathology and a causal role in various diseases, which is gaining interest in pharmacological and medical fields. We need to find a way or method to activate mitophagy pathways without harming the body's normal functioning to remove the dysfunctional mitochondria causing disease pathologies. However, very few human studies on mitophagy modulators have been reported.

Quercetin is a natural flavonoid found abundantly in apples, honey, raspberries, onions, red grapes, cherries, citrus fruits, and green leafy vegetables (Hollman, P C, and M B Katan., 1999). It is used to treat various cardiovascular diseases, cancer, and neurodegenerative diseases (Serban, Maria-Corina, et al., 2016; Musthafa M. E., Akbar M., 2016; Hashemzaei, Mahmoud, et al., 2017). *Periplaneta americana* (*P. americana*), popularly known as the cockroach, has been used extensively as medicine in China for more than 2000 years due to its positive benefits on blood circulation, blood stasis, digestion, and diuresis. Increased H9C2 cell viability, decreased inflammatory mediator levels, and activation of Pink/Parkin mediated mitophagy are the outcomes of the research of *P. americana* extract on lipopolysaccharide-(LPS-) caused cardiomyocyte damage. (Li, Jie, et al., 2019). Citrus peels contain polymethoxyflavone quercetogetin (Gattuso, Giuseppe, et al., 2007). Quercetogetin inhibits apoptosis by lowering the expression of cleaved caspase 3, 8, and 9 along with the generation of ROS and the mitophagy proteins DRP1 and pink1 in a study using Beas-2B and NHBE human bronchial epithelial cells. (Son, Eun Suk, et al., 2018). *Pomegranate* and *Ginkgo biloba* extracts have been investigated primarily in vitro as mitophagy modulators. (Tan, Sijie, et al., 2019; Wang, Wenjun, et al., 2019). To improve mitophagy, notoginsenoside R1 (NGR1), a saponin isolated from *Panax notoginseng*, increases the levels of PINK1, Parkin, and the ratio of LC3-II/LC-I in the retinas of leptin receptor KO mice (db/db mice), a type 2 diabetes model. (Zhou, Ping, et al., 2019). Salidroside (Sal) is a phenylpropanoid glycoside extracted from *Rhodiola that promotes mitophagy, parkin expression, scavenger ROS, and inhibits apoptosis. It can treat neural diseases like spinal cord injury (*Zhang, Zengjie, et al., 2018; Wang, Chenggui, et al., 2018). A steroidal alkaloid called tomatidine has been discovered in tomato leaves and skins. Studies on Caenorhabditis elegans have demonstrated that the skinhead-1 protein (SKN-1)/nuclear factor erythroid 2-(NFE2-) related factor 2 (Nrf2) pathway is activated, sustaining muscle cell homeostasis and mitophagy, and extending life span (Fang, Evandro F et al., 2017; Caprioli, Giovanni, et al., 2015). Metformin was discovered to promote parkin-mediated mitophagy in an obese mice model by accelerating the breakdown of mitofusin and inhibiting the interaction between cytosolic p53 and parkin (Song, Young Mi et al., 2016).

Many pharmacological medicines or chemicals are also used to modulate mitophagy. Muscle atrophy in streptozocin-induced diabetic mice is treated with resveratrol. Due to increased mitochondrial biogenesis and a decrease in mitophagy, resveratrol consumption improves muscular performance in tests of rip strength and treadmill running. Resveratrol increased the expression of nuclear respiratory factor- (NRF-) 1, peroxisome proliferator-activated receptor gamma coactivator 1-alpha (PGC-1α), cytochrome c oxidase (Cox) IV, and mitochondrial transcription factor A (mtTFA). It reduced the expression of mitochondrial fission 1 protein (FIS1), mitochondrial fission regulatory proteins (e.g., phospho-dynamin-related protein 1 (p-DRP1)and mitochondrial fission factor (MFF)) and mitochondrial fusion regulatory proteins (e.g., p-DRP1 (Ser637), mitofusin-2 (MFN2), and OPA1 mitochondrial dynamin-like GTPase (OPA1) (Wang, Dongtao, et al., 2018). It is also used to treat Alzheimer's disease, reducing *caused cell death and* mitochondrial dysfunction (Wang, Hui, et al., 2018). Diabetes has long been treated using berberine, the primary active ingredient of the traditional Chinese remedies Coptis Root and Cortex Phellodendri. In a study, berberine (100 nM, for 30 min) decreased H9C2 cell hypertrophy by balancing the fusion and fission of mitochondria. (Hang, Weijian, et al., 2018).

When given before a cardiac ischemia episode, resveratrol glycoside SIRT1 activators have a potent antioxidative action and cardioprotective effect (Gurusamy, Narasimman, et al., 2010). Another medication for treating atherosclerosis is melatonin, which activates the Parkin/SIRT3/FOXO3a pathway to encourage mitophagy in atherogenic mouse models. (Ma, Sai, et al., 2018). It can activate parkin-mediated mitophagy in Parkinson's disease and be therapeutic (Díaz-Casado, María E et al., 2016). Metformin is famous for its anti-hyperglycaemic effect, and its ability to activate mitophagy has been used to treat various diseases especially type 2 DCM. It activates mitophagy via the AMPK pathway by increasing the expression of PINK1, parkin, Nix, and LC3 genes (Bhansali, Shipra, et al., 2020). The ability of nitric oxide to control mitochondrial dynamics and biogenesis is also well documented. According to research, NO can cause parkin to move from the nucleus to the mitochondria of dopaminergic neuronal cells (Han, Ji-Young, et al., 2015). Dopaminergic neurons can benefit from NO therapy by having their ETC enzyme levels return (Kim, Kyung-Hee et al., 2012).

Due to its capacity to activate to promote mitophagy even in the absence of PINK1 and Parkin, targeting de-ubiquitinating enzymes (DUBs) is currently of considerable interest. USP30, a cysteine protease found in the OMM, is one of the most interesting DUB targets because it inhibits parkin-mediated mitophagy (Bingol, Baris, et al., 2014).

Conclusion

TThe function of mitophagy in disorders related to old age and aging is well recognized. It is crucial for maintaining oxidative stress, calcium homeostasis, dysfunctional mitochondria, and other molecular events that lead to the occurrence of diseases. Both overexpression and reduced expression of mitophagy are harmful to the cell, making mitophagy a double-edged sword. In some diseases increased level of mitophagy causes extreme mitochondrial fission responsible for the pathogenesis of the disease, while in some decreased level of mitophagy causes a build-up of dysregulated mitochondria and ROS in the cell that eventually impacts the cellular machinery negatively. Therefore, mitophagy has proved itself as a possible therapeutic target for several age-related illnesses. Various tools have been developed for the treatment and activation of mitophagy. Many natural compounds and chemicals are developed that can regulate mitophagy in a correct appropriate manner. Lots of drugs are developing day by day that can regulate mitophagy.

There are still many unrevealed mitophagy druggable targets along the mitophagy pathway that can be used to treat impaired mitophagy. Research is required for proper information about the mechanism of action of discovered drugs to make their proper clinical use and trials.

Much of the data is available on the role of pink1/parkin-mediated mitophagy in the pathogenesis of the disease, but the information about other mitophagy pathways is still unrevealed. There is still no permanent cure for AD, PD DCM, and heart failure, which is required because of their high mortality and morbidity rates.

References

Abel, O., Powell, J. F., Andersen, P. M., & Al-Chalabi, A. (2012). ALSoD: A user-friendly online bioinformatics tool for amyotrophic lateral sclerosis genetics. *Human mutation*. 33(9), 1345–1351.

Ashrafi, G., & Schwarz, T. L. (2013). The pathways of mitophagy for quality control and clearance of mitochondria. *Cell death and differentiation*. 20(1), 31–42.

Baricault, L., Ségui, B., Guégand, L., Olichon, A., Valette, A., Larminat, F., & Lenaers, G. (2007). OPA1 cleavage depends on decreased mitochondrial ATP levels and bivalent metals. *Experimental cell research*. 313(17), 3800–3808.

Barsoum, M. J., Yuan, H., Gerencser, A. A., Liot, G., Kushnareva, Y., Gräber, S., Kovacs, I., Lee, W. D., Waggoner, J., Cui, J., White, A. D., Bossy, B., Martinou, J. C., Youle, R. J., Lipton, S. A., Ellisman, M. H., Perkins, G. A., & Bossy-Wetzel, E. (2006). Nitric oxide-induced mitochondrial fission is regulated by dynamin-related GTPases in neurons. *The EMBO journal*. 25(16), 3900–3911.

Bayne, A. C., Mockett, R. J., Orr, W. C., & Sohal, R. S. (2005). Enhanced catabolism of mitochondrial superoxide/hydrogen peroxide and aging in transgenic Drosophila. *The Biochemical journal*. 391(Pt 2), 277–284.

Benjamin, E. J., Muntner, P., Alonso, A., Bittencourt, M. S., Callaway, C. W., Carson, A. P., Chamberlain, A. M., Chang, A. R., Cheng, S., Das, S. R., Delling, F. N., Djousse, L., Elkind, M., Ferguson, J. F., Fornage, M., Jordan, L. C., Khan, S. S., Kissela, B. M., Knutson, K. L., Kwan, T. W., American Heart Association Council on Epidemiology and Prevention Statistics Committee and Stroke Statistics Subcommittee (2019). Heart Disease and Stroke Statistics-2019 Update: A Report From the American Heart Association. *Circulation*. 139(10), e56–e528.

Bhansali, S., Bhansali, A., Dutta, P., Walia, R., & Dhawan, V. (2020). Metformin upregulates mitophagy in patients with T2DM: A randomized placebo-controlled study. *Journal of cellular and molecular medicine*. 24(5), 2832–2846.

Bleazard, W., McCaffery, J. M., King, E. J., Bale, S., Mozdy, A., Tieu, Q., Nunnari, J., & Shaw, J. M. (1999). The dynamin-related GTPase Dnm1 regulates mitochondrial fission in yeast. *Nature cell biology*. 1(5), 298–304.

Bokov, A., Chaudhuri, A., & Richardson, A. (2004). The role of oxidative damage and stress in aging. *Mechanisms of ageing and development*. 125(10-11), 811–826.

Brandt, T., Cavellini, L., Kühlbrandt, W., & Cohen, M. M. (2016). A mitofusin-dependent docking ring complex triggers mitochondrial fusion in vitro. *eLife*. 5, e14618.

Bua, E., Johnson, J., Herbst, A., Delong, B., McKenzie, D., Salamat, S., & Aiken, J. M. (2006). Mitochondrial DNA-deletion mutations accumulate intracellularly to detrimental levels in aged human skeletal muscle fibers. *American journal of human genetics*. 79(3), 469–480.

Bugger, H., Schwarzer, M., Chen, D., Schrepper, A., Amorim, P. A., Schoepe, M., Nguyen, T. D., Mohr, F. W., Khalimonchuk, O., Weimer, B. C., & Doenst, T. (2010). Proteomic remodelling of mitochondrial oxidative pathways in pressure overload-induced heart failure. *Cardiovascular research*. 85(2), 376–384.

Cagalinec, M., Safiulina, D., Liiv, M., Liiv, J., Choubey, V., Wareski, P., Veksler, V., & Kaasik, A. (2013). Principles of the mitochondrial fusion and fission cycle in neurons. *Journal of cell science*. 126(Pt 10), 2187–2197.

Caprioli, G., Cahill, M., Logrippo, S., & James, K. (2015). Elucidation of the mass fragmentation pathways of tomatidine and β1-hydroxytomatine using orbitrap mass spectrometry. *Natural product communications*. 10(4), 575–576.

Castro, M., Suarez, E., Kraiselburd, E., Isidro, A., Paz, J., Ferder, L., & Ayala-Torres, S. (2012). Aging increases mitochondrial DNA damage and oxidative stress in liver of rhesus monkeys. *Experimental gerontology*. 47(1), 29–37.

Chakrabarti, R., Ji, W. K., Stan, R. V., de Juan Sanz, J., Ryan, T. A., & Higgs, H. N. (2018). INF2-mediated actin polymerization at the ER stimulates mitochondrial calcium uptake, inner membrane constriction, and division. *The Journal of cell biology*. 217(1), 251–268.

Chan D. C. (2006). Mitochondria: dynamic organelles in disease, aging, and development. *Cell*. 125(7), 1241–1252.

Chan D. C. (2012). Fusion and fission: interlinked processes critical for mitochondrial health. *Annual review of genetics*. 46, 265–287.

Chaugule, V. K., Burchell, L., Barber, K. R., Sidhu, A., Leslie, S. J., Shaw, G. S., & Walden, H. (2011). Autoregulation of Parkin activity through its ubiquitin-like domain. *The EMBO journal*. 30(14), 2853–2867.

Chen, G., Han, Z., Feng, D., Chen, Y., Chen, L., Wu, H., Huang, L., Zhou, C., Cai, X., Fu, C., Duan, L., Wang, X., Liu, L., Liu, X., Shen, Y., Zhu, Y., & Chen, Q. (2014). A regulatory signaling loop comprising the PGAM5 phosphatase and CK2 controls receptor-mediated mitophagy. *Molecular cell*. 54(3), 362–377.

Chen, H., & Chan, D. C. (2009). Mitochondrial dynamics--fusion, fission, movement, and mitophagy--in neurodegenerative diseases. *Human molecular genetics*. 18(R2), R169–R176.

Chen, H., Detmer, S. A., Ewald, A. J., Griffin, E. E., Fraser, S. E., & Chan, D. C. (2003). Mitofusins Mfn1 and Mfn2 coordinately regulate mitochondrial fusion and are essential for embryonic development. *The Journal of cell biology*. 160(2), 189–200.

Chen, J. Q., Yager, J. D., & Russo, J. (2005). Regulation of mitochondrial respiratory chain structure and function by estrogens/estrogen receptors and potential physiological/pathophysiological implications. *Biochimica et biophysica acta*. 1746(1), 1–17.

Chen, K. H., Guo, X., Ma, D., Guo, Y., Li, Q., Yang, D., Li, P., Qiu, X., Wen, S., Xiao, R. P., & Tang, J. (2004). Dysregulation of HSG triggers vascular proliferative disorders. *Nature cell biology*. 6(9), 872–883.

Chen, L., Na, R., Boldt, E., & Ran, Q. (2015). NLRP3 inflammasome activation by mitochondrial reactive oxygen species plays a key role in long-term cognitive impairment induced by paraquat exposure. *Neurobiology of aging*. 36(9), 2533–2543.

Chen, M., Chen, Z., Wang, Y., Tan, Z., Zhu, C., Li, Y., Han, Z., Chen, L., Gao, R., Liu, L., & Chen, Q. (2016). Mitophagy receptor FUNDC1 regulates mitochondrial dynamics and mitophagy. *Autophagy*. 12(4), 689–702.

Cho, B., Cho, H. M., Jo, Y., Kim, H. D., Song, M., Moon, C., Kim, H., Kim, K., Sesaki, H., Rhyu, I. J., Kim, H., & Sun, W. (2017). Constriction of the mitochondrial inner compartment is a priming event for mitochondrial division. *Nature communications*. 8, 15754.

Choi, S. Y., Huang, P., Jenkins, G. M., Chan, D. C., Schiller, J., & Frohman, M. A. (2006). A common lipid links Mfn-mediated mitochondrial fusion and SNARE-regulated exocytosis. *Nature cell biology*. 8(11), 1255–1262.

Chouchani, E. T., Pell, V. R., Gaude, E., Aksentijević, D., Sundier, S. Y., Robb, E. L., Logan, A., Nadtochiy, S. M., Ord, E., Smith, A. C., Eyassu, F., Shirley, R., Hu, C. H., Dare, A. J., James, A. M., Rogatti, S., Hartley, R. C., Eaton, S., Costa, A., Brookes, P. S., … Murphy, M. P. (2014). Ischaemic accumulation of succinate controls reperfusion injury through mitochondrial ROS. *Nature*. 515(7527), 431–435.

Chu C. T. (2019). Multiple pathways for mitophagy: A neurodegenerative conundrum for Parkinson's disease. *Neuroscience letters*. 697, 66–71.

Cipolat, S., Martins de Brito, O., Dal Zilio, B., & Scorrano, L. (2004). OPA1 requires mitofusin 1 to promote mitochondrial fusion. *Proceedings of the National Academy of Sciences of the United States of America*. 101(45), 15927–15932.

Cirulli, E. T., Lasseigne, B. N., Petrovski, S., Sapp, P. C., Dion, P. A., Leblond, C. S., Couthouis, J., Lu, Y. F., Wang, Q., Krueger, B. J., Ren, Z., Keebler, J., Han, Y., Levy, S. E., Boone, B. E., Wimbish, J. R., Waite,

L. L., Jones, A. L., Carulli, J. P., Day-Williams, A. G., Goldstein, D. B. (2015). Exome sequencing in amyotrophic lateral sclerosis identifies risk genes and pathways. *Science (New York, N.Y).* 347(6229), 1436–1441.

Clark, I. E., Dodson, M. W., Jiang, C., Cao, J. H., Huh, J. R., Seol, J. H., Yoo, S. J., Hay, B. A., & Guo, M. (2006). Drosophila pink1 is required for mitochondrial function and interacts genetically with parkin. *Nature.* 441(7097), 1162–1166.

Cohen, M. M., Leboucher, G. P., Livnat-Levanon, N., Glickman, M. H., & Weissman, A. M. (2008). Ubiquitin-proteasome-dependent degradation of a mitofusin, a critical regulator of mitochondrial fusion. *Molecular biology of the cell.* 19(6), 2457–2464.

Cortopassi, G. A., & Arnheim, N. (1990). Detection of a specific mitochondrial DNA deletion in tissues of older humans. *Nucleic acids research.* 18(23), 6927–6933.

de Brito, O. M., & Scorrano, L. (2008). Mitofusin 2 tethers endoplasmic reticulum to mitochondria. *Nature.* 456(7222), 605–610.

De Strooper, B., & Karran, E. (2016). The Cellular Phase of Alzheimer's Disease. *Cell.* 164(4), 603–615.

Deb, P., Sharma, S., & Hassan, K. M. (2010). Pathophysiologic mechanisms of acute ischemic stroke: An overview with emphasis on therapeutic significance beyond thrombolysis. *Pathophysiology: the official journal of the International Society for Pathophysiology.* 17(3), 197–218.

Decker, R. S., & Wildenthal, K. (1980). Lysosomal alterations in hypoxic and reoxygenated hearts. I. Ultrastructural and cytochemical changes. *The American journal of pathology.* 98(2), 425–444.

Delettre, C., Lenaers, G., Griffoin, J. M., Gigarel, N., Lorenzo, C., Belenguer, P., Pelloquin, L., Grosgeorge, J., Turc-Carel, C., Perret, E., Astarie-Dequeker, C., Lasquellec, L., Arnaud, B., Ducommun, B., Kaplan, J., & Hamel, C. P. (2000). Nuclear gene OPA1, encoding a mitochondrial dynamin-related protein, is mutated in dominant optic atrophy. *Nature genetics.* 26(2), 207–210.

Deng, H., Dodson, M. W., Huang, H., & Guo, M. (2008). The Parkinson's disease genes pink1 and parkin promote mitochondrial fission and/or inhibit fusion in Drosophila. *Proceedings of the National Academy of Sciences of the United States of America.* 105(38), 14503–14508.

DeVay, R. M., Dominguez-Ramirez, L., Lackner, L. L., Hoppins, S., Stahlberg, H., & Nunnari, J. (2009). Coassembly of Mgm1 isoforms requires cardiolipin and mediates mitochondrial inner membrane fusion. *The Journal of cell biology.* 186(6), 793–803.

Doblado, L., Lueck, C., Rey, C., Samhan-Arias, A. K., Prieto, I., Stacchiotti, A., & Monsalve, M. (2021). Mitophagy in Human Diseases. *International journal of molecular sciences.* 22(8), 3903.

Douarre, C., Sourbier, C., Dalla Rosa, I., Brata Das, B., Redon, C. E., Zhang, H., Neckers, L., & Pommier, Y. (2012). Mitochondrial topoisomerase I is critical for mitochondrial integrity and cellular energy metabolism. *PloS one.* 7(7), e41094.

Du, C., Fang, M., Li, Y., Li, L., & Wang, X. (2000). Smac, a mitochondrial protein that promotes cytochrome c-dependent caspase activation by eliminating IAP inhibition. *Cell.* 102(1), 33–42.

Esposito, L., Raber, J., Kekonius, L., Yan, F., Yu, G. Q., Bien-Ly, N., Puoliväli, J., Scearce-Levie, K., Masliah, E., & Mucke, L. (2006). Reduction in mitochondrial superoxide dismutase modulates Alzheimer's disease-like pathology and accelerates the onset of behavioral changes in human amyloid precursor protein transgenic mice. *The Journal of neuroscience: the official journal of the Society for Neuroscience.* 26(19), 5167–5179.

Essa, M. M., Akbar, M., & Guillemin, G. (Eds.). (2016). *The Benefits of Natural Products for Neurodegenerative Diseases.* Cham, Switzerland: Springer.

Fang, E. F., Waltz, T. B., Kassahun, H., Lu, Q., Kerr, J. S., Morevati, M., Fivenson, E. M., Wollman, B. N., Marosi, K., Wilson, M. A., Iser, W. B., Eckley, D. M., Zhang, Y., Lehrmann, E., Goldberg, I. G., Scheibye-Knudsen, M., Mattson, M. P., Nilsen, H., Bohr, V. A., & Becker, K. G. (2017). Tomatidine enhances lifespan and healthspan in C. elegans through mitophagy induction via the SKN-1/Nrf2 pathway. *Scientific reports.* 7, 46208.

Flannery, P. J., & Trushina, E. (2019). Mitochondrial dynamics and transport in Alzheimer's disease. *Molecular and cellular neurosciences.* 98, 109–120.

Frey, T. G., & Mannella, C. A. (2000). The internal structure of mitochondria. *Trends in biochemical sciences.* 25(7), 319–324.

Friedman, J. R., Lackner, L. L., West, M., DiBenedetto, J. R., Nunnari, J., & Voeltz, G. K. (2011). ER tubules mark sites of mitochondrial division. *Science (New York, N.Y.)*. 334(6054), 358–362.

Fröhlich, C., Grabiger, S., Schwefel, D., Faelber, K., Rosenbaum, E., Mears, J., Rocks, O., & Daumke, O. (2013). Structural insights into oligomerization and mitochondrial remodelling of dynamin 1-like protein. *The EMBO journal*. 32(9), 1280–1292.

Fu, J. D., & Yang, H. T. (2006). Developmental regulation of intracellular calcium homeostasis in early cardiac myocytes. *Sheng li xue bao: [Acta physiologica Sinica]*. 58(2), 95–103.

Galloway, C. A., Lee, H., Nejjar, S., Jhun, B. S., Yu, T., Hsu, W., & Yoon, Y. (2012). Transgenic control of mitochondrial fission induces mitochondrial uncoupling and relieves diabetic oxidative stress. *Diabetes*. 61(8), 2093–2104.

Gandre-Babbe, S., & van der Bliek, A. M. (2008). The novel tail-anchored membrane protein Mff controls mitochondrial and peroxisomal fission in mammalian cells. *Molecular biology of the cell*. 19(6), 2402–2412.

Gattuso, G., Barreca, D., Gargiulli, C., Leuzzi, U., & Caristi, C. (2007). Flavonoid composition of Citrus juices. *Molecules (Basel, Switzerland)*. 12(8), 1641–1673.

Green, D. R., & Kroemer, G. (2004). The pathophysiology of mitochondrial cell death. *Science (New York, N.Y.)*. 305(5684), 626–629.

Griffin, E. E., Graumann, J., & Chan, D. C. (2005). The WD40 protein Caf4p is a component of the mitochondrial fission machinery and recruits Dnm1p to mitochondria. *The Journal of cell biology*. 170(2), 237–248.

Gurusamy, N., Lekli, I., Mukherjee, S., Ray, D., Ahsan, M. K., Gherghiceanu, M., Popescu, L. M., & Das, D. K. (2010). Cardioprotection by resveratrol: a novel mechanism via autophagy involving the mTORC2 pathway. *Cardiovascular research*. 86(1), 103–112.

Haddad, D., & Nakamura, K. (2015). Understanding the susceptibility of dopamine neurons to mitochondrial stressors in Parkinson's disease. *FEBS letters*. 589(24 Pt A), 3702–3713.

Ham, P. B., 3rd, & Raju, R. (2017). Mitochondrial function in hypoxic ischemic injury and influence of aging. *Progress in neurobiology*. 157, 92–116.

Han, J. Y., Kang, M. J., Kim, K. H., Han, P. L., Kim, H. S., Ha, J. Y., & Son, J. H. (2015). Nitric oxide induction of Parkin translocation in PTEN-induced putative kinase 1 (PINK1) deficiency: functional role of neuronal nitric oxide synthase during mitophagy. *The Journal of biological chemistry*. 290(16), 10325–10335.

Hang, W., He, B., Chen, J., Xia, L., Wen, B., Liang, T., Wang, X., Zhang, Q., Wu, Y., Chen, Q., & Chen, J. (2018). Berberine Ameliorates High Glucose-Induced Cardiomyocyte Injury via AMPK Signaling Activation to Stimulate Mitochondrial Biogenesis and Restore Autophagic Flux. *Frontiers in pharmacology*. 9, 1121.

HARMAN D. (1956). Aging: a theory based on free radical and radiation chemistry. *Journal of gerontology*. 11(3), 298–300.

Harman D. (1972). The biologic clock: the mitochondria? *Journal of the American Geriatrics Society*. 20(4), 145–147.

Harman D. (1981). The aging process. *Proceedings of the National Academy of Sciences of the United States of America*. 78(11), 7124–7128.

Hashemzaei, M., Delarami Far, A., Yari, A., Heravi, R. E., Tabrizian, K., Taghdisi, S. M., Sadegh, S. E., Tsarouhas, K., Kouretas, D., Tzanakakis, G., Nikitovic, D., Anisimov, N. Y., Spandidos, D. A., Tsatsakis, A. M., & Rezaee, R. (2017). Anticancer and apoptosis-inducing effects of quercetin in vitro and in vivo. *Oncology reports*. 38(2), 819–828.

Hollensworth, S. B., Shen, C., Sim, J. E., Spitz, D. R., Wilson, G. L., & LeDoux, S. P. (2000). Glial cell type-specific responses to menadione-induced oxidative stress. *Free radical biology & medicine*. 28(8), 1161–1174.

Hollman, P. C., & Katan, M. B. (1999). Dietary flavonoids: intake, health effects and bioavailability. *Food and chemical toxicology: an international journal published for the British Industrial Biological Research Association*. 37(9-10), 937–942.

Hwang, S., Disatnik, M. H., & Mochly-Rosen, D. (2015). Impaired GAPDH-induced mitophagy contributes to the pathology of Huntington's disease. *EMBO molecular medicine*. 7(10), 1307–1326.

Ingerman, E., Perkins, E. M., Marino, M., Mears, J. A., McCaffery, J. M., Hinshaw, J. E., & Nunnari, J. (2005). Dnm1 forms spirals that are structurally tailored to fit mitochondria. *The Journal of cell biology.* 170(7), 1021–1027.

Ishihara, N., Eura, Y., & Mihara, K. (2004). Mitofusin 1 and 2 play distinct roles in mitochondrial fusion reactions via GTPase activity. *Journal of cell science.* 117(Pt 26), 6535–6546.

Ishihara, N., Fujita, Y., Oka, T., & Mihara, K. (2006). Regulation of mitochondrial morphology through proteolytic cleavage of OPA1. *The EMBO journal.* 25(13), 2966–2977.

Jang, J. Y., Blum, A., Liu, J., & Finkel, T. (2018). The role of mitochondria in aging. *The Journal of clinical investigation.* 128(9), 3662–3670.

Jin K. (2010). Modern Biological Theories of Aging. *Aging and disease,* 1(2), 72–74.

Johnson, A. A., Akman, K., Calimport, S. R., Wuttke, D., Stolzing, A., & de Magalhães, J. P. (2012). The role of DNA methylation in aging, rejuvenation, and age-related disease. *Rejuvenation research.* 15(5), 483–494.

Joshi, G., Sultana, R., Perluigi, M., & Butterfield, D. A. (2005). In vivo protection of synaptosomes from oxidative stress mediated by Fe2+/H2O2 or 2,2-azobis-(2-amidinopropane) dihydrochloride by the glutathione mimetic tricyclodecan-9-yl-xanthogenate. *Free radical biology & medicine.* 38(8), 1023–1031.

Kamer, K. J., & Mootha, V. K. (2015). The molecular era of the mitochondrial calcium uniporter. Nature reviews. *Molecular cell biology.* 16(9), 545–553.

Kane, L. A., Lazarou, M., Fogel, A. I., Li, Y., Yamano, K., Sarraf, S. A., Banerjee, S., & Youle, R. J. (2014). PINK1 phosphorylates ubiquitin to activate Parkin E3 ubiquitin ligase activity. *The Journal of cell biology.* 205(2), 143–153.

Karbowski, M., Lee, Y. J., Gaume, B., Jeong, S. Y., Frank, S., Nechushtan, A., Santel, A., Fuller, M., Smith, C. L., & Youle, R. J. (2002). Spatial and temporal association of Bax with mitochondrial fission sites, Drp1, and Mfn2 during apoptosis. *The Journal of cell biology.* 159(6), 931–938.

Kattoor, A. J., Pothineni, N., Palagiri, D., & Mehta, J. L. (2017). Oxidative Stress in Atherosclerosis. *Current atherosclerosis reports.* 19(11), 42.

Kazlauskaite, A., Kondapalli, C., Gourlay, R., Campbell, D. G., Ritorto, M. S., Hofmann, K., Alessi, D. R., Knebel, A., Trost, M., & Muqit, M. M. (2014). Parkin is activated by PINK1-dependent phosphorylation of ubiquitin at Ser65. *The Biochemical journal.* 460(1), 127–139.

Kazlauskaite, A., Martínez-Torres, R. J., Wilkie, S., Kumar, A., Peltier, J., Gonzalez, A., Johnson, C., Zhang, J., Hope, A. G., Peggie, M., Trost, M., van Aalten, D. M., Alessi, D. R., Prescott, A. R., Knebel, A., Walden, H., & Muqit, M. M. (2015). Binding to serine 65-phosphorylated ubiquitin primes Parkin for optimal PINK1-dependent phosphorylation and activation. *EMBO reports.* 16(8), 939–954.

Khalil, B., El Fissi, N., Aouane, A., Cabirol-Pol, M. J., Rival, T., & Liévens, J. C. (2015). PINK1-induced mitophagy promotes neuroprotection in Huntington's disease. *Cell death & disease.* 6(1), e1617.

Kim, J., Kundu, M., Viollet, B., & Guan, K. L. (2011). AMPK and mTOR regulate autophagy through direct phosphorylation of Ulk1. *Nature cell biology.* 13(2), 132–141.

Kim, K. H., Song, K., Yoon, S. H., Shehzad, O., Kim, Y. S., & Son, J. H. (2012). Rescue of PINK1 protein null-specific mitochondrial complex IV deficits by ginsenoside Re activation of nitric oxide signaling. *The Journal of biological chemistry.* 287(53), 44109–44120.

Kitada, T., Asakawa, S., Hattori, N., Matsumine, H., Yamamura, Y., Minoshima, S., Yokochi, M., Mizuno, Y., & Shimizu, N. (1998). Mutations in the parkin gene cause autosomal recessive juvenile parkinsonism. *Nature.* 392(6676), 605–608.

Klein, C., & Westenberger, A. (2012). *Genetics of Parkinson's disease. Cold Spring Harbor perspectives in medicine.* 2(1), a008888.

Kocahan, S., & Doğan, Z. (2017). Mechanisms of Alzheimer's Disease Pathogenesis and Prevention: The Brain, Neural Pathology, N-methyl-D-aspartate Receptors, Tau Protein and Other Risk Factors. *Clinical psychopharmacology and neuroscience: the official scientific journal of the Korean College of Neuropsychopharmacology.* 15(1), 1–8.

Koirala, S., Bui, H. T., Schubert, H. L., Eckert, D. M., Hill, C. P., Kay, M. S., & Shaw, J. M. (2010). Molecular architecture of a dynamin adaptor: implications for assembly of mitochondrial fission complexes. *The Journal of cell biology.* 191(6), 1127–1139.

Koirala, S., Guo, Q., Kalia, R., Bui, H. T., Eckert, D. M., Frost, A., & Shaw, J. M. (2013). Interchangeable adaptors regulate mitochondrial dynamin assembly for membrane scission. *Proceedings of the National Academy of Sciences of the United States of America.* 110(15), E1342–E1351.

Kraus, F., & Ryan, M. T. (2017). The constriction and scission machineries involved in mitochondrial fission. *Journal of cell science.* 130(18), 2953–2960.

Kubli, D. A., Zhang, X., Lee, Y., Hanna, R. A., Quinsay, M. N., Nguyen, C. K., Jimenez, R., Petrosyan, S., Murphy, A. N., & Gustafsson, A. B. (2013). Parkin protein deficiency exacerbates cardiac injury and reduces survival following myocardial infarction. *The Journal of biological chemistry.* 288(2), 915–926.

Kundu, M., Lindsten, T., Yang, C. Y., Wu, J., Zhao, F., Zhang, J., Selak, M. A., Ney, P. A., & Thompson, C. B. (2008). Ulk1 plays a critical role in the autophagic clearance of mitochondria and ribosomes during reticulocyte maturation. *Blood.* 112(4), 1493–1502.

Labrousse, A. M., Zappaterra, M. D., Rube, D. A., & van der Bliek, A. M. (1999). C. elegans dynamin-related protein DRP-1 controls severing of the mitochondrial outer membrane. *Molecular cell.* 4(5), 815–826.

Lagier-Tourenne, C., Polymenidou, M., Hutt, K. R., Vu, A. Q., Baughn, M., Huelga, S. C., Clutario, K. M., Ling, S. C., Liang, T. Y., Mazur, C., Wancewicz, E., Kim, A. S., Watt, A., Freier, S., Hicks, G. G., Donohue, J. P., Shiue, L., Bennett, C. F., Ravits, J., Cleveland, D. W., Yeo, G. W. (2012). Divergent roles of ALS-linked proteins FUS/TLS and TDP-43 intersect in processing long pre-mRNAs. *Nature neuroscience.* 15(11), 1488–1497.

Lazarou, M., Sliter, D. A., Kane, L. A., Sarraf, S. A., Wang, C., Burman, J. L., Sideris, D. P., Fogel, A. I., & Youle, R. J. (2015). The ubiquitin kinase PINK1 recruits autophagy receptors to induce mitophagy. *Nature.* 524(7565), 309–314.

Lee, J. E., Westrate, L. M., Wu, H., Page, C., & Voeltz, G. K. (2016). Multiple dynamin family members collaborate to drive mitochondrial division. *Nature.* 540(7631), 139–143.

Lee, Y. J., Jeong, S. Y., Karbowski, M., Smith, C. L., & Youle, R. J. (2004). Roles of the mammalian mitochondrial fission and fusion mediators Fis1, Drp1, and Opa1 in apoptosis. *Molecular biology of the cell.* 15(11), 5001–5011.

Lee, Y., Lee, H. Y., Hanna, R. A., & Gustafsson, Å. B. (2011). Mitochondrial autophagy by Bnip3 involves Drp1-mediated mitochondrial fission and recruitment of Parkin in cardiac myocytes. American journal of physiology. *Heart and circulatory physiology.* 301(5), H1924–H1931.

Lewis, S. C., Uchiyama, L. F., & Nunnari, J. (2016). ER-mitochondria contacts couple mtDNA synthesis with mitochondrial division in human cells. *Science (New York, N.Y.).* 353(6296), aaf5549.

Li, J., Shi, W., Zhang, J., & Ren, L. (2019). To Explore the Protective Mechanism of PTEN-Induced Kinase 1 (PINK1)/Parkin Mitophagy-Mediated Extract of Periplaneta Americana on Lipopolysaccharide-Induced Cardiomyocyte Injury. *Medical science monitor: international medical journal of experimental and clinical research.* 25, 1383–1391.

Li, X. H., Du, L. L., Cheng, X. S., Jiang, X., Zhang, Y., Lv, B. L., Liu, R., Wang, J. Z., & Zhou, X. W. (2013). Glycation exacerbates the neuronal toxicity of β-amyloid. *Cell death & disease.* 4(6), e673.

Li, X. J., Li, S. H., Sharp, A. H., Nucifora, F. C., Jr, Schilling, G., Lanahan, A., Worley, P., Snyder, S. H., & Ross, C. A. (1995). A huntingtin-associated protein enriched in brain with implications for pathology. *Nature.* 378(6555), 398–402.

Liang, Q., Carlson, E. C., Donthi, R. V., Kralik, P. M., Shen, X., & Epstein, P. N. (2002). Overexpression of metallothionein reduces diabetic cardiomyopathy. *Diabetes.* 51(1), 174–181.

Lim, S., Lee, S. Y., Seo, H. H., Ham, O., Lee, C., Park, J. H., Lee, J., Seung, M., Yun, I., Han, S. M., Lee, S., Choi, E., & Hwang, K. C. (2015). Regulation of mitochondrial morphology by positive feedback interaction between PKCδ and Drp1 in vascular smooth muscle cell. *Journal of cellular biochemistry.* 116(4), 648–660.

Lim, Y., Rubio-Peña, K., Sobraske, P. J., Molina, P. A., Brookes, P. S., Galy, V., & Nehrke, K. (2019). Fndc-1 contributes to paternal mitochondria elimination in C. elegans. *Developmental biology.* 454(1), 15–20.

Liu, L., Feng, D., Chen, G., Chen, M., Zheng, Q., Song, P., Ma, Q., Zhu, C., Wang, R., Qi, W., Huang, L., Xue, P., Li, B., Wang, X., Jin, H., Wang, J., Yang, F., Liu, P., Zhu, Y., Sui, S., Chen, Q. (2012). Mitochondrial outer-membrane protein FUNDC1 mediates hypoxia-induced mitophagy in mammalian cells. *Nature cell biology.* 14(2), 177–185.

Lopes da Fonseca, T., & Outeiro, T. F. (2014). ATP13A2 and Alpha-synuclein: a Metal Taste in Autophagy. *Experimental neurobiology*. 23(4), 314–323.

Magalingam, K. B., Radhakrishnan, A., Ping, N. S., & Haleagrahara, N. (2018). Current Concepts of Neurodegenerative Mechanisms in Alzheimer's Disease. *BioMed research international*. 2018, 3740461.

Makino, A., Scott, B. T., & Dillmann, W. H. (2010). Mitochondrial fragmentation and superoxide anion production in coronary endothelial cells from a mouse model of type 1 diabetes. *Diabetologia*. 53(8), 1783–1794.

Malka, F., Guillery, O., Cifuentes-Diaz, C., Guillou, E., Belenguer, P., Lombès, A., & Rojo, M. (2005). Separate fusion of outer and inner mitochondrial membranes. *EMBO reports*. 6(9), 853–859.

Malpartida, A. B., Williamson, M., Narendra, D. P., Wade-Martins, R., & Ryan, B. J. (2021). Mitochondrial Dysfunction and Mitophagy in Parkinson's Disease: From Mechanism to Therapy. *Trends in biochemical sciences*. 46(4), 329–343.

Manczak, M., & Reddy, P. H. (2012). Abnormal interaction between the mitochondrial fission protein Drp1 and hyperphosphorylated tau in Alzheimer's disease neurons: implications for mitochondrial dysfunction and neuronal damage. *Human molecular genetics*. 21(11), 2538–2547.

Manczak, M., Calkins, M. J., & Reddy, P. H. (2011). Impaired mitochondrial dynamics and abnormal interaction of amyloid beta with mitochondrial protein Drp1 in neurons from patients with Alzheimer's disease: implications for neuronal damage. *Human molecular genetics*. 20(13), 2495–2509.

Martorell-Riera, A., Segarra-Mondejar, M., Muñoz, J. P., Ginet, V., Olloquequi, J., Pérez-Clausell, J., Palacín, M., Reina, M., Puyal, J., Zorzano, A., & Soriano, F. X. (2014). Mfn2 downregulation in excitotoxicity causes mitochondrial dysfunction and delayed neuronal death. *The EMBO journal*. 33(20), 2388–2407.

McGavock, J. M., Lingvay, I., Zib, I., Tillery, T., Salas, N., Unger, R., Levine, B. D., Raskin, P., Victor, R. G., & Szczepaniak, L. S. (2007). Cardiac steatosis in diabetes mellitus: a 1H-magnetic resonance spectroscopy study. *Circulation*. 116(10), 1170–1175.

Mears, J. A., Lackner, L. L., Fang, S., Ingerman, E., Nunnari, J., & Hinshaw, J. E. (2011). Conformational changes in Dnm1 support a contractile mechanism for mitochondrial fission. *Nature structural & molecular biology*. 18(1), 20–26.

Mears, J. A., Lackner, L. L., Fang, S., Ingerman, E., Nunnari, J., & Hinshaw, J. E. (2011). Conformational changes in Dnm1 support a contractile mechanism for mitochondrial fission. *Nature structural & molecular biology*. 18(1), 20–26.

Meglei, G., & McQuibban, G. A. (2009). The dynamin-related protein Mgm1p assembles into oligomers and hydrolyzes GTP to function in mitochondrial membrane fusion. *Biochemistry*. 48(8), 1774–1784.

Miquel, J., Economos, A. C., Fleming, J., & Johnson, J. E., Jr (1980). Mitochondrial role in cell aging. *Experimental gerontology*. 15(6), 575–591.

Mkrtchyan, G. V., Abdelmohsen, K., Andreux, P., Bagdonaite, I., Barzilai, N., Brunak, S., Cabreiro, F., de Cabo, R., Campisi, J., Cuervo, A. M., Demaria, M., Ewald, C. Y., Fang, E. F., Faragher, R., Ferrucci, L., Freund, A., Silva-García, C. G., Georgievskaya, A., Gladyshev, V. N., Glass, D. J., Gorbunova, V., de Grey, A., He, W. W., Hoeijmakers, J., Hoffmann, E., Horvath, S., Houtkooper, R. H., Jensen, M. K., Jensen, M. B., Kane, A., Kassem, M., de Keizer, P., Kennedy, B., Karsenty, G., Lamming, D. W., Lee, K. F., MacAulay, N., Mamoshina, P., Mellon, J., Molenaars, M., Moskalev, A., Mund, A., Niedernhofer, L., Osborne, B., Pak, H. H., Parkhitko, A., Raimundo, N., Rando, T. A., Rasmussen, L. J., Reis, C., Riedel, C. G., Franco-Romero, A., Schumacher, B., Sinclair, D. A., Suh, Y., Taub, P. R., Toiber, D., Treebak, J. T., Valenzano, D. R., Verdin, E., Vijg, J., Young, S., Zhang, L., Bakula, D., Zhavoronkov, A., Scheibye-Knudsen, M. "ARDD 2020: from aging mechanisms to interventions." *Aging* vol. 12,24 (2020): 24484-24503.

Moors, T. E., Paciotti, S., Ingrassia, A., Quadri, M., Breedveld, G., Tasegian, A., Chiasserini, D., Eusebi, P., Duran-Pacheco, G., Kremer, T., Calabresi, P., Bonifati, V., Parnetti, L., Beccari, T., & van de Berg, W. (2019). Characterization of Brain Lysosomal Activities in GBA-Related and Sporadic Parkinson's Disease and Dementia with Lewy Bodies. *Molecular neurobiology*. 56(2), 1344–1355.

Moors, T., Paciotti, S., Chiasserini, D., Calabresi, P., Parnetti, L., Beccari, T., & van de Berg, W. D. (2016). Lysosomal Dysfunction and α-Synuclein Aggregation in Parkinson's Disease: Diagnostic Links. *Movement disorders: official journal of the Movement Disorder Society*. 31(6), 791–801.

Morales, P. E., Arias-Durán, C., Ávalos-Guajardo, Y., Aedo, G., Verdejo, H. E., Parra, V., & Lavandero, S. (2020). Emerging role of mitophagy in cardiovascular physiology and pathology. *Molecular aspects of medicine*. 71, 100822.

Morciano, G., Patergnani, S., Bonora, M., Pedriali, G., Tarocco, A., Bouhamida, E., Marchi, S., Ancora, G., Anania, G., Wieckowski, M. R., Giorgi, C., & Pinton, P. (2020). Mitophagy in Cardiovascular Diseases. *Journal of clinical medicine*. 9(3), 892.

Mozdy, A. D., McCaffery, J. M., & Shaw, J. M. (2000). Dnm1p GTPase-mediated mitochondrial fission is a multi-step process requiring the novel integral membrane component Fis1p. *The Journal of cell biology*. 151(2), 367–380.

Nikoletopoulou, V., Markaki, M., Palikaras, K., & Tavernarakis, N. (2013). Crosstalk between apoptosis, necrosis and autophagy. *Biochimica et biophysica acta*. 1833(12), 3448–3459.

Nishizawa Y. (2001). Glutamate release and neuronal damage in ischemia. *Life sciences*. 69(4), 369–381.

Novak, I., Kirkin, V., McEwan, D. G., Zhang, J., Wild, P., Rozenknop, A., Rogov, V., Löhr, F., Popovic, D., Occhipinti, A., Reichert, A. S., Terzic, J., Dötsch, V., Ney, P. A., & Dikic, I. (2010). Nix is a selective autophagy receptor for mitochondrial clearance. *EMBO reports*. 11(1), 45–51.

Nunnari, J., & Suomalainen, A. (2012). Mitochondria: in sickness and in health. *Cell*. 148(6), 1145–1159.

Nuriel, T., Peng, K. Y., Ashok, A., Dillman, A. A., Figueroa, H. Y., Apuzzo, J., Ambat, J., Levy, E., Cookson, M. R., Mathews, P. M., & Duff, K. E. (2017). The Endosomal-Lysosomal Pathway Is Dysregulated by APOE4 Expression in Vivo. *Frontiers in neuroscience*. 11, 702.

Olichon, A., Emorine, L. J., Descoins, E., Pelloquin, L., Brichese, L., Gas, N., Guillou, E., Delettre, C., Valette, A., Hamel, C. P., Ducommun, B., Lenaers, G., & Belenguer, P. (2002). The human dynamin-related protein OPA1 is anchored to the mitochondrial inner membrane facing the inter-membrane space. *FEBS letters*. 523(1-3), 171–176.

Onyango, I. G., Khan, S. M., & Bennett, J. P., Jr (2017). Mitochondria in the pathophysiology of Alzheimer's and Parkinson's diseases. *Frontiers in bioscience (Landmark edition)*. 22(5), 854–872.

Orr, W. C., & Sohal, R. S. (1994). Extension of life-span by overexpression of superoxide dismutase and catalase in Drosophila melanogaster. *Science (New York, N.Y.)*. 263(5150), 1128–1130.

Palikaras, K., & Tavernarakis, N. (2012). Mitophagy in neurodegeneration and aging. *Frontiers in genetics*. 3, 297.

Palmer, C. S., Osellame, L. D., Laine, D., Koutsopoulos, O. S., Frazier, A. E., & Ryan, M. T. (2011). MiD49 and MiD51, new components of the mitochondrial fission machinery. *EMBO reports*. 12(6), 565–573.

Pickrell, A. M., & Youle, R. J. (2015). The roles of PINK1, parkin, and mitochondrial fidelity in Parkinson's disease. *Neuron*. 85(2), 257–273.

Ponikowski, P., Voors, A. A., Anker, S. D., Bueno, H., Cleland, J., Coats, A., Falk, V., González-Juanatey, J. R., Harjola, V. P., Jankowska, E. A., Jessup, M., Linde, C., Nihoyannopoulos, P., Parissis, J. T., Pieske, B., Riley, J. P., Rosano, G., Ruilope, L. M., Ruschitzka, F., Rutten, F. H., ESC Scientific Document Group (2016). 2016 ESC Guidelines for the diagnosis and treatment of acute and chronic heart failure: The Task Force for the diagnosis and treatment of acute and chronic heart failure of the European Society of Cardiology (ESC)Developed with the special contribution of the Heart Failure Association (HFA) of the ESC. *European heart journal*. 37(27), 2129–2200.

Porter, R. K., & Brand, M. D. (1995). Mitochondrial proton conductance and H+/O ratio are independent of electron transport rate in isolated hepatocytes. *The Biochemical journal*. 310 Pt 2), 379–382.

Prabhakaran, D., Jeemon, P., & Roy, A. (2016). Cardiovascular Diseases in India: Current Epidemiology and Future Directions. *Circulation*. 133(16), 1605–1620.

Qi, X., Qvit, N., Su, Y. C., & Mochly-Rosen, D. (2013). A novel Drp1 inhibitor diminishes aberrant mitochondrial fission and neurotoxicity. *Journal of cell science*. 126(Pt 3), 789–802.

Rambold, A. S., & Pearce, E. L. (2018). Mitochondrial Dynamics at the Interface of Immune Cell Metabolism and Function. *Trends in immunology*. 39(1), 6–18.

Reddy P. H. (2009). Amyloid beta, mitochondrial structural and functional dynamics in Alzheimer's disease. *Experimental neurology*. 218(2), 286–292.

Reddy, P. H., & Beal, M. F. (2008). Amyloid beta, mitochondrial dysfunction and synaptic damage: implications for cognitive decline in aging and Alzheimer's disease. *Trends in molecular medicine*. 14(2), 45–53.

Redza-Dutordoir, M., & Averill-Bates, D. A. (2016). Activation of apoptosis signalling pathways by reactive oxygen species. *Biochimica et biophysica acta,* 1863(12), 2977–2992.

Richter C. (1992). Reactive oxygen and DNA damage in mitochondria. *Mutation research*. 275(3-6), 249–255.

Riehle, C., Wende, A. R., Sena, S., Pires, K. M., Pereira, R. O., Zhu, Y., Bugger, H., Frank, D., Bevins, J., Chen, D., Perry, C. N., Dong, X. C., Valdez, S., Rech, M., Sheng, X., Weimer, B. C., Gottlieb, R. A., White, M. F., & Abel, E. D. (2013). Insulin receptor substrate signaling suppresses neonatal autophagy in the heart. *The Journal of clinical investigation*. 123(12), 5319–5333.

Rijzewijk, L. J., van der Meer, R. W., Smit, J. W., Diamant, M., Bax, J. J., Hammer, S., Romijn, J. A., de Roos, A., & Lamb, H. J. (2008). Myocardial steatosis is an independent predictor of diastolic dysfunction in type 2 diabetes mellitus. *Journal of the American College of Cardiology*. 52(22), 1793–1799.

Rojo, M., Legros, F., Chateau, D., & Lombès, A. (2002). Membrane topology and mitochondrial targeting of mitofusins, ubiquitous mammalian homologs of the transmembrane GTPase Fzo. *Journal of cell science*. 115(Pt 8), 1663–1674.

Ross W. N. (2012). Understanding calcium waves and sparks in central neurons. *Nature reviews. Neuroscience*. 13(3), 157–168.

Ross, C. A., & Tabrizi, S. J. (2011). Huntington's disease: from molecular pathogenesis to clinical treatment. *The Lancet. Neurology*. 10(1), 83–98.

Rujiviphat, J., Meglei, G., Rubinstein, J. L., & McQuibban, G. A. (2009). Phospholipid association is essential for dynamin-related protein Mgm1 to function in mitochondrial membrane fusion. *The Journal of biological chemistry*. 284(42), 28682–28686.

Russo, S. B., Baicu, C. F., Van Laer, A., Geng, T., Kasiganesan, H., Zile, M. R., & Cowart, L. A. (2012). Ceramide synthase 5 mediates lipid-induced autophagy and hypertrophy in cardiomyocytes. *The Journal of clinical investigation*. 122(11), 3919–3930.

Sas, K., Robotka, H., Toldi, J., & Vécsei, L. (2007). Mitochondria, metabolic disturbances, oxidative stress and the kynurenine system, with focus on neurodegenerative disorders. *Journal of the neurological sciences*. 257(1-2), 221–239.

Sauvé, V., Lilov, A., Seirafi, M., Vranas, M., Rasool, S., Kozlov, G., Sprules, T., Wang, J., Trempe, J. F., & Gehring, K. (2015). A Ubl/ubiquitin switch in the activation of Parkin. *The EMBO journal*. 34(20), 2492–2505.

Schapira, A. H., Cooper, J. M., Dexter, D., Jenner, P., Clark, J. B., & Marsden, C. D. (1989). Mitochondrial complex I deficiency in Parkinson's disease. *Lancet (London, England)*. 1(8649), 1269.

Seabright, A. P., Fine, N., Barlow, J. P., Lord, S. O., Musa, I., Gray, A., Bryant, J. A., Banzhaf, M., Lavery, G. G., Hardie, D. G., Hodson, D. J., Philp, A., & Lai, Y. C. (2020). AMPK activation induces mitophagy and promotes mitochondrial fission while activating TBK1 in a PINK1-Parkin independent manner. *FASEB journal: official publication of the Federation of American Societies for Experimental Biology*. 34(5), 6284–6301.

Selivanov, V. A., Votyakova, T. V., Pivtoraiko, V. N., Zeak, J., Sukhomlin, T., Trucco, M., Roca, J., & Cascante, M. (2011). Reactive oxygen species production by forward and reverse electron fluxes in the mitochondrial respiratory chain. *PLoS computational biology*. 7(3), e1001115.

Selkoe D. J. (2002). Alzheimer's disease is a synaptic failure. *Science (New York, N.Y.)*. 298(5594), 789–791.

Serban, M. C., Sahebkar, A., Zanchetti, A., Mikhailidis, D. P., Howard, G., Antal, D., Andrica, F., Ahmed, A., Aronow, W. S., Muntner, P., Lip, G. Y., Graham, I., Wong, N., Rysz, J., Banach, M., & Lipid and Blood Pressure Meta-analysis Collaboration (LBPMC) Group (2016). Effects of Quercetin on Blood Pressure: A Systematic Review and Meta-Analysis of Randomized Controlled Trials. *Journal of the American Heart Association*. 5(7), e002713.

Sergin, I., Bhattacharya, S., Emanuel, R., Esen, E., Stokes, C. J., Evans, T. D., Arif, B., Curci, J. A., & Razani, B. (2016). Inclusion bodies enriched for p62 and polyubiquitinated proteins in macrophages protect against atherosclerosis. *Science signaling*. 9(409), ra2.

Shen, X., Zheng, S., Metreveli, N. S., & Epstein, P. N. (2006). Protection of cardiac mitochondria by overexpression of MnSOD reduces diabetic cardiomyopathy. *Diabetes*. 55(3), 798–805.

Shirendeb, U., Reddy, A. P., Manczak, M., Calkins, M. J., Mao, P., Tagle, D. A., & Reddy, P. H. (2011). Abnormal mitochondrial dynamics, mitochondrial loss and mutant huntingtin oligomers in Huntington's disease: implications for selective neuronal damage. *Human molecular genetics*. 20(7), 1438–1455.

Shires, S. E., & Gustafsson, Å. B. (2015). Mitophagy and heart failure. *Journal of molecular medicine (Berlin, Germany).* 93(3), 253–262.

Smith, E. F., Shaw, P. J., & De Vos, K. J. (2019). The role of mitochondria in amyotrophic lateral sclerosis. *Neuroscience letters.* 710, 132933.

Sohal, R. S., & Orr, W. C. (2012). The redox stress hypothesis of aging. *Free radical biology & medicine.* 52(3), 539–555.

Sohal, R. S., Agarwal, S., Dubey, A., & Orr, W. C. (1993). Protein oxidative damage is associated with life expectancy of houseflies. *Proceedings of the National Academy of Sciences of the United States of America.* 90(15), 7255–7259.

Son, E. S., Kim, S. H., Ryter, S. W., Yeo, E. J., Kyung, S. Y., Kim, Y. J., Jeong, S. H., Lee, C. S., & Park, J. W. (2018). Quercetogetin protects against cigarette smoke extract-induced apoptosis in epithelial cells by inhibiting mitophagy. *Toxicology in vitro: an international journal published in association with BIBRA.* 48, 170–178.

Song, Y. M., Lee, W. K., Lee, Y. H., Kang, E. S., Cha, B. S., & Lee, B. W. (2016). Metformin Restores Parkin-Mediated Mitophagy, Suppressed by Cytosolic p53. *International journal of molecular sciences.* 17(1), 122.

Song, Z., Chen, H., Fiket, M., Alexander, C., & Chan, D. C. (2007). OPA1 processing controls mitochondrial fusion and is regulated by mRNA splicing, membrane potential, and Yme1L. *The Journal of cell biology.* 178(5), 749–755.

Šonský, I., Vodička, P., Vodičková Kepková, K., & Hansíková, H. (2021). Mitophagy in Huntington's disease. *Neurochemistry international.* 149, 105147.

Sowter, H. M., Ratcliffe, P. J., Watson, P., Greenberg, A. H., & Harris, A. L. (2001). HIF-1-dependent regulation of hypoxic induction of the cell death factors BNIP3 and NIX in human tumors. *Cancer research.* 61(18), 6669–6673.

Stadtman E. R. (1992). Protein oxidation and aging. *Science* (New York, N.Y.). 257(5074), 1220–1224.

Stewart, V. C., & Heales, S. J. (2003). Nitric oxide-induced mitochondrial dysfunction: implications for neurodegeneration. *Free radical biology & medicine.* 34(3), 287–303.

Stock, A. J., Kasus-Jacobi, A., Wren, J. D., Sjoelund, V. H., Prestwich, G. D., & Pereira, H. A. (2016). The Role of Neutrophil Proteins on the Amyloid Beta-RAGE Axis. *PloS one.* 11(9), e0163330.

Stojanovski, D., Koutsopoulos, O. S., Okamoto, K., & Ryan, M. T. (2004). Levels of human Fis1 at the mitochondrial outer membrane regulate mitochondrial morphology. *Journal of cell science.* 117(Pt 7), 1201–1210.

Sun, M. S., Jin, H., Sun, X., Huang, S., Zhang, F. L., Guo, Z. N., & Yang, Y. (2018). Free Radical Damage in Ischemia-Reperfusion Injury: An Obstacle in Acute Ischemic Stroke after Revascularization Therapy. *Oxidative medicine and cellular longevity.* 2018, 3804979.

Susin, S. A., Lorenzo, H. K., Zamzami, N., Marzo, I., Snow, B. E., Brothers, G. M., Mangion, J., Jacotot, E., Costantini, P., Loeffler, M., Larochette, N., Goodlett, D. R., Aebersold, R., Siderovski, D. P., Penninger, J. M., & Kroemer, G. (1999). Molecular characterization of mitochondrial apoptosis-inducing factor. *Nature.* 397(6718), 441–446.

Svenning, S., & Johansen, T. (2013). Selective autophagy. *Essays in biochemistry.* 55, 79–92.

Tan, S., Yu, C. Y., Sim, Z. W., Low, Z. S., Lee, B., See, F., Min, N., Gautam, A., Chu, J., Ng, K. W., & Wong, E. (2019). Pomegranate activates TFEB to promote autophagy-lysosomal fitness and mitophagy. *Scientific reports.* 9(1), 727.

Tanji, K., Mori, F., Kakita, A., Takahashi, H., & Wakabayashi, K. (2011). Alteration of autophagosomal proteins (LC3, GABARAP and GATE-16) in Lewy body disease. *Neurobiology of disease.* 43(3), 690–697.

Taylor, J. P., Brown, R. H., Jr, & Cleveland, D. W. (2016). Decoding ALS: from genes to mechanism. *Nature.* 539(7628), 197–206.

Thai, P. N., Seidlmayer, L. K., Miller, C., Ferrero, M., Dorn, G. W., II, Schaefer, S., Bers, D. M., & Dedkova, E. N. (2019). Mitochondrial Quality Control in Aging and Heart Failure: Influence of Ketone Bodies and Mitofusin-Stabilizing Peptides. *Frontiers in physiology.* 10, 382.

Thompson, J. W., Wei, J., Appau, K., Wang, H., Yu, H., Spiga, M. G., Graham, R. M., & Webster, K. A. (2015). Bnip3 Binds and Activates p300: Possible Role in Cardiac Transcription and Myocyte Morphology. *PloS one*. 10(8), e0136847.

Tieu, Q., & Nunnari, J. (2000). Mdv1p is a WD repeat protein that interacts with the dynamin-related GTPase, Dnm1p, to trigger mitochondrial division. *The Journal of cell biology*. 151(2), 353–366.

Tobore T. O. (2019). On the central role of mitochondria dysfunction and oxidative stress in Alzheimer's disease. *Neurological sciences: official journal of the Italian Neurological Society and of the Italian Society of Clinical Neurophysiology*. 40(8), 1527–1540.

Toescu E. C. (2005). *Normal brain ageing: models and mechanisms. Philosophical transactions of the Royal Society of London. Series B, Biological sciences*. 360(1464), 2347–2354.

Trifunovic, A., Wredenberg, A., Falkenberg, M., Spelbrink, J. N., Rovio, A. T., Bruder, C. E, Bohlooly-Y. M, Gidlöf, S., Oldfors, A., Wibom, R., Törnell, J., Jacobs. H. T., Larsson, N. G. (2004). Premature ageing in mice expressing defective mitochondrial DNA polymerase. *Nature*. 429, 417–423.

Valente, E. M., Abou-Sleiman, P. M., Caputo, V., Muqit, M. M., Harvey, K., Gispert, S., Ali, Z., Del Turco, D., Bentivoglio, A. R., Healy, D. G., Albanese, A., Nussbaum, R., González-Maldonado, R., Deller, T., Salvi, S., Cortelli, P., Gilks, W. P., Latchman, D. S., Harvey, R. J., Dallapiccola, B., Wood, N. W. (2004). Hereditary early-onset Parkinson's disease caused by mutations in PINK1. *Science (New York, N.Y.)*. 304(5674), 1158–1160.

Van Houten, B., Woshner, V., & Santos, J. H. (2006). Role of mitochondrial DNA in toxic responses to oxidative stress. *DNA repair*. 5(2), 145–152.

Viña, J., Borras, C., Abdelaziz, K. M., Garcia-Valles, R., & Gomez-Cabrera, M. C. (2013). The free radical theory of aging revisited: the cell signaling disruption theory of aging. *Antioxidants & redox signaling*. 19(8), 779–787.

Vindis, C., Elbaz, M., Escargueil-Blanc, I., Augé, N., Heniquez, A., Thiers, J. C., Nègre-Salvayre, A., & Salvayre, R. (2005). Two distinct calcium-dependent mitochondrial pathways are involved in oxidized LDL-induced apoptosis. *Arteriosclerosis, thrombosis, and vascular biology*. 25(3), 639–645.

Wang, C., & Youle, R. J. (2009). The role of mitochondria in apoptosis. *Annual review of genetics*. 43, 95–118.

Wang, D. B., Kinoshita, C., Kinoshita, Y., & Morrison, R. S. (2014). p53 and mitochondrial function in neurons. *Biochimica et biophysica acta*. 1842(8), 1186–1197.

Wang, D., Sun, H., Song, G., Yang, Y., Zou, X., Han, P., & Li, S. (2018). Resveratrol Improves Muscle Atrophy by Modulating Mitochondrial Quality Control in STZ-Induced Diabetic Mice. *Molecular nutrition & food research*. 62(9), e1700941.

Wang, H., Jiang, T., Li, W., Gao, N., & Zhang, T. (2018). Resveratrol attenuates oxidative damage through activating mitophagy in an in vitro model of Alzheimer's disease. *Toxicology letters*. 282, 100–108.

Wang, Q., Zhang, M., Torres, G., Wu, S., Ouyang, C., Xie, Z., & Zou, M. H. (2017). Metformin Suppresses Diabetes-Accelerated Atherosclerosis via the Inhibition of Drp1-Mediated Mitochondrial Fission. *Diabetes*. 66(1), 193–205.

Wang, S., Zhao, Z., Feng, X., Cheng, Z., Xiong, Z., Wang, T., Lin, J., Zhang, M., Hu, J., Fan, Y., Reiter, R. J., Wang, H., & Sun, D. (2018). Melatonin activates Parkin translocation and rescues the impaired mitophagy activity of diabetic cardiomyopathy through Mst1 inhibition. *Journal of cellular and molecular medicine*. 22(10), 5132–5144.

Wang, W., Wang, M., Ruan, Y., Tan, J., Wang, H., Yang, T., Li, J., & Zhou, Q. (2019). Ginkgolic Acids Impair Mitochondrial Function by Decreasing Mitochondrial Biogenesis and Promoting FUNDC1-Dependent Mitophagy. *Journal of agricultural and food chemistry*. 67(36), 10097–10106.

Wang, Y., Martinez-Vicente, M., Krüger, U., Kaushik, S., Wong, E., Mandelkow, E. M., Cuervo, A. M., & Mandelkow, E. (2009). Tau fragmentation, aggregation and clearance: the dual role of lysosomal processing. *Human molecular genetics*. 18(21), 4153–4170.

Wauer, T., Simicek, M., Schubert, A., & Komander, D. (2015). Mechanism of phospho-ubiquitin-induced PARKIN activation. *Nature*. 524(7565), 370–374.

Westermann B. (2008). Molecular machinery of mitochondrial fusion and fission. *The Journal of biological chemistry*. 283(20), 13501–13505.

Wills, J., Jones, J., Haggerty, T., Duka, V., Joyce, J. N., & Sidhu, A. (2010). Elevated tauopathy and alpha-synuclein pathology in postmortem Parkinson's disease brains with and without dementia. *Experimental neurology*. 225(1), 210–218.

Wu, H., Xue, D., Chen, G., Han, Z., Huang, L., Zhu, C., Wang, X., Jin, H., Wang, J., Zhu, Y., Liu, L., & Chen, Q. (2014). The BCL2L1 and PGAM5 axis defines hypoxia-induced receptor-mediated mitophagy. *Autophagy*. 10(10), 1712–1725.

Wu, M. Y., Li, C. J., Hou, M. F., & Chu, P. Y. (2017). New Insights into the Role of Inflammation in the Pathogenesis of Atherosclerosis. *International journal of molecular sciences*. 18(10), 2034.

Wu, S., Lu, Q., Ding, Y., Wu, Y., Qiu, Y., Wang, P., Mao, X., Huang, K., Xie, Z., & Zou, M. H. (2019). Hyperglycemia-Driven Inhibition of AMP-Activated Protein Kinase α2 Induces Diabetic Cardiomyopathy by Promoting Mitochondria-Associated Endoplasmic Reticulum Membranes In Vivo. *Circulation*. 139(16), 1913–1936.

Xie, Z., Lau, K., Eby, B., Lozano, P., He, C., Pennington, B., Li, H., Rathi, S., Dong, Y., Tian, R., Kem, D., & Zou, M. H. (2011). Improvement of cardiac functions by chronic metformin treatment is associated with enhanced cardiac autophagy in diabetic OVE26 mice. *Diabetes*. 60(6), 1770–1778.

Xu, X., Kobayashi, S., Chen, K., Timm, D., Volden, P., Huang, Y., Gulick, J., Yue, Z., Robbins, J., Epstein, P. N., & Liang, Q. (2013). Diminished autophagy limits cardiac injury in mouse models of type 1 diabetes. *The Journal of biological chemistry*. 288(25), 18077–18092.

Yamano, K., Matsuda, N., & Tanaka, K. (2016). The ubiquitin signal and autophagy: an orchestrated dance leading to mitochondrial degradation. *EMBO reports*. 17(3), 300–316.

Ye, X., Sun, X., Starovoytov, V., & Cai, Q. (2015). Parkin-mediated mitophagy in mutant hAPP neurons and Alzheimer's disease patient brains. *Human molecular genetics*. 24(10), 2938–2951.

Yoon, Y., Galloway, C. A., Jhun, B. S., & Yu, T. (2011). Mitochondrial dynamics in diabetes. *Antioxidants & redox signaling*. 14(3), 439–457.

Yoon, Y., Krueger, E. W., Oswald, B. J., & McNiven, M. A. (2003). The mitochondrial protein hFis1 regulates mitochondrial fission in mammalian cells through an interaction with the dynamin-like protein DLP1. *Molecular and cellular biology*. 23(15), 5409–5420.

Yu, T., Robotham, J. L., & Yoon, Y. (2006). Increased production of reactive oxygen species in hyperglycemic conditions requires dynamic change of mitochondrial morphology. *Proceedings of the National Academy of Sciences of the United States of America*. 103(8), 2653–2658.

Zhang, J., & Ney, P. A. (2009). Role of BNIP3 and NIX in cell death, autophagy, and mitophagy. *Cell death and differentiation*. 16(7), 939–946.

Zhang, J., Loyd, M. R., Randall, M. S., Waddell, M. B., Kriwacki, R. W., & Ney, P. A. (2012). A short linear motif in BNIP3L (NIX) mediates mitochondrial clearance in reticulocytes. *Autophagy*. 8(9), 1325–1332.

Zhang, X., Yan, H., Yuan, Y., Gao, J., Shen, Z., Cheng, Y., Shen, Y., Wang, R. R., Wang, X., Hu, W. W., Wang, G., & Chen, Z. (2013). Cerebral ischemia-reperfusion-induced autophagy protects against neuronal injury by mitochondrial clearance. *Autophagy*. 9(9), 1321–1333.

Zhou, P., Xie, W., Meng, X., Zhai, Y., Dong, X., Zhang, X., Sun, G., & Sun, X. (2019). Notoginsenoside R1 Ameliorates Diabetic Retinopathy through PINK1-Dependent Activation of Mitophagy. *Cells*. 8(3), 213.

Zuo, W., Zhang, S., Xia, C. Y., Guo, X. F., He, W. B., & Chen, N. H. (2014). Mitochondria autophagy is induced after hypoxic/ischemic stress in a Drp1 dependent manner: the role of inhibition of Drp1 in ischemic brain damage. *Neuropharmacology*. 86, 103–115.

Chapter 10

Triggering Mitophagy: A Promising Approach to Overcome Chemoresistance in Cancer Therapy

Pragya Singh
Ankita Tayal
and Shalini Mani*
Department of Biotechnology, Jaypee Institute of Information Technology, Noida, Uttar Pradesh, India

Abstract

Cancer is an abnormal cell condition where cells divide uncontrollably and damage the tissues. Despite the availability of various treatments for cancer, it remains one of the leading causes of death. The main reason is the developed drug resistance to chemotherapy and most medications used for treatment. Therefore, there is a need to identify new targets and develop new treatments for drug–resistant carcinomas. Recent studies show mitochondria, due to their importance in developing cancer cells, as an emerging target for cancer therapeutics that can also help alleviate chemoresistance. Mitophagy, the removal of damaged mitochondria, a process involved in mitochondrial homeostasis, is linked with cancer cell proliferation and death. Therefore, identifying the role of mitophagy and mitochondrial functionality in different stages and types of cancer will aid in developing optimal therapies that regulate mitophagy.

Keywords: chemoresistance, mitochondrial homeostasis, mitophagy, mitochondrial dysfunction, cancer

Introduction

Cancer is a set of illnesses known to include the uncontrollable division of body cells that manifest over time. It starts when the cell rebels against the limitations placed on normal cell division, i.e., metastasis, and proceeds according to its plan. Our understanding of cancer has changed due to the research conducted in the past 30 years, primarily due to the advancements

* Corresponding Author's Email: shalini.mani@jiit.ac.in.

In: Mitophagy and Its Role in Health and Disease
Editors: Shalini Mani and Samiksha Wasnik
ISBN: 979-8-89113-470-6

in the tools used in molecular biology. The predominantly used therapies for treating cancer include surgery, immunotherapy, radiation therapy, gene therapy, laser therapy, combination therapy, and targeted therapy, with chemotherapy being the most popular and promising option. Even after all the advancements in cancer research, drug resistance to chemotherapy remains a significant issue and is to blame for most patient relapses and higher mortality rates. Nearly all medications used to treat the deadliest carcinomas have been reported to have drug resistance. Researching and developing new treatments is imperative due to the rising occurrence of these drug-resistant carcinomas. In eukaryotes, mitochondria are quite a crucial organelle.

Recent research has shown that mitochondria play a significant role in cancer development, can be used as targets for chemotherapy, and can help alleviate chemoresistance. To maintain their homeostasis and control their quality, mitochondria use a variety of processes. Maintaining mitochondrial homeostasis depends on mitophagy's continuous fusion, fission, and removal of damaged mitochondria.

It has been observed that mitochondrial dysfunction and injury are caused by hypoxia, starvation, energy stress, cell senescence, ROS, aberrant protein aggregation, and drug exposure. An increased mitochondrial mass is typically found in cancer cells, showing increased mitochondrial biogenesis, which is also associated with increased chemoresistance. A double membrane that encloses the mitochondria during mitophagy causes autophagosomes. After the mitochondrial disintegration, the lysosome and the resulting mitophagy-lysosome fuse to allow for the recycling of metabolites.

Excessive mitophagy can cause the loss of functioning mitochondria, which disrupts the metabolic energy demand and culminates in mitochondrial death via mitophagy. In contrast, mitophagy can help cells survive by eliminating damaged mitochondria for better defense against hostile environments. Hence, inducers and inhibitors of mitophagy are equally potent as prospective therapies for cancer. We now know that mitophagy plays a dual role in cancer therapy after evaluating numerous studies on particular chemotherapeutic agents, their sensitivity in cancer therapy, and the participation of mitochondrial dynamics in its treatment. Designing the best therapies to control mitophagy requires advances in understanding carcinogenesis and chemoresistance in the context of mitochondrial functionality for cancer cells in different stages and types of cancer.

Cancer

Tumors are formed when the cells proliferate uncontrollably and lose the usual constraints on cell division, i.e., metastasis. This is how cancer develops. Cancer endangers the human body when tumor growth affects essential organs and tissues.

The hyperactivity of growth-promoting pathway-associated proteins occurs due to a mutated form of proto-oncogenes called the oncogenes, leading to rapid, uncontrollable cell multiplication because of growth factors. The nearby cells undergo expansion due to growth factors. During this process, there is also the risk of dividing cells, creating more growth factors. Understanding putative hallmarks of cancer growth and treatment has made notable progress in recent years (Understanding Cancer, NIH Curriculum Supplement Series, 2007).

Radiation therapy, surgery, chemotherapy, immunotherapy, and hormone therapy are all examples of present treatments for cancer. Radiation therapy primarily takes away the cell's ability to divide and proliferate (Bernier et al., 2004). Targeted therapy focuses on one specific area, such as the intracellular organelles or tumor vasculature, while sparing the surrounding tissue (Bazak et al., 2015). Immunotherapy controls immune responses in various directions using medications (such as immunosuppressants), biologicals, vitamins and minerals, transplantation, and immunizations (Amjad et al., 2022).

Chemotherapy is used to halt tumor growth and cell division, which prevents cells from undergoing metastasis. Chemotherapeutic medications can be administered intravenously, orally, subcutaneously, intramuscularly, or intrathecally. By interfering with DNA, RNA, or protein synthesis, traditional chemotherapy medications lead to malignant cells showing dysfunction in the preformed molecules (Urruticoechea et al., 2010).

Chemoresistance and Multiple-Drug Resistance

Nearly all medications used to treat the deadliest tumors have drug resistance. Despite advancements in cancer treatments, chemotherapeutic drug resistance remains a significant issue. It has been noted that drug resistance can either be acquired or innate (Longley et al., 2005).

Because of inherent resistance-mediating elements in the tumor mass, intrinsic chemoresistance, as its name suggests, is an innate property of cancer cells that reduces the efficiency of the treatment even before it is applied. Contrarily, acquired resistance, which appears after treatment administration, is typically due to mutations or changes in the working of the adaptive response machinery, which range from the upregulation of therapeutic targets, drug inactivation, overexpression of drug transporters (MDR pumps), encouragement of repair of damaged DNA, changes in metabolism, and the alternative survival pathways being awakened to the upregulation of drug transporters (Housman et al., 2014). Drug resistance is directly or indirectly responsible for more than half of cancer patients' deaths. Data extracted from the COSMIC database shows the resistance of some major chemotherapy drugs such as imatinib, sorafenib, pembrolizumab, cizotinib, and many more in which some of the resistance developed are unique, and some are developed through mutation.

Figure 1 shows how many commonly known chemotherapy drugs such as cisplatin, doxorubicin, oxaliplatin, paclitaxel, carboplatin, and vinblastine already have their resistance identified, which further affirms the gravity of the problem of chemoresistance. The pathways that lead to resistance to chemotherapy include drug inactivation, multidrug resistance, cell death inhibition (apoptosis suppression), pharmacological targets, modifications to drug metabolism, including epigenetic modifications, enhanced DNA repair, and gene amplification. Drug resistance-related tumor invasion and metastasis account for 90% of chemotherapy failures. It has been shown that cancer relapses caused by multidrug resistance have occurred even after surgical removal. Drug resistance in cancer is mediated by several processes, including alteration in drug targets, activated pro-survival pathways, and improper induction of cell death (Holohan et al., 2013).

The management of anticancer drug resistance has not significantly improved clinical results in the last few decades. Therefore, there is a pressing need to increase anticancer medication sensitivity by identifying fresh suitable targets (Yan, Li, 2018). Drug resistance mechanisms must be defined to understand the cause of this therapeutic failure, and it is crucial to find new targets to overcome chemoresistance and enhance the effectiveness of chemotherapy.

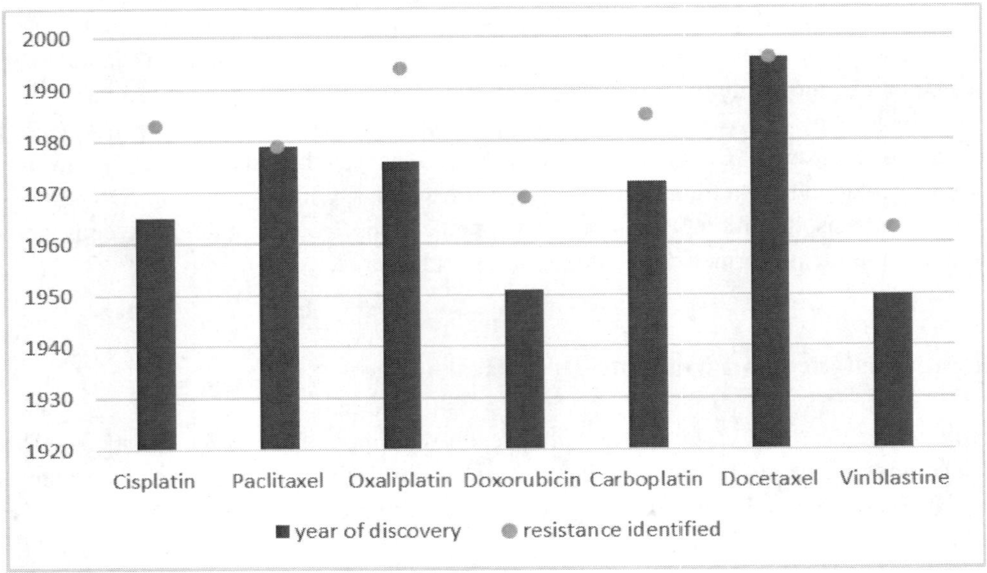

Figure 1. Year of discovery of chemotherapeutic drugs and their approximate year of identification of resistance.

Mitochondria: Not Just the Powerhouse of the Cell

By managing the synthesis and dispersion, and the distribution of energy through oxidative phosphorylation (OXPHOS), which is relative to oxygen and the availability of substrate., The mitochondria perform a crucial role in cells. This organelle has been connected to numerous aspects of carcinogenesis and tumor growth because of its crucial cellular function. Various chemotherapy medications specifically target mitochondria directly or indirectly through the circuits controlling mitochondrial function. As a result, medications that target mitochondria have become promising means of interfering with tumor adaptations and eradicating cancer cells.

Mitochondrial Metabolism in Cancer-Resistant Cells

Because glycolytic cancer stem cells differentiate and multiply, the metabolic condition of cancer cells is essential for comprehending the mechanisms underlying treatment resistance, as it is possible to see a change from anaerobic to aerobic metabolism (Lyakhovich et al., 2016).

The characteristic metabolic flexibility of cancer is based, at the molecular level, on the reconnection of the various metabolic pathways. This is accomplished through the creation or degradation of essential proteins, which enable metabolites to switch routes in response to the needs of the cells. The Warburg effect involves the cells using a high amount of glucose when oxygen is present and is the first of the recognized processes of this metabolic remodeling. Through a process known as aerobic glycolysis, lactate is created and secreted as a result (Warburg, 1956).

Although this method is ineffective at producing energy in the form of ATP, tumor cells typically absorb glucose at a greater rate, resulting in a net ATP generation comparable to that of OXPHOS, which is brought on by various mutations in respiratory complexes (Hanahan et al., 2011)

Cancer cells have low ROS levels due to metabolic reprogramming required to maintain self-renewal and improve the antioxidant defense mechanism (Lyakhovich et al., 2016). Therefore, glycolytic cancerous cells are typically acclimated to a characteristic low oxygen environment present in multiple types of cancer, where the glycolytic metabolism predominates to make up for a deficiency in mitochondria.

Lung, breast, and sarcoma cancer stem cells exhibit increased levels ofALDH1 expression (aldehyde dehydrogenase-1), which is indicative of enhanced chemotherapeutic drug resistance (Lohberger et al., 2012), (Jiang et al., 2009), (Tanei et al., 2009). In both cell lines and samples taken from patients, high mitochondrial mass levels correlate with high ALDH activity, which may indicate higher mitochondrial biogenesis or slower mitochondrial degeneration. Doxycycline, an inhibitor of mitochondrial biogenesis, thus targets ALDH+ breast CSCs (Lin et al., 2018).

The majority of intracellular RNOS is produced by mitochondria, which also produce superoxide (O_2), peroxides (H_2O_2, RO_2H), and free radicals (Sosa et al., 2013). A limited portion of these RNOS is created under typical circumstances due to OXPHOS, either as signal transduction mediators or as components of immune response systems. These defenses are thwarted by stressful circumstances, harming the mitochondria. The amount of RNOS produced by damaged mitochondria also increases as oxidative damage progresses (Zorov et al., 2014).

After receiving a combination of radiotherapy and chemotherapy, CSCs displayed better resistance to DNA damage due to this elevated production of RNOS protection genes (Diehn et al., 2009).

We can now better comprehend the various regulatory processes that the mitochondria undergo to maintain mitochondrial and cellular activity since we better understand the metabolic alterations that the mitochondria undergo in a treatment-resistant cancer cell.

Regulation of Mitochondrial Dynamics

The mitochondria continually involve fission, fusion, transfer, and autophagy to preserve mitochondrial and cellular integrity. A collection of conserved GTPases controls how the mitochondrial functions (hydrolase enzymes binding to GTP and hydrolyzing it into GDP). By isolating a small portion of the mitochondria, the fission that occurs during mitosis, also known as mitofission (Figure 2), aids in the segregation of mtDNA and eliminates all the faulty mitochondria. In the meantime, mitofusion (Figure 3), also known as mitochondrial fusion, is crucial when there is a high ATP demand and attempts to prevent organelle death. Resistance transfer through mitochondria has been seen to occur through particular channels, including gap junctions, extracellular vesicles (EVs, which include exosomesand microvesicles), and tunneling nanotubes (TNTs, made of long actin fibers that help in contacting nonadjacent cells). This dynamic process is called MitoTransfer (Figure 4). Mitophagy, which is the autophagy of the mitochondria and is regulated by two mechanisms that will be covered in depth, is one of the most significant regulatory activity events in mitochondria.

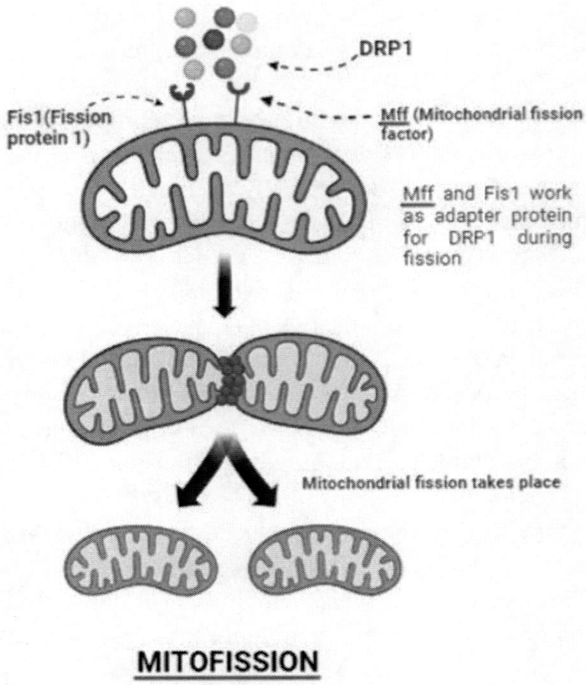

Figure 2. Schematic showing the mitochondrial fission.

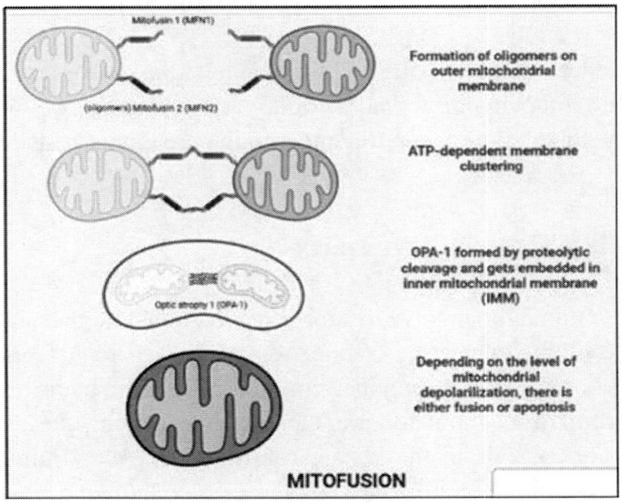

Figure 3. Schematic showing the Mitochondrial Fusion.

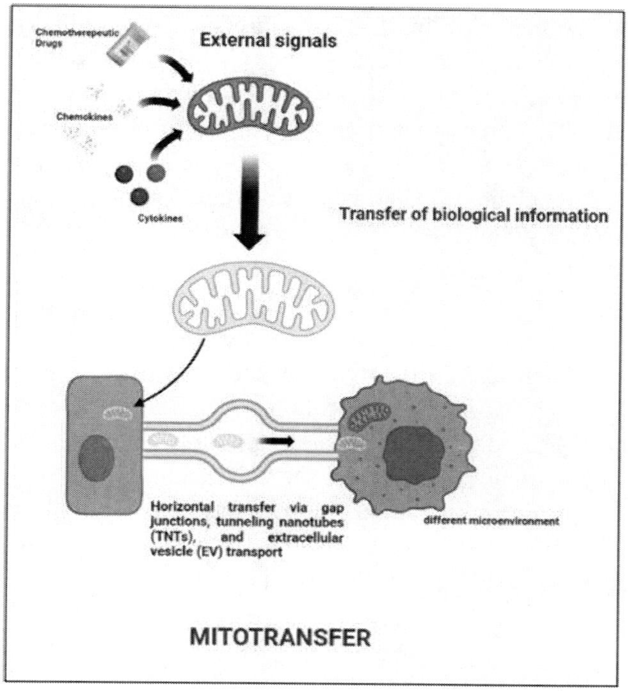

Figure 4. Schematic showing the mitochondrial transfer.

Mitophagy

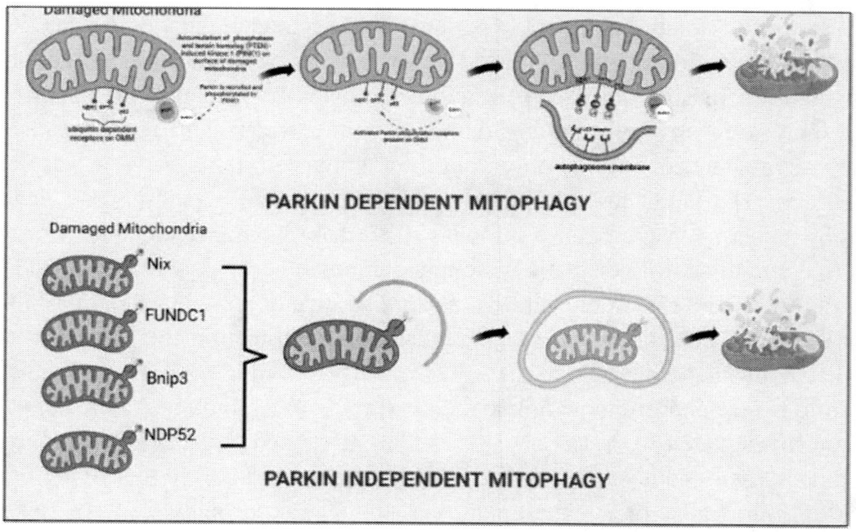

Figure 5. Schematic showing two different types of mitophagy: PARKIN-dependent and PARKIN-independent.

Mitophagy uses autophagy to eliminate superfluous or damaged mitochondria in response to diverse stressful stimuli. Instead of canonical autophagy, mitophagy only targets mitochondria that specific mitophagy receptors have identified. The receptorsystem primarily consists of FUNDC1, NDP52, or Nix receptors on the surface of the mitochondria, as well as the Parkin/PTEN-induced putative kinase 1 (PINK1) receptor system and the microtubule-associated proteins 1 light chain 3 (LC3)-interacting region (LIR)-containing receptor system. It is heavily controlled by several NIP3-like protein X (NIX)/BNIP3 signaling pathways (Figure 5).

Mitophagy Manipulation/ Modulation as a Potential Cancer Therapy Target

Multiple investigations have discovered several variables that control mitophagy and play a role in cancer cell migration, carcinogenesis, and cell death during the past few decades. The analysis of the signal pathways that control mitophagy is now vitally necessary to pinpoint possible treatment targets. Here, we review how to comprehend mitophagy concerning carcinogenesis, metastasis, therapy, and drug resistance.

Mitophagy Inhibition and Chemoresistance Case Studies

It is frequently assumed that mitophagy brought on by chemotherapy medications is a defense mechanism used by cancer cells and a contributing factor in chemotherapy resistance. Praharaj P.P. and others (2015) and Su YC et al. (2016) found that Galectin-1 facilitates tumor growth by driving cancer cell adhesion, migration, and proliferation. They also found that it caused chemoresistance to cisplatin treatment in patients with hepatocellular carcinoma (HCC). It was discovered that cisplatin treatment-induced mitochondrial potential loss and apoptosis might be decreased by increasing mitophagy, which is mediated by galectin-1. Galectin-1-triggered autophagy may aid cells in removing damaged mitochondria and resisting chemotherapy treatments, as evidenced by the observation of enhanced mitophagy in cisplatin-treated hepatoma cells. The study also found that cisplatin-induced hepatoma cell death may be countered by galectin-1 via inducing autophagy and that this improved anticancer impact was shown in BALB/c mice, with cisplatin treatment combined with thiodigalactoside (TDG) being related with inhibition. This demonstrated the significance of soluble galectin-1 in tumors and the effectiveness of galectin-1 inhibitor and cisplatin combination therapy for cooperatively lowering hepatoma chemoresistance.

p62'srole in mitophagy and its link to the sensitivity of cisplatin in A2780 cells for ovarian cancer were investigated by Yu et al., 2021. They theorized that one of the reasons for the cisplatin's decreased apoptosis rate in the UBA group, in addition to the higher efficiency of mitochondrial removal in their experiment, was the increased binding of HK2 and VDAC1. The work that described how the UBA domain of p62 affects the mitochondrial location of HK2 and alters the cisplatin sensitivity of ovarian cancer cells provided a lead for developing mitophagy inhibitors and HK2-targeted inhibitors for enhancing cisplatin sensitivity.

In hypoxia-induced sorafenib-resistant HCC cells, Wu et al2020's research revealed that the ATAD3A-PINK1/PARKIN axis controlled hyperactivated mitophagy. Through the mitophagy route, it was proposed that ATAD3A was crucial in developing sorafenib resistance brought on by hypoxia. Resistance to sorafenib and hypoxia-induced mitophagy was correlated with gains and losses of ATAD3A. Under hypoxia, miR-210-5P increased and helped control sorafenib resistance. According to the study, ATAD3A was a valuable target for overcoming sorafenib resistance and a predictor of sorafenib response and clinical outcomes.

In sixty oral squamous cell carcinoma (OSCC) samples, Naik PP et al.,2017 conducted a retrospective study of CD44, ABC1, and ADAM17 about the various clinicopathological variables and their connection. Their findings showed that resistant FaDu cells had a larger autophagic flux than their parental counterparts. According to the study, autophagy suppression caused CD44 expression in CDDP-resistant FaDu cells to fall, suggesting a possible connection between CD44 expression and autophagy-mediated chemoresistance. However, additional research is needed to better understand the therapeutic potential of explicitly targeting mitophagy instead of bulk autophagy. The team also demonstrated that oral cancer may develop chemoresistance due to mitophagy.

A mitophagy inhibitor, cyclosporin A (CsA), significantly increased the susceptibility of PIWIL1-overexpressing multiple myeloma (MM) cells to chemotherapeutic medicines, according to research by Wang et al.,2022 supporting the function of mitophagy in causing MM chemoresistance. They discovered PIWIL1's latest functional mechanism, which promotes resistance to chemotherapy drugs by triggering mitophagy and the population of myeloma stem cells via AKT-mTOR and PARKIN-dependent mitophagy pathways. However, further research must be validated to establish the precise molecular target in PIWIL1 controlling the mentioned traits. It is necessary to conduct more research to learn more about the downstream molecular targets of PIWIL1 that are regulated by mitophagy and myeloma stem cells. These can make it easier to create cutting-edge MM treatment plans.

The particularly intriguing work of He et al., 2021 found that mitophagosome development activated the protective mitophagy caused by TMZ in glioma cells, so the BNIP3 assisted in preparing damaged mitochondria for autophagic removal. Analysis revealed that FOXO3a, a crucial regulator of cell survival, shielded glioma cells against TMZ-induced DNA DSBs by encouraging mitophagy mediated by BNIP3. Ultimately, the study hypothesized that TMZ treatment sensitivity would increase in glioma cells by inhibiting FOXO3a.

Mitophagy Activation in Certain Cases to Remove Chemoresistance

Although activating mitophagy has been suggested as a potential strategy to treat cancer, mitophagy typically acts as an antagonist in cases of chemoresistance in cancer therapy.

Results from Ma et al., 2020 demonstrated a concept that was the preclinical evidence for combination therapy that targeted mitophagy to enhance the effectiveness of chemotherapy by increasing hepatocellular carcinoma apoptosis through DRP1 inhibition.DRP1's function in HCC mitophagy induced by cisplatin was investigated with a particular dynamin-related protein 1 (DRP1) inhibitor called Mdivi 1. *In vitro* and *in vivo* testing was performed to determine the synergistic impact of cisplatin and the DRP1 inhibitor on HCC. As a result, cisplatin activated DRP1 to cause mitophagy in survived HCC cells.

The coupled administration of melatonin and sorafenib boosted the expression of parkin 6 h after usage, but sorafenib alone had no effect, according to a study by Prieto-Domnguez et al., 2016. The current study results show that co-administration of melatonin and sorafenib causes early mitophagic response in HCC cells, which is associated with overflow of ROS and depolarization of mitochondrial membranes, which promote apoptosis and decreased cell viability. The study also found that sorafenib and melatonin administration together improved sensitivity in HCC cell lines, suggesting that melatonin can be a valuable adjuvant for sorafenib treatment of HCC.

In the study by Zheng et al.,2018, most patients who received the drug Bortezomib, used in chemotherapy to treat multiple myelomas, developed acquired resistance to it. According to reports, thioredoxin levels are linked to the development of cancer. In bortezomib-resistant myeloma cells, it was shown that thioredoxin expression dramatically increased, which was associated with a rise in the concentration of bortezomib. The fact that PX12's effects on myeloma cells that have bortezomib resistance were inhibited by thioredoxin and prevented by blocking mitophagy is particularly significant because it validated that for the induction of cell death in bortezomib-resistant myeloma cells by PX12, mitophagy was a crucial molecular mechanism. It was proposed through this study that Thioredoxin was a possible target for clinical therapies against multiple myeloma.

In their study, Dany et al. (2016) showed how to activate mitophagy by specifically targeting mitochondria with the drug LCL-461, a highly soluble C18-ceramide analog that accumulates in mitochondria of acute myeloid leukemia (AML) and causes AML cells resistant to crenolanib both *in situ* and *in vivo* to undergo lethal mitophagy. Ceramide was crucial in controling autophagy. They opined that further research was still required to understand how FLT3-targeted therapy affected how AML responded fully.

Conclusion

Drug resistance significantly reduces the efficacy of chemotherapy. Due to complicated drug resistance mechanisms, modulating anticancer drug resistance has had minimal impact on improving clinical outcomes during the past few decades. Using autophagy, the crucial mitochondrial quality control mechanism known as the mitophagy selectively destroys the surplus or defective mitochondria. There is mounting evidence that the modulation of mitophagy contributes to the emergence of cancerous cells and drug resistance in various tumor types. Initially, mitophagy was believed to be a suppressor of oncogenesis, preserving cell homeostasis and preventing cells from becoming oncogenic. However, by destroying damaged mitochondria and lowering mitochondrial ROS, mitophagy also, in some cases, promote the endurance of the cancer cells under cytotoxic stress. As a result, there is a debate over whether to induce or inhibit mitophagy in cases of cancer medication resistance. In this review, we discuss the most recent data on the dual role of mitophagy in cancer therapy resistance. Even though there was no agreement on whether mitophagy suppresses or promotes cancer medication resistance in particular cancer types, the best course of action for each patient can be determined based on the cancer type and anticancer agent. Following a review of numerous studies regarding the induction of mitophagy and its benefit in modifying the sensitivity of cancer treatment, additional research will be required to determine whether these studies can also be applied to various types of carcinomas. Inhibiting mitophagy is probably more effective when combined with chemotherapy drugs in specific cancers.

References

Amjad, M. T., Chidharla, A., & Kasi, A. (2022). Cancer Chemotherapy. In *StatPearls*. StatPearls Publishing.

Bazak R, Houri M, El Achy S, et al. (2015). Cancer active targeting by nanoparticles: a comprehensive review of the literature. *J Cancer Res Clin Oncol.* 141(5):769–784.

Bernier J, Hall, E. J., Giaccia, A.(2004). Radiation oncology: a century of achievements. *Natur. e*4:737–747.

Diehn M., Cho R. W., Lobo, N. A., Kalisky, T., Dorie, M. J., Kulp, A. N., Qian. D., Lam, J. S., Ailles, L. E., Wong, M., Joshua, B., Kaplan, M. J., Wapnir, I., Dirbas, F. M., Somlo, G., Garberoglio, C., Paz, B., Shen, J., Lau, S. K., Quake, S. R., Brown, J. M., Weissman, I. L, Clarke, M. F.(2009). Association of reactive oxygen species levels and radioresistance in cancer stem cells. *Nature.* 458:780–783.

Hanahan, D., Weinberg, R. A. (2011). Hallmarks of Cancer: The Next Generation. *Cell.* 144:646–674.

He, C., Lu, S., Wang, X. Z., Wang, C. C., Wang, L., Liang, S. P., Luo, T. F., Wang, Z. C., Piao, M. H., Chi, G. F., & Ge, P. F. (2021). FOXO3a protects glioma cells against temozolomide-induced DNA double-strand breaks via promotion of BNIP3-mediated mitophagy. *Acta pharmacologica Sinica*, *42*(8), 1324–1337.

Holohan, C., Van Schaeybroeck, S., Longley, D. B., Johnston, P. G. (2013). Cancer drug resistance: an evolving paradigm. *Nat Rev Cancer* 13: 714-726.

Housman, G., Byler, S., Heerboth, S., Lapinska, K., Longacre, M., Snyder, N., Sarkar, S. (2014). Drug Resistance in Cancer: An Overview. *Cancers.* 6:1769–1792.

Jiang F., Qiu Q., Khanna A., Todd N. W., Deepak, J., Xing, L., Wang, H., Liu, Z., Su, Y., Stass, S. A., Katz, R. L.(2009). Aldehyde dehydrogenase 1 is a tumor stem cell-associated marker in lung cancer. *Mol. Cancer Res.* 7:330–338.

Lin, C., Lo, M., Moody, R. R., Stevers, N. O., Tinsley, S. L., Sun, D. (2018). Doxycycline targets aldehyde dehydrogenase-positive breast cancer stem cells. *Oncol. Rep.* 39:3041–3047.

Lohberger, B., Rinner, B., Stuendl, N., Absenger, M., Liegl-Atzwanger, B., Walzer, S.M., Windhager, R., Leithner, A.(2012). Aldehyde dehydrogenase 1, a potential marker for cancer stem cells in human sarcoma. *PLoS ONE.* 7:e43664.

Longley, D. B., Johnston, P. G. (2005). Molecular Mechanisms of Drug Resistance. *J. Pathol.* 205:275–292.

Lyakhovich, A., Lleonart, M. E. (2016). Bypassing Mechanisms of Mitochondria-Mediated Cancer Stem Cells Resistance to Chemo- and Radiotherapy. *Oxidative Med. Cell. Longev.* 2016:1716341.

Ma, M., Lin, X. H., Liu, H. H., Zhang, R., & Chen, R. X. (2020). Suppression of DRP1-mediated mitophagy increases the apoptosis of hepatocellular carcinoma cells in the setting of chemotherapy. *Oncology reports*, *43*(3), 1010–1018.

Dany, M., Gencer, S., Nganga, R., Thomas, R. J., Oleinik, N., Baron, K. D., Szulc, Z. M., Ruvolo, P., Kornblau, S., Andreeff, M., Ogretmen, B.(2016). Targeting FLT3-ITD signaling mediates ceramide-dependent mitophagy and attenuates drug resistance in AML. *Blood* 128 (15): 1944–1958.

National Institutes of Health (US); Biological Sciences Curriculum Study. NIH Curriculum Supplement Series [Internet]. Bethesda (MD): National Institutes of Health (US); 2007. Understanding Cancer. Available from: https://www.ncbi.nlm.nih.gov/books/NBK20362/.

Praharaj, P. P., Naik, P. P., Panigrahi, D. P., Bhol, C. S., Mahapatra, K. K., Patra, S., Sethi, G., Bhutia, S.K.(2019). Intricate Role of Mitochondrial Lipid in Mitophagy and Mitochondrial Apoptosis: Its Implication in Cancer Therapeutics. *Cell Mol. Life Sci.* 76:1641–1652.

Prieto-Domínguez, N., Ordóñez, R., Fernández, A., Méndez-Blanco, C., Baulies, A., Garcia-Ruiz, C., Fernández-Checa, J. C., Mauriz, J. L., & González-Gallego, J. (2016). Melatonin-induced increase in sensitivity of human hepatocellular carcinoma cells to sorafenib is associated with reactive oxygen species production and mitophagy. *Journal of pineal research*, *61*(3), 396–407.

Sosa, V., Moliné, T., Somoza, R., Paciucci, R., Kondoh, H., Lleonart, M.E. (2013). Oxidative stress and cancer: An overview. *Ageing Res. Rev.* 12:376–390.

Su, Y. C., Davuluri, G. V., Chen, C. H., Shiau, D. C., Chen, C. C., Chen, C. L., Lin, Y. S., & Chang, C. P. (2016). Galectin-1-Induced Autophagy Facilitates Cisplatin Resistance of Hepatocellular Carcinoma. *PloS one*, *11*(2), e0148408.

Tanei, T., Morimoto, K., Shimazu, K., Kim, S. J., Tanji, Y., Taguchi, T., Tamaki, Y., Noguchi, S. (2009). Association of Breast Cancer Stem Cells Identified by Aldehyde Dehydrogenase 1 Expression with

Resistance to Sequential Paclitaxel and Epirubicin-Based Chemotherapy for Breast Cancers. *Clin. Cancer Res.* 15:4234–4241.

Urruticoechea, A., Alemany, R., Balart, J., Villanueva, A., Viñals, F., & Capellá, G. (2010). Recent advances in cancer therapy: an overview. *Current pharmaceutical design*, *16*(1), 3–10.

Wang, Y., Yao, L., Teng, Y., Yin, H., & Wu, Q. (2022). PIWIL1 Drives Chemoresistance in Multiple Myeloma by Modulating Mitophagy and the Myeloma Stem Cell Population. *Frontiers in oncology*, *11*, 783583.

Warburg, O.(1956). On the Origin of Cancer Cells. *Science.*123:309–314.

Yan, C., & Li, T. S. (2018). Dual Role of Mitophagy in Cancer Drug Resistance. *Anticancer research*, *38*(2), 617–621.

Yu, S., Yan, X., Tian, R., Xu, L., Zhao, Y., Sun, L., & Su, J. (2021). An Experimentally Induced Mutation in the UBA Domain of p62 Changes the Sensitivity of Cisplatin by Up-Regulating HK2 Localisation on the Mitochondria and Increasing Mitophagy in A2780 Ovarian Cancer Cells. *International journal of molecular sciences*, *22*(8), 3983.

Zheng, Z., Fan, S., Zheng, J., Huang, W., Gasparetto, C., Chao, N. J., Hu, J., & Kang, Y. (2018). Inhibition of thioredoxin activates mitophagy and overcomes adaptive bortezomib resistance in multiple myeloma. *Journal of hematology & oncology*, *11*(1), 29.

Zorov, D. B., Juhaszova, M., Sollott, S. J. (2014). Mitochondrial reactive oxygen species (ROS) and ROS-induced ROS release. *Physiol. Rev.*94:909–950.

Chapter 11

A Workout for the Brain to Trigger Mitophagy

Aditi Kaushik
Saboor Khan
Ankita Tayal
and Shalini Mani*

Department of Biotechnology, Jaypee Institute of Information Technology, Noida, Uttar Pradesh, India

Abstract

Mitophagy is the process by which damaged or dysfunctional mitochondria are removed from cells through a selective autophagic process. It plays a crucial role in maintaining cellular homeostasis and preventing the accumulation of damaged mitochondria, which can lead to cellular dysfunction and disease. There is limited research on the direct effect of yoga on mitophagy. However, studies have shown that yoga can increase autophagy in general, which may indirectly impact mitophagy. Yoga has been shown to stimulate the autophagy pathway by activating the AMPK/mTOR pathway, which regulates energy homeostasis and cellular metabolism. Research has also suggested that yoga may increase mitochondrial biogenesis, the process by which new mitochondria are produced in cells. Increased mitochondrial biogenesis can help maintain healthy mitochondrial function and reduce the accumulation of damaged mitochondria. In addition, yoga reduces oxidative stress, which can damage mitochondria and lead to mitophagy. By reducing oxidative stress, yoga may help prevent the accumulation of damaged mitochondria and promote healthy mitochondrial function. Overall, while there is limited research on the relationship between yoga and mitophagy, there is evidence to suggest that yoga can positively impact mitochondrial health and cellular homeostasis, which may indirectly impact mitophagy.

Keywords: mitophagy, AMPK/mTOR pathway, energy homeostasis

Introduction

Neurodegenerative diseases are a group of disorders that mainly affect the neurons in the human brain, causing them to deteriorate and die gradually. These illnesses significantly threaten

* Corresponding Author's Email: shalini.mani@jiit.ac.in.

In: Mitophagy and Its Role in Health and Disease
Editors: Shalini Mani and Samiksha Wasnik
ISBN: 979-8-89113-470-6

human health, mainly due to the growing older population. The nervous system, which comprises the brain and spinal cord, comprises neurons that do not typically regenerate or replace themselves. Hence, the body cannot repair or replace them when they die or sustain injuries. Neurodegenerative diseases are severe and incurable, impairing movement, speech, breathing, mental functioning (dementia), and motor coordination (ataxia). They affect numerous families, and coping with them is challenging for patients and their loved ones (Gitler AD et al., 2017).

Neurodegenerative diseases have diverse pathophysiologies, including but not limited to the following examples:

a) Alzheimer's disease (AD) and other dementias
b) Parkinson's disease (PD)
c) Huntington's disease (HD)
d) Spinocerebellar ataxia (SCA), etc.

Neurodegenerative diseases have various causes, including inherited genetic alterations. In these cases, parents pass on a defective gene to their offspring, leading to conditions like Huntington's disease, certain forms of Alzheimer's disease, and motor neuron disease.

In most cases, a combination of genetic and environmental factors contributes to the development of neurodegenerative diseases, making it difficult to predict who may contract a disease. Specific genetic abnormalities have been identified for some diseases, although the genetic impacts on neurodegenerative disorders are not yet fully understood. Environmental factors such as chronic exposure to poisons, chemicals, and pesticides can also contribute to developing diseases like Parkinson's disease. Dementia affects more than 350,000 Australians, and around 80,000 people have Parkinson's disease. As the population ages, these numbers are expected to increase, leading to rising healthcare concerns regarding neurodegenerative illnesses. Alzheimer's disease affected an estimated 5.4 million Americans in 2016, and it is anticipated that 930,000 Americans might be affected by Parkinson's disease in 2020.

Yoga is becoming a popular form of self-care that can improve the quality of life (QOL). Research has explored the use of yoga as a supplementary health therapy for neurological ailments like multiple sclerosis (MS), fibromyalgia (FM), and diseases in the peripheral nervous system (Mishra SK et al., 2012). A study on yoga treatment for musculoskeletal diseases found significant functional improvements, including decreased pain and increased agility (Ward L et al., 2013). Recent research suggests that yoga is one of many complementary and alternative medicine (CAM) modalities recommended for neurological problems, but each modality has different benefits (Wells RE et al., 2010). This chapter discusses how yoga can improve QOL for patients with neurodegenerative illnesses but does not examine the impact of other CAM modalities on neurological disorders. Certain styles of yoga may not be beneficial for specific medical conditions. Practices emphasizing continuous movement (*vinyasa*), execution in extreme heat (*Bikram*), or repetitive, intense breathing may be ill-advised for clients or patients with neurological health challenges. Gentle or therapeutic yoga styles, customized for each client or patient's abilities, are better suited for such cases (Bartlett SJ et al., 2013).

Yoga and Mitophagy

Yoga is a traditional practice that improves physical and mental well-being, prevents diseases, and may be used as an adjunct therapy for serious illnesses. Yoga has been shown to positively affect the epigenome of sperm and the integrity of mitochondrial and nuclear DNA. It is a science that influences various physiological processes, including primary metabolism, epigenetics, repair of damaged DNA, oxidative reactions, senescence, organ system maintenance, and a well-defined psycho-neuro-immune axis (Gautam S. et al., 2021).

Over the last decade, yoga has gained immense popularity and remains a top fitness trend. Practitioners report significant improvements in their health, and the practice is highly adaptable to meet the needs of individuals or large groups. Yoga is a practical option for people of all ages, fitness levels, and abilities to increase physical activity and overall well-being. While there is evidence linking yoga to positive outcomes such as QOL, and perceived stress, there is limited evidence to support the effect of yoga on objectively measured physiological indicators of health, such as inflammation or glucose. Therefore, there is a lack of reliable and conclusive information about the potential impact of yoga on health outcomes (Bethany Forseth et al., 2021).

Yoga promotes self-regulatory systems by reducing physical, psychic, and psychological stress, using various high- and low-level networks. It is often referred to as "meditation in action" as it enhances focused attention during various practices such as asanas (physical postures), pranayama (breathing methods), dhyana (meditation), and savasana (relaxation techniques). Scientific research has linked yoga to many objectively measurable benefits, including stress and anxiety reduction, protection against depression, improved sleep quality, decreased sympathetic distress, lower blood pressure, reduced inflammation, and overall quality of life. Yoga operates through various mechanisms, such as inducing a relaxation response, increasing endorphin levels, improving cardiac-vagal tone, decreasing sympathetic overactivity, lowering intraocular pressure, and preventing additional retinal ganglion cell damage (Gautam S. et al., 2021).

Scientific studies have demonstrated that yoga can enhance cognitive skills, including memory, cerebral cortex thickness, brain connectivity, electroencephalogram (EEG) improvements, and neuroplasticity. Yoga also plays a role in preserving genomic stability and chromosome integrity by maintaining telomere length, upregulating the telomerase activity, and increasing sirtuin expression levels. This helps to slow down cellular aging. In addition, yoga downregulates pro-inflammatory cytokines, restores transcripts involved in DNA repair, and regulates the cell cycle checkpoint (Gautam S. et al., 2021).

Gautam S. et al. (2021) reported that regular daily yoga practice could result in an increase in several mitochondrial parameters, including mitochondrial membrane potential, cytochrome C oxidase (COX-II), nicotinamide adenine dinucleotide (NAD+), mitochondrial copy number, and the expression of genes responsible for maintaining mitochondrial integrity and promoting mitochondrial biogenesis, such as AMPK, IGF1R, PRC-1, TFAM, SIRT-1, TIMP-1, and KLOTHO. Furthermore, practicing yoga can also elevate melatonin levels, a powerful antioxidant with a high concentration within the mitochondria, which is essential in reducing oxidative damage to mitochondrial DNA and preventing mutations.

In addition, yoga improves mitochondrial integrity and optimizes markers related to oxidative stress, altering the levels of transcripts responsible for maintaining mitochondrial health. As a result, it helps to maintain healthy levels of free radicals, increase the number of mitochondria, improve COX-II activity, elevate NAD+ levels, reduce disease activity, and mitigate its effects on physical health.

Neurodegenerative Diseases and Mitophagy

Mitochondria play a crucial role in the oxidative phosphorylation system, which generates adenosine triphosphate (ATP), maintains cellular homeostasis, regulates Ca2+, and facilitates redox signaling. To preserve mitochondrial homeostasis, cells have several interconnected pathways that work together to remove damaged mitochondria. Mitophagy, a specialized form of autophagy, is the only known pathway that selectively eliminates defective mitochondria (Lou G et al., 2020).

A study using a mouse model of Purkinje cell degeneration observed that altered mitochondria caused excessive damage and death of neuronal cells in the cerebellum. These findings suggest that insufficient and uncontrolled mitophagy can have harmful effects (Kamat PK et al., 2014). Since neuronal cells rely heavily on mitochondrial function, reduced or inadequate mitophagy function is now believed to play a crucial role in various neurodegenerative diseases, including Alzheimer's disease, Parkinson's disease, Huntington's disease, and amyotrophic lateral sclerosis (ALS). This impaired mitochondrial function makes neurons vulnerable to oxidative stress and bioenergetic effects, indicating that mitochondrial defects significantly impact neuronal survival (Palomo GM et al., 2015).

To understand the effect of mitophagy disorders on neuronal cells leading to oxidative damage and neurodegeneration diseases, we will discuss the following disease in relation to malfunctioning of the mitochondrial machinery.

Alzheimer's Disease (AD)

Alzheimer's disease (AD) is a prevalent neurodegenerative disorder that causes memory loss and cognitive impairment. Substantial evidence indicates that changes in mitochondrial machinery, motility, morphology, and activity are directly related to AD and are crucial early events in the disease's development (Fang, E.F. et al., 2019). High levels of 4-hydroxy nonenal (4-HNE) and malondialdehyde (MA) in the brains of AD patients have been reported to cause injury mediated by reactive oxygen species (ROS) (Butterfield DA et al., 2002). Resveratrol mitigates Aβ1-42-induced cell death and increases mitophagy, while upregulation of Bcl-1 reduces Aβ aggregation and neurodegeneration (Lee JH et al., 2010; Wang H et al., 2018). Additionally, recent studies have found that PS-1 mutations indirectly affect mitochondrial function by impairing mitophagy machinery (García-Escudero V et al., 2013). Several proteins related to mitophagy, such as Bcl2L13, BNIP3L/NIX, PINK1, p-TBK1, FUNDC1, p-ULK1 (ser555), AMBRA1, and MUL1, are downregulated in the brains of AD patients (Fang, E.F. et al., 2019).

Parkinson's Disease (PD)

PD is a recurrent, and debilitating neurodegenerative disease that is characterized by the gradual loss of dopaminergic neurons, leading to the presence of four primary symptoms, including rigidity, bradykinesia, and postural instability (Giorgi et al., 2021) (Balestrino et al., 2020). Studies examining PINK1 and Parkin mutations have shown that mutations such as R42P, R275W, A240R, and T415N can interfere with the release of p62 and HDAC6, which in turn hinders the removal of impaired mitochondria (Lee et al., 2010). The protein LRRK2 has been found to contribute to mitochondrial clearance by setting up a complex with Miro1 (Hsieh et al., 2016). Additionally, research has shown that DJ1 and α-synuclein can impact mitophagy by causing mitochondrial fragmentation.

Amyotrophic Lateral Sclerosis (ALS)

ALS is a fatal condition where motor neurons experience localized degeneration and protein aggregation. Various proteins, including OPTN, TDP-43, TBK1, FUS, p62, and PIPK1, have been identified as contributing to the progression of ALS. These proteins lead to the accumulation of swollen and aggregated mitochondria, resulting in increased production of ROS, impaired mitochondrial dynamics, and reduced calcium buffering capacity. It has been suggested that both over- and under-stimulation of the mitophagy machinery can play a role in the development of ALS (Ruffoli et al., 2015), (Smith et al., 2019).

The Effect of Yoga on the Progression of Neurodegenerative Diseases

Yoga is an ancient Indian mind-body approach considered a form of complementary and alternative medicine (CAM) (Wahbeh et al., 2008) and is not associated with any religious beliefs. It is an effective mind-body medicine for preventing and treating neurological disorders (Mishra SK et al., 2012). Studies have shown that the practice of yoga can trigger mitophagy, which has positive effects on the progression of neurological disorders. Several neurological disorders have been reviewed in this context.

Alzheimer's Disease

AD, a condition that results in dementia, memory loss, cognitive decline, and neuropsychiatric symptoms, has been the focus of several studies exploring the benefits of yoga and mindful meditation in improving AD pathophysiology (MacLeod et al., 2015; Ashford JW et al., 2015; McCaffrey R et al., 2014). One study found that an 8-week yoga program improved basic cognitive tests such as the Six-Minute Walk Test in adults with moderate and severe AD (McCaffrey R et al., 2014). The Basal forebrain complex (BFC) neurons, which are affected by the severe reduction of nerve growth factor (NGF) in AD patients, can be protected by increasing NGF levels through triggering mitophagy (Martorana F et al., 2018; Balasubramanian S et al., 2015). Moreover, evidence suggests that Alzheimer's is linked to a reduction in cerebral γ-aminobutyric acid (GABA) neurons, which is associated with higher levels of anxiety (Mohr et al., 1986; Kosel M et al., 2004). Interestingly, performing yoga and meditation stimulates the prefrontal cortex, which activates the thalamus's reticular nucleus and produces GABA (Cornwall J et al., 1988).

A recent study aimed to explore the relationship between the neurotransmitters released during yoga and their ability to prevent cellular death. To do this, human induced pluripotent stem cells were treated with various factors such as nerve growth factor, bone morphogenetic factor 9, and purmorphamine to generate basal forebrain cholinergic neurons, which were then used to create an in-vitro model for Alzheimer's disease damaged cells. Several neurotransmitters were tested individually and in combination to assess their neuroprotective effects. These included nerve growth factors, dopamine, histamine, serotonin, and acetylcholine. The combination of serotonin and dopamine, and serotonin and histamine, effectively rescue cholinergic function, resulting in increased levels of choline acetyltransferase expression. These results suggest that either activating serotonin receptors to inhibit the production of Aβ oligomers or inhibiting the synthesis of nitric oxide could be responsible for the positive effects in this study (Hassan A. et al., 2020).

Parkinson's Disease

Parkinson's disease is the second most prevalent neurodegenerative disorder characterized by the progressive loss of dopaminergic neurons (Li S et al., 2016), resulting in various physical and cognitive disturbances (Dong J et al., 2016). Several studies have found that practicing yoga can improve physical functions, such as bradykinesia, muscle and strength (Ni M et al., 2016), rigidity, stability (Sumec R et al., 2015), Unified Parkinson Disease Rating Scale (UPDRS) (Ni M et al., 2016) and tremors (Sharma NK et al., 2015), and provide an improvement with depression (Boulgarides LK et al., 2014), QOL (Moriello G et al., 2013) and the quality of life in patients with Parkinson's disease. Additionally, a case study conducted over 24 weeks found that combining yoga and physical therapy exercises followed by home exercise plans effectively improved the mobility, muscle strength, and balance of a patient with Parkinson's disease (Ni et al., 2016). Parkinson's shows a significant decrease in dopamine, homovanillic acid, serotonin, γ-aminobutyric acid (GABA), acetylcholine, and noradrenaline neurotransmitters. As discussed before, yoga increases the production of GABA and restores the function of the neuronal cells (proven in vitro). The combination of serotonin and dopamine works similarly to the one described in AD. Thus, yoga helps stabilize the body's extremities, strengthen core muscles, and provide additional strength, power, flexibility, and balance (Ribeiro S et al., 2014).

Amyotrophic Lateral Sclerosis (ALS)

Individuals with amyotrophic lateral sclerosis (ALS), a condition that affects motor neurons, encounter muscle cramps throughout the intensifying stages of the disease. A study has shown that practicing yoga can effectively manage cramps in ALS patients (Ribeiro Iyengar S et al., 2015). In another case study, an ALS patient developed pressure ulcers in the ischial area after being immobile for a year. However, a 3-week course of Iyengar yoga therapy proved effective in repositioning the patient, reducing pain, and healing pressure ulcers (Ribeiro S et al., 2014).

Conclusion

The relationship between yoga and cognitive function has been a topic of concern in exercise neuroscience, although research in this area is still in its early stages and does not adhere to the strict procedures used in exercise cognition studies. To better understand the extent and nature of neurobiological changes, it is crucial to conduct systematic randomized trials of yoga, compare its effects to other exercise-based therapies, and conduct long-term studies of individuals who practice yoga. The dorsolateral prefrontal cortex and hippocampus undergo significant changes with aging. Studies indicate that yoga may affect the functional connectivity of the default mode network (DMN), the dorsolateral prefrontal cortex's activity while performing analytical tasks, and the formation of these brain regions. Therefore, yoga and other behavioral therapies may mitigate age- and neurodegenerative-related declines.

References

Ashford JW, Mahoney L, Burkett T. (2015) A role for complementary and integrative medicine in Alzheimer's disease prevention. *J Alzheimers Dis*, 48(1):13–4.

Balasubramanian S, Mintzer JE, Wahlquist AE. (2015) Induction of salivary nerve growth factor by yogic breathing: a randomized controlled trial. *Int Psychogeriatr*; 27(1):168–70.

Balestrino, R., and Schapira, A. H. V. (2020). Parkinson Disease. *Eur. J. Neurol*, 27 (1), 27–42.

Bartlett SJ, Moonaz SH, Mill C, Bernatsky S, Bingham CO., (2013) III Yoga in rheumatic diseases. *CurrRheumatol Rep*.;15(12):387.

Bethany Forseth, Michele Polfuss, Michael Brondino, Michael W. Lawlor, Margaret J. Beatka, Mariah J. Prom, Janis Eells, Jeri-Ann Lyons,(2021) Association between yoga, physiologic and psychologic health: A cross-sectional study, *Complementary Therapies in Clinical Practice.*

Boulgarides LK, Barakatt E, ColemanSalgado B. (2014) Measuring the effect of an eight-week adaptive yoga program on the physical and psychological status of individuals with Parkinson's disease. *A pilot studies. Int J Yoga Therapy*, 24:31–41.

Butterfield DA, Lauderback CM (2002) Lipid peroxidation and protein oxidation in Alzheimer's disease brain: potential causes and consequences involving amyloid β-peptide-associated free radical oxidative stress 1, 2. *Free Radic Biol Med,* 32:1050-1060.

Cornwall J, Phillipson O. (1988) Mediodorsal and reticular thalamic nuclei receive collateral axons from prefrontal cortex and laterodorsal tegmental nucleus in the rat. *Neuroscience letters*, 88:121–126.

Dong J, Cui Y, Li S, Le, W.. (2016) Current pharmaceutical treatments and alternative therapies of parkinson's disease. *Curr Neuropharmacol*, 14(4):339–55.

Fang, EF, Hou Y, Palikaras K, Adriaanse BA, Kerr JS, Yang B, Lautrup S, Hasan-Olive MM, Caponio D, Dan X, Rocktäschel P, Croteau DL, Akbari M, Greig NH, Fladby T, Nilsen H, Cader MZ, Mattson MP, Tavernarakis N, Bohr VA. (2019) Mitophagy inhibits amyloid-β and tau pathology and reverses cognitive deficits in models of Alzheimer's disease. *Nature Neuroscience,* 22 (3), 401-412.

García-Escudero V, Martín-Maestro P, Perry G, Avila J (2013) Deconstructing mitochondrial dysfunction in Alzheimer disease. *Oxid Med Cell Longev,* 2013:162152.

Gautam S, Saxena R, Dada T, Dada R. (2021) Yoga-Impact on Mitochondrial Health: Clinical Consequences. *Ann Neurosci*, 28(3-4):114-116.

Giorgi, C.; Bouhamida, E.; Danese, A.; Previati, M.; Pinton, P.; Patergnani, S. (2021) Relevance of Autophagy and Mitophagy Dynamics and Markers in Neurodegenerative Diseases. *Biomedicines,* 9, 149.

Gitler AD, Dhillon P, Shorter J. (2017). Neurodegenerative disease: models, mechanisms, and a new hope. *Dis Model Mech.*

Hassan A, Robinson M, Willerth SM (2020). Determining the mechanism behind yoga's effects on preventing the symptoms of Alzheimer's disease. *Neural Regen Res*, 15(2):261-262.

Hawk, M.A, Gorsuch CL, Fagan P, Lee C, Kim SE, Hamann JC, Mason JA, Weigel KJ, Tsegaye MA, Shen L, Shuff S, Zuo J, Hu S, Jiang L, Chapman S, Leevy WM, DeBerardinis RJ, Overholtzer M, Schafer ZT. (2018). RIPK1-mediated induction of mitophagy compromises the viability of extracellular-matrix-detached cells. *Nat Cell Biol,* 20 (3), 272-284.

Hsieh, CH., Shaltouki, A., Gonzalez, AE., Bettencourt da Cruz, A., Burbulla, LF., St Lawrence, E., Schüle B, Krainc D, Palmer TD, Wang X. (2016). Functional Impairment in Miro Degradation and Mitophagy Is a Shared Feature in Familial and Sporadic Parkinson's Disease. *Cell Stem Cell, 19* (6), 709–724.

Kamat PK, Kalani A, Kyles P, Tyagi SC, Tyagi N (2014) Autophagy of mitochondria: a promising therapeutic target for neurodegenerative disease. *Cell Biochem Biophys*, 70:707-719.

Kosel M, Rudolph U, Wielepp P, Luginbuhl M, Schmitt W, Fisch HU, Schlaepfer TE. (2004) Diminished GABAA Receptor Binding Capacity and a DNA Base Substitution in a Patient with Treatment-Resistant Depression and Anxiety. *Neuropsychopharmacology*; 29:1762–1762.

Lee, J. Y., Nagano, Y., Taylor, J. P., Lim, K. L., and Yao, T. P. (2010). Disease- causing Mutations in Parkin Impair Mitochondrial Ubiquitination, Aggregation, and HDAC6-dependent Mitophagy. *J. Cel. Biol. 189* (4), 671–679.

Lee JH, Yu WH, Kumar A, Lee S, Mohan PS, Peterhoff CM, Wolfe DM, Martinez-Vicente M, Massey AC, Sovak G. (2010) Lysosomal proteolysis and autophagy require presenilin 1 and are disrupted by Alzheimer-related PS1 mutations. *Cell,* 141:1146-1158.

Li S, Dong J, Cheng C, Weidong L. (2016) Therapies for diseases: alternatives to current pharmacological interventions. *J Neural Transm* (Vienna);123(11):1279–99.

Lou G, Palikaras K, Lautrup S, Scheibye-Knudsen M, Tavernarakis N, Fang EF (2020). Mitophagy and Neuroprotection. *Trends Mol Med*; 26(1):8-20.

MacLeod, R., Hillert, E. K., Cameron, R. T., & Baillie, G. S. (2015). The role and therapeutic targeting of α-, β-and γ-secretase in Alzheimer's disease. *Future science OA*, 1(3).

Martorana F, Gaglio D, Bianco MR, Aprea F, Virtuoso A, Bonanomi M, Alberghina L, Papa M, Colangelo AM. (2018) Differentiation by nerve growth factor (NGF) involves mechanisms of crosstalk between energy homeostasis and mitochondrial remodeling. *Cell Death Dis*, 9(3):391.

McCaffrey R, Park J, Newman D, Hagen D. (2014) The effect of chair yoga in older adults with moderate and severe Alzheimer's disease. *Res Gerontol Nurs*; 7 (4):171–7.

Mishra SK, Singh P, Bunch SJ, Zhang R. (2012) The therapeutic value of yoga in neurological disorders. *Ann Indian Acad Neurol*, 15(4):247–254.

Mohr, Erich; Bruno, Giuseppe; Foster, Norman; Gillespie, Marjorie; Cox, Christiane; Hare, Theodore A.; Tamminga, Carol; Fedio, Paul; Chase, Thomas N. (1986). GABA-Agonist Therapy for Alzheimer's Disease. *Clinical Neuropharmacology*, 9(3), 257–263.

Moore, A.S. and Holzbaur, E.L. (2016) Dynamic recruitment and activation of ALS-associated TBK1 with its target optineurin are required for efficient mitophagy. *Proc Natl Acad Sci, U S A* 113 (24), E3349-58.

Moriello G, Denio C, Abraham M, DeFrancesco D, Townsley J (2013); Incorporating yoga into an intense physical therapy program in someone with Parkinson's disease: a case report. *J Bodyw Mov Ther,* 17(4):408–17.

Ni, Meng; Mooney, Kiersten; Signorile, Joseph F. (2016). Controlled pilot study of the effects of power yoga in Parkinson's disease. *Complementary Therapies in Medicine*, 25(), 126–131.

Ni M, Mooney K, Signorile JF. (2016) Controlled pilot study of the effects of power yoga in Parkinson's disease. *Complement Ther Med;* 25:126–31.

Ni M, Signorile JF, Mooney K, Balachandran A, Potiaumpai M, Luca C, Moore JG, Kuenze CM, Eltoukhy M, Perry AC. (2016) Comparative effect of power training and high-speed yoga on motor function in older patients with Parkinson disease. *Arch Phys Med Rehabil*; 97(3):345–54. e15.

Palomo GM, Manfredi G (2015) Exploring new pathways of neurode-generation in ALS: The role of mitochondria quality control. *Brain Res,* 1607:36-46.

Ribeiro S. Iyengar (2014) Yoga therapy as an intervention for cramp management in individuals with amyotrophic lateral sclerosis: three case reports. *J Altern Complement Med,* 20(4):322–6.

Ribeiro Iyengar S. (2015) Yoga therapy intervention for ischial pressure ulcers in a patient with amyotrophic lateral sclerosis: a case study. *J Altern Complement Med*; 21(9):578–82.

Ruffoli, R., Bartalucci, A., Frati, A., and Fornai, F. (2015). Ultrastructural Studies of ALS Mitochondria Connect Altered Function and Permeability with Defects of Mitophagy and Mitochondriogenesis. *Front. Cel. Neurosci, 9*, 341.

Sharma NK, Robbins K, Wagner K, Colgrove YM. (2015) A randomized controlled pilot study of the therapeutic effects of yoga in people with Parkinson's disease. *Int J Yoga*;8 (1):74–9.

Smith, E. F., Shaw, P. J., and De Vos, K. J. (2019). The Role of Mitochondria in Amyotrophic Lateral Sclerosis. *Neurosci. Lett*, 710, 132933.

Šumec R, Filip P, Sheardová K, Bareš M. (2015) psychological benefits of nonpharmacological methods aimed for improving balance in Parkinson's disease: a systematic review. *Behav Neurol*: 620674.

Wahbeh, H.; Elsas, S.-M; Oken, B. S. (2008). Mind-body interventions: Applications in neurology. *Neurology*, 70(24), 2321–2328.

Wang H, Jiang T, Li W, Gao N, Zhang T (2018) Resveratrol attenuates oxidative damage through activating mitophagy in an in vitro model of Alzheimer's disease. *Toxicol Lett* 282:100-108.

Ward L, Stebbings S, Cherkin D, Baxter GD. (2013) Yoga for functional ability, pain and psychosocial outcomes in musculoskeletal conditions: a systematic review and meta-analysis. *Musculoskeletal Care*.; 11(4):203–217.

WEHI Neurological Disorders Available from: https://www.wehi.edu.au/research-diseases/development-and-ageing/neurodegenerative-disorders

Weishaupt JH, Hyman T, Dikic I. (2016) Common Molecular Pathways in Amyotrophic Lateral Sclerosis and Frontotemporal Dementia. *Trends Mol Med 22* (9), 769-783.

Wells RE, Phillips RS, Schachter SC, McCarthy EP. (2010) Complementary and alternative medicine use among US adults with common neurological conditions. *J Neurol.;* 257(11):1822–1831.

Chapter 12

Role of Herbs on Mitophagy to Treat Neurodegenerative Disorders

Kaushlendra Kumar[1]
Ashok Kumar Tiwari[2]
and Rachana[2,*]
[1]Department of Biotechnology, Delhi Technological University, New Delhi, India
[2]Department of Biotechnology, Jaypee Institute of Information Technology, Noida, Uttar Pradesh, India

Abstract

Autophagy is an evolutionary self-preservation mechanism that allows the body to eliminate unhealthy cells or organelles and recycle some of them in cellular cleaning and repair. Cellular breakdown and recycling processes are very much conserved in all eukaryotes. Mitophagy is called when unhealthy mitochondria are removed from the cells during this process. During mitophagy, the mitochondria are first marked with various molecular signatures, and this step gets them selected for their removal from the cellular environment. Furthermore, a few "particular receptors on the mitochondrial outer membrane," or the ubiquitin molecules, bind to various other surface receptors and enhance mitophagy via the formation of autophagosomes around them. Autophagosome formation is initiated from the Golgi body, endoplasmic reticulum, or plasma membrane. Activation of various proteins like ULK1, ATG13, Fip200, and ATG101 or receptors like the ULK1 complex leads to the attachment of the Beclin 1 Complex to the phagophore, which ultimately results in the activation of more proteins toward autophagosome formation. Mitophagic receptors such as Parkin/PINK1, ULK1, Nix, and Bnip3 have been reported to be the prime receptor for mitophagy. Mitophagy is an essential process for life, but sometimes, it is dysregulated due to various factors like oxidative stress, inflammation, mutation, and aging, which may lead to various diseases, including diseases of the central nervous system (CNS) such as Alzheimer's (AD), Parkinson's diseases (PD) and Huntington's (HD). These diseases are reported to be associated with many genes, proteins, enzymes, and receptors, which are also found to be related to the mitophagy pathway. Parkinson's disease is reported to be associated with the gene mutations found in PARK9, PARK7, PARK1/4, PARK6, PARK17, and PARK8. In AD, the abnormal interaction of protein tau and Aβ is observed, which is usually removed by autophagy. In the case of Huntington's disease, the SQSTM1/p62 interaction decreased, which further affected the

* Corresponding Author's Email: rachana.dr@iitbombay.org.

In: Mitophagy and Its Role in Health and Disease
Editors: Shalini Mani and Samiksha Wasnik
ISBN: 979-8-89113-470-6

autophagy and mitophagy processes. Several therapeutic molecules, including the molecules obtained from the herbs, have been tried to balance the processes of autophagy and mitophagy. This article reviews the importance of mitophagy in cases of several nervous system disorders. Also, it intends to discuss the roles of various herbs in managing mitophagy and treating these central nervous system disorders by targeting various receptors or proteins that are part of the mitophagy pathway.

Keywords: central nervous system disorders, mitophagy, autophagy, PINK, Aβ, ULK1, SQSTM1/p62, Alzheimer's disease, Huntington's disease (HD), Parkinson's disease (PD)

Introduction

Once a cell decides to remove specific molecules and/or organelles from the cytoplasm, it induces autophagy via priming the damaged molecule(s)/organelle(s) for removal. Thus, autophagy is a process that breaks down damaged organelles and molecules within the cell by the structure called autophagosome (Fimia et al., 2013). It is the cytoprotective action that leads to cellular cleansing and recycling of cell organelles and nutrients. Within the process of autophagy, mitophagy is the mechanism through which autophagy selectively destroys damaged mitochondria (Tolkovsky, 2009; Fimia et al., 2013). Several trigger factors induce autophagy, such as mutations in the mitochondrial DNA or protein receptors, which might ultimately result in mitochondrial dysfunction, thus initiating the process of mitophagy (Ding and Yin 2012). Evidence indicates that mitophagy is essential for several pivotal physiological processes, such as paternal mitochondrial degradation, terminal differentiation of RBCs, neurodegenerative disease (NS) prevention, and ischemia-reperfusion (Williams et al. 2017). Mitophagy maintains the mitochondrial quantity and quality in the cells by removing unhealthy or extra mitochondria that could produce ROS, the reactive oxygen species, and other toxic chemicals that direct various harmful activities leading to cell death (Zhang et al., 2017).AD, HD, PD, and many other neural diseases are considered to be neurodegenerative disorders. These are distinguished from many other neurological disorders as they carry defective mitochondria characteristics, including their morphology, dynamics, function, and the process of mitophagy (Cen et al., 2021). Although the specific causes of these disorders are unknown, mitophagy stimulation has been found to rebuild the healthy quality of proteins and a balanced mitochondrial environment and has been proposed for the symptomatic therapy of neurological disorders (Cen et al., 2021). In the current manuscript, the authors introduce autophagy and mitophagy and their roles in various neurodegenerative diseases (NDs) and review the role of various herbs against these neurodegenerative diseases regarding their effectiveness in regulating mitophagy.

Mechanisms of Autophagy

The autophagy pathway is divided into five phases, i.e., initiation, nucleation, elongation, maturation, and fusion (Figure 1 and 2). Autophagy begins by forming a c-shaped phagophore from either the endoplasmic reticulum, plasma membrane, or Golgi body (Levine and Kroemer, 2019). When nutrients are deprived, the initiation phase of autophagy starts with the translocation of various protein receptors such as ULK1, ATG13, Fip200, and ATG101 (also known as ULK1 complex) to form phagophore (Dikic and Elazar, 2018). The exact location of the ULK1 complex is not precise, but various studies show that it forms a structure near the endoplasmic reticulum during starvation. These proteins of the ULK1 complex are responsible for further activation of the autophagic membrane nucleation phase (Figure 1) (Hosokawa et al., 2009). The ULK1 complex is reported to have a central role in the autophagy process as it is known to combine the signals from upstream signaling molecules like AMPK and mTOR to the downstream autophagy pathway (Parzych and Klionsky, 2014). The starvation condition deactivates the mTOR protein and leads to the dephosphorylation of ULK1 and ATG13, and then, ULK1 kinase is activated to initiate the autophagy process. Additionally, during starvation conditions, AMPK (AMP-Activated Protein Kinase) inhibits the mTORC1 via mTOR negative regulator activation and phosphorylation of Tuberculosis Sclerosis Complex2 (TSC2) (Pang et al. 2021). An increase in the AMP: ATP and ADP: ATP ratios are immediately sensed by AMPK, activating AMPK. These controls have been shown to restrict cell cycle progression and protein synthesis regulating cell size and preventing apoptosis downstream of mTORC1 suppression (Pang et al., 2021).

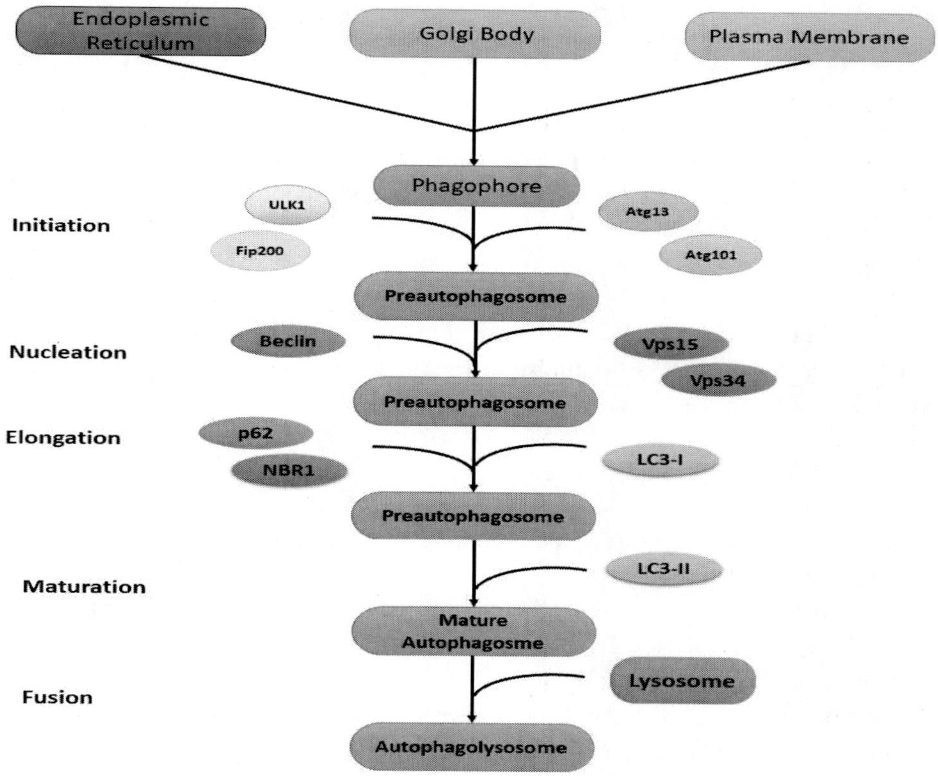

Figure 1. Flow chart of the autophagy pathway: It starts with the formation of photophores from the ER, plasma membrane, and Golgi body. They are divided into five phases, i.e., initiation, nucleation, elongation, maturation, and fusion. It starts with the formation of phagophores from the endoplasmic reticulum, Golgi body, and plasma membrane. It starts with the translocation of various protein receptors, ULK1 is the central receptor for phagophore initiation. ULK1 helps in phagophore formation, and it activates the protein Beclin with some protein Vps15 and Vps34; this protein recruits another protein called LC3-I, and after that, it recruits a protein called LC3-II, and the whole this process makes the phagophore into a mature autophagosome, which is further fused with a lysosome to form a complete structure called Autophagolysosome.

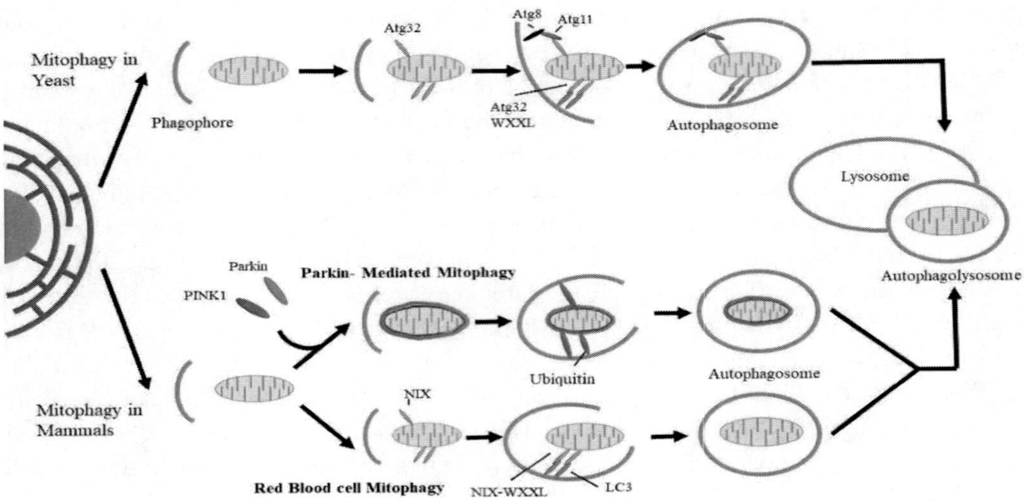

Figure 2. A schematic representation of mitophagy in yeast and mammals, a comparison.

As aforementioned, ULK1 (in the dephosphorylated state) plays a significant role in the initiation process. Further, ULK1 activates the Beclin1 protein, a multifunctional protein that acts as a scaffold for many binding partners such as Vps34 and Vps15. Beclin is present in phagophore, the cytomembrane formed from either the Golgi body, endoplasmic reticulum, or plasma membrane through the phosphorylation of beclin1 by ULK1, which along with some more proteins like Vps34 and Vps15, is located on the beclin complex. Phosphorylation of Beclin1 results in increasing the activity of Vps34 and Vps15 (Mcalpine et al., 2013; Cao and Klionsky, 2007). Vps15 of the beclin complex is necessary for autophagy, which is brought on by different types of stress, such as nutrition starvation, hypoxia, and oxidative stress. The Beclin complex recruits phosphatidylinositol, which is phosphorylated by the Vps34, to form phosphatidylinositol 3-phosphate. It helps in autophagosome formation, and results in the recruitment of protein LC3-1 (1A/1B-light chain 3), p62 (ubiquitin cargo binding protein), and NBR1 (autophagy cargo receptor) to the autophagosome (Lin and Hurley 2016; Ding and Yin 2012). In the nucleation phase, other receptors like Atg5, Atg12, and Atg16, present on the phagophore surface, also help target the particular substrate for the autophagy process(Ding and Yin 2012; Lin and Hurley 2016).

In the elongation phase, p62 and NBR1 are those proteins that act as partial selective receptors for different substrates. LC3-I is cleaved from LC3 by a cysteine protease Atg. Atg4 is essential for autophagy because it first modifies Atg8 at its C-terminus (Mcalpine et al. 2013). p62 and NBR1 both have a UBA domain and bind to ubiquitinated proteins as well as a peptide motif (LIR), which binds specifically to Atg8-family (LC3/GABARAP), which is located on the developing autophagosome membrane. Once LC3-I is targeted to autophagosome formation, it is activated by Atg7, which is bound with Atg3 present on pre-autophagosome, and it conjugates with phosphoethanolamine to form LC3-II, and all these protein receptors help in the formation and elongation of the autophagosome (Yang and Klionsky 2010; Nakatogawa et al., 2009; Kelekar, 2006). This mature autophagosome obtained further gets fused with lysosome by the proteins called SNARE and Rab7, and it is then converted into autolysosome. Further, the membranes fused, and the content in the autophagosome is released to the lysosome, and the substrate is degraded in the acidic lysosome by proteases (Onishi et al., 2021).

Mechanism of Mitophagy

Mitophagy in Yeast

In yeast, the functioning of mitochondria, morphology, and autophagic activity depends upon the growth conditions, such as: if glucose is present or yeast is obtaining the energy by anaerobic respiration (Tolkovsky, 2009). Initiation of autophagy occurs when there is less amount of amino acid and glucose in the medium: the starvation conditions (Fukuda and Kanki, 2018). The Mechanistic Target of Rapamycin Complex 1 (mTORC1): The protein complex controls the synthesis of proteins and serves as a food, energy, and redox sensor, activated when glucose and amino acids are highly available. When nutrition levels drop significantly, the mTOR pathway is blocked, and ULK1, FIP200, and Atg13 produce downstream autophagy effector proteins like berlin, Vps15, and Vps34, which trigger autophagy (Mcalpine et al., 2013).

Many researchers have found that the Atg32 autophagy protein is essential for mitophagy (Okamoto et al., 2009) (Figure 2). Atg32 is crucial for cargo loading and the formation of an autophagosome, which functions as a scaffold protein to draw in additional autophagy proteins and build an autophagosome (Kanki et al., 2009). Atg32 protein binds to Atg11 localized at PAS (pre-autophagosomal structure) (required for cargo loading), and Atg11 binds to Atg8 Protein; Atg32 has a WXXL-like Atg8-binding motif in its cytosolic domain that must attach to Atg8 to occur mitophagy (Figure 2) (Kanki et al., 2009; Okamoto et al., 2009; Kondo-Okamoto et al., 2012). Atg8 is connected to phagophore (cytomembrane derived from endoplasmic reticulum) to recruit the mitochondria into autophagosomes (Antón et al. 2016). Expression of Atg32 increases in oxidizing conditions, suggesting that it may also play a role in mitochondrial quality control; if the mitochondria are healthy, there is no chance of autophagy (Innokentev et al., 2020).

Mitophagy in Mammals

Parkin and PINK1 Dependent Pathway

Different pathways regulate the mitophagy process in mammals, and the most well-studied pathways for mitophagy are regulated by, i.e., PINK1 and PD-related protein Parkin (Wang et al., 2019). Although the PINK1 gene is distributed throughout all organs, normal cells typically maintain this protein at an undetectable level in the cytosol. The PINK1 N-terminal mitochondria targeting sequence (MTS) is transported across healthy mitochondria's outer and inner membranes after targeting (Valente et al., 2004). The schematics (Figure 3) show that PINK1 is translocated to the mitochondria. It attaches to the TOM (Translocase of Outer Mitochondrial Membrane) complex of the outer membrane of mitochondria and transfers to the inner mitochondrial membrane called TIM (Translocase of Inner Mitochondrial Membrane) complex, and when PINK1 gets attached to the inner membrane and gets trimmed by the protein called Presenilin-Associated Rhomboid-Like protein (PARL) a mitochondrial protein helps in apoptosis and signal transduction. After that, it gets dissociated from mitochondria and degraded by the proteasomes inside the cell. To stop the mitophagy of healthy mitochondria, this mechanism is maintained by deficient endogenous PINK1 levels in a polarized mitochondrion (Whitworth et al., 2008: Deas et al., 2011; Shi et al., 2011). In damaged mitochondria, PINK1 attaches to the mitochondrial outer membrane but cannot enter the inner membrane of the mitochondria because the critical proton gradient maintained by the electron transport chain is absent due to abnormal conditions of mitochondria (Figure 3). PINK1 regulates the ubiquitin enzyme, which phosphorylates it on ser65, helping recruit parkin. Both are used as a feed-forward method to activate parkin. Ubiquitin-binding autophagy receptors can be activated by Parkin after it assembles ubiquitin chains on the outer membrane of mitochondria. The ubiquitin chain signals the autophagosome to bind with them, and the mitophagy process starts. The ubiquitin chain forms a structure called "p62 autophagy adapter," which contains the K-63 ubiquitin-binding domain and LC3-1 binding domain which helps in the attachment of autophagosomes with the mitochondria. On the outer membrane of harmed mitochondria, PINK1 builds up and activates "Parkin's E3 ubiquitin ligase" activity and brings Parkin to the faulty mitochondria. Then, Parkin causes selective autophagy by ubiquitinating proteins on the outer mitochondrial membrane (Wade Harper, Ordureau, and Heo, 2018; Kane et al., 2014).

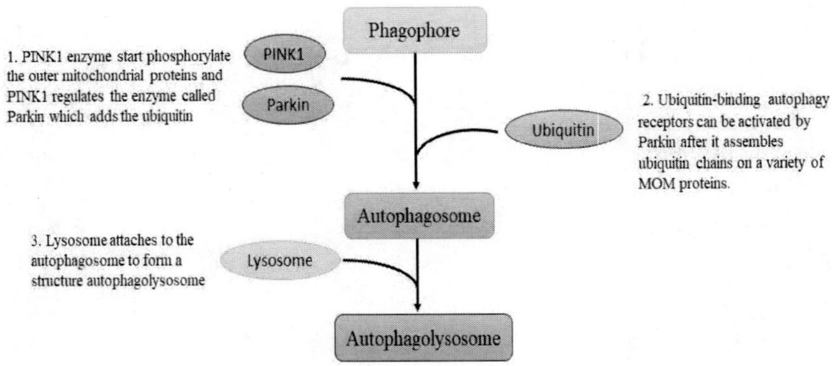

Figure 3. A sequence of events during autophagy in mammals.

Mitophagy in Red Blood Cells

Another pathway of mitophagy is majorly studied in mammalian red blood cells; there is a lack of mitochondria in mature RBCs of mammals. This is accomplished through mitophagy only as the reticulocytes mature. This procedure involves the NIP3-like protein X (NIX/BNIP3L), one of the selective autophagy receptors, called Nix (BNIP3L), involved in several human illnesses. When BNIP3L/NIX attaches to mitochondria and the phagophore membrane, it facilitates mitochondrial autophagosome engulfment and subsequent transport to lysosomes for degradation (Aerbajinai et al., 2003). It is in OMM. It also contains a WXXL-like motif in the cytoplasmic region, the same as in yeast cells. Furthermore, it attaches to the Atg8 orthologue, GABA receptor-mediated proteins, LC3 and LC3 homolog. LC3/GABARAP proteins are on the autophagosome membrane via the conserved LC3-interacting region (Figure 4) (Schwarten et al., 2009; Novak et al., 2010). This helps mitochondria to recruit phagophores, and the receptors and proteins present on phagophores help in the formation of autophagosomes which leads to mitophagy (Palikaras et al., 2018; 2015). ULK1 and ATG7, which are present in phagophores, are also essential autophagic proteins for mitophagy in RBC (reticulocytes) (Mortensen et al., 2010; Kundu et al., 2008; J. Zhang et al., 2009).

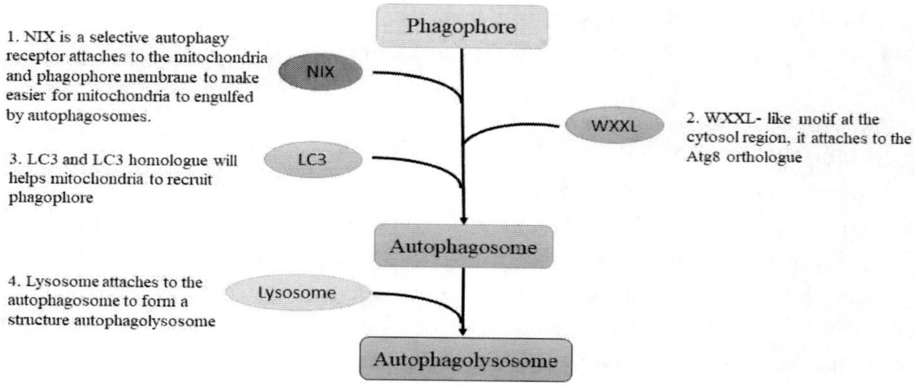

Figure 4. The sequence of events during mitophagy in RBCs.

The above explanation of autophagy and mitophagy represents the role of various proteins and genes in these processes. Several neural diseases also affect these proteins and genes, indicating that they play essential roles. This further implies that if we balance their function via certain drugs/molecules, we might be able to control the progression of the diseases. The following is the description of various essential herbs and their molecules observed to modulate the behaviors of the proteins and genes.

Natural Products and Their Effect on Mitophagy for Neurodegenerative Diseases

Herbs have been excellent alternatives to pharmaceutical medications. The following describes a few essential herbs/their extracts/purified molecules that have been reported to modulate mitophagy/autophagy in neural diseases.

Natural Extracts and Formulations Used to Improve Mitophagy for Neurological Disorders

Panaxnotoginseng

*Panaxnotoginseng*is a Chinese herb mainly used for improved blood circulation and activation, disappearing blood stasis, and treating nervous disorders. The significant proportion that helps treat AD from this herb is Panax *notoginseng* saponins (PNS). Some studies have shown that PNS can enhance memory and cognition for AD treatment (Zhang et al., 2020).

In a study by Jiang et al. (2012), PC12 cells were treated with Aβ25-35 (Amyloid β-protein fragment 25-35) to create an AD model and were further treated with Panax *notoginseng (PNS).* Transmission electron microscopy analysis of these cells revealed that PNS enhanced the development of autolysosomes, making it more straightforward for the damaged mitochondria to be engulfed by the autophagosome in Aβ-treated PC12 cells. Western blotting assessed the expression of p62, LC3-I, and LC3-II (autophagy marker proteins), and it was found that the PNS treatment drastically decreased the p62 protein while significantly enhancing the LC3-II/LC3-I ratio this way, PNS treatment could enhance the autophagic activity. Interestingly, PNS treatment led to a drop in the p-mTOR/mTOR ratio, indicating a lesser degree of mTOR activation, the key factor regulating autophagy that can increase cellular autophagy in AD-like circumstances.

Ginkgo Biloba

Ginkgo biloba is often called the "Maidenhair tree," commonly found in China. *Ginkgo biloba* seeds have been used to treat NDs for many years (Mei et al., 2017). It shows various properties like anti-aging, anti-inflammatory and neuroprotective. *Ginkgo biloba* extract has been observed to provide neuroprotection in many animals suffering from Parkinson's disease and AD (Yadav et al., 2013; El-Ghazaly et al., 2015). Ginkgolic acid (GA) is an active substance extracted from the nuts, leaves, and testa of *Ginkgo biloba* herbs which shows effectiveness in neurodegenerative disorders.

In the HeLa cells treated with GA, researchers found massive upregulation in LC3B-II lipidation (the autophagy marker) and some LC3 puncta (an indicator of autophagosome formation) (Wang et al., 2019). LC3 was lipidated during autophagosome formation, and this LC3-phospholipid conjugate (LC3-II) was observed on autophagosomes and autolysosomes. They proposed that lysosomal hydrolases may destroy intra-autophagosomal LC3-II and showed an increase in LC3-II accumulation as a hallmark of autophagy. Furthermore, compared to GA therapy alone, CQ (lysosomal inhibitor) and GA combination enhanced LC3B-II much more,indicating that GA boosted the autophagy flux. They also observed the co-localization of mitochondria and the lysosomes/autophagosomes.

In addition, compared to GA therapy alone, the combination of CQ (lysosomal inhibitors) and GA raised the level of LC3B-II much more, indicating that GA boosted the autophagic flux. Similarly, in lung cancer cells, GA has been demonstrated to prevent the epithelial-to-mesenchymal transition and to reduce colon cancer cell proliferation, migration, and invasion through inhibition of the mTOR signaling pathway, which is the primary negative regulator of autophagy. In addition, the GA-induced loss of mitochondrial mass was shown to be reversed by chemically or genetically induced autophagy suppression. These studies indicate that GA can trigger mitophagy (Wang et al., 2019).

Withaniasomnifera

Withaniasomnifera (Ws), often known as Ashwagandha, is an essential Indian herb that can prevent many illnesses(Mirjalili et al., 2009). It is reported to be a nerve tonic and has aphrodisiac properties that enhance memory and learning capability. Ws root has many pharmacological properties, like anti-inflammatory (Sun et al., 2016), antioxidant (Sun et al., 2016), anxiolytic-depressant, and memory-enhancing properties (Shivamurthy et al., 2016). According to various studies, withaferin-A, a significant component of Ws extract, has been shown to induce autophagy (Hahm and Singh, 2013).

In another study by (Kumar and Kumar, 2009), the root extract of Ws was shown to modulate mitochondrial damage, biochemical parameters, and behavioral outcomes in 3-nitropropionic acid-induced models of HD. The study showed that the Chronic Ws therapy effectively balanced the increased nitrite levels and lipid peroxidation and restored the levels of antioxidant enzymes in the brain's striatum and cortex, pointing to the potential antioxidant activity of Ws, which is correlated with HD by various scientists.

In a preclinical study on rats, it has been figured out that with anodes-A and C, the active phytochemicals of WS, bound to the active motif of Aβ (Aβ25–35) distinctively and might protect cells from the toxicity caused by Aβ-induced avoiding fibril formation, which is known to decrease the autophagy and mitophagy process and hence, might help in the prevention of AD(Jayaprakasam et al., 2010).

Curcuma Longa

Turmeric, also known as *Curcuma longa*, has been used for various health benefits for years. *Curcuma longa* is mainly used for a healthy nervous system, immune system, blood vessels, and heart. It acts as a neuroprotective agent against neuronal death, damage in BBB (blood-brain barrier), and brain aging (Balaet al., 2006; Omosa, 2017).

Curcumin promotes mitophagy (Cao et al., 2020). According to researchers, *Curcumin longa* may control mitophagy and remove damaged or unhealthy mitochondria by modifying signaling pathways relevant to autophagy (de Oliveira et al., 2016; Wu et al., 2011)

In a study on "PTEN-induced putative kinase 1 (PINK1) Knock Down SH-SY5Y cellular model" of PD, the results highlighted the protective effects of curcumin against the abnormal mitochondria and apoptosis in PINK1-deficient and paraquat (a PD inducing compound) exposed cells. It has been hypothesized that curcumin activates PINK1 activity and prevents neuron death due to abnormal mitochondria (Van der Merwe et al., 2017).

Artemisia Argyi Leaf Extract

The medicinal qualities of plants in the *Artemisia* genus have traditionally been used against liver, spleen, kidney infections and neurological disorders (Hussain, 2020; Wu et al., 2022). *Artemisia argyi* has also been shown to have anti-inflammatory and antioxidant properties (Kang et al., 2016).

In a study on the "1-methyl-4-phenyl-pyridinium (MPP+)-induced PD model" established in human neuroblastoma cell line -SH-SY5Y, as well as on a "1-methyl-4-phenyl-1,2,3,6-tetrahydro-pyridine (MPTP)-induced PD model" in C57BL/6 J mice, *Artemisia argyi*could protect from the disease by activation ofthe cell survival pathway called as PI3K/AKT/β-catenin pathway (Wu et al., 2022). Mitochondrial damage caused by MPP+ was overcome by upregulating mitophagy and downregulating the mitochondrial fission regulator -phospho-dynamin-related protein 1 (pDRP1) (Phosphorylated at ser616), in SH-SY5Y cells after the treatment with *Artemisia argyi*leaf extract. A study showed that *Artemisia argyi*could protect against cell viability loss caused by MPP+,and theaccumulation of mitochondrial and total ROSwas also reduced. Furthermore, the expression of the TRPML1 protein wasincreased, which induced autophagy,facilitating the clearance of α-synuclein (Wu et al., 2022).

In another study (Lu et al. 2022), it was reported that eupatilin, a major flavonoid of *A. argyi, could* prevent the increase in glutamate concentrations and excitotoxicity caused by kainic acid (KA) in the cortex of rats. Eupatilin preserves normal levels of glutaminase (responsible for theproduction of glutamate in neurons) and excitatory amino acid transporter1/3 (EAAT1/3), the proteins which cause glutamate uptake in the synaptic cleft. Excess levels ofglutamate, the primary excitatory neurotransmitter in the CNS, causes excessive activation of glutamate receptors like α-amino-3-hydroxy-5-methyl4-

isoxazolepropionic acid (AMPA) andN-methyl-D-aspartate (NMDA) receptors. All this ultimately overloadsthe level of Ca^{2+}, increases protease activation, and elevates free radical and mitochondrial dysfunction leading to eventual neural death.

Danggui-Shaoyao San

Dangguishaoyaosan (DSS), the Chinese herbal formula, also known as, Dangguijakyak-san (DJS) or Toki-shakuyaku-san (TJ-23), has been recommended in Traditional Chinese Medicine to treat AD. It is a combination of *Angelica sinensis (Oliv.) Diels., Ligusticum chuanxiong* Hort., *Paeonia lactiflora* pall., *Poria cocos* (Schw.) Wolf, *Alisma orientalis* (Sam.) Juzep., *Atractylodesmacrocephala*Koidz., It helps improve memory dysfunction, protects from aging and regulate monoamine neurotransmitters' metabolism (Xin Fu et al., 2016; Z. Song et al., 2021).

In a study, DSS has been found to influence free radical-mediated neuropathy via anti-inflammatory, antioxidant, and anti-apoptosis effects in hippocampus cells (Song et al., 2021). Furthermore, it has also been shown to facilitate modulations of the central monoamine neurotransmitter system, consequently ameliorating central cholinergic nervous system dysfunction and scopolamine's reduction in acetylcholine levels (Xin Fu et al., 2016). DSS improves dopaminergic, adrenergic, and serotonergic nervous system function. It also has been shown to alleviate cognitive deficiencies in Alzheimer's disease (AD) patients, suggesting it might be effective as a disease therapy (Xin Fu et al., 2022).

Tao hong Si wu Decoction

Traditional Chinese medicine Tao hong Si wu Decoction (THSWD) comprises six herbs, including *Angelica sinensis* (Oliv.) Diels, *Carthamus tinctorius L, Conioselinumanthriscoides, Prunus persica* (L.) *Rehmanniaglutinosa* (Gaertn), Batsch and *Paeonia lactiflora* Pall (Huang et al., 2021). THSWD shows neuroprotective and anti-dementia effects by inhibiting neuronal apoptosis, preventing the death of neuronal cells, regulating brain neurotransmitters, and improving cerebral blood flow in cerebral ischemia (Fang-Fang et al., 2020; Han et al., 2015).

Researchers used transmission electron microscopy to examine the ultrastructural mitochondrial phenotype to determine how THSWD therapy affected the activation of mitochondrial autophagy in a rat model. Treatment with THSWD in a rat model reduced the mitochondrial morphological change, whereas, Mdivi-1 therapy reversed THSWD's helpful benefits. After THSWD therapy, proteins related to mitochondrial autophagy were elevated. After that, they looked into the levels of expression for the autophagy-related proteins LC3B and Beclin1. Compared to the MCAO/R-treated group, the THSWD-treated group shows a higher conversion of LC3-I and Beclin to LC3-II. Mdivi-1 (an autophagy inhibitor) significantly reduced the Beclin1 and LC3-II/LC3-I expression (Ji et al., 2022).

Polygala Saponins

Radix polygalae, a Chinese medication, is a component of polygala saponins (PSS), which are extracted from the root of *Polygala tenuifolia* herb and show good properties like anti-inflammatory, neuroprotective, and antidepressant in treating disease (Deng et al., 2020; Shin et al. 2014; Zhao et al., 2020).

According to a study, PSS activated mitophagy in microglial BV-2 cells as an increase in the co-localization of LC3 and mitochondria was used to detect as well as an increase in the number of autophagic vacuoles encircling the mitochondria (Qiu et al., 2022). Further investigation also discovered that PSS increased SHP-2, a cytoplasmic SH2 domain that contains tyrosine phosphatase protein, and activated the PINK1/parkin & and AMPK/mTOR signaling pathways. PSS dramatically reduced NLRP3 inflammasome activation in BV-2 cells stimulated by A53T-α-synuclein, Aβ (1-42), or Q74. This effect was diminished by the autophagy inhibitor bafilomycin A1 and the antioxidant SHP099 (an SHP-2 inhibitor). Additionally, it was discovered that PSS encouraged the autophagic destruction of NLRP3 inflammasomes by co-localizing LC3 and ASC. PSS also reduced the apoptosis of PC-12 cells exposed to conditioned media (Qiu et al., 2022).

Dendrobium nobileLindl. Alkaloids

Dendrobium nobile is a vital herb known for antiaging and neuroprotection for years. The Dendrobium nobile ingredients used for neurodegenerative diseases are the *Dendrobium nobile*Lindl Alkaloids (DNLA). In a study, it was shown that DNLA protects the brain against Manganese-induced neurotoxicity. Manganese and mitochondrial interactions are well established for causing the fall off of biological energy and an increase in oxidative stress. The author identified that it functions through PINK1/Parkin-mediated autophagy (Xiaolong Fu et al., 2021). The cells treated with Mn had upregulated levels of LC3-II and p62. While the rise in p62 protein may have signaled a decline in autophagic turnover, the rise in LC3-II protein was a hallmark of higher autophagosome production (Padmakrishnan et al., 2019). The elevated p62 protein expression showed that Manganese had inhibitory effects on autophagic flux. Studies suggested that DNLA promotes autophagosome breakdown and autophagic flux, reducing the toxicity caused by Mn, by reducing LC3-II and p62 levels in cells after DNLA pre-treatment (Gugnani et al., 2018).

The Parkin and PINK1 mutations discussed above have been proven to result in mitochondrial dysfunction and may be responsible for the development of PD (Zeng et al., 2018; Song et al., 2017). Pre-treatment of cells with DNLA has been demonstrated to significantly increase the expression of PINK1 and Parkin, resulting in increased cell survival. This study showed that DNLA can activate the PINK1 and Parkin-mediated mitophagy and exhibits neuroprotective properties against manganese-induced cytotoxicity (Xiaolong et al., 2021).

Purified Fractions or Pure Molecules Isolated from Herbs

Astragaloside IV

The herb *Astragalus membranaceus* contains a substance called astragaloside IV (AS-IV), which is recognized for its neuroprotective properties (Wen et al., 2018; Chang et al., 2016; Jiang et al., 2019; Zhang and Deng, 2019). Evidence suggests its effectiveness against Parkinson's disease. Some previous studies suggested that *in vitro,* astragaloside IV can boost dopamine neuron tolerance to neurotoxins and decrease apoptosis (Chan et al., 2009).

In a study, researchers investigated the mechanism of astrocyte senescence inhibition by AS-IV; they examined the autophagy-associated protein level in astrocytes. The LC3-II expression was lowered in senescent astrocytes, and the p62 level was much higher as we compared it to the non-senescent astrocytes. Moreover, the expression levels of Parkin and PINK1 (mitophagy-related protein) had significantly decreased. On the other hand, senescent astrocytes show higher expression of TOM20 (Translocase of the Outer Membrane)than non-senescent astrocytes in mitochondrial protein extracts, which suggests a low mitophagy flux. AS-IV treatment in senescent astrocytes was suggested to restore the mitophagy level, improve mitochondrial membrane potential, and reduce the ROS level in mitochondria. This result showed that AS-IV can promote mitophagy and improve the function of mitochondria (Xia et al., 2020).

Resveratrol

Resveratrol is a natural phytoconstituent in numerous fruit skins and plants, including grapes, blueberries, and raspberries (Ma and Li, 2020). It has good antioxidant properties (Wang et al., 2018) and prevents neurodegenerative diseases like AD, HD, and PD.

A study by Wang et al. (2018) revealed that resveratrol could attenuate the cell death induced by Aβ1-42 and activate the mitophagy process by inducing the amount of acidic vesicular organelles, the ratio of LC3-II/LC3-I, expression of Parkin and Beclin-1, and the presence of LC3 and TOMM20 together in Aβ1-42-treated PC12 cells (Wang et al., 2018). It was suggested by Wu et al., that critical mechanisms underlying the protective effects of resveratrol may involve mitophagy. Resveratrol shielded human peritoneal mesothelial cells (HPMCs) from ROS-NLRP3-mediated inflammatory damage by promoting mitophagy (Wu. et al., 2016). According to the study, 3-MA(3-methyladenine) dramatically reduced the mitophagy generated by resveratrol and reduced the protective effects of resveratrol, indicating that mitophagy played a crucial part in the protective impact of resveratrol (Wang et al., 2018).

Silibinin

Silybum marianum, generally found in North Africa and Southern Europe, has an active ingredient called silibinin. It is used as a trauma therapy in Germany and shows protective potential on PD (Biedermann et al., 2014).

Liu et al., in 2021, worked on an MPTP-induced PD model to evaluate the motor protective effect of silibinin. The results of the study showed that silibinin inhibits lipid peroxidation *in vivo. However,* it also prevents dopaminergic neurodegeneration and boosts the production of the pro-inflammatory cytokines (IL-1β and TNFα) and the antioxidant response, but also by raising the mediators of mitophagy PINK1 and Parkin, and lowering the toxicity of the α-synuclein and "NLR family pyrin domain containing 3" (NLRP3) activation. The neuroprotective effects of silibinin are further supported by the restoration of TH expression in the striatum and substantia nigra (SN) (Liu et al., 2021). Silibinin dramatically decreased MPTP-induced dopaminergic neuronal degeneration and motor dysfunction in a different investigation. The results indicate that silibinin has advantages in MPTP-induced models of Parkinson's disease because it makes the mitochondrial membrane potential more stable (Lee et al., 2015).

Eclalbasaponin l

Eclalbasaponin l (ECl) is an active compound obtained from the *Aralia elata* herb (Wang et al., 2017). It is abundant in the essential components of the human body, including fat, vitamins, minerals, and plant fiber (Shikov et al., 2016).

Wang group's study in 2019 suggested that ECl can increase mitophagy by increasing LC3-I to LC3-II conversion and a high level of LC3, Parkin, and PINK expression in the cells (Wang et al., 2019). In an earlier study by Wang et al. from 2018, it was concluded that through the stimulation of the Nrf2/HO-1 pathway, EC1 reduces oxidative damage brought on by H_2O_2, increasing the activities of glutathione peroxidase (GSH-Px) and superoxide dismutase (SOD). Malondialdehyde (MDA) levels must be decreased to reactivate the antioxidant system. Because of its ability to promote the expression of antioxidant enzymes, Nrf2 activation can be viewed mainly as a target for preventing oxidative stress. Notably, employing Baf or 3-MA to block autophagy reduces EC1's capacity to lessen H_2O_2-induced oxidative damage. This suggests that EC1 is a promising novel medication for treating nervous system disorders linked to oxidative stress, which works via autophagy activation, and antioxidative mechanisms in neural cell death.

Wogonin

The pharmacological activities of *Scutellariabaicalensis Georgi have* been described, including anti-viral, anti-bacterial, and anti-inflammatory properties(Yue et al., 2012; Wardenburg et al., 2007). Wogonin, the primary component of *Scutellariabaicalensis*georgi, has shown various biological properties, including anti-inflammatory and anti-tumor effects (Zhao et al., 2014; Chen et al., 2013). It was discovered that wogonin contributes to neuronal defense (Lin, 2011). According to the study by Li et al. (2005) using human tissue, the mTOR pathway was found to be more activated in damaged brain regions in Alzheimer's disease brains compared to healthy brains (Li et al., 2005). It was shown that the therapy with the mTOR inhibitor rapamycin enhanced autophagy markers in neurons, decreased the level of Aβ plaque, and reversed the memory deficits in AD mice models (Caccamo et al., 2010; Spilman et al., 2010).

A study suggested that wogonin could induce autophagy and the maturation of autophagosomes by increasing the concentration of LC3-II in SH-SY5Y cells and primary cortical astrocytes. The phosphorylation of ULK1, mTOR, and P70S6K was also examined because these proteins are crucial for mTOR-mediated autophagy. The outcomes showed that in SH-SY5Y cells and primary cortical astrocytes, the phosphorylation levels of these proteins were inhibited (Zhu and Wang 2015). Studies have shown that when the mTOR signaling activity is elevated, the phosphorylation level of GSK3β (on serine 9) is reduced. As a result, tau protein would typically have a high level of phosphorylation (Caccamo et al., 2013; Tang et al., 2013). Researchers looked into how wogonin affected the phosphorylation of GSK3ß and tau and found that wogonin could both decrease tau phosphorylation in SH-SY5Y cells and enhance GSK3β phosphorylation (Zhu and Wang, 2015).

Palmatine

An isoquinoline alkaloid known as palmatine is found in the medicinal herb *Fibraurearecisa* Pierre. Palmatine has been therapeutically used to treat pelvic inflammation, stomach pain, chronic endometritis, enteritis, and Alzheimer's disease (AD) (Fangand Hongyu, 2016; Pei et al., 2022). It has many pharmacological attributes, including anti-inflammation and antibacterial (Yan et al., 2017). Additionally, it is essential in triggering mitophagy and treating ND.

A study showed the effect of palmatine on mitophagy in DSS mice. The findings showed that DSS mice had reduced levels of PINK1 and increased levels of p62, while Parkin and LC3-II reductions failed to reach statistical significance. However, palmatine reduced the P62 level and encouraged the transformation of LC3-I into LC3-II and the production of PINK1/Parkin (Mai et al., 2019). The following table summarizes the studies discussed above.

Table 1. A summary of studies showing the effect of different natural compounds on triggering mitophagy

S.no.	Herbs	Main Component	Targets
1	*Panax notoginseng*	Panax Notoginseng Saponin	Decreased p62 protein and increased LC3-II/LC3-I ratio protect from Aβ-induced injury by inhibiting mTOR activation.
2	*Ginkgo biloba*	*Ginko biloba* Extract and Ginkgolic acid	The massive increase in LC3-II and lower FUN14 domain containing 1(FUNDC1) lipidation and some LC3 puncta, Boost the autophagy flux.
3	*Dangguishaoyaosan*	Dangguishaoyaosan extract	The expression levels of Parkin, PINK1, and LC3-I /LC3-II increased, and p62 decreased.
4	*Withaniasomnifera*	Withaferin-A	P62 level decreased autophagy-related protein 5 (ATG5), and because the level is increased, LC3-II expression increased and promoted autophagy.
5	*Curcuma longa*	curcuma longa extract	Boost the expression of PINK1 and Parkin, promote mitophagy
6	*Artemisia argyi*	Artemisia leaf extract	Dynamin-related protein-2 (DRP2) and Phospho-DRP1 (pDRP1) both downregulated.
7	Tao hongsiwu decoction	It is made up of *Angelicae Sinensis* Radix, *Paeoniaeradix* Alba, *Praeparata*, *Chuanxiong rhizoma*, *Persicaeseman*, *Rehmanniaeradix* and *Carthamiflos.*	High conversion of LC3-I and beclin to LC3-II. Mdivi-1 autophagy inhibitor significantly reduced Beclin and LC3-II/LC3-I expression.
8	The skin of blueberries, grapes, raspberries	Resveratrol is extracted from the skin of blueberries, grapes, and raspberries.	Attenuates Aβ1-42-induced mitophagy and cell death activated by inducing the acidic vesicular organelles amount, LC3-II/LC3-I ratio, parkin and Beclin-1.
9	*Polygala tenuifolia*	Polygala saponin Chinese medicine Radix polygala extracted from the root of	Activate mitophagy by increasing the autophagic vacuoles quality, PINK1/parkin and AMPK/mTOR are activated and reduced the NLRP3 inflammasome activation
10	*Dendrodiumnobile*	*Dendrodiumnobile*lindl. alkaloids	Improves autophagic flux, ameliorating Mn-elected toxicity by reducing p62 and LC3-II levels, activating the PINK1 and Parkin-mediated mitophagy.
11	*Astragalousmembranaceus*	Astragaloside IV	Restore mitophagy level and improve mitochondrial membrane potential and may decrease ROS level,
12	*Silybum marianum*	Silibinin	Parkin and Pink1 are recovered, the LC3-II/LC3-I ratio decreases, and mitophagy.
13	*Aralia Elata*	Eclalbasaponin I	Increase LC3-I to LC3-II conversion, regulate the oxidative stress-induced mitophagy, and inhibit apoptosis.
14	Scutellariabaicalensisgeorgi	Wogonin	Decreased tau and GSK3β Scutellariabaicalensisgeorgi phosphorylation and increased level of LC3-II
15	*Fibraurearecisa*pierre	Palmatine	Reduced p62 protein and accelerated the transformation of LC3-I to LC3-II, increased PINK1/Parkin expression.

Conclusion

The natural causes of neurodegenerative diseases (ND) are not fully understood, and research is still going on to figure out the root causes. It has been observed from the above discussion that the proteins involved in mitophagy and autophagy are seen to be affected during CNS disorders. Parkin/PINK1, ULK1, Nix, and Bnip3 are the primary receptors of mitophagy. Dysregulation of mitophagy in neural tissue can be initiated due to factors such as oxidative stress, inflammation, and aging and may lead to CNS disorders like Alzheimer's disease, Huntington's disease, and Parkinson's disease.

It can be clearly understood from this discussion that the level of expression of specific proteins associated with and the signaling mechanism during mitophagy can be balanced to control these diseases. The reasons for abnormal mitophagy and autophagy are found to be different for different neurodegenerative diseases. For example, in Alzheimer's disease, abnormal interactions of P-Tau, $A\beta$, and Drp1 have been reported. In Parkinson's disease, abnormal mitophagy is due to the mutation in the Parkin gene, and Huntington's disease, decreased interaction of mitochondria with SQSTM1/p62 autophagy receptors occurs.

While reviewing various herbs utilized against several CNS disorders, it has been observed that many of them, as mentioned above, target various proteins involved in autophagy and mitophagy, membrane potential control, production of ROS, and mitogenesis, among other aspects of mitochondrial biology. It can be said that the above natural products can regulate mitophagy through various mechanisms and molecular targets, which raises the possibility of using these substances as therapeutics for various neurodegenerative disorders caused by defective mitophagy. It is suggested that if various components taken from different herbs may be combined to form formulations, different proteins involved in mitophagy can be targeted simultaneously, which may be able to design new and efficacious therapeutics against CNS disorders.

References

Aerbajinai, W., Giattina, M., Lee, Y. T., Raffeld, M. and Miller, J. L. (2003). The proapoptotic factor Nix is coexpressed with Bcl-xL during terminal erythroid differentiation. *Blood*, *102*(2), pp.712-717.

Antón, Z., Landajuela, A., Hervás, J. H., Montes, L. R., Hernández-Tiedra, S., Velasco, G., Goni, F. M. and Alonso, A. (2016). Human Atg8-cardiolipin interactions in mitophagy: Specific properties of LC3B, GABARAPL2 and GABARAP. *Autophagy*, *12*(12), pp.2386-2403.

Bala, K., Tripathy, B. C. and Sharma, D. (2006). Neuroprotective and anti-aging effects of curcumin in aged rat brain regions. *Biogerontology*, *7*, pp.81-89.

Biedermann, D., Vavříková, E., Cvak, L. and Křen, V., (2014). Chemistry of silybin. *Natural product reports*, *31*(9), pp.1138-1157.

Caccamo, A., Magrì, A., Medina, D. X., Wisely, E. V., López-Aranda, M. F., Silva, A. J. and Oddo, S. (2013). mTOR regulates tau phosphorylation and degradation: implications for Alzheimer disease and other tauopathies. *Aging cell*, *12*(3), pp.370-380.

Caccamo, A., Majumder, S., Richardson, A., Strong, R. and Oddo, S. (2010). Molecular interplay between mammalian target of rapamycin (mTOR), amyloid-β, and Tau: effects on cognitive impairments. *Journal of Biological Chemistry*, *285*(17), pp.13107-13120.

Cao, S., Wang, C., Yan, J., Li, X., Wen, J. and Hu, C. (2020). Curcumin ameliorates oxidative stress-induced intestinal barrier injury and mitochondrial damage by promoting Parkin-dependent mitophagy through the AMPK-TFEB signal pathway. *Free Radical Biology and Medicine*, *147*, pp.8-22.

Cao, Y. and Klionsky, D. J. (2007). Physiological functions of Atg6/Beclin 1: a unique autophagy-related protein. *Cell research*, *17*(10), pp.839-849.

Cen, X., Zhang, M., Zhou, M., Ye, L. and Xia, H. (2021). Mitophagy regulates neurodegenerative diseases. *Cells*, *10*(8), p.1876.

Chan, W. S., Durairajan, S. S. K., Lu, J. H., Wang, Y., Xie, L. X., Kum, W. F., Koo, I., Yung, K. K. L. and Li, M. (2009). Neuroprotective effects of Astragaloside IV in 6-hydroxydopamine-treated primary nigral cell culture. *Neurochemistry International*, *55*(6), pp.414-422.

Chang, C. P., Liu, Y. F., Lin, H. J., Hsu, C. C., Cheng, B. C., Liu, W. P., Lin, M. T., Hsu, S. F., Chang, L. S. and Lin, K. C. (2016). Beneficial effect of astragaloside on Alzheimer's disease condition using cultured primary cortical cells under β-amyloid exposure. *Molecular neurobiology*, *53*, pp.7329-7340.

Chen, X. M., Bai, Y., Zhong, Y. J., Xie, X. L., Long, H. W., Yang, Y. Y., Wu, S. G., Jia, Q. and Wang, X. H. (2013). Wogonin has multiple anti-cancer effects by regulating c-Myc/SKP2/Fbw7α and HDAC1/HDAC2 pathways and inducing apoptosis in human lung adenocarcinoma cell line A549. *PLoS One*, *8*(11), p.e79201.

Cj, P., Hv, E., Vijayakurup, V., R Menon, G., Nair, S. and Gopala, S. (2019). High LC3/Beclin expression correlates with poor survival in glioma: a definitive role for autophagy as evidenced by in vitro autophagic flux. *Pathology & Oncology Research*, *25*, pp.137-148.

de Oliveira, M. R., Jardim, F. R., Setzer, W. N., Nabavi, S. M. and Nabavi, S. F. (2016). Curcumin, mitochondrial biogenesis, and mitophagy: Exploring recent data and indicating future needs. *Biotechnology advances*, *34*(5), pp.813-826.

Deas, E., Plun-Favreau, H., Gandhi, S., Desmond, H., Kjaer, S., Loh, S. H., Renton, A. E., Harvey, R. J., Whitworth, A. J., Martins, L. M. and Abramov, A. Y. (2011). PINK1 cleavage at position A103 by the mitochondrial protease PARL. *Human molecular genetics*, *20*(5), pp.867-879.

Deng, X., Zhao, S., Liu, X., Han, L., Wang, R., Hao, H., Jiao, Y., Han, S. and Bai, C. (2020). Polygala tenuifolia: a source for anti-Alzheimer's disease drugs. *Pharmaceutical Biology*, *58*(1), pp.410-416.

Dikic, I. and Elazar, Z. (2018). Mechanism and medical implications of mammalian autophagy. *Nature reviews Molecular cell biology*, *19*(6), pp.349-364.

Ding, W. X. and Yin, X. M. (2012). Mitophagy: mechanisms, pathophysiological roles, and analysis. *Biological chemistry*, *393*(7), pp.547-564.

El-Ghazaly, M. A., Sadik, N. A., Rashed, E. R. and Abd-El-Fattah, A. A. (2015). Neuroprotective effect of EGb761® and low-dose whole-body γ-irradiation in a rat model of Parkinson's disease. *Toxicology and industrial health*, *31*(12), pp.1128-1143.

Fang, Q. and LI, H. (2016). Effect of palmatine sequential combination of exercise therapy on serum inflammatory markers in pelvic inflammation and its clinical effect. *Chinese Journal of Biochemical Pharmaceutics*, pp.164-166.

Chen, F. F., Wang, M. M., Xia, W. W., Peng, D. Y, Han, L. (2020). Tao-Hong-Si-Wu Decoction promotes angiogenesis after cerebral ischaemia in rats via platelet microparticles. *Chinese journal of natural medicines*, *18*(8), pp.620-627.

Fimia, G. M., Kroemer, G. and Piacentini, M. (2013). Molecular mechanisms of selective autophagy. *Cell Death & Differentiation*, *20*(1), pp.1-2.

Fu, X., Chen, S., Wang, X., Shen, Y., Zeng, R., Wu, Q., Lu, Y., Shi, J. and Zhou, S. (2021). Dendrobium nobileLindl. alkaloids alleviate Mn-induced neurotoxicity via PINK1/Parkin-mediated mitophagy in PC12 cells. *Biochemistry and Biophysics Reports*, *26*, p.100877.

Fu, X., Liu, Q., Sun, X., Chang, H., Liu, Y. and Han, J. (2022). Research Advances in the Treatment of Alzheimer's Disease with Polysaccharides of Danggui-Shaoyao-San. *Journal of Alzheimer's Disease*, *85*(1), pp.7-19.

Fu, X., Wang, Q., Wang, Z., Kuang, H. and Jiang, P. (2016). Danggui-Shaoyao-San: new hope for Alzheimer's disease. *Aging and disease*, *7*(4), p.502.

Fukuda, T. and Kanki, T. (2018). Mechanisms and physiological roles of mitophagy in yeast. *Molecules and Cells*, *41*(1), p.35.

Gugnani, K. S., Vu, N., Rondón-Ortiz, A. N., Böhlke, M., Maher, T. J. and Pino-Figueroa, A. J. (2018). Neuroprotective activity of macamides on manganese-induced mitochondrial disruption in U-87 MG glioblastoma cells. *Toxicology and applied pharmacology*, *340*, pp.67-76.

Hahm, E. R. and Singh, S. V. (2013). Autophagy fails to alter withaferin A-mediated lethality in human breast cancer cells. *Current cancer drug targets*, *13*(6), pp.640-650.

Han, L., Ji, Z., Chen, W., Yin, D., Xu, F., Li, S., Chen, F., Zhu, G. and Peng, D. (2015). Protective effects of tao-Hong-si-wu decoction on memory impairment and hippocampal damage in animal model of vascular dementia. *Evidence-Based Complementary and Alternative Medicine*.

Harper, J. W., Ordureau, A. and Heo, J. M. (2018). Building and decoding ubiquitin chains for mitophagy. *Nature reviews Molecular cell biology*, *19*(2), pp.93-108.

Hosokawa, N., Sasaki, T., Iemura, S. I., Natsume, T., Hara, T. and Mizushima, N. (2009). Atg101, a novel mammalian autophagy protein interacting with Atg13. *Autophagy*, *5*(7), pp.973-979.

Huang, S., Chen, Y., Pan, L., Fei, C., Wang, N., Chu, F., Peng, D., Duan, X. and Wang, Y. (2021). Exploration of the potential mechanism of Tao Hong Si Wu decoction for the treatment of breast cancer based on network pharmacology and in vitro experimental verification. *Frontiers in Oncology*, *11*, p.731522.

Hussain, A., (2020). The genus Artemisia (Asteraceae): A review on its ethnomedicinal prominence and taxonomy with emphasis on foliar anatomy, morphology, and molecular phylogeny. *Proceedings of the Pakistan Academy of Sciences: B. Life and Environmental Sciences*, *57*(1), pp.1-28.

Innokentev, A., Furukawa, K., Fukuda, T., Saigusa, T., Inoue, K., Yamashita, S. I. and Kanki, T. (2020). Association and dissociation between the mitochondrial Far complex and Atg32 regulate mitophagy. *Elife*, *9*, p.e63694.

Jayaprakasam, B., Padmanabhan, K. and Nair, M. G. (2010). Withanamides in Withaniasomnifera fruit protect PC-12 cells from β-amyloid responsible for Alzheimer's disease. *Phytotherapy Research*, *24*(6), pp.859-863.

Ji, Z. J., Shi, Y., Li, X., Hou, R., Yang, Y., Liu, Z. Q., Duan, X. C., Liu, Q., Chen, W. D. and Peng, D. Y. (2022). Neuroprotective effect of TaohongSiwu decoction on cerebral ischemia/reperfusion injury via mitophagy-NLRP3 inflammasome pathway. *Frontiers in Pharmacology*, *13*.

Jiang, Y., Li, S., Xie, X., Li, H., Huang, P., Li, B., Huo, L., Zhong, J., Li, Y. and Xia, X. (2021). Exploring the Mechanism of Panax notoginseng Saponins against Alzheimer's Disease by Network Pharmacology and Experimental Validation. *Evidence-Based Complementary and Alternative Medicine*, *2021*.

Kane, L. A., Lazarou, M., Fogel, A. I., Li, Y., Yamano, K., Sarraf, S. A., Banerjee, S. and Youle, R. J. (2014). PINK1 phosphorylates ubiquitin to activate Parkin E3 ubiquitin ligase activity. *Journal of Cell Biology*, *205*(2), pp.143-153.

Kang, S. H., Kim, K., Lee, J. H., Ahn, B. O., Won, S. Y., Sohn, S. H. and Kim, J. S. (2016). The complete chloroplast genome sequence of medicinal plant, Artemisia argyi. *Mitochondrial DNA Part B*, *1*(1), pp.257-258.

Kanki, T., Wang, K., Cao, Y., Baba, M. and Klionsky, D. J. (2009). Atg32 is a mitochondrial protein that confers selectivity during mitophagy. *Developmental cell*, *17*(1), pp.98-109.

Kelekar, A. (2006). Autophagy. *Annals of the New York Academy of Sciences*, *1066*(1), pp.259-271.

Kondo-Okamoto, N., Noda, N. N., Suzuki, S. W., Nakatogawa, H., Takahashi, I., Matsunami, M., Hashimoto, A., Inagaki, F., Ohsumi, Y. and Okamoto, K. (2012). Autophagy-related protein 32 acts as autophagic degron and directly initiates mitophagy. *Journal of Biological Chemistry*, *287*(13), pp.10631-10638.

Kumar, P. and Kumar, A. (2009). Possible neuroprotective effect of Withaniasomnifera root extract against 3-nitropropionic acid-induced behavioral, biochemical, and mitochondrial dysfunction in an animal model of Huntington's disease. *Journal of medicinal food*, *12*(3), pp.591-600.

Kundu, M., Lindsten, T., Yang, C. Y., Wu, J., Zhao, F., Zhang, J., Selak, M. A., Ney, P. A. and Thompson, C. B. (2008). Ulk1 plays a critical role in the autophagic clearance of mitochondria and ribosomes during reticulocyte maturation. *Blood, The Journal of the American Society of Hematology*, *112*(4), pp.1493-1502.

Lee, Y., Park, H. R., Chun, H. J. and Lee, J. (2015). Silibinin prevents dopaminergic neuronal loss in a mouse model of Parkinson's disease via mitochondrial stabilization. *Journal of Neuroscience Research*, *93*(5), pp.755-765.

Levine, B. and Kroemer, G. (2019). Biological functions of autophagy genes: a disease perspective. *Cell*, *176*(1-2), pp.11-42.

Li, X., Alafuzoff, I., Soininen, H., Winblad, B. and Pei, J. J. (2005). Levels of mTOR and its downstream targets 4E-BP1, eEF2, and eEF2 kinase in relationships with tau in Alzheimer's disease brain. *The FEBS journal*, *272*(16), pp.4211-4220.

Lin, B. (2011). Polyphenols and neuroprotection against ischemia and neurodegeneration. *Mini reviews in medicinal chemistry*, *11*(14), pp.1222-1238.

Lin, M. G. and Hurley, J. H. (2016). Structure and function of the ULK1 complex in autophagy. *Current opinion in cell biology*, *39*, pp.61-68.

Liu, X., Liu, W., Wang, C., Chen, Y., Liu, P., Hayashi, T., Mizuno, K., Hattori, S., Fujisaki, H. and Ikejima, T. (2021). Silibinin attenuates motor dysfunction in a mouse model of Parkinson's disease by suppression of oxidative stress and neuroinflammation along with promotion of mitophagy. *Physiology & Behavior*, *239*, p.113510.

Ma, B. N. and Li, X. J. (2020). Resveratrol extracted from Chinese herbal medicines: a novel therapeutic strategy for lung diseases. *Chinese Herbal Medicines*, *12*(4), pp.349-358.

Mai, C. T., Wu, M. M., Wang, C. L., Su, Z. R., Cheng, Y. Y. and Zhang, X. J. (2019). Palmatine attenuated dextran sulfate sodium (DSS)-induced colitis via promoting mitophagy-mediated NLRP3 inflammasome inactivation. *Molecular immunology*, *105*, pp.76-85.

McAlpine, F., Williamson, L. E., Tooze, S. A. and Chan, E. Y. (2013). Regulation of nutrient-sensitive autophagy by uncoordinated 51-like kinases 1 and 2. *Autophagy*, *9*(3), pp.361-373.

Mei, N., Guo, X., Ren, Z., Kobayashi, D., Wada, K. and Guo, L. (2017). Review of Ginkgo biloba-induced toxicity, from experimental studies to human case reports. *Journal of Environmental Science and Health, Part C*, *35*(1), pp.1-28.

Mirjalili, M. H., Moyano, E., Bonfill, M., Cusido, R. M. and Palazón, J. (2009). Steroidal lactones from Withaniasomnifera, an ancient plant for novel medicine. *Molecules*, *14*(7), pp.2373-2393.

Mortensen, M., Ferguson, D. J. P., Edelmann, M., Kessler, B., Morten, K. J., Komatsu, M. and Simon, A. K. (2010). Loss of autophagy in erythroid cells leads to defective removal of mitochondria and severe anemia in vivo. *Proceedings of the National Academy of Sciences*, *107*(2), pp.832-837.

Nakatogawa, H., Suzuki, K., Kamada, Y. and Ohsumi, Y. (2009). Dynamics and diversity in autophagy mechanisms: lessons from yeast. *Nature reviews Molecular cell biology*, *10*(7), pp.458-467.

Novak, I., Kirkin, V., McEwan, D. G., Zhang, J., Wild, P., Rozenknop, A., Rogov, V., Löhr, F., Popovic, D., Occhipinti, A. and Reichert, A. S. (2010). Nix is a selective autophagy receptor for mitochondrial clearance. *EMBO reports*, *11*(1), pp.45-51.

Okamoto, K., Kondo-Okamoto, N. and Ohsumi, Y. (2009). Mitochondria-anchored receptor Atg32 mediates degradation of mitochondria via selective autophagy. *Developmental cell*, *17*(1), pp.87-97.

Onishi, M., Yamano, K., Sato, M., Matsuda, N. and Okamoto, K. (2021). Molecular mechanisms and physiological functions of mitophagy. *The EMBO journal*, *40*(3), p.e104705.

Palikaras, K., Lionaki, E. and Tavernarakis, N. (2015). Coupling mitogenesis and mitophagy for longevity. *Autophagy*, *11*(8), pp.1428-1430.

Pang, X., Zhang, X., Jiang, Y., Su, Q., Li, Q. and Li, Z. (2021). Autophagy: Mechanisms and therapeutic potential of flavonoids in cancer. *Biomolecules*, *11*(2), p.135.

Parzych, K. R. and Klionsky, D. J. (2014). An overview of autophagy: morphology, mechanism, and regulation. *Antioxidants & redox signaling*, *20*(3), pp.460-473.

Pei, H., Zeng, J., Chen, W., He, Z. and Du, R. (2022). Network pharmacology and molecular docking integrated strategy to investigate the pharmacological mechanism of palmatine in Alzheimer's disease. *Journal of Biochemical and Molecular Toxicology*, *36*(11), p.e23200.

Qiu, W. Q., Ai, W., Zhu, F. D., Zhang, Y., Guo, M. S., Law, B. Y. K., Wu, J. M., Wong, V. K. W., Tang, Y., Yu, L. and Chen, Q. (2022). Polygala saponins inhibit NLRP3 inflammasome-mediated neuroinflammation via SHP-2-Mediated mitophagy. *Free Radical Biology and Medicine*, *179*, pp.76-94.

Schwarten, M., Mohrlüder, J., Ma, P., Stoldt, M., Thielmann, Y., Stangler, T., Hersch, N., Hoffmann, B., Merkel, R. and Willbold, D. (2009). Nix directly binds to GABARAP: a possible crosstalk between apoptosis and autophagy. *Autophagy*, *5*(5), pp.690-698.

Shi, G., Lee, J. R., Grimes, D. A., Racacho, L., Ye, D., Yang, H., Ross, O. A., Farrer, M., McQuibban, G. A. and Bulman, D. E. (2011). Functional alteration of PARL contributes to mitochondrial dysregulation in Parkinson's disease. *Human molecular genetics*, *20*(10), pp.1966-1974.

Shikov, A. N., Pozharitskaya, O. N. and Makarov, V. G., (2016). Aralia elata var. mandshurica (Rupr. & Maxim.) J. Wen: An overview of pharmacological studies. *Phytomedicine*, *23*(12), pp.1409-1421.

Shin, I. J., Son, S. U., Park, H., Kim, Y., Park, S. H., Swanberg, K., Shin, J. Y., Ha, S. K., Cho, Y., Bang, S. Y. and Lew, J. H. (2014). Preclinical evidence of rapid-onset antidepressant-like effect in Radix Polygalae extract. *PloS one*, *9*(2), p.e88617.

Shivamurthy, S., Manchukonda, R. S. and Ramadas, D. (2016). Evaluation of learning and memory enhancing activities of protein extract of Withaniasomnifera (Ashwagandha) in Wistar albino rats. *Int J Basic Clin Pharmacol*, *5*(2), pp.453-457.

Song, D., Ma, J., Chen, L., Guo, C., Zhang, Y., Chen, T., Zhang, S., Zhu, Z., Tian, L. and Niu, P. (2017). FOXO3 promoted mitophagy via nuclear retention induced by manganese chloride in SH-SY5Y cells. *Metallomics*, 9(9), pp.1251-1259.

Song, Z., Luo, D., Wang, Y., Zheng, Y., Chen, P., Xia, X., He, C., Yu, W., Li, P., Xiao, C. and Cheng, S. (2021). Neuroprotective effect of dangguishaoyaosan via the mitophagy-apoptosis pathway in a rat model of alzheimer's disease. *Evidence-Based Complementary and Alternative Medicine*, *2021*, pp.1-11.

Spilman, P., Podlutskaya, N., Hart, M. J., Debnath, J., Gorostiza, O., Bredesen, D., Richardson, A., Strong, R. and Galvan, V. (2010). Inhibition of mTOR by rapamycin abolishes cognitive deficits and reduces amyloid-β levels in a mouse model of Alzheimer's disease. *PloS one*, *5*(4), p.e9979.

Sun, G. Y., Li, R., Cui, J., Hannink, M., Gu, Z., Fritsche, K. L., Lubahn, D. B. and Simonyi, A., (2016). Withaniasomnifera and its withanolides attenuate oxidative and inflammatory responses and up-regulate antioxidant responses in BV-2 microglial cells. *Neuromolecular medicine*, *18*, pp.241-252.

Tang, Z., Bereczki, E., Zhang, H., Wang, S., Li, C., Ji, X., Branca, R. M., Lehtiö, J., Guan, Z., Filipcik, P. and Xu, S. (2013). Mammalian target of rapamycin (mTor) mediates tau protein dyshomeostasis: implication for Alzheimer disease. *Journal of Biological Chemistry*, *288*(22), pp.15556-15570.

Tolkovsky, A. M. (2009). Mitophagy. *Biochimica et Biophysica Acta (BBA)-Molecular Cell Research*, *1793*(9), pp.1508-1515.

Valente, E. M., Abou-Sleiman, P. M., Caputo, V., Muqit, M. M., Harvey, K., Gispert, S., Ali, Z., Del Turco, D., Bentivoglio, A. R., Healy, D. G. and Albanese, A. (2004). Hereditary early-onset Parkinson's disease caused by mutations in PINK1. *Science*, *304*(5674), pp.1158-1160.

van der Merwe, C., van Dyk, H. C., Engelbrecht, L., van der Westhuizen, F. H., Kinnear, C., Loos, B. and Bardien, S. (2017). Curcumin rescues a PINK1 knock down SH-SY5Y cellular model of Parkinson's disease from mitochondrial dysfunction and cell death. *Molecular neurobiology*, *54*, pp.2752-2762.

Wang, H., Jiang, T., Li, W., Gao, N. A. and Zhang, T. (2018). Resveratrol attenuates oxidative damage through activating mitophagy in an in vitro model of Alzheimer's disease. *Toxicology letters*, *282*, pp.100-108.

Wang, P., Zhu, L., Sun, D., Gan, F., Gao, S., Yin, Y. and Chen, L. (2017). Natural products as modulator of autophagy with potential clinical prospects. *Apoptosis*, *22*, pp.325-356.

Wang, W., Wang, M., Ruan, Y., Tan, J., Wang, H., Yang, T., Li, J. and Zhou, Q. (2019). Ginkgolic acids impair mitochondrial function by decreasing mitochondrial biogenesis and promoting FUNDC1-dependent mitophagy. *Journal of agricultural and food chemistry*, *67*(36), pp.10097-10106.

Wang, W., Yao, G. D., Shang, X. Y., Zhang, Y. Y., Song, X. Y., Hayashi, T., Zhang, Y. and Song, S. J. (2019). Eclalbasaponin I causes mitophagy to repress oxidative stress-induced apoptosis via activation of p38 and ERK in SH-SY5Y cells. *Free Radical Research*, *53*(6), pp.655-668.

Wang, Y., Liu, N. and Lu, B. (2019). Mechanisms and roles of mitophagy in neurodegenerative diseases. *CNS neuroscience & therapeutics*, *25*(7), pp.859-875.

Wardenburg, J. B., Patel, R. J. and Schneewind, O. (2007). Surface proteins and exotoxins are required for the pathogenesis of Staphylococcus aureus pneumonia. *Infection and immunity*, *75*(2), pp.1040-1044.

Wen, W., Chen, J., Ding, L., Luo, X., Zheng, X., Dai, Q., Gu, Q., Liu, C., Liang, M., Guo, X. and Liu, P. (2018). Astragaloside exerts anti-photoaging effects in UVB-induced premature senescence of rat dermal fibroblasts through enhanced autophagy. *Archives of biochemistry and biophysics*, *657*, pp.31-40.

Whitworth, A. J., Lee, J. R., Ho, V. M. W., Flick, R., Chowdhury, R. and McQuibban, G. A. (2008). Rhomboid-7 and HtrA2/Omi act in a common pathway with the Parkinson's disease factors Pink1 and Parkin. *Disease models & mechanisms*, *1*(2-3), pp.168-174.

Williams, J. A., Zhao, K., Jin, S. and Ding, W. X. (2017). New methods for monitoring mitochondrial biogenesis and mitophagy in vitro and in vivo. *Experimental Biology and Medicine*, *242*(8), pp.781-787.

Wu, J., Li, X., Zhu, G., Zhang, Y., He, M. and Zhang, J. (2016). The role of Resveratrol-induced mitophagy/autophagy in peritoneal mesothelial cells inflammatory injury via NLRP3 inflammasome activation triggered by mitochondrial ROS. *Experimental Cell Research*, *341*(1), pp.42-53.

Wu, L. K., Agarwal, S., Kuo, C. H., Kung, Y. L., Day, C. H., Lin, P. Y., Lin, S. Z., Hsieh, D. J. Y., Huang, C. Y. and Chiang, C. Y. (2022). Artemisia Leaf Extract protects against neuron toxicity by TRPML1 activation and promoting autophagy/mitophagy clearance in both in vitro and in vivo models of MPP+/MPTP-induced Parkinson's disease. *Phytomedicine*, *104*, p.154250.

Wu, Q., Hou, Y. L., Sun, B., Zhuang, K., Zhang, H. C. and Jin, D. J. (2011). Curcumin induce apoptosis of CNE-2z cells via caspase-dependent mitochondrial intrinsic pathway. *Afr J Pharm Pharmacol*, *7*(5), pp.1748-1756.

Xia, M. L., Xie, X. H., Ding, J. H., Du, R. H. and Hu, G. (2020). Astragaloside IV inhibits astrocyte senescence: implication in Parkinson's disease. *Journal of Neuroinflammation*, *17*(1), pp.1-13.

Yadav, S. K., Prakash, J., Chouhan, S. and Singh, S. P. (2013). Mucuna pruriens seed extract reduces oxidative stress in nigrostriatal tissue and improves neurobehavioral activity in paraquat-induced Parkinsonian mouse model. *Neurochemistry international*, *62*(8), pp.1039-1047.

Yan, B., Wang, D., Dong, S., Cheng, Z., Na, L., Sang, M., Yang, H., Yang, Z., Zhang, S. and Yan, Z. (2017). Palmatine inhibits TRIF-dependent NF-κB pathway against inflammation induced by LPS in goat endometrial epithelial cells. *International immunopharmacology*, *45*, pp.194-200.

Yang, Z. and Klionsky, D. J. (2010). Eaten alive: a history of macroautophagy. *Nature cell biology*, *12*(9), pp.814-822.

Yue, G. G., Chan, B. C., Kwok, H. F., To, M. H., Hon, K. L., Fung, K. P., Lau, C. B. and Leung, P. C., (2012). Screening for anti-inflammatory and bronchorelaxant activities of 12 commonly used Chinese herbal medicines. *Phytotherapy research*, *26*(6), pp.915-925.

Zeng, X. S., Geng, W.S., Jia, J. J., Chen, L. and Zhang, P. P. (2018). Cellular and molecular basis of neurodegeneration in Parkinson disease. *Frontiers in aging neuroscience*, *10*, p.109.

Zhang, J., Randall, M. S., Loyd, M. R., Dorsey, F.C., Kundu, M., Cleveland, J. L. and Ney, P. A. (2009). Mitochondrial clearance is regulated by Atg7-dependent and-independent mechanisms during reticulocyte maturation. *Blood, The Journal of the American Society of Hematology*, *114*(1), pp.157-164.

Zhang, T., Dong, K., Xiao, L., Li, G., and Zhang, Z. (2020). Effects of Co-administration of icariin and Panax notoginseng saponins on intestinal microbiota and hippocampal protein expression in a mouse model of Alzheimer's disease. *Neuropsychiatric Disease and Treatment*, pp.2169-2179.

Zhang, W., Siraj, S., Zhang, R. and Chen, Q. (2017). Mitophagy receptor FUNDC1 regulates mitochondrial homeostasis and protects the heart from I/R injury. *Autophagy*, *13*(6), pp.1080-1081.

Zhao, K., Song, X., Huang, Y., Yao, J., Zhou, M., Li, Z., You, Q., Guo, Q. and Lu, N. (2014). Wogonin inhibits LPS-induced tumor angiogenesis via suppressing PI3K/Akt/NF-κBsignaling. *European Journal of Pharmacology*, *737*, pp.57-69.

Zhu, Y. and Wang, J. (2015). Wogonin increases β-amyloid clearance and inhibits tau phosphorylation via inhibition of mammalian target of rapamycin: potential drug to treat Alzheimer's disease. *Neurological Sciences*, *36*, pp.1181-1188.

Chapter 13

Therapeutic Tools for Mitophagy

Vinayak Agarwal
Divya Jindal
and Manisha Singh*

Department of Biotechnology, Jaypee Institute of Information Technology, Noida, Uttar Pradesh, India

Abstract

Mitochondria are necessary organelles that control cellular homeostasis and induce apoptosis. The removal of mitochondria via autophagy, also known as mitophagy, is thus essential for maintaining proper cellular functions. Indeed, it has recently been proposed that mitophagy plays a critical role in the terminal differentiation of various pathologies such as neurodegenerative disorders (NDD), cardiovascular injuries, hepatic diseases, muscular dystrophy, aging, and metabolic diseases. The removal of damaged mitochondria by autophagy requires two primary stages, encompassing the activation of general autophagy and priming of damaged mitochondria for preferential autophagic recognition. State-of-the-art advancements in mitophagy research report comprehensively that priming facilitated by mitochondria stems from the Pink1-Parkin independent or dependent signaling pathway, as well as several biomarker marker proteins that impart their expression in mitophagy activation or inhibition. In this chapter, we summarize our current understanding of mitophagy mechanisms while also discussing the pathophysiological roles of mitophagy in detail.

Keywords: mitophagy, PINK1 protein, parkin protein, autophagy, mitochondrial damage, USP30 inhibitors

Introduction

Featuring their DNA and the capacity to generate ATP, mitochondria are vital organelles that power several mitochondrial functions. Mitochondria rely on oxidative phosphorylation (OXPHOS) to produce energy. Several characteristics of the OXPHOS machinery are encoded

* Corresponding Author's Email: manishasingh1295@gmail.com.

In: Mitophagy and Its Role in Health and Disease
Editors: Shalini Mani and Samiksha Wasnik
ISBN: 979-8-89113-470-6

in both nuclear and mitochondrial DNA (mtDNA). Mitophagy can respond to several stressful stimuli, including nutritional deprivation and oxidative stress, along with the strategic elimination of mitochondria (Palikaras K et al., 2015). Several known mechanisms control mitophagy, but PTEN-induced putative kinase 1 (PINK1) and the E3-ubiquitin ligase Parkin are indeed the best-studied types. Parkinson's disease (PD) autosomal recessive versions have been linked to mutations in both the PINK1 and Parkin (PARK2) genes (Pickles S et al., 2018).

Whenever the mitochondria are not experiencing stress, PINK1 is imported into the cell through the TOM/TIM complex in a way that is dependent on the membrane potential of the cell, thereby causing PINK1 to be broken down by proteolytic enzymes (Deas et al., 2011). Following release to the cytosol, the N-terminally shortened PINK1 is destroyed either by proteasomes. The transit of PINK1 throughout the mitochondrial membrane is disrupted by mitochondrial membrane potential loss, which causes aggregation of uncleaved PINK1 just at the outer mitochondrial membrane. Then, either by directly phosphorylating the Parkin Ub-like (UBL) domain or by phosphorylating ubiquitin, PINK1 controls the activation and recruitment of the cytosolic Parkin. Once active, Parkin accelerates the ubiquitination of several substrates, creating new substrates for Pink1 and resulting in a positive intake forward mechanism (Ordureau et al., 2014). In recent years, numerous substrates have ubiquitinated rendering PINK1/Parkin-mediated signaling pathways, including mitochondrial outside membrane proteins and autophagy receptors (Sarraf et al., 2013). Essential mitochondrial fusion proteins, mitofusin 1 and 2 (Mfn1 and Mfn2), for example, play a vital role in degradation in a PINK1/parkin-dependent pathway which ends up making mitochondria accessible for degradation and precludes the fusion of damaged mitochondria with healthy networks. Contingent double-knockout of Mfn1 and Mfn2 in mice prompts mitochondrial dysfunction, and Mfn2-depleted cardiomyocytes do not often attract Parkin to the entire mitochondrial outer membrane as a direct consequence. Other mitochondrial proteins, including Miro1 and VDAC1, have been reported to have a similar priming role. Parkin has recently been found to target the apoptotic protein BAK, further linking Parkin-mediated mitophagy to the control of cellular death. The activation of autophagy substrate receptors such as p62, Optineurin, and NDP52 at the mitochondrial membrane by ubiquitination processes triggered by PINK1 and Parkin facilitates the engulfment of mitochondria by autophagosomes. Notably, functional PINK1-Parkin-mediated mitophagy depends on transcriptional control. Cellular stress, such as reactive oxygen species or ethanol exposure, for example, stimulates PINK1-Parkin-mediated mitophagy, which yields in the nuclear translocation of several transcription factors, which would include transcription factor EB (TFEB), and nuclear respiratory factors (NRFs), which regulate the expression of mitochondrial, autophagy and lysosomal genes. Parkin expression is tightly regulated by stress pathways, notably unfolded protein response pathways and activating transcription factor 4 (ATF4). All of this emphasizes the many different therapeutic approaches to combat the PINK1-Parkin signaling pathway. Channels employed by mitophagy are Pink1/Parkin-independent and mostly depend on receptor proteins to facilitate the assembly of LC3/GABAR APs for mitochondrial removal. During reticulocyte differentiation in mammals, the BCL2-related protein NIX (also referred to as BNIP3L) causes mitophagy(Sarraf SA et al., 2013). In agreement with this, anemia and reticulocytosis appear in NIX knockout mice.

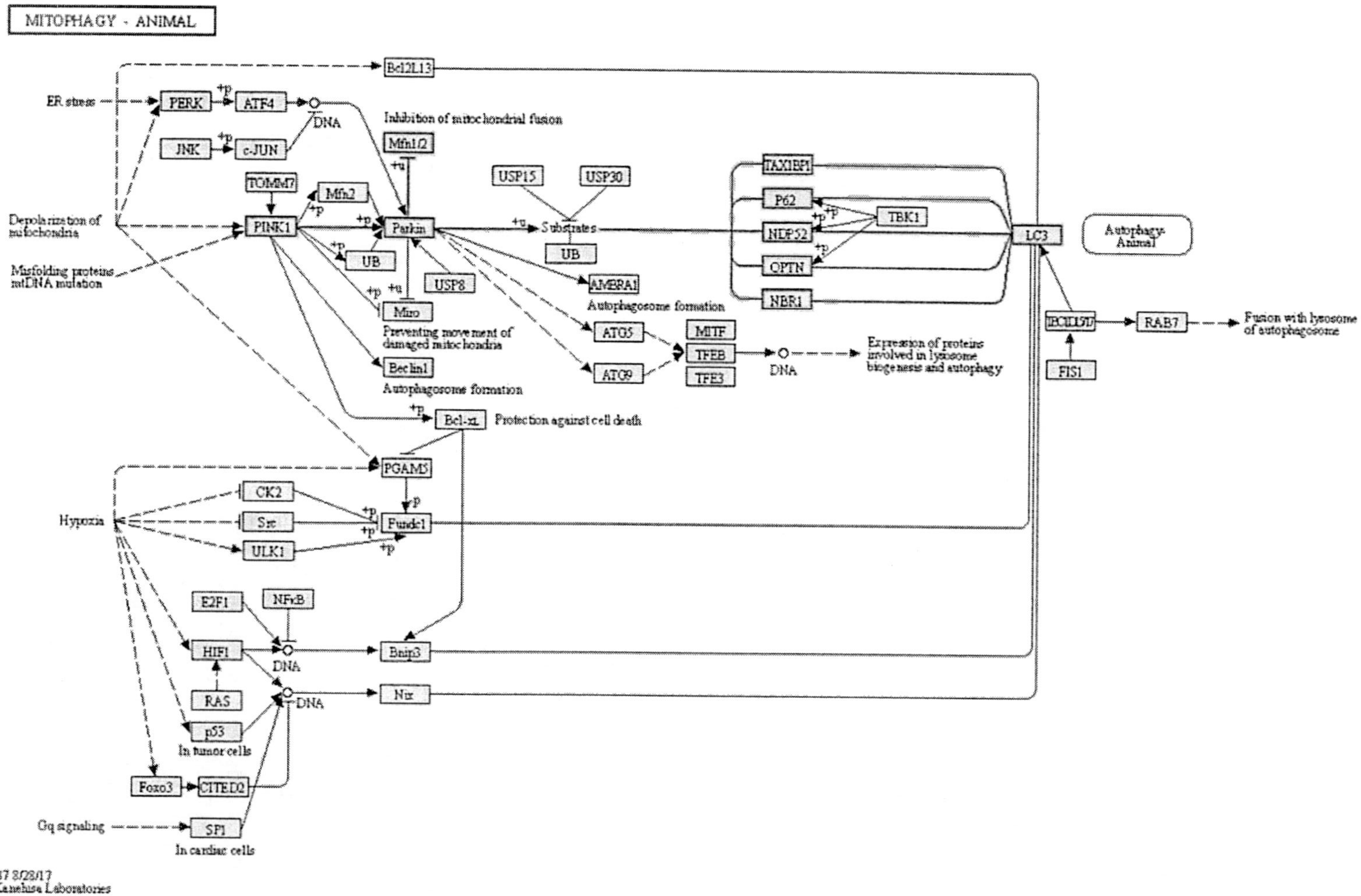

Figure 1. Mitophagy pathways; PINK1-Parkin mediated & PINK-Parkin dependent mitophagy.

Reexpressing LIR-mutant NIX in inadequate NIX reticulocytes partially reversed the identified phenotype, indicating LC3-independent or even autophagy-independent systems for mitochondrial removal in reticulocyte differentiation. The LC3-interacting (LIR) motif mediates the communication of NIX with LC3 protein members. By encouraging mitochondrial fission, FUNDC1, another protein with a LIRmotif, controls mitophagy in hypoxic environments. FUNDC1 and NIX, but not Pink1 or Parkin, are increased to maintain a functioning mitochondrial network during cardiac precursor cell development. So too, is mitophagy, controlled in a lineage-dependent manner. Piecemeal autophagy or vesicles produced from mitochondria can mediate the local removal of mitochondrial sub-domains. It is believed that PINK1/Parkin regulates the production of mitochondrial-derived vesicles but not the classical autophagy process. Instead, the build-up of unfolded protein mitochondrial protein aggregates triggers the targeted mobilization of Parkin and autophagy proteins, promoting the destruction of mitochondrial subdomains. Recently, it was discovered that mitochondrial sub-regions are degraded by LC3C and p62 by a PINK1/Parkin-independent piecemeal mitophagy. Further research is required to determine their physiological significance and whether the protein architecture for such mitochondrial degradation biochemical mechanism overlaps with the traditional mitophagy pathways (Figure 1) (Hirota Y et al., 2015).

Biomarkers Involved in Mitophagy

Among several biomarkers, pS65-Ub is a potential biomarker that can be used to evaluate the expression of PINK1 and Parkin effectively. A particular pS65-Ub antibody demonstrates a buildup of pS65-Ub on mitochondria inside the brains of aged human subjects and patients having Lewy body disease, and it may be utilized for an ELISA in human biofluids, such as cerebrospinal fluid. As an alternative, phosphoproteomics methods may be effective for finding pS65-Ub in adult biofluids. Mass spectrometry can also detect pS65-Ub. Low levels of pS65-Ub may identify patients with deficient mitophagy in their blood or cerebrospinal fluid, and an elevation in pS65-Ub operating as mitophagy stimulator therapy may offer a pharmacodynamic read-out. Several biological markers are being established to monitor mitochondrial health, including a novel PET tracer that tends to affect the control signal to mitochondrial complex one in 1-methyl-4-phenyl-1,2,3,6-tetrahydropyridine (MPTP)-treated monkeys and is currently being investigated in human clinical trials by the pre-competitive consortium MIND-MAPS. If effective, this PET tracer may aid in stratifying patients diagnosed with mitochondrial complex I deficit and indirectly gauging the sensitivity to mitophagy-enhancing small molecules as the intended result of these treatments.

In other words, good health mitochondria with standard mitochondrial complex I levels. Multiple fluid indicators of mitochondrial DNA (mtDNA) integrity have been cited, including mtDNA ploidy, deletion, methylation, point mutations, and degradation (Nishida Y et al., 2009). Unaffected by age or gender, a significant drop in mtDNA ploidy is seen in the blood cells of PD patients and correlates with the abnormalities found in the postmortem brain. In the cerebral fluid, lower cell-free pulsatile mtDNA levels are seen in PD in its early stages. Further studies are required to grasp the temporal dynamics of alteration in cell-free, entirely propagated mtDNA levels for biomarker applications.

Nonetheless, symptomatic PD subjects carrying LRRK2 gene mutation had higher levels of cell-free circulating mtDNA than asymptomatic LRRK2 mutation carriers, idiopathic PD patients, and safeguards. While there was no link between age, clinical presentation, gender, severity, and methylated mtDNA was identified as significantly curtailed in the cerebral fluid of PD patients. Compared to control samples, post-mortem PD brains displayed a decline in methylation in almost all CpG and non-CpG sites in the D-loop region of the mtDNA, which is crucial for both mitochondrial transcription and replication. However, additional research is required to confirm the concordance between blood and the brain. The number of mutations linked to various clinical symptoms has increased due to mtDNA sequence research; however, particular mutations linked to PD have not yet been identified. The substantia nigra has a substantial amount of cardiolipins, and mono-oxygenated, mitochondria-specific, and plasma levels of polyunsaturated cardiolipins have been elevated in rats exposed to rotenone, an operative used primarily on animal models of PD. These increased numbers of biomarkers associated with mitochondrial integrity have recently been investigated in animal models of PD but will need beyond that and impartial validation in human postmortem brain tissue and biofluids. When combined, on-target PINK1/Parkin activation markers like pS65-Ub, PET imaging, and/or biofluid measurements of mitochondrial viability may make it possible to evaluate small-molecule mitophagy enhancers in a medical setting see Table 1 (Youle RJ and Narendra DP., 2011).

Table 1. Biomarker proteins associated with the regulation of mitophagy

Biomarker Protein	Expression in Mitophagy	References
Positive Regulators for Mitophagy		
1. PARKIN – DEPENDENT MITOPHAGY MEDIATORS		
Parkin	Mitophagy induced by Outer Mitochondrial Membrane (OMM) proteins	(Chen Y and Dorn GW., 2013)
PINK1	Mitophagy induced by OMM along with phosphorylated ubiquitin protein	(Cereghetti GM et al., 2008)
p62, NBR1	LIR attached to LC3 and Ubiquitin-binding protein induces mitophagy	(Kondapalli C et al., 2012)
HDAC6	Promotes autophagosome-lysosome fusion	(Lee JY et al., 2010)
LC3 receptor	PINK1 dependent assembled to incur damage to mitochondria	(Hasson SA et al., 2013; Ribas V et al., 2016)
TAXBP1	Partially involved in LC3 receptor-mediated mitophagy	(Hasson SA et al., 2013; Sandoval H et al., 2008)
HSPA1L	Promotes translocation of Parkin protein into the OMM	(Zhang H et al., 2008)
BAG4	Halt the transposition of Parkin to OMM	(Chu CT et al., 2013)
HSP72	Positively influence the recruitment of Parkin	(Wu W et al., 2014)
2. PARKIN- INDEPENDENT MITOPHAGY MEDIATORS		
NIX	High expression in the concluding stages of erythroid maturation leads to mitochondrial degradation.	(Murakawa T et al., 2015)
Bnip3	Under a hypoxic environment, it positively influences mitophagy via LC3 interaction.	(Chu CT et al., 2013; Murakawa T et al., 2015)
Cardiolipin	Initiates mitophagy upon mitochondrial injury	(Murakawa T et al., 2015)
FUNDC1	Interaction of phosphorylated ULK1 at Ser^{17} with LC3 to promote mitophagy	(Nakamura N and Hirose S, 2008)
Bcl2-L-13	Atg32 is a mammalian homolog that causes mitophagy & and mitochondrial fragmentation.	(Sasarman F et al., 2010)
Negative Regulators for Mitophagy		
USP30	Rescue mitochondria from the abnormal phenotype by its expression	(Sasarman F et al., 2010)
LRPPRC (C19ORF5)	It is part of a large protein in mitochondria; this complex regulates post-transcriptional gene expression.	(Kerr JS et al., 2017)

Mitophagy-Governed Pathology

Mmitophagy is perhaps a preferential autophagic process that removes dysfunctional mitochondria and is required for cellular homeostasis. It is powered up by inner membrane depolarization and is crucial in highly diversified post-mitotic cells strongly reliant on aerobic metabolism, including neurons, muscle cells & and hepatocytes. It has been proposed that both defective and prolonged mitophagy add value to age-related NDDs such as Alzheimer's disease (AD) & and PD, metabolic disorders, muscular dystrophy, myocardial illness, and vascular complications leading to diabetes and liver disease, among many others. Pharmacological and dietary initiatives that reinstate mitophagy equilibrium and expedite the obliteration of irreparably mitochondrial injury could thus be used as potential treatments for various chronic diseases. Despite remarkable advancements in this field, primarily extracted directly from *in vitro* and model organism models, human functionalities predicated depending upon regulatory oversight of mitochondrial performance in patients, have yet to be authorized (Youle RJ and Narendra DP., 2011).

Neurodegenerative Disorders

Alzheimer's Disease

Studies on post-mortem cellular moiety of brain tissue and AD patients who are still alive have shown that neurons in the afflicted brain regions have decreased mitochondrial function. Neuronal uptake of radiolabeled glucose is reduced in PET brain scans, and biochemical tests show the efficiency of mitochondrial enzymes responsible for oxidative phosphorylation, so the TCA cycle is diminished(Youle RJ and Narendra DP., 2011). The 40–42 amino acid Aβ is partially located inside the lattice domain of the β – β-precursor protein (APP), a transmembrane protein(Miller S and Muqit MM, 2019). Three enzymes—α-, β-, and γ-secretase—cleave APP. α-secretase cleaves at the membrane surface to release a secreted version of APP, whereas β- and γ-secretase cleave APP somewhere at the N and C termini of Aβ to generate intact Aβ. Early-onset dominantly inherited AD can be triggered by genetic polymorphisms in either APP or the γ-secretase enzymes presenilin 1 (PS1)(Um JH and Yun J, 2017). In this form of AD, the mutations induce a faster age-dependent formation of Ab1-42, which undergoes robust self-aggregation and causes neurotoxicity. Increasing data suggest that amyloidogenic APP synthesis and related neuroinflammatory processes are facilitated by mitochondrial dysfunction and accompanying oxidative stress brought on by aging, sedentary lifestyles, or high-calorie intake(Doblado L et al., 2021). When given a diet heavy in sucrose, APP-mutant transgenic mice show rapid Aβ accumulation and cognitive decline(Jung, Hye Seung and Myung-Shik Lee, 2010).

Parkinson's Disease

Parkinson's disease family cases have been separately related to mutations in PARK2 (Parkin) and PARK6 (PINK1), which connect mitophagy issues to the deterioration of dopaminergic neurons 1 (Chen G et al., 2020). Dopaminergic neurons from people with familial or sporadic PD appear to be more sensitive towards respiratory chain inhibitors and contain more mtDNA mutations, which cause mitochondrial malfunction (Nah J et al., 2011) by inhibiting general autophagy, other PD-related mutations, such as those in α-synuclein that form α-synuclein aggregates, indirectly cause a block in mitophagy. These results suggest that mitophagy plays a significant role in PD prevention overall (Ma X et al., 2020).

Metabolic Diseases and Diabetes

Diabetes mellitus, metabolic disorders, or type 2 diabetes have all been related to mitochondrial malfunction, brought on by inherited and inherited mtDNA mutations when people age (Ma X et al., 2020). The onset of type 2 diabetes in the elderly is substantially associated with the age-dependent decline of mitochondrial function, even though particular mitophagy deficiencies have not been directly linked to insulin resistance(Sauvé V et al., 2022). It would be expected that mitophagy-related problems in mitochondrial metabolism, such as impaired fatty acid oxidation that results in lipid accumulation in the liver or muscle, would affect insulin signaling and induce diabetes (Kumar A. et al., 2017).

Muscle Dystrophy and Ageing

Most tissues have a decline in autophagy rates with aging, which reduces mitochondrial turnover and impairment of mitochondrial function(Andreux PA et al., 2019). The phenomenon has been associated with diminished respiration, excessive ROS formation, and age-related injury to various tissues, including the pancreas, cardiac muscle, liver, and skeletal muscle (Popov LD, 2020). Since mitochondria are the principal generator of ROS in the cell, any disruption of mitochondrial quality management, such as decreased mitophagy, is predicted to accelerate aging by the "free radical hypothesis of aging"(Okatsu K et al., 2018).

Myocardial Injury

Mmyocardial ischemia induces the buildup of CO_2 and bicarbonate, which impedes mitophagy during reperfusion and promotes myocardial injury. The instability of the mitochondrial performance regulatory systems, notably mitophagy, and mitochondrial dynamics, has indeed been associated with an array of forms of cardiac pathology, including ischemic cardiovascular disease, cardiac hypertrophy, diabetic cardiomyopathy, and cardiac arrest (Manning G et al., 2002). Bicarbonate does not reduce proteasome or autophagy activity, indicating that bicarbonate inhibits mitophagy-specific mitochondrial breakdown pathways. It is still unclear how bicarbonate suppresses mitophagy at the molecular level. About ischemia/reperfusion in the heart, PGAM5, a mitochondrial protein associated with the RIP1/RIP3/MLKL complex, is crucial in triggering PINK1-dependent mitophagy. Due to the genetic inhibition of PGAM5 during ischemia/reperfusion, damaged mitochondria build up, oxidative stress increases, and necroptosis in the heart is exacerbated(Manning G et al., 2002; Chaugule VK et al., 2011).

Hepatic Disease

Previous research demonstrating that 1-AT deficiency causes mitochondrial malfunction by obstructing mitochondrial degradation by autophagy led to the initial discovery of the relationship between mitochondrial autophagy, liver damage, and mitophagy. A serum glycoprotein called 1-AT serves as an inhibitor of harmful neutrophil proteases. Several naturally occurring mutants of 1-AT, such as the S variant (where the glutamic acid [Glu] at residue 264 [Glu264] is metamorphosed to valine [Val]) and the Z variant (where the glutamic acid [Glu] at residue 264 [Glu264] is mutated to lysine [Lys]), have been discovered and demonstrated to contribute to the pathogenicity in association to human diseases, including chronic liver-associated diseases. Inclusion bodies formed due to homozygosity for the Z variation of 1AT (also known as 1ATZ) are maintained in the ER, which hinders 1AT's ability to be secreted into the bloodstream and damages the liver. Numerous investigations have shown that removing 1-ATZ protein aggregates may play a role in autophagy in preventing liver damage. It has been shown that genetically silencing ATGs, including ATG5, prevents the autophagic pathway from working correctly, which prevents the breakdown of 1-ATZ in yeast cells. It has been demonstrated that rapamycin and other autophagy-enhancing medications increase autophagy, which in turn promotes the clearance of 1-ATZ to repair liver damage brought on by 1-AT shortage. According to one study, 1-ATZ aggregation causes mitochondrial dysfunction inside the hepatic tissue of patients who lack 1-AT. Later, it was discovered that liver tissue samples from individuals with 1-AT deficiency and livers from 1-ATZ transgenic mice exhibit induced mitochondrial injury-associated mitophagy.

Furthermore, it has been demonstrated that, in the context of I/R- and A/R-induced hepatic damage, deficiencies in mitophagy increase the MPT and bring about mitochondrial dysfunction. The reduction of MFN2 deacetylation (at the Lys655 and Lys662 residues) and the impairment of mitochondrial autophagy caused by I/R-induced SIRT1 down-regulation were shown to cause mitochondrial malfunction and liver damage. The restoration of mitochondrial autophagy and the suppression of mitochondrial defects and cell death after I/R were demonstrated to occur when SIRT1 was overexpressed by adenoviral gene transfer and its activity was increased by the pharmacological activator SIRT1702 (Chaugule VK et al., 2011).

Therapeutic Intervention Modulating Mitophagy

PINK1 has been sourced and sliced by the protease PARL in healthy mitochondria. USP30 removes ubiquitin residues from OMM proteins. Mitochondrial impairment can occur as a result of both natural and experimental processes. MPP+ (the active form of MPTP) and rotenone were used to inhibit mitochondrial complex I (MC1). When mitochondrial damage occurs, PINK1 migrates to the OMM, is activated by autophosphorylation, and engages Parkin via Ser65 phosphorylation. Parkin ubiquitinates many OMM proteins and several ubiquitin residues, which are then phosphorylated on Ser65 by PINK1, creating a positive feed-forward loop. USP30 does not eliminate the long phosphorylated ubiquitin chains on OMM proteins, culminating in mitochondrion engulfment by autophagosomal membranes and lysosome recycling. Mitophagy-enhancing therapies either increase PINK1 and Parkin expression or inhibit USP30 (Chaugule VK et al., 2011; Kondapalli C et al., 2012).

Mitophagy Activators

Humans have an early onset of PD explained by the primary phenotypes obtained by the deletion of PINK1 or Parkin in rat models or human iPSC-derived dopaminergic neurons. Using a comprehensive experimental framework, the study shows that PINK1/Parkin-regulated mitophagy subdues *in vivo* innate immunity, which causes impairment in mitophagy and gives rise to enhanced expression due to interferon genes (STING) pathway booster, potentially causing an inflammatory phenotype that could very well cause cognitive collapse. The above study indicates that a broader perspective than dopaminergic neurodegeneration might be needed to comprehend the ramifications of dysfunctional mitophagy entirely. Importantly, patients having Parkin mutations also have elements of innate immune activation. PINK1 and Parkin were demonstrated to drive mitochondrial antigen presentation and initiate adaptive immune reactions in a rat cell line of macrophages through mitochondrial-derived vesicles, supporting a function for PINK1 and Parkin in immunological responses, early in disease development, intrinsic immunity, and inflammation change in PINK1 KO mice. By administering lipopolysaccharide injections, nigral degeneration caused by inflammation is more susceptible when parkin KO is present. It has been suggested that inflammation triggers subsequent inflammatory processes by increasing mitochondrial damage. Therefore, it is possible that immunological responses and mitochondrial malfunction work in concert to cause PD and that numerous rodent models only replicate a component of the systemic malformations (Kondapalli C et al., 2012).

PINK1 Activators

Even though PINK1 is unique across species, insect PINK1 homologs, such as Pediculushumanuscorporis PhPINK1 and Tribolium castaneum TcPINK1, have been discovered and can be communicated in significant quantities in E. Coli to investigate the PINK1 framework and substrate recognition frameworks, and the ramifications of the autosomal recessive mutations that cause Parkinson's disease. PINK1 differs from other protein kinases due to three distinct insertions within the kinase domain, resulting in a C-terminal region with no homology to any other known protein. Nearly 30 human PINK1 (hPINK1) mutations, most of which are projected to terminate the C-terminal region or are located primarily inside the kinase domain, have been found in patients (Kondapalli C et al., 2012; Kumar A. et al., 2015).

When mitochondria become depolarized, PINK1 rapidly builds up on the OMM and joins the translocase of an outer membrane (TOM) complex to form a vast 700-kDa multimeric complex, which in turn causes PINK1 to be auto-phosphorylated and activated. For hPINK1, 3 autophosphorylation areas have been identified: Thr257, Ser228, and Ser402, the latter two affecting Parkin and the following ubiquitin phosphorylation. Ser402 is found inside the conserved promoter regions of hPINK1 and is expected to stabilize the active conformation and increase kinase activity following autophosphorylation by interacting with residues inside the catalytic motif. Parkin recruitment is reduced by Ser402 and autophosphorylation activation. However, Ser228 autophosphorylation is not required for Parkin recruitment or the start of mitophagy, despite increasing substrate phosphorylation *in vitro*. Instead, the structure of insect PINK1 suggests that Ser202 (phPINK) or Ser205 (TcPINK1) phosphorylation in the N-lobe, which corresponds to human Ser228 and Ser228, respectively, is essential for kinase activity and integrity as well as for ubiquitin attachment to the Globe. By promoting dimerization or increasing autophosphorylation by adhering to allosteric regulatory regions, small molecules can activate PINK1 more effectively. As an alternative, using ubiquitin conformational modulators to boost ubiquitin phosphorylation levels is suggested. The ATP analog kinetin triphosphate found in WO-2015/123365, is a neo-substrate used in a third method to boost PINK1 activity *in vitro*. Future studies would therefore clarify whether any of these strategies can produce effective organic compounds for translation in animal studies and eventually in humans. No kinetin implications were detectable in PINK1 KO rats after oral ingestion for up to six months, but no dopaminergic neurodegeneration was observed in this study, leaving the question open as to how far kinetin could mitigate pathophysiology caused by PINK1 KO (Kumar A et al., 2015).

Parkin Activators

Parkin is a member of the RBR [RING-in-between-RING (IBR)-RING] class of E3 ligases, which includes RING1 domains for E2 recruitment, RING2 domains for catalyzing ubiquitin transfer, and IBR domains for connecting. Five domains make Parkin: Ubl (ubiquitin-like domain), RING0, RING1, IBR, & and RING2. Parkin is the only RBR ligase with the Ubl& and RING0 domains. However, all RBR ligases share the RBR module. More than a hundred different mutations were identified to cause deletions, truncations, and exon duplications in Parkin-associated PD, all impairing Parkin function. The fact that Parkin mutations linked to PD are discovered in all five domains, as well as at domain interfaces and linker regions, suggests that the structural stability of each domain is crucial to the development of PD. Parkin's crystal structures reveal three ways it assumes an autoinhibited conformation under baseline circumstances. First, the RING0 domain substantially obscures the catalytic cysteine (Cys431). Second, the predicted E2 binding on RING1 is blocked by a short helical Parkin repressor element (REP) and the Ubl domain. Third, the transfer of ubiquitin in the auto-inhibited configuration of Parkin is prevented by the distance of over 30 between the RING2 catalytic cysteine and anticipated E2 catalytic cysteine. Recent structural investigations have demonstrated the mechanisms of Parkin stimulation and the place on the RING0 domain where the phosphorylated Ubl interacts.

Additionally, a preserved linker region that connects the RING0 and Ubl domains serves as an activating component that promotes the emergence of the RING2 motif. These novel structures give designers of Parkin activators atomic insights on how to thwart autoinhibitory mechanisms and promote catalytic activity. It is currently unknown whether ubiquitin transfer occurs due to a conformational shift or due to numerous Parkin molecules working together. An UbFluor sensor has indeed been devised to quantify changes in Parkin activity in response to phosphorylation precisely, allosteric stimulation by pUb, protein substrates, and activating structural mutations, assisting in developing small chemical activators. In addition to assays for finding such molecules, WO-2018/023029 reported small molecule Parkin activators that are functional in *vivo* experiments, although there are presently no cellular or *in vivo* studies with Parkin activators (Komander D et al., 2009).

Mitophagy Inhibitors

Using the V-ATPase (vacuolar-type H+ -ATPase) antagonist bafilomycin A1 or the lysosomotropic drugs chloroquine and hydroxychloroquine is the most extensively used method for inhibiting the process of mitophagy *in vitro*. Mitophagy has also been stopped by substances that block the production of autophagosomes, which also suppress general autophagy. The most well-known example is the class III PI3K inhibitor 3-methyladenine. Despite being particularly good at blocking mitophagy, these methods are mainly unspecific and impact some other autophagic and lysosomal processes (Cornelissen T et al., 2014).

USP30 Inhibitors

Deubiquitinating enzymes (DUBs), which catalyze ubiquitin removal from substrates, control the activity of E3 ubiquitin ligases such as Parkin. Modulating DUBs implicated in mitophagy is an exciting strategy for creating small-molecule transcription factors of mitophagy due to the difficulties in raising PINK1 or Parkin activity, as described above. It has been demonstrated that two DUBs, USP15, and USP30, prevent Parkin-mediated mitochondrial ubiquitination and mitophagy in Drosophila and human cells *in vitro* (Cornelissen T et al., 2018). By reducing ubiquitin residues off Parkin and stopping its auto-inhibition, it has been demonstrated that another DUB, USP8, may trigger mitophagy. Due to its localization at the OMM, which preferentially hydrolyzes Lys6 and Lys11-linked ubiquitin chains, USP30 is a particularly appealing therapeutic target to promote mitochondrial ubiquitination. In their initial research, Bingol et al. demonstrated that USP30 KO promotes mitophagy in human cell lines that overexpress Parkin and can reverse parkinsonian phenotypes in Drosophila expressing PINK1 or Parkin mutants. Numerous topologies of USP30 have indeed been solved, adding to our understanding of how it works. USP30 is hypothesized to have a homeostatic role by eliminating and repurposing ubiquitin residues at OMM. USP30 may aid in limiting Parkin activity and mitophagy to the area of mitochondrial damage when it happens. Unresolved is whether USP30 works before or after PINK1 activation. It was demonstrated in cells *in vitro* using the mtKeima and mitoQ reporter systems that the reduction of USP30 enhances mitophagy but not when PINK1 is absent. According to these results, USP30 may function upstream of PINK1 and stop tonically active mitophagy. The same researchers demonstrated that USP30 deprivation also enhances pexophagy by removing peroxisomes from cells, indicating a hitherto unrecognized regulatory function of USP30 in pexophagy. There is no published information on the effects of USP30 KO in mice (Cunningham CN et al., 2015).

Peptide Inhibitors of Receptor-Mediated Mitophagy

Novel methods for blocking mitophagy have been described, but their use has been sparse so far. The latest studies have primarily concentrated on peptide antagonists of receptor-mediated mitophagy that are premised on the LIR of FUNDC1. These sequences contain the Ser13 residue, for which the phosphorylation status controls FUNDC1's interaction with LC3 and consequently its capacity to direct mitochondria toward autophagosomes. The mitophagy receptor's binding to LC3 was demonstrated to be disrupted by the LIR mimetic peptide 9 QDYESDDDSYEVLDLTE25 a Kd of 5.8 M, which would be five times lower than the counterpart containing pSer13. The pairing of the unphosphorylated peptide inhibitors to a

TAT-derived sequence developed an intriguing cell-permeable probe that could decrease FCCP-induced mitophagy without impacting overall autophagy, highlighting the ability of such compounds to affect the operation in a specific way (Kageyama Y et al., 2014).

Suppression of Mitochondrial Fragmentation

Reducing mitochondrial fragments, a phenomenon linked to mitophagy is a distinct strategy for inhibiting the process. When there is mitochondrial stress, substances that limit the action of Drp1 can stop the division of the mitochondria and possibly stop the start of mitophagy. Mitochondrial division inhibitor (mdivi), an allosteric regulator of Drp1, was discovered through phenotypic screening in yeast cells. Mdivi was demonstrated to have cardio-protective benefits *in vivo* by preventing aberrant mitophagy and suppressing mitochondrial fission, at least in part. Another instance is the Drp1 modulator P110, which comprises the evolutionarily conserved Drp1 region 110 (49DLLPRGT55) coupled to a cell-permeable TAT peptide-derived sequence. By preferentially breaking Drp1's PPI with the OMM-localized protein Fis1, P110 inhibits Drp1 by lowering its enzymatic expression and preventing its translocation to mitochondria. Interestingly, P110 inhibits aberrant mitochondrial fragmentation brought on by several mitochondrial stressors, notably FCCP and MPP+, but does not decline mitochondrial fission under the basal state. To our knowledge, no research on P110's effects on autophagy has been published yet, so it is unknown whether the compound can negatively affect the process (Givvimani S et al., 2012).

Conclusion

Mitochondrial quality regulation is essential for the body's healthy operation in stress responses. Due to the high incidence of mitochondrial dysfunction in age-related and heart failure, a good comprehension of the biochemical pathways that establish extensive mitochondrial testing in cardiomyocytes, how they are severely impacted throughout development, aging, or stress, and how they can continue to improve is critical, which could lead to the development of new therapeutics for patients. To a certain point, researchers believe that the following issues are notably pressing.

New studies have shown that mitophagy operates outside of the context of severe mitochondrial stress, for example, during resting physiology, guaranteeing the integrity of both the mitochondrial network and contributing to the maintenance of cellular homeostasis. The drawbacks of provoking mitophagy via pharmacologically induced acute depolarization illustrate the importance of alternative methods to not only help regulate the operation but also eventually use it for therapeutic purposes. On the other hand, as recently published successful examples show, an improved comprehension of the signaling pathways involved in mitophagy regulation will provide possibilities for specific target pharmacological interventions via the advancement of functionalized chemical probes.

Finally, molecular interventions that specifically regulate mitophagy levels should be identified. It has been postulated that PINK1, Parkin, and USP30 are viable targets for mitophagy modulation. Mitophagy may be stimulated by interventions that promote the upregulation of intrinsic PINK1 and Parkin or the downregulation of USP30. Mitophagy may be stimulated by interventions that stimulate the mitochondrial unfolded protein response. In any case, secondary metabolites for these potential targets must still be researched for a smooth distribution while avoiding side effects. Researchers believe in better comprehending the role of mitochondrial quality regulatory systems about a diverse selection of pathologies such as myocardial injuries, NDDs, hepatic diseases, aging, muscular dystrophy, and metabolic diseases by combining knowledge gained from various aspects of the investigations.

References

Andreux PA, Blanco-Bose W, Ryu D, Burdet F, Ibberson M, Aebischer P, Auwerx J, Singh A, Rinsch C. (2019). The mitophagy activator urolithin A is safe and induces a molecular signature of improved mitochondrial and cellular health in humans. *Nature Metabolism*. 1(6):595-603.

Cereghetti GM, Stangherlin A, De Brito OM, Chang CR, Blackstone C, Bernardi P, Scorrano L. (2008). Dephosphorylation by calcineurin regulates the translocation of Drp1 to mitochondria. *Proceedings of the National Academy of Sciences*. 105(41):15803-8.

Chaugule VK, Burchell L, Barber KR, Sidhu A, Leslie SJ, Shaw GS, Walden H. (2011). Autoregulation of Parkin activity through its ubiquitin-like domain. *The EMBO journal*. 30(14):2853-67.

Chen G, Kroemer G, Kepp O. (2020). Mitophagy: an emerging role in aging and age-associated diseases. *Frontiers in cell and developmental biology*. 8:200.

Chen Y, Dorn GW. (2013). PINK1-phosphorylated mitofusin 2 is a Parkin receptor for culling damaged mitochondria. *Science*. 340(6131):471-5.

Chu CT, Ji J, Dagda RK, Jiang JF, Tyurina YY, Kapralov AA, Tyurin VA, Yanamala N, Shrivastava IH, Mohammadyani D, Qiang Wang KZ. (2013). Cardiolipin externalization to the outer mitochondrial membrane acts as an elimination signal for mitophagy in neuronal cells. *Nature cell biology*. 15(10):1197-205.

Cornelissen T, Haddad D, Wauters F, Van Humbeeck C, Mandemakers W, Koentjoro B, Sue C, Gevaert K, De Strooper B, Verstreken P, Vandenberghe W. (2014). The deubiquitinase USP15 antagonizes Parkin-mediated mitochondrial ubiquitination and mitophagy. *Human molecular genetics*. 23(19):5227-42.

Cornelissen T, Verstreken P, Vandenberghe W. (2018). Imaging mitophagy in the fruit fly. *Autophagy*. 14(9):1656-7.

Cunningham CN, Baughman JM, Phu L, Tea JS, Yu C, Coons M, Kirkpatrick DS, Bingol B, Corn JE. (2015). USP30 and parkin homeostatically regulate atypical ubiquitin chains on mitochondria. *Nature cell biology*. 17(2):160-9.

Deas E, Plun-Favreau H, Gandhi S, Desmond H, Kjaer S, Loh SH, Renton AE, Harvey RJ, Whitworth AJ, Martins LM, Abramov AY. (2011). PINK1 cleavage at position A103 by the mitochondrial protease PARL. *Human molecular genetics*. 20(5):867-79.

Doblado L, Lueck C, Rey C, Samhan-Arias AK, Prieto I, Stacchiotti A, Monsalve M. (2021). Mitophagy in human diseases. *International journal of molecular sciences*. 22(8):3903.

Givvimani S, Munjal C, Tyagi N, Sen U, Metreveli N, Tyagi SC. (2012). Mitochondrial division/mitophagy inhibitor (Mdivi) ameliorates pressure overload induced heart failure. *PloS one*. 7(3): e32388.

Hasson SA, Kane LA, Yamano K, Huang CH, Sliter DA, Buehler E, Wang C, Heman-Ackah SM, Hessa T, Guha R, Martin SE. (2013). High-content genome-wide RNAi screens identify regulators of parkin upstream of mitophagy. *Nature*. 504(7479):291-5.

Hirota Y, Yamashita SI, Kurihara Y, Jin X, Aihara M, Saigusa T, Kang D, Kanki T. (2015). Mitophagy is primarily due to alternative autophagy and requires the MAPK1 and MAPK14 signaling pathways. *Autophagy*. 11(2):332-43.

Jung, Hye Seung, and Myung-Shik Lee. (2010). Role of autophagy in diabetes and mitochondria. *Annals of the New York Academy of Sciences*. 1201.1 (2010): 79-83.

Kageyama Y, Hoshijima M, Seo K, Bedja D, Sysa-Shah P, Andrabi SA, Chen W, Höke A, Dawson VL, Dawson TM, Gabrielson K. (2014). Parkin-independent mitophagy requires D rp1 and maintains the integrity of mammalian heart and brain. *The EMBO journal*. 33(23):2798-813.

Kerr JS, Adriaanse BA, Greig NH, Mattson MP, Cader MZ, Bohr VA, Fang EF. (2017). Mitophagy and Alzheimer's disease: cellular and molecular mechanisms. *Trends in neurosciences.* 40(3):151-66.

Komander D, Clague MJ, Urbé S. (2009). Breaking the chains: structure and function of the deubiquitinases. *Nature reviews Molecular cell biology*. 10(8):550-63.

Kondapalli C, Kazlauskaite A, Zhang N, Woodroof HI, Campbell DG, Gourlay R, Burchell L, Walden H, Macartney TJ, Deak M, Knebel A. (2012). PINK1 is activated by mitochondrial membrane potential depolarization and stimulates Parkin E3 ligase activity by phosphorylating Serine 65. *Open biology*. 2(5):120080.

Kondapalli C, Kazlauskaite A, Zhang N, Woodroof HI, Campbell DG, Gourlay R, Burchell L, Walden H, Macartney TJ, Deak M, Knebel A. (2012). PINK1 is activated by mitochondrial membrane potential depolarization and stimulates Parkin E3 ligase activity by phosphorylating Serine 65. *Open biology*. 2(5):120080.

Kumar A, Aguirre JD, Condos TE, Martinez-Torres RJ, Chaugule VK, Toth R, Sundaramoorthy R, Mercier P, Knebel A, Spratt DE, Barber KR. (2015). Disruption of the autoinhibited state primes the E3 ligase parkin for activation and catalysis. *The EMBO journal*. 34(20):2506-21.

Kumar A, Tamjar J, Waddell AD, Woodroof HI, Raimi OG, Shaw AM, Peggie M, Muqit MM, van Aalten DM. (2017). Structure of PINK1 and mechanisms of Parkinson's disease-associated mutations. *Elife*. 6: e29985.

Lee JY, Nagano Y, Taylor JP, Lim KL, Yao TP. (2010). Disease-causing mutations in parkin impair mitochondrial ubiquitination, aggregation, and HDAC6-dependent mitophagy. *Journal of Cell Biology*. 189(4):671-9.

Ma X, McKeen T, Zhang J, Ding WX. (2020). Role and mechanisms of mitophagy in liver diseases. *Cells*. 9(4):837.

Manning G, Whyte DB, Martinez R, Hunter T, Sudarsanam S. (2002). The protein kinase complement of the human genome. *Science*. 298(5600):1912-34.

Miller S, Muqit MM. (2019). Therapeutic approaches to enhance PINK1/Parkin mediated mitophagy for the treatment of Parkinson's disease. *Neuroscience letters*. 705:7-13.

Murakawa T, Yamaguchi O, Hashimoto A, Hikoso S, Takeda T, Oka T, Yasui H, Ueda H, Akazawa Y, Nakayama H, Taneike M. (2015). Bcl-2-like protein 13 is a mammalian Atg32 homologue that mediates mitophagy and mitochondrial fragmentation. *Nature communications*. 6(1):1-4.

Nah J, Miyamoto S, Sadoshima J. (2011). Mitophagy as a protective mechanism against myocardial stress. *Comprehensive Physiology*. 7(4):1407-24.

Nakamura N, Hirose S. (2008). Regulation of mitochondrial morphology by USP30, a deubiquitinating enzyme present in the mitochondrial outer membrane. *Molecular biology of the cell*. 19(5):1903-11

Nishida Y, Arakawa S, Fujitani K, Yamaguchi H, Mizuta T, Kanaseki T, Komatsu M, Otsu K, Tsujimoto Y, Shimizu S. (2009). Discovery of Atg5/Atg7-independent alternative macroautophagy. *Nature*. 461(7264):654-8.

Okatsu K, Sato Y, Yamano K, Matsuda N, Negishi L, Takahashi A, Yamagata A, Goto-Ito S, Mishima M, Ito Y, Oka T. (2018). Structural insights into ubiquitin phosphorylation by PINK1. *Scientific reports*. 8(1):1-0.

Ordureau A, Sarraf SA, Duda DM, Heo JM, Jedrychowski MP, Sviderskiy VO, Olszewski JL, Koerber JT, Xie T, Beausoleil SA, Wells JA. (2014). Quantitative proteomics reveal a feedforward mechanism for mitochondrial PARKIN translocation and ubiquitin chain synthesis. *Molecular cell*. 56(3):360-75.

Palikaras K, Daskalaki I, Markaki M, Tavernarakis N. (2017). Mitophagy and age-related pathologies: Development of new therapeutics by targeting mitochondrial turnover. *Pharmacology & therapeutics*. 178:157-74.

Palikaras K, Lionaki E, Tavernarakis N. (2015). Coordination of mitophagy and mitochondrial biogenesis during ageing in C. elegans. *Nature*. 521(7553):525-8.

Pickles S, Vigié P, Youle RJ. (2018). Mitophagy and quality control mechanisms in mitochondrial maintenance. *Current Biology*. 28(4): R170-85.

Popov LD. (2020). Mitochondrial biogenesis: An update. *Journal of Cellular and Molecular Medicine*. 24(9):4892-9.

Ribas V, Drew BG, Zhou Z, Phun J, Kalajian NY, Soleymani T, Daraei P, Widjaja K, Wanagat J, de Aguiar Vallim TQ, Fluitt AH. (2016). Skeletal muscle action of estrogen receptor α is critical for the maintenance of mitochondrial function and metabolic homeostasis in females. *Science translational medicine*. 8(334):334ra54-.

Sandoval H, Thiagarajan P, Dasgupta SK, Schumacher A, Prchal JT, Chen M, Wang J. (2008). Essential role for Nix in autophagic maturation of erythroid cells. *Nature*. 454(7201):232-5.

Sarraf SA, Raman M, Guarani-Pereira V, Sowa ME, Huttlin EL, Gygi SP, Harper JW. (2013). Landscape of the PARKIN-dependent ubiquitylome in response to mitochondrial depolarization. *Nature*. 496(7445):372-6.

Sasarman F, Brunel-Guitton C, Antonicka H, Wai T, Shoubridge EA, LSFC Consortium. (2010). LRPPRC and SLIRP interact in a ribonucleoprotein complex that regulates posttranscriptional gene expression in mitochondria. *Molecular biology of the cell*. 21(8):1315-23.

Sauvé V, Sung G, MacDougall EJ, Kozlov G, Saran A, Fakih R, Fon EA, Gehring K. (2022). Structural basis for feedforward control in the PINK1/Parkin pathway. *The EMBO Journal*. e109460.

Tourrière H, Chebli K, Zekri L, Courselaud B, Blanchard JM, Bertrand E, Tazi J. (2003). The RasGAP-associated endoribonuclease G3BP assembles stress granules. *The Journal of cell biology*.160(6):823-31.

Um JH, Yun J. (2017). Emerging role of mitophagy in human diseases and physiology. *BMB reports*. 50(6):299.

Wu W, Tian W, Hu Z, Chen G, Huang L, Li W, Zhang X, Xue P, Zhou C, Liu L, Zhu Y. (2014). ULK 1 translocates to mitochondria and phosphorylates FUNDC 1 to regulate mitophagy. *EMBO reports*. 15(5):566-75.

Youle RJ, Narendra DP. (2011). Mechanisms of mitophagy. *Nature reviews Molecular cell biology*. 12(1):9-14.

Zhang H, Bosch-Marce M, Shimoda LA, Tan YS, Baek JH, Wesley JB, Gonzalez FJ, Semenza GL. (2008). Mitochondrial autophagy is an HIF-1-dependent adaptive metabolic response to hypoxia. *Journal of Biological Chemistry*. 283(16):10892-903.

Chapter 14

Therapeutic Interventions in Mitophagy

Arup Kumar Misra[1,*]
and Siddhartha Dutta[2]
[1]Department of Pharmacology, All India Institute of Medical Sciences, Mangalagiri, Andhra Pradesh, India
[2]Department of Pharmacology, All India Institute of Medical Sciences, Rajkot, Gujarat, India

Abstract

The mitochondrion is an organelle essential for energy synthesis, the breakdown of cytoplasmic proteins, and cell death. A unique autophagic process called mitophagy removes damaged mitochondria and maintains the body's equilibrium intact. According to recent studies, several physiological functions, including cellular differentiation, nerve protection, and homeostasis, appear to be influenced by mitophagy. The processes of mitophagy, mitochondrial fusion, and mitochondrial fission—known as "mitochondrial dynamics"—that control the quantity, quality, and architecture of mitochondria—have subsequently been attributed to different pathologies. During various pathological transformations of the underlying disease, changes in mitochondrial morphology may serve as a pathway for bioenergetic adaptation, which may link mitochondrial dynamics to the equilibrium between energy demand and nutrition supply. Mitophagy, which depends on the fission/fusion cycle, is crucial to mitochondrial dynamics because it eliminates unhealthy and malfunctioning mitochondria. This chapter discusses potential pharmaceutical and investigational products that can modulate autophagy as fresh approaches to fight various biological diseases.

Keywords: mitophagy, mitochondrial dynamics, inflammation, oxidative stress, cell death, activators, inhibitors, modulators

Introduction

The primary sites for cellular respiration and energy production are mitochondria covered by the bilayer membranes, mitochondrial inner membrane (IMM), and mitochondrial outer membrane (OMM) (Friedman and Nunnari, 2014). The cristae is a highly significant structure

[*] Corresponding Author's Email: arup2003m@gmail.com.

In: Mitophagy and Its Role in Health and Disease
Editors: Shalini Mani and Samiksha Wasnik
ISBN: 979-8-89113-470-6

folded into the inner membrane and contains ATP synthase and several different cytochromes (Kuehlbrandt, 2015). The brain consumes 20% of all human energy despite making up only 2% of the body's weight (Raichle and Gusnard, 2002). Using glucose and preserving mitochondrial function is necessary for this massive energy expenditure (Castro et al., 2018). As a result, maintaining the mitochondrial network's viability and activity is crucial to maintaining the nervous system's homeostasis (Swerdlow and Wilkins, 2020).

Defect mitochondria can be hazardous by over-producing ROS, consuming ATP via the reversal of ATP synthase, and interfering with several other metabolic processes (Youle and van der Bliek, 2012).

Mitophagy is essential for removing outdated or damaged mitochondria to preserve the integrity of the mitochondrial pool for cellular homeostasis, which affects many physiological and pathological processes (Fang et al., 2019). Autophagy, which rigorously regulates the quality and number of mitochondria, targets defective mitochondria for phagocytosis and death (Wang et al., 2019). Maintaining the integrity of mitochondria and the balance of metabolism necessitates mitophagy. Homeostasis and mitophagy regulate mitochondrial biogenesis, which is essential for cellular and organismal function. A mismatch among them might result in mitochondria increasing oxygen demand and excessive ROS production, all of which can result in cellular deterioration and try to open up cell death pathways. Numerous mechanisms, including the ubiquitin-mediated system and the mitochondrial receptor-mediated pathway, are a part of the mitophagy activation process (Murata et al., 2015).

Functional Association of Mitochondrial Dynamics and Mitophagy

Changes in mitochondrial structure and size are connected to the fission and fusion of the mitochondria, which depend on the needs of the cell (Trotta and Chipuk, 2017). Since mammalian cell autophagosomes range in size from 500 to 1500 nm (Wang and Klionsky, 2003), it seems sensible to assume that a large mitochondrial network should be divided through fission before mitophagy. In that regard, it has been noted that mitophagy necessitates either the degradation of mitochondrial fusion-associated components or factors related to mitochondrial fission (Shirihai et al., 2015). As a result, proteins involved in mitochondrial dynamics are suggested to work with the LC3 adapters/receptors in mitophagy. In other words, overexpressed FUNDC1 and BCL2L13 interact with the mitochondrial fission protein Drp1, which is required for mitochondrial fragmentation and Parkin-mediated mitophagy in both FUNDC1 and BNIP3 (Lee et al., 2011; Wu et al., 2016). However, a recent study found that mitophagy is unnecessary for mitochondrial fission (Yamashita and Kanki, 2017). Therefore, more research on Drp1's function in mitophagy is required. OPA1, a mitochondrial fusion protein, might additionally serve as a mitophagic component.

Through its interaction with OPA1, BNIP3 causes the fragmentation of the mitochondria (Landes et al., 2010). OPA1 long isoform deletion from PHB2 results in aberrant cristae formation and fragmentation of the mitochondria (Merkwirth et al., 2008). According to a study, for the inner membrane of the mitochondria to fuse, there must be a significant amount of cardiolipin in the inner membrane and that cardiolipin must bind to OPA1 (Ban et al., 2017). Additionally, following contact with FIS1, TBC1D15 facilitates mitochondrial autophagic encapsulation upon Parkin activation (Yamano et al., 2014). Therefore, it is thought that due to the interaction of LC3 adapter/receptors with mitochondrial dynamics factors, mitophagy is related to mitochondrial dynamics.

Mitophagy and Human Diseases

Numerous clinical illnesses are linked to abnormalities in the mitophagy and/or mitochondrial quality control; several neurodegenerative diseases are among the best-researched examples (Rodolfo et al., 2017). Parkin and PINK1 are hereditary contributors to Parkinson's disease and operate as neuroprotective mitophagy regulators (Ashrafi et al., 2014). According to Ye et al. (2015), Parkin is implicated in Alzheimer's disease as the disease proceeds through the depletion of mitochondria, and its overexpression minimizes the signs associated with it (Martin-Maestro et al., 2016). With Huntington's disease, mitophagy is disrupted. Mutant huntingtin operates as a platform to facilitate selective autophagy or impair mitochondrial transport to the lysosome (Martinez-Vicente et al., 2010). (Wong and Holzbaur, 2014). Although it is unclear what role autophagy plays in ALS, mitochondrial dysfunction is also seen in this condition (Rodolfo et al., 2017). As part of the ALS process, mutant superoxide dismutase-1 (SOD1) fragments the mitochondria and blocks retrograde axonal transport in the mitochondria (Magrane et al., 2014). Furthermore, mutant SOD1 binds to p62, OPTN, and LC3 (Rodolfo et al., 2017). Therefore, decreased autophagosome ubiquitinated mitochondrial targeting may be a pathological component of ALS.

Cancer and mitophagy go hand in hand (Drake et al., 2017). Inhibiting mitophagy lowers the stemness of cells during cell self-renewal, and in cancer cells, metabolic reprogramming is associated with decreased mitophagy and increased mitochondrial biogenesis (Vazquez et al., 2013) (Drake et al., 2017). P62-mediated removal of damaged mitochondria reduces inflammation by lowering the NLRP3 inflammasome's activation in macrophages, which is correlated with inflammation-induced cell death (Zhong et al., 2016). The poly (ADP-ribose) polymerase-1 (PARP1)-SIRT1-BNIP3/NIX pathway induces mitophagy when DNA is damaged to prevent the production of ROS or cell-dangerous inflammasomes (Drake et al., 2017).

Variables affecting mitophagy in cardiovascular diseases can either be beneficial or harmful. It has been demonstrated that in the absence of BNIP3, FUNDC1, MFN1, and MFN2 have the opposite effect, and Parkin and PGAM5 protect against heart damage. Parkin and PINK1 also play a preventative role in atherosclerosis (Bravo-San Pedro et al., 2017). Similarly, it is unclear what role mitophagy plays in lung illness. The lack of PINK1 protects against mitophagy-dependent necroptosis in chronic obstructive pulmonary disease (COPD) but may result in idiopathic pulmonary fibrosis (IPF) (Mizumura et al., 2014). In human bronchial epithelial cells, mitochondrial ROS regulates senescence. Mitophagy lowers these levels. Consequently, the type of cell determines how mitophagy functions in lung illness (Cloonan and Choi, 2016).

Renal impairment occurs before mitochondrial dysfunction in acute kidney injury. In cases of acute kidney injury and metabolic acidosis, a factor contributing to chronic kidney disease, mitophagy protects renal cells from being damaged. The kidneys are harmed by a high-calorie diet, which also reduces autophagy and mitophagy and is linked to distorted mitochondrial morphology (Parikh et al., 2015). Thyroid hormone (TH) improves the consequences of nonalcoholic fatty liver disease (NAFLD) by promoting mitophagy and mitochondrial biogenesis. They increase the abundance of mRNA for ULK1, NIX, BNIP3, LC3, and p62 (Sinha and Yen, 2016). In addition, mitophagy helps reduce NAFLD when AMPK is activated (Smith et al., 2016). There are, of course, current, multiple investigations that highlight the relevance of mitophagy in the etiology of other maladies as well.

Alzheimer's Disease and Therapeutic Interventions

The principal hallmark of Alzheimer's dementia (AD), a devastating neurodegenerative condition, is the steady decline in cognitive function. Neurons of AD patients have a significant accumulation of extracellular Aβ plaques, MAPT/Tau tangles, and damaged mitochondria. Current findings imply that mitochondrial dysfunction may be the critical pathophysiological factor in AD because the accumulation of damaged mitochondria occurs before the accumulation of Aβ plaques and accelerates the deposition of Aβ. One possible approach to treating AD is to encourage restoring mitochondrial activity. The two processes that make up mitochondrial homeostasis, mitophagy, and mitochondrial biogenesis, are tightly regulated, and both are weakened in AD patients. As a cycle-regulated process, mitochondrial homeostasis, and increasing mitophagy can promote the production of mitochondria.

The study employed a high-throughput method for finding mitophagy inducers (Cen et al., 2020). As a reliable mitophagy sensor that has been well-established, a stable expression cell line using the mt-Keima protein was used to determine the presence of mitophagy, and a high throughput imaging technique was used to calculate the proportion of mitochondria that reach lysosomes. FDA-approved or potential medications were evaluated and discovered that several apoptosis inducers, including BH3 mimics, can successfully induce mitophagy. Most of these medications prevent the activity of BCL2-family proteins, which permits apoptosis. Interestingly, the MCL1 inhibitor UMI-77 can activate mitophagy, showing that MCL1 might be involved in this process.

Although most known mitophagy inducers, notably CCCP, that have been investigated in previous research typically operate by damaging mitochondria, compelling cells to adopt mitophagy, the safety of these compounds is the primary concern. There was no evidence that UMI-77 was damaging to mitochondria. On the other hand, UMI-77 can lead to the selective destruction of damaged mitochondria, indicating its potential for clinical usage. MCL1 functions in numerous biological processes as an anti-apoptotic protein. Previous research indicates that MCL1 deletion in adult myocytes leads to mitochondrial malfunction, reduced PINK1-PRKN/PARK2 signaling, and impaired autophagy start. With its putative function in establishing synapses and embryonic development, MCL1 may regulate lifespan and several cellular functions, including mitochondrial dynamics. Designed explicitly for MCL1, UMI-77 blocks MCL1 and BAX-BAK1 from interacting, allowing BAX-BAK1 to cause apoptosis. However, at the sublethal level of 5 M, it could not trigger apoptosis, suggesting that MCL1 has an additional, unidentified function.

UMI-77 can successfully treat the cognitive dysfunction shown in APP-PSEN1/PS1 mice and induce mitophagy in vivo. This is demonstrated by the favorable results of APPPSEN1 animals in the water maze test and the reduction of inflammatory cytokine levels and extracellular Aβ plaque deposition in the brain. Moreover, the mitochondrial morphology of brain neurons is restored. Patients with AD create more reactive oxygen species (ROS) than healthy individuals due to risk factors such as genetic mutations, aging, and environmental variables. This results in mitochondrial malfunction, Aβ plaque formation, and the production of inflammatory cytokines.

Nevertheless, Aβ can enter cells and accumulate in mitochondria. This interaction alters the dynamics of mitochondria, lowers their activity, and raises their levels of ROS. UMI-77 reduces the generation of Aβ plaques and inflammatory cytokines by eliminating the damaged mitochondria, which raises the possibility that the mitochondrial damage may play a significant role in the onset of AD. It also shows that improving mitochondrial activity and promoting mitophagy are possible therapeutic strategies for treating Alzheimer's disease (Cen et al., 2021). The FDA-approved drug candidate UMI-77, a BH3 mimic, can induce mitophagy by releasing MCL1, which may assist in reversing the pathophysiology of Alzheimer's disease. MCL1 is a novel mitophagy receptor protein.

Parkinson's Disease and Therapeutic Interventions

Parkinson's disease (PD) genetics and pathophysiology suggest that mitochondria are involved in the condition's etiology. Elegant research over the last two decades has elucidated the intricate molecular signaling underlying mitophagy, a process of mitochondrial quality control that entails identifying and expulsing damaged mitochondria from the cell. Both sporadic and familial PD show mitochondrial deficiencies, more notably impaired mitophagy. Early-onset Parkinson's disease (PD) is thought to be caused by loss-of-function mutations in the crucial mitophagy regulators PINK1 and Parkin. To create a PD treatment that can modify the disease, it is pharmacologically important to understand mitophagy and how to speed up the removal of damaged mitochondria. Despite significant knowledge of both PINK1-Parkin-dependent and -independent mitochondrial quality control mechanisms, the therapeutic potential of targeting mitophagy has not yet been extensively researched.

Pathways dependent and independent of PINK1-Parkin present several possible biological intervention points to enhance mitophagy. Several chemicals are already accessible that disrupt biology by inhibiting or activating a particular target. These compounds help address the idea's validity and, in some situations, serve as a foundation for therapy development. Some of them are discussed below:

Protease

OMA1

A crucial mitophagy-initiating event is the arrest of PINK1 import caused by mitochondrial membrane depolarization. Numerous PD-linked PINK1 variants have failed to stabilize OMA1, an IMM-embedded metalloprotease, imports and cleaves proteins. By facilitating PINK1 import arrest preceding OMA1 identification, Tom7, a small auxiliary protein of the TOM complex, has been postulated to regulate the lateral release of conformationally kinase-active PINK1 from the TOM40 channel (Sekine S et al., 2019). The import of PD-related PINK1 mutations, which are stopped by depolarization at the OMM, is restored when OMA1 is suppressed. Additionally, OMA1/cells showed a modest increase in PINK1 stabilization caused by mitochondrial membrane depolarization, irrespective of modifications in PINK1 steady-state import (Sekine S et al., 2019). These findings, along with proof that OMA1 plays a protective effect in a Phb2-deficient (Phb2NKO) neurodegenerative paradigm that only harms forebrain neurons (Korwitz A et al., 2016), imply that OMA1 inhibitors may be useful as therapeutic agents.

USP33/VDU1

Deubiquitinases are a family of proteins that may remove ubiquitin chains from substrates (DUBs). USP33 largely resists parkin autoubiquitination, an OMM localized DUB, at Lys435. USP33 silencing enhances Parkin stability and mitophagy by boosting K63-linked ubiquitin chain formation on Parkin and accelerating depolarization-dependent mitochondrial translocation. Unexpectedly, in the substantia nigra and amygdala, USP33 levels are elevated in two areas of the central nervous system (CNS) that are affected by Parkinson's disease (PD) (Niu K et al., 2020).

Deubiquitinases (DUBs)

Many DUBs, including USP30, USP15 (Cornelissen T et al., 2014; Cornelissen T et al., 2018), USP8 (Durcan TM et al., 2014), and a splice version of USP35, affect mitophagy (Wang Y et al., 2015).

Phosphatase

PINK1-inhibiting protein phosphatase type II (PPEF2) dephosphorylates ubiquitin and prevents PINK1-mediated mitophagy. Phospho-Ser65-ubiquitin levels rise due to PPEF2 silencing, and both Parkin- and Parkin-independent basal and stress-induced mitophagy are enhanced. It is interesting to note that PPEF2 and PINK1 have been found to regulate many proteins negatively. These findings suggest potential functions as an antiapoptotic phosphatase, the regulation of mitochondrial biogenesis, and mitophagy in tissues and cells with reduced Parkin expression (Wall CE et al., 2019). The fact that phosphorylated ubiquitin linkages are more resistant to DUB-mediated cleavage suggests that dephosphorylation of ubiquitin may be an essential regulator in controlling rates of mitophagy (Wauer T et al., 2015).

Traumatic Brain Injury (TBI) and Therapeutic Interventions

Acute structural injury and brain dysfunction caused by external stimuli are called traumatic brain injury (TBI) (Bailes JE and Borlongan CV, 2020). Typically, TBI begins as a primary injury brought on by external factors before developing over time into a secondary injury involving cell death, oxidative stress, and inflammatory response. These biological mechanisms interact with mitophagy during the occurrence and progression of TBI to control mitochondrial quality, which may impact nerve cell survival and degeneration.

Traumatic brain injury results in more severe oxidative damage, an inflammatory response, and cell death and is associated with mitochondrial dysfunction. Mitochondrial damage results in a considerable release of reactive oxygen species and promotes the formation of pro-inflammatory molecules, which cause inflammation and oxidative stress. Moreover, damaged mitochondria cause the activation of mitophagy. If mitophagy is inadequate to trigger the cell death process, an imbalance in the delicate equilibrium between mitophagy, ROS production, and the inflammatory response may start, drive, or speed up the process. Therapeutic targets that promote mitophagy to reduce downstream cascades such as oxidative damage, inflammatory response, and cell death may reduce neurological functioning after TBI.

Agents Acting on the Inflammatory Pathway and Mitophagy

As the awareness of the mitochondrial damage caused by TBI has grown recently, improving brain-mitochondrial function has emerged as a novel therapeutic target. Inhibiting the posttraumatic inflammatory response was traditionally the primary goal of pharmacological therapy for TBI (Filev AD et al., 2021). Some medications, however, stop the inflammation brought on by the mitophagy pathway and improve mitochondrial function caused by TBI. These are described below:

Ren et al. discovered that Resolvin D1 uses mitophagy to eliminate damaged mitochondria and prevent neuroinflammation (Ren YZ et al., 2020). Similarly, through controlling neuroinflammation and mitophagy, the IL-10 and p53 activators pifithrin-μ and pifithrin-α can additionally provide neuroprotection (Maiti P et al., 2019 and Yang LY et al., 2020). Interestingly, it was discovered that the prevention of NLRP3-mediated inflammatory bodies' activation might be used with rapamycin-induced mitophagy, and the combined therapy had a more significant neuroprotective impact (Chen Y et al., 2019). Lin et al. have observed that the inhibitory effect of melatonin on inflammation is controlled by activating mitophagy and eliminating only damaged mitochondria. Since there is concrete evidence linking mitophagy with inflammation (Lin C et al., 2016), inhibiting it will considerably worsen the inflammation brought on by TBI.

The NLRP3 inflammasome and PINK1-mediated mitophagy interaction may further enhance neuroprotection by inhibiting NLRP3 inflammasome activation post-TBI (Chen Y et al., 2019). Fisetin was shown by Ding et al. to inhibit inflammatory activation in sepsis-associated encephalopathy by promoting mitophagy (Ding H et al., 2022). The idea is that FUNDC1 could be a new therapeutic target for the treatment of ICH as it is a mitophagy receptor that lowers inflammation caused by ICH by stimulating mitophagy (Zheng S et al., 2021). Our data collectively demonstrate that mitochondrial damage can trigger proinflammatory pathways, which can accumulate due to inflammation.

Agents Acting on the Oxidative Stress and Mitophagy

Increased energy requirements after TBI result in excessive ROS generation, which damages and dysfunctions mitochondrial DNA (Wang Y et al., 2019). When mitophagy targets damaged mitochondria, an endogenous neuroprotective process, the damaged mitochondria can be swiftly removed, preserving neurons' ability to function normally (Okamoto K et al., 2014). However, stressful surroundings and stimulation circumstances also affect how much mitophagy is activated. Targeting damaged mitochondria with moderate mitophagy helps to preserve their functionality and promotes the survival of neurons (Youle RJ and Narendra DP, 2011). Damaged mitochondria are not correctly removed when mitophagy is insufficient, and ROS continues to build up. As a result, mitochondrial proteins and DNA are oxidized, aggravating the damage to the mitochondria (Fleury C et al., 2002). Too many mitochondria are removed by excessive mitophagy (Chu CT et al., 2014). As mitochondria are essential for neurons, excessive mitochondrial clearance will promote nerve cell death and harm brain function (Kim I et al., 2007).

According to Wu et al., by inhibiting both mitophagy and mitochondrial division, mitochondrial division inhibitor 1 (Mdivi-1) can decrease the damage to the blood-brain barrier and cell death brought on by TBI (Wu Q et al., 2018). Niu et al. showed that mitophagy plays a protective role in injured mitochondria by establishing that Mdivi-1 might exacerbate the damage caused to TBI mice (Niu F et al., 2019). The results could differ depending on the degree of mitophagy, as was already indicated. The mitochondrial structure and function after TBI may improve by increasing moderate mitophagy or decreasing excessive mitophagy (Saito T and Sadoshima J, 2015). This could also possibly promote the survival of neurons. In addition, mitophagy routes vary spatially and kinetically in neurons and immortalized cells, indicating that these two cell types may have several distinct purposes and outcomes. Because they commonly employ local repair or partial degradation processes, nerve cells have a higher threshold for mitophagy than other types of cells (Chu CT et al., 2014).

In 2020, Zakarya et al. observed that the superoxide dismutase simulant 5,6-dicarboxy-1,1,3,3-tetraethylisoindolin-2-oxyl (DCTEIO) could enhance tissue repair and neural function in TBI rats by conserving the mitochondria's structure and function by scavenging ROS, minimizing oxidative stress, and encouraging mitophagy. In animal research, when treated with DCTEIO immediately after injury, the damage range of brain tissue decreased after 24 hours and six weeks after injury (Zakarya R et al., 2020). By demonstrating that triiodothyronine (T3) improved the clearance of damaged mitochondria by boosting PINK1 and significantly reduced the generation of ROS, Lin et al. revealed that T3 reduced neuronal mortality through mitophagy (Lin C et al., 2020). According to these results, mitophagy prevents the accumulation of ROS and damaged mitochondria.

Agents Acting on the Cell Death and Mitophagy

Neurons only experience cell death when the cumulative effects of numerous stimuli exceed their potential for recovery (Chi H et al., 2018). A traumatic event, such as a TBI, prompts immediate neuronal cell death, which is frequently observed in neurodegenerative illnesses, and a fast drop in energy production in the damaged neurons. Necrosis and programmed cell death are the two stages of neuronal cell death following TBI. Any chance of minimizing neuronal necrosis brought on by membrane disruption and irreversible metabolic abnormalities appears to depend on the early implementation of neuroprotective treatments. This second wave of neuronal programmed cell death appears during a window of opportunity for effective targeted therapy (Lenzlinger PM et al., 2001 Zhang X et al., 2005). This chapter summarises the possible therapeutic role in modifying cell death to enhance neurological outcomes.

Caspase- and Caspase-9 are apoptosis markers activated by Mdivi-1, which Niu et al. showed to prevent mitophagy activation selectively. This suggests that mitophagy significantly reduces the cell apoptosis brought on by TBI (Niu F et al., 2019). TUNEL-positive structures and the number of LC3 puncta in cells decreased; Wu et al. found that Mdivi-1 had similar effects on TBI, suggesting that autophagy may be a factor in these effects. Cardiolipin-dependent mitophagy, an endogenous neuroprotective mechanism that is triggered by traumatic brain injury, and prevents neuronal death and behavioral impairments, was demonstrated by Chao et al. (Chao H et al., 2019). By inhibiting pink1-mediated mitophagy, Lin et al. found that T3 therapy may provide a TBI treatment plan (Lin C et al., 2020).

Pyroptosis is accompanied by inflammation and brought on by proinflammatory signals (Bergsbaken T et al., 2009). Current studies indicate that mitophagy acts as a negative regulator of pyroptosis because it prevents oxygen deprivation or hypercapnia from promoting the process in microglia (Ding HG et al., 2020, Hu Z et al., 2021). Chen et al. showed that suppressing pyroptosis activation post-TBI increases the neuroprotection caused by rapamycin-induced mitophagy (Chen Y et al., 2019). In both depression and Parkinson's disease (PD) mouse models, quercetin inhibits the activation of pyroptosis in microglia via increasing mitophagy (Han X et al., 2021). Recent research has identified a novel type of cell death called ferroptosis, frequently accompanied by significant iron build-up and lipid peroxidation (Li J et al., 2020). A deficiency of ferritin promotes osteoblastic ferroptosis through the PINK1/Parkin-mediated mitophagy pathway, according to studies on type 2 diabetes osteoporosis (Wang X et al., 2022). However, there is no information on the connection between mitophagy and other cell degeneration processes, such as ferroptosis in the brain. A more precise knowledge of the association between ferroptosis and various illnesses may generate original thoughts for developing therapies that deliberately address ferroptosis.

Therapeutic targets that enhance mitophagy attenuate downstream cascades such as oxidative damage, inflammatory response, and cell death, which may alleviate neurological impairment following TBI.

Cancer and Therapeutic Interventions

Mitophagy performs a variety of roles in the development of cancer and carcinogenesis. Cancer cells' status and subtypes significantly impact whether they act as tumor promoters or suppressors (Fei P et al., 2004; Kulikov A et al., 2017). Mitophagy can eliminate diseased or dysfunctional mitochondria to preserve the right balance between the number and the quality of mitochondria. Healthy mitochondria produce fewer ROS after mitophagy, which minimizes their ability to trigger tumors (Waris G and Ahsan H, 2006). On the other hand, once the tumors have developed, mitophagy can function to have a cytoprotective mechanism to promote tumor progression against chemotherapy-induced apoptosis (Yan C et al., 2017, Im E et al., 2016). The aggregation of damaged mitochondria is controlled by functional mitophagy, which also prevents cancer growth. The PINK/Parkin pathway can induce mitophagy in liver cells, eliminate damaged mitochondria, and reduce ROS generation. The loss or mutation of genes like PARK2 and BNIP3, which can encourage carcinogenesis and the spread of cancer, can result in mitophagy suppression (Poulogiannis G. et al., 2010, Cesari R. et al., 2003). Parkin-encoding loss-of-function mutations in the PARK2 gene have been found in human colorectal cancer. Parkin can be overexpressed, which can prevent colorectal cancer cells from proliferating. Intestinal cancer development is increased in PARK2 heterozygous deletion mice (Poulogiannis G et al., 2010). Mitophagy that FUNDC1 mediates also prevents inflammasome activation and guards against the development of liver cancer. As defective mitochondria build up in liver cells following the targeted reduction of FUNDC1, inflammasome activation, and carcinogenesis are triggered (Li W et al., 2019).

Mitophagy and Drug Resistance

Some chemotherapy medications cause mitochondrial malfunction, release harmful by-products such as ROS, and alter normal metabolic processes (Carew J et al., 2003, Yao N et al., 2019). Cytoprotective mechanisms such as mitophagy assist the body to adapt to chemotherapy. As a result, focusing on mitochondria is thought to be a promising anticancer treatment.

A platinum-based drug with extensive use, cisplatin exhibits anticancer efficacy against several malignancies. Cancer cells eventually develop cisplatin resistance. As a result, getting around medication resistance is difficult (Liu Y et al., 2020, Li Z et al., 2019). Caveolin-1 (Cav-1)/Parkin-mediated mitophagy is implicated in the chemoresistance of the non-small cell lung cancer cell line A549. The cav-1-knockdown A549 cells were found to be more susceptible to cisplatin due to the downregulated Rho-associated coiled-coil-containing protein kinase 1 (ROCK1), which in turn diminishes Parkin-mediated mitophagy (Liu Y et al., 2020). The cisplatin resistance of these cells relies on a protein called apurinic endonuclease 1 (APE1), which is overexpressed in A549 cells and triggers Parkin-mediated mitophagy. As per Li Z et al., the inhibition of APE1 promotes cell death and restores cisplatin sensitivity. Furthermore, the E3 ubiquitin ligase ARIH1 is required for PINK1-dependent mitophagy to begin without Parkin (Li Z et al., 2019). A549 cell death brought on by cisplatin is prevented by ARIH-induced mitophagy. The same dose of cisplatin dominates cell proliferation in ARIH1 knockout cells (Villa E et al., 2017).

By inducing mitochondrial apoptosis, the novel betulinic acid (BA) derivative B5G1 displays a decisive anticancer action against cancer cells resistant to many drugs. B5G1 can, however, trigger mitophagy by stimulating PINK1 and eventually recruiting Parkin. When mitophagy is inhibited by mitochondrial division inhibitor 1 (mdivi-1) or bafilomycin, drug-resistant cancer cells become susceptible to B5G1 (Yao N et al., 2019).

Doxorubicin, a DNA-damaging substance, significantly impacts cell survival because it causes cell death and mitochondrial malfunction. Nevertheless, damaged mitochondria are cleared through BNIP3L-mediated mitophagy in colorectal cancer to reduce oxidative stress and promote cell survival. Its susceptibility to doxorubicin is considerably increased by the suppression of mitophagy caused by BNIP3L deletion (Yan C et al., 2017).

Mitophagy as a Therapeutic Target for Cancer

Although mitophagy helps cells survive stress by adapting, it can also kill cells if it clears too many mitochondria. Therefore, mitophagy inducers and inhibitors may be therapeutic in cancer therapy.

A critical factor in reducing cancer cell drug resistance is the prevention of mitophagy (Carew J et al., 2003; Chen L. et al., 2018). Chemotherapy for cervical cancer is less effective due to cisplatin drug resistance. A well-known endogenous indoleamine and antioxidant, melatonin (N-acetyl-5-methoxytryptamine), lessens hypoxia-ischemia damage and enhances sleep. Additionally, it can stop the growth of tumors and stop mitophagy. Parkin is regulated by melatonin, which worsens the apoptosis of cervical cancer cells by reducing c-Jun N-terminal kinase (JNK) and mitophagy (Chen L et al., 2018).

Due to innate mitophagy and autophagy, chemotherapy medicines such as cisplatin do not effectively eradicate cancer cells in hepatic carcinoma. To improve mitophagy, cisplatin stimulates dynamin-related protein 1 (DRP1). Instead of immediately inducing apoptosis, mdivi-1 or bafilomycin, inhibitors of lysosome function, enhance the cisplatin responsiveness of liver cancer cells (Ma M et al., 2020). Liensinine, a mitophagy inhibitor, can significantly boost breast cancer cells' sensitivity to cisplatin. Liensinine inhibits mitophagy by limiting excessive autophagosome accumulation, autophagosome-lysosome fusion, and the emergence of multiple essential lysosomal hydrolases (Zhou J et al., 2015).

Additionally, improved mitophagy may offer a potential plan for therapeutic intervention in various malignancies. Ketoconazole, an oral antifungal drug, enhances PINK1/Parkin-mediated mitophagy by inhibiting COX-2 synthesis, which causes hepatic cancer cells to induce apoptosis (Chen Y et al., 2019).

Furthermore, sorafenib stimulates the PINK1/Parkin-controlled process of mitophagy, which leads to the death of liver cancer cells. It helps to stabilize PINK1 on the mitochondrial outer membrane to activate mitophagy and attracts Parkin to impaired mitochondria (Zhang C et al., 2017).

Two mitochondria-targeted drugs, Mito-CP and Mito-Metformin, release ULK1 from mTOR-mediated control, reduce mitochondrial membrane potential, and hinder the growth of colorectal cancer cells by mitophagy (Boyle K et al., 2018).

Kidney Diseases and Therapeutic Interventions

Global health concerns and significant financial burdens are associated with renal disorders, including acute and chronic kidney damage and declining renal functioning (Lameire NH et al., 2013 Duann P et al., 2016). Harmful ROS and RNS are produced when the disease manifests, contributing to acute renal injury and chronic kidney disease. This oxidative stress injury may be driven on by ischemia/reperfusion, an energy depletion spurred on by an issue involving ATP energization or mitochondrial biogenesis, or a failure to clear damaged mitochondria (mitophagy). Potential causes of the disorders above include the build-up of inflammatory cytokines, the stimulation of immune cells, apoptosis, and tissue damage. The renal phenotypes of mitochondrial dysfunction can include podocytopathy, nephrotic syndromes, cystic kidney disease, proximal tubular dysfunction, and tubular interstitial disease. Along with increasing mortality and morbidity rates, all types of renal dysfunction may increase cardiovascular disease risk.

Research projects have focused on this development toward renal illnesses because protecting mitochondria may be a potentially beneficial treatment method (Tabara LC et al., 2014 Kezic A et al., 2016). These substances include counteracting mitochondrial oxidants, fostering mitochondrial biogenesis and ATP generation, controlling ROS metabolism, protecting cardiolipins, inhibiting mPTP, or inhibiting mitochondrial fragmentation, as described below.

Mitochondria-Targeted Antioxidants

MitoQ, MitoTEMPO, MitoE, Mito-CP, SkQ1, and SkQR1 constitute the few mitochondrial-targeted antioxidants to minimize mitochondrial oxidative stress. These compounds deliver existing redox agents to the mitochondrial matrix by conjugating them with the TPP+ (triphenyl alkyl phosphonium cation) moiety (Kezic A et al., 2016). Based on membrane potential, these ROS scavenger substances could penetrate the membrane's lipid bilayer and accumulate at the matrix (m). MitoQ is safe in clinical studies for fatty acid disease, Parkinson's disease, and chronic kidney disease. In conclusion, several animal models have shown that antioxidants that target mitochondria are an efficient way to prevent or lessen many kidney illnesses.

The Activator of Mitochondria Biogenesis

Acute organ injury recovery requires considerable metabolic and energy demands, which calls for the induction of mitochondrial biogenesis. The AMPK/SIRT/PGC-1 axis is essential for the synthesis of mitochondria. PGC-1 is a crucial element that regulates NAD synthesis in a coordinated manner, lowers oxidative stress, and speeds up recovery from AKI (Tran MT et al., 2016). SIRT1, a NAD-dependent deacetylase, positively regulates PGC-1 expression and activity. These substances include formoterol, AICAR, and resveratrol. Resveratrol therapy optimizes mitochondrial mass, improves mitochondrial respiratory function, and minimizes oxidative stress in a rat hemorrhagic shock model (Wang H et al., 2015). Additionally, in a model involving septic AKI induced by lipopolysaccharide, resveratrol modifies the immune response by reducing inflammation triggered by macrophages (Chen L et al., 2015). In a rat model of I/R-induced AKI, pre-treatment with the AMPK activator AICAR reduced nitrosative stress and improved renal function (Lempiainen J et al., 2012). It also mitigated damage and tubular necrosis. Increasing the number of mtDNA copies, oxygen consumption rate, and PGC-1 synthesis, the solid 2-adrenoreceptor agonist formoterol stimulates mitochondrial biogenesis. An animal model of I/R-induced AKI is accomplished by improving renal function, preserving impaired renal tubules, and limiting necrosis (Jesinkey SR et al., 2014). In conclusion, therapies that influence NAD regulation and mitochondrial biogenesis offer considerable promise for treating renal disease, but further research is needed to determine how well they will work in practice.

Modulation of ATP Synthesis and ROS Metabolism

Synthetic indole acetic acid plant hormone derivative mitochondria acid 5 (MA-5) has been demonstrated to enhance ATP synthesis. Leigh syndrome and MELAS are two examples of mitochondrial illnesses reported to benefit from MA-5 treatment (Suzuki T et al., 2015). There is evidence that complex V at the mitochondrial crista junction is assembled and oligomerized by MA-5, maintaining the dynamics of the mitochondria and preventing fragmentation. These recent discoveries offer fresh means of protecting the heart and kidneys (Suzuki T et al., 2015; Suzuki T et al., 2016).

Cardiolipin Protections

Cardiolipin governs the structural and functional flexibility of IMMs. Hence, its peroxidation or loss has been linked to aging and many types of AKI and CKD. An original finding was the development of a cardiolipin-target chemical to enhance ETC effectiveness and restore cellular bioenergetics. The cardiolipin-targeted SS-31 peptide, also known as MTP-131 or Bendavia, stabilizes cardiolipin in AKI and CKD models, scavenges mitochondrial ROS, regulates cytC activity, and inhibits the mitochondrial permeability transition (MPT) pore (Szeto HH et al., 2014; and Szeto HH and Birk AV, 2014). Better renal hemodynamics, function, and mitochondrial protective effects were seen after treatment with SS-31 in pigs with stenotic kidney injury (Eirin A et al., 2014). SS-31 has entered numerous clinical trials for treating AKI renal microvascular dysfunction in hypertension and I/R damage of myocardial infarction (Szeto HH et al., 2014).

mPTP Inhibitors

The Ca2+-dependent nonselective IMM localized mitochondrial permeability transition pore (mPTP), which is a Ca2+-dependent pore, opens under various pathological conditions, such as Ca2+ overload or oxidative stress. The opening of the mPTP appears to be what causes the IMM to become permeable to tiny molecules (1,500 Da in molecular weight), lose its proton motive force, disconnect OXPHOS, and swell up mitochondrially, which is commonly followed by AKI (Wang Z et al., 2013). Cyclosporin A (CsA) is an effective inhibitor of mPTP through its interactions with cyclophilin D, a structural component and mediator of mPTP important for controlling Ca2+- and ROS-dependent opening. In a clinical investigation, the benefits of small doses of CsA (at submicromolar concentration), which inhibit mPTP opening and mitochondrial swelling, are being examined in acute myocardial infarction reperfusion injury (NCT01595958). CsA reduced podocyte damage through mitochondrial protection in an animal model of adriamycin-induced podocyte injury (Guan N et al., 2015). A clinical trial (NCT02397213) is currently being conducted to evaluate CsA renal protection in cardiac surgery (potential AKI condition). CsA is renotoxic at high levels, so its use in medicine is restricted.

Research suggests that the enzyme glycogen synthase kinase (GSK)-3 connects all the signaling pathways that regulate the mitochondrial permeability transition (MPT). To stop MPT and increase cell viability during AKI, the thiadiazolidinone derivative and specific GSK-3 inhibitor TDZD-8 function as non-ATP competitive inhibitors (Wang Z et al., 2013 Wang Z et al., 2015).

Fission Inhibitors

Mitochondrial fusion is required for the uniform distribution of mtDNA, matrix proteins, and lipids throughout all merged mitochondria. Mitophagy clearance of defective mitochondria and mitochondrial expansion following mitosis rely on mitochondrial fission. Mdivi-1 (mitochondrial fission inhibitor-1) is a newly discovered tiny medication that selectively and irreversibly inhibits the assembly and GTPase activity of the fission protein DRP1 (Cassidy-Stone A et al., 2008). Mdivi-1 causes the rapid and reversible formation of fused mitochondria in various preclinical animal studies, including I/R injury to the heart, liver, kidney, and brain (Rosdah AA et al., 2016). Mdivi-1 simultaneously increases ATP synthesis and protects against mPTP opening in acute cardiorenal syndrome (Sumida M et al., 2015). It also reduces ROS stress, tubular epithelial cell apoptosis, and rhabdomyolysis-induced AKI in rats (Tang WX et al., 2013). Mdivi-1 has a variety of functions that could either be cytoprotective or cytotoxic, and there is still no evidence that it protects against renotoxicity in humans despite having therapeutic promise (Rosdah AA et al., 2016).

K(ATP) Channel Opener

A smooth muscle vasodilator, levosimendan, increases Ca2+ sensitivity and is clinically used to treat heart failure. In cardiomyocytes suffering from ischemic heart disease, levosimendan has positive pleiotropic effects, including protecting the mitochondria (Papp Z et al., 2012). A mitochondrial ATP-sensitive potassium channel that aids cardiomyocytes in conserving mitochondrial energy may play a role in the molecular mechanisms of levosimendan. Pre-clinical and clinical investigations also revealed beneficial circulation benefits in the brain, lungs, and renoprotection under potentially lethal stress circumstances. Levosimendan is the subject of a clinical investigation currently being conducted (NCT01720030) for AKI.

Mitochondrial-targeting medicines have been demonstrated to be effective interventions for maintaining mitochondrial architecture and functioning in several animal models of kidney injury. MitoQ and SS-31 have demonstrated promising promise against specific kidney injuries and illnesses in clinical trials.

Lung Diseases and Therapeutic Interventions

There is growing evidence that changes in autophagy may contribute to the pathophysiology of several chronic diseases, including chronic obstructive pulmonary disease and asthma (COPD). Increased autophagy promotes type 2 immune responses and eosinophilic inflammation in asthma, whereas decreased autophagy may benefit neutrophilic asthma. Smoking may activate autophagy in acute exposure, leading to ciliary dysfunction and death of airway epithelial cells. Studies have shown that autophagy is impaired in stable COPD, with reduced autophagic flux and accumulation of abnormal mitochondria (defective mitophagy) linked to cellular senescence. As autophagy can be increased or diminished depending on the cellular environment and different cell types, it is challenging to target autophagy therapeutically. Rapamycin, metformin, carbamazepine, cardiac glycosides, and statins are only a few current medications that may trigger autophagy; others, like chloroquine, may suppress it. Although more selective medications are being developed, these nonspecific medications are still used to treat asthma and COPD today.

Autophagy-specific medications have not yet been created. The fact that different cell types may be affected by autophagy differently, making the total impact of a medication uncertain, is a significant issue. Several such mouse models are available (Kuma A et al., 2017). Cell-specific deletion of autophagy-related genes can somewhat alleviate this. Both increases and decreases in autophagy have been associated with asthma and COPD. This may depend on the type of cells, the intensity and duration of a particular cellular stress, and different asthma and COPD pheno-endotypes. Inhibitors should be helpful because, for instance, autophagy appears to be elevated in eosinophilic inflammation in asthma. It is impeded, nevertheless, in severe neutrophilic asthma, where activators could be required. Smoking and acute stress may promote autophagy in COPD, even though autophagy is usually inhibited in chronic illnesses. Combination therapies are now possible because it has been discovered that several medication classes act at various locations along the autophagic pathway and can promote or reduce autophagy (Galluzzi L et al., 2017).

Furthermore, dietary adjustments and lifestyle modifications may also have a positive impact on autophagy and serve as supplementary strategies. Given that most medications utilized have other effects on the cell, it may be challenging to anticipate and understand the effects of therapeutic interventions. For instance, cellular senescence is strongly linked to autophagy and is inhibited by the autophagy-activating drug rapamycin (Aman Y et al., 2021).

Autophagy Activators

To treat stable COPD and severe non-T2 asthma, it may be necessary to use several medications found to boost autophagy when it is decreased by disease.

Rapamycin and related rapalogues, including temsirolimus, everolimus, and umerolimus, which inhibit mTORC1, cause autophagy. In mice, emphysema develops quickly when mTORC1 is overexpressed in alveolar epithelial cells because mTOR is active in the lungs of people with COPD (Mitani A. et al., 2016 and Houssaini A. et al., 2018). Rapamycin reduces the increased mTOR activation in COPD lung endothelial cells and inhibits the generation of proinflammatory mediators like IL-6 and CXCL8. Oral rapamycin delays the development of emphysema in mice with mTORC1 overexpression after exposure to cigarette smoke (Houssaini et al., 2018). The benefits of rapamycin are associated with improved mitochondrial activity and a reduction in cellular senescence. An increase in autophagy is most likely how these advantages are mediated. Rapamycin, on the other hand, had a suppressive impact on Th17 cells in mice that had been sensitized to ovalbumin and exposed to it, which was consistent with a defect in autophagy in neutrophilic asthma (Wu J et al., 2020). Rapamycin is also effective in a typical mouse model of allergic asthma, with a drop in lung eosinophils and a decrease in eosinophil differentiation (Hua W et al., 2015).

AMPK activators, which inhibit mTORC1 as well, can also cause autophagy to become active. Even though other AMPK activators have been developed (Kim J et al., 2016), only metformin—routinely used to treat type 2 diabetes—has been tested in clinical studies. Through the activation of AMPK, metformin lowers asthma exacerbations in asthmatic patients (Wu TD et al., 2019) and reduces allergic-induced structural changes and inflammation in mice with sensitivity to allergens (Ma W et al., 2022). Metformin also shields mice from emphysema and structural changes in the airways brought on by cigarette smoke, and it has been linked in epidemiological research to a slower progression of COPD. However, these studies have not yet made a strong connection between metformin's beneficial effects and the activation of autophagy; instead, other adverse effects of the medication may be to blame.

Anticonvulsant carbamazepine induces autophagy by suppressing PI3K activity (Galluzzi L et al., 2017). In mice exposed to cigarette smoke, carbamazepine reduces aggresome buildup in the lungs and prevents the development of emphysema (Tran I et al., 2015). There are no known clinical trials for COPD with carbamazepine. Digoxin and other cardiac glycosides can treat defective autophagy in airway epithelial cells that show delayed autophagosome maturation after smoking by reducing BICD1 and p62 accumulation (Mercado N et al., 2019). The tiny chemical SMER28 promotes autophagy in these cells without the assistance of mTORC1 (Mercado N et al., 2019).

Depending on the cellular stimulus and the cell type, atorvastatin activates or prevents excessive autophagy (Ashrafizadeh M et al., 2020). In mice exposed to ovalbumin, simvastatin enhances the expression of autophagy proteins like ATG5, LC3B, and Beclin-1, as well as the number of autophagosomes in lung tissue (Gu W et al., 2017). Simvastatin simultaneously lowers the extracellular matrix and inhibits the expression of IL-4, IL-5, and IL-13. 3-MA can counteract these statin-related effects. Statins may help asthma patients better regulate their symptoms (Sunata K et al., 2022). The anti-inflammatory benefits of an inhaled corticosteroid may be enhanced by a statin's influence on autophagy, which could benefit asthmatic patients (Maneechotesuwan K et al., 2021). Statins may similarly help COPD patients by reducing neutrophilic inflammation (Maneechotesuwan K et al., 2015).

While oxidative stress appears to be a primary mechanism that affects autophagy in COPD and asthma, antioxidants may not be the best way to restore average autophagic flux. The hunt is on for antioxidants that are more efficient and more tolerated because present antioxidants, including N-acetylcysteine, are ineffective in COPD (Barnes PJ et al., 2019).

Autophagy Inhibitors

Excessive autophagy is a pathophysiological component of airway sickness; autophagy inhibitors may be therapeutic. Therefore, this might be a strategy for reducing autophagy and potentiating eosinophilic inflammation in asthma. When 3-MA inhibits autophagy, the anti-inflammatory cytokine IL-10 is generated in alveolar macrophages from asthma patients (Maneechotesuwan K et al., 2021). Due to toxicity concerns, such as the activation of apoptosis, 3-MA is not appropriate for therapeutic dosing. The formation of autophagosomes is decreased because 3-MA suppresses Class III PI3Ks (Chicote J et al., 2020). As autophagy inhibition reduces the surge in autophagy observed in animal models and airway epithelial cells after acute exposure to cigarette smoke, it may be beneficial for treating acute exacerbations of COPD. Because other studies indicate that autophagy is impaired in COPD, it is unclear whether autophagy should be promoted or repressed. Different cell types and types of cellular stress, including short-term high-level exposure to cigarette smoke or long-term low-level exposure, probably have diverse impacts on autophagy pathways.

The well-tolerated drugs chloroquine and hydroxychloroquine inhibit lysosomal function to restrict autophagy, but they also have a variety of other pharmacological actions, rendering them highly non-specific. Despite developing Lys5, a more targeted lysosomal inhibitor, no airway disease trials have been conducted.

Two further specialized autophagy inhibitors are now being investigated as cancer therapeutics, the ULK-1 inhibitors SBI-0206965 and MRT68921 (Galluzzi L et al., 2017). Effective autophagy inhibitors such as ULK-1 inhibitors (Egan DF et al., 2015) and specific inhibitors of VPS34, a Class III PI3K, are also being investigated and developed as cancer therapies (Dowdle WE et al., 2014). The autophagy-modulating properties of several naturally occurring substances have also been discovered, including traditional Chinese herbal treatments used to treat asthma (Zhang Y et al., 2021). Autophagy activators such as metformin and rapamycin and medications that specifically stimulate the PINK1/Parkin pathway may target defective mitophagy (Iorio R et al., 2021). The current study also found that cannabis promotes mitophagy and stimulates PINK1/Parkin signaling (Ramirez A et al., 2022). Elamipretide, a tetrapeptide that targets the mitochondria and stabilizes cardiolipin while preserving mitochondrial activity, is now being researched for heart failure and mitochondrial disorders (Karaa et al., 2020).

Cardiac glycosides, frequently used to treat cardiovascular problems, activate autophagy even if they can have inhibitory effects. There is a concentrated attempt to find more selective medications because autophagy is crucial to cancer, cardiovascular disease, and neurological disorders. Rapamycin and its analogs, metformin, and chloroquine, have all been discovered to affect autophagy, although they also have multiple additional effects and are therefore not explicitly autophagy-targeting medications (Martins WK et al., 2021).

Diabetes Mellitus and Therapeutic Interventions

The association between the prevalence of type 2 diabetes and diseases such as inflammatory bowel disease, non-alcoholic steatohepatitis, and even a range of cancers like colon and liver cancer substantially strain our society's healthcare system. The balance of cellular mineral elements, energy management, and lipid metabolism are all regulated by autophagy, a vital factor in metabolic homeostasis. Metabolic homeostasis and type 2 diabetes regulation are significantly influenced by mitophagy, according to a wealth of research. Insulin resistance has been linked to faulty mitophagy, a fact that has long been known. Moreover, the development of insulin resistance results in several diseases, such as nephropathy, retinopathy, and cardiovascular disease.

By this, type 2 diabetes will have a proven and effective therapeutic target in restoring mitophagy. Recent research has shown that several phytochemicals can prevent the development of diabetes by preventing the onset of mitophagy, which stops cell malfunction. Given the above occurrence, mitophagy inducers should be viewed as potential and cutting-edge therapeutic medicines for treating diabetes. An increasing amount of research has shown that autophagy modulators may affect autophagy, making them possible treatments for diabetes mellitus.

Natural Products

Natural products control mitophagy to restore the dynamics of mitochondria in several ways. Traditionally used in Danshen, traditional Chinese medicine, antioxidant salvianolic acid B is a naturally occurring family of 1-benzofurans. Salvinolic acid B inhibited Rho-associated protein kinase-mediated mitophagy in the context of exposure to high glucose conditions and oxidized low-density lipoproteins by reducing the expression of fission proteins such as Dynamin-1-like protein (DRP 1) and mitochondrial fission 1 protein (FIS 1) and causing an endothelial cell to die as a result (Wang X et al., 2020). Another single molecule, ginseng-sanqi-chuanxiong (GSC), has inhibitory effects on the cardiovascular defenses connected with diabetes that are sparked by increased glucose and palmitate. Through stimulation of the AMPK pathway, ginseng-Sanqi-Chuanxiong extracts increased autophagosome production and mitophagy to remove mitochondrial ROS (Ko YS et al., 2020).

Curcumin

In various tissues, curcumin reduces lipid peroxidation and oxidative stress (Ak T and Gulçin I, 2008). Curcumin's antioxidant benefits are credited with improving mitochondrial biogenesis, decreasing ROS, and upregulating antioxidant enzymes. By simultaneously activating AMPK/PGC-1alpha, mitochondrial regulator proteins, mitochondrial transcription factor A, and nuclear respiratory factor 1 in adipocytes, curcumin increased mitochondrial biogenesis (Lone J et al., 2016). Increased levels of mtDNA and proteins involved in mitochondrial functions, such as the master regulator of mitochondrial biogenesis PGC-1alpha and the uncoupling protein 1 (UCP1) for proton transfer and ATP synthesis in the matrix of the mitochondrion, were observed after curcumin administration in an in vivo model. Clinical studies have demonstrated that giving prediabetes subjects curcumin (500 mg daily for 90 days) improved their body mass index and blood glucose levels (Karandish M et al., 2021). In addition, during 12 weeks, 80 mg of curcumin daily reduced insulin resistance and LDL cholesterol and improved overall antioxidant capacity compared to placebo groups (Mokhtari M et al., 2021).

Caffeine

Green tea, coffee, and chocolate all include caffeine, a key ingredient of xanthine alkaloids. In hepatic tissues, lipophagy has been shown to lower the autophagy-dependent lipid content (Zhu X et al., 2017). Additionally, caffeine has been shown to promote mitophagy, which increases fatty acid oxidation and mitochondrial synthesis while potentiating the rate of mitochondrial breakdown in skeletal muscle (Enyart DS et al., 2020).

Quercetin

Sophora japonica L. and Platycladusorientalis (L.) Fanco contains a crucial ingredient called quercetin. There is evidence that quercetin has therapeutic potential for treating diabetic neuropathy. Quercetin decreased ROS production, ATP generation, and mitochondrial damage by promoting the AMPK/PGC-1 pathway in the high-glucose-exposed rat Schwann cell line and STZ-induced diabetic rats (Zhang Q et al., 2021). Quercetin increased the activity of the mitochondrial complex I and made it easier for PCG-1alpha and PPAR to bind, which regulates how much energy is produced and used in metabolic disorders. The integrity and redox status of the intracellular mitochondria in the hepatocytes were also enhanced. As a result, as well as cellular NAD+/NADH. Additionally, it has been found that quercetin and its derivative quercetin-3-O-glucuronide can lower triglyceride accumulation by altering fatty acid synthase, which also activates AMPK to promote the restoration of mitochondrial mass and biogenesis through control of mitophagy (Herranz-Lopez M et al., 2020). Clinical research found that a four-week daily regimen of 500 mg of quercetin lowered plasma uric acid, a risk factor for type 2 diabetes (Shi Y and Williamson G, 2016).

Berberine

Berberine, a quaternary ammonia compound, is in the traditional Chinese herbs Coptis chinensis Franch and PhellodendronChinense Schneid. Berberine corrected mitochondrial dysfunction, fatty acid oxidation failure, and ROS generation in diabetes in mouse models of diabetic kidney disease. Ultimately, berberine balanced the mitochondrial energy homeostasis by amplifying the peroxisome proliferator-activated receptor coactivator-1 (PGC-1) signaling pathway (Qin X et al., 2020). Berberine alleviates diabetic kidney disease, high glucose-induced mitochondrial fission and fusion abnormalities, and mitogenesis. By activating mitophagy in a way dependent on the AMPK signaling pathway, berberine reduced the number of mitochondria and cardiomyocyte damage (Hang W et al., 2018). By regulating the gut flora, Berberine (0.6 g twice daily for 12 weeks) reduced glycated hemoglobin, blood triglycerides, and total cholesterol in T2DM patients.

Vitamins

A vitamin B12 deficiency is common in type 2 diabetics and may indicate that type 2 diabetes has been diagnosed (Shivaprasad C et al., 2020). Vitamin D3 has also been linked to diabetes. Diabetes risk has been reported to decrease with vitamin D3 supplementation (Dawson-Hughes B et al., 2020). Metformin and vitamin D3 combination therapy for diabetes reduced reactive oxygen species and inflammation (Cojic M et al., 2020). A different study also showed that vitamin D3 lessened the harmful effects of diabetes. High-dose cholecalciferol supplementation of 40,000 IU/week for 24 weeks was associated with improvements in clinical manifestation, cutaneous microcirculation, and inflammatory markers in people with T2DM with peripheral neuropathy (Karonova T et al., 2020). Using vitamin D3 supplements at a dose of 4000 IU per day for 24 months did not reduce the risk of developing diabetes more than a placebo, according to a different clinical study (Pittas AG et al., 2019). It is possible that the variance results from the varied vitamin doses utilized in the two tests. Moreover, combining vitamin therapy with anti-diabetic medicine may have more benefits for treating T2DM than alone. In a clinical experiment, it was discovered that type 2 diabetes mellitus patients who received pioglitazone (45 mg/day) and vitamin E (400 IU twice a day) together had lower non-alcoholic fatty liver disease activity ratings than those who received only vitamin E (Bril F et al., 2019).

These substances have been shown to benefit the onset and progression of type 2 diabetes mellitus. However, long-term use of these natural treatments for type 2 diabetes mellitus may have adverse side effects. A major organelle in cellular energy expenditure is mitochondrial dynamics, too. Cellular homeostasis will suffer if one aspect of mitochondrial dynamics is completely inhibited. Hence, the partial reduction of fission and restoration of mitophagy is the more practical goal of type 2 diabetes mellitus therapy.

Cardiovascular Diseases and Aging with Therapeutic Interventions

Ccardiac hypertrophy helps the heart meet a higher oxygen demand and is usually initiated as an adaptive response to hemodynamic stress or myocardial injury. Although initially beneficial, hypertrophy can eventually contribute to the onset of heart disease by worsening interstitial fibrosis and impairing ventricular function. The initiation and progression of pathological remodeling now heavily rely on metabolic changes. Because of the myocardium's high level of oxidative tissue, mitochondria are crucial for maintaining the heart's top performance. Recently, "mitochondrial dynamics" the processes of mitochondrial fusion, fission, biogenesis, and mitophagy that affect mitochondrial morphology, quality, and abundance, have been connected to cardiovascular disease. Alterations in mitochondrial morphology may act as a bioenergetic adaptation mechanism during the pathological remodeling of the heart. Another crucial role of mitochondrial dynamics is the removal of unhealthy and dysfunctional mitochondria by mitophagy, which depends on the fission/fusion cycle.

Although calorie restriction appears to be the most effective strategy for increasing longevity, boosting autophagy, and lowering age-related comorbidities (Alfaras I et al., 2016), it might not be the ideal option for the elderly due to immunodeficiency and malnutrition. To extend healthy lifespans and/or prevent cardiac aging, several anti-aging medicines are now accessible (Alfaras I et al., 2016 Saraswat K and Rizvi SI, 2017). Many organic compounds, including rapamycin, the AMPK/Sirt1 activators resveratrol, and metformin, as well as derivatives of polyphenols, may operate as caloric-restriction mimics to lengthen healthy aging (Alfaras I et al., 2016 and Cabo R and Navas P, 2016).

Currently, available information discusses the importance of autophagy dysregulation in cardiac illnesses as well as the significance of rapamycin, resveratrol, metformin, nicotinamide, and NAD+ precursors for treating aging and cardiac anomalies that are related to autophagy (Nair S and Ren J, 2012 and Shirakabe A et al., 2016). We will also discuss recent epigenetic interventions and non-pharmacological strategies like calorie restriction.

Rapamycin and mTOR Inhibitors

A mTOR inhibitor and autophagy inducer that has received FDA approval is rapamycin (Kennedy BK and Lamming DW, 2016). The two present forms of mTOR have different structures and biological functions. Rapamycin sensitivity and the need for mTORC1 in autophagy regulation. Mammalian lethal with SEC13 protein 8 (mLST8), Raptor, mTOR, and a proline-rich Akt substrate of 40 kDa make up this protein (mTORC1). Rictor, mSin1, and mLST8 make up mTORC2, which is rapamycin-insensitive (Kennedy BK and Lamming DW, 2016). Clinical uses for rapamycin and its analogs, or rapalogs, like RAD001, include the treatment of particular cancers, blockage of cardiac stents, and preventing organ rejection after kidney transplantation. C. elegans, Drosophila, and mice all have longer lifespans when mTOR is inhibited (Kennedy BK and Lamming DW, 2016; Weichhart T, 2018).

Rapamycin treatment in later life can restore aging-related heart dysfunction (Flynn JM et al., 2013). Rapamycin has been demonstrated to have beneficial effects on cardiac aging due to its enhancement of mitochondrial content, suppression of inflammation, and reversal of the age-related metabolic switch from fatty acid oxidation to glycolysis (Chiao YA and Rabinovitch PS, 2015 and Flynn JM et al., 2013). It is possible to move from "an irreversible arrest into a reversible state" using rapamycin by relieving cell cycle inhibition, which prevents cells from being forced to multiply if the cell cycle is stopped (Demidenko ZN et al., 2009). Intriguingly, young animals treated with rapamycin for three months had similar outcomes (Neff F et al., 2013), indicating that rapamycin has effects independent of age.

The unfolded protein response (UPRmt) of the mitochondria and aging may both be affected by changes in the ratio of mitochondrial to nuclear oxidative phosphorylation, which rapamycin may be able to alter (Houtkooper RH et al., 2013). While mTOR inhibition is helpful for the therapy of cardiac aging, the entire deletion of mTORC1 may cause cardiomyopathy, a diminished hypertrophic response, and accelerated heart failure (Zhang D et al., 2010). The adaptive hypertrophic response, commonly regarded as adaptive, often occurs in response to pressure overload or hemodynamic stress in aging.

Resveratrol and Sirtuin Activators

The class III histone deacetylases known as NAD+-dependent sirtuins are lauded as "the fountain of youth" because they significantly impact longevity and aberrant aging. Resveratrol, the first naturally occurring flavonoid, is found in red wine, mulberries, peanuts, and rhubarb. It encourages Sirtuin activation in several pathological situations to have favorable effects on the heart and fight aging (Arafa MH et al., 2014; Lin CH et al., 2014). Foxo3 phosphorylation, mitochondrial homeostasis, cAMP phosphodiesterase inhibition, and Sirt1/PI3K/Akt-mediated resveratrol prevent cardiac aging and improve cardiac function (Zhang Y et al., 2014). Resveratrol's ability to prevent ventricular hypertrophy was eliminated by sirt3 impairment (Chen T et al., 2015). To improve cardiac performance and lifespan, Longevinex, a commercially available resveratrol formulation, also stimulates autophagy, upregulates Sirt1 and Sirt3, and nuclear translocates FOXOs (Mukherjee S et al., 2010). Resveratrol and sirtuins may increase longevity and cardiac health by controlling protein hypoacetylation, a crucial stage in the autophagy process (Mukherjee S et al., 2010). In addition to Sirtuins, resveratrol is notable for suppressing Class I PI3K, stimulated by Akt-insulin signaling, competing with ATP for the catalytic site (Frojdo S et al., 2007).

Berberine, AMPK Activators, and Metformin

According to research, the anti-diabetic drug metformin has been demonstrated to directly phosphorylate Ser633 and Ser1177 and boost AMPK (De Haes W et al., 2014). Metformin lengthens life span and health span in C. elegans and mice through mitohormetic regulation of ROS production, AMPK-mediated stimulation of autophagy, and the elevation of NAD+ levels (De Haes W et al., 2014 and Martin-Montalvo A et al., 2013). According to data from our lab, metformin medication temporarily appeared to alter the aging phenotype of the heart in mice (Turdi S et al., 2010). Results from the United Kingdom Prospective Diabetes Study show that compared to sulfonylurea, insulin, or dietary management, metformin reduces 10-year mortality in myocardial infarction, stroke, or all causes (Tzoulaki I et al., 2009).

Berberine, a naturally occurring substance, activates AMPK and reduces blood LDL cholesterol (Lee YS et al., 2006). Through the suppression of mTOR, berberine increases autophagy, which improves heart function by reducing cardiac remodeling, collagen deposition, apoptosis, and fibrosis (Li MH et al., 2014). In long-term follow-up trials, berberine improves cardiac function and reduces patient mortality in patients with chronic congestive heart failure (Zeng XH et al., 2003). According to more data, berberine delays H2O2-induced senescence in senescent cells while increasing NAD+ levels and autophagic flux (Han X et al., 2016). By reducing neuroinflammation in aging, berberine also reduces postoperative cognitive deficits (Zhang Z et al., 2016). Sirt1, PGC-1, and ATP synthesis are other signaling mechanisms linked to berberine's anti-aging effects, AMPK, and inducing autophagy (Yu Y et al., 2018).

Nicotinamide and NAD+ Precursors

NAD+levels decrease with age due to age-related CD38 overexpression or downregulation of nicotinamide phosphoribosyl transferase (Nampt), a vital enzyme for NAD+ synthesis (Shirakabe A et al., 2016). Neurodegenerative disorders and cardiovascular aging are characterized by NAD+ depletion, which can be treated with medication to increase cellular NAD+ levels (Fang EF et al., 2017). By activating Sirtuins like Sirt1 and Sirt3, which result in protein deacetylation and the activation of autophagy, NAD+ precursors like nicotinamide mononucleotide and nicotinamide riboside or upregulating Nampt may slow down cardiac aging (Shirakabe A et al., 2016). A targeted NAD metabolome study showed decreased levels of nicotinamide mononucleotide salvage in nicotinamide mononucleotide-treated animals, albeit this effect was reversed by overexpression of de novo NAD production enzymes. NAD+ levels may not be significantly increased by nicotinamide mononucleotide, but it may promote Sirtuin target acetylation and slow cardiac aging without survival advantages (Mitchell SJ et al., 2018).

Spermidine

Spermidine, a polyamine, is involved in several biological processes, including the activation of autophagy, DNA stability, transcription, translation, and demise. Spermidine, an autophagy-inducing compound, is a neurodegenerative, cognitive, memory, and motor dysfunction therapeutic that delays aging. As a result, spermidine may have a preventative effect on aging hearts and show therapeutic potential in cardiac aging. Other studies have shown that supplementation with spermidine-rich plant extract is safe and well-tolerated in mice and the elderly, making longer-term spermidine intervention in people possible (Schwarz C et al., 2018). Dietary polyamines (spermidine in particular) offer cardiovascular benefits through mitochondrial homeostasis, anti-inflammation, and delayed stem cell senescence (Madeo F et al., 2018).

Osteopontin (OPN) Inhibitor

These researchers examined the effect of surgical VAT removal and the small-molecule OPN inhibitor Agelastatin A on heart aging. Excessive adiposity and extracellular matrix, particularly OPN (a marker for cardiac problems including calcification), are usually associated with cardiac aging. Interestingly, the consequences of myocardial fibrosis in aging hearts, matched by OPN insufficiency, were improved by VAT removal, significantly decreasing circulating levels of TGF and OPN. VAT ablation and OPN deficiency accelerated the senescence of cardiac fibroblasts, which constrained their activation. Agelastatin A restored heart fibrosis and dysfunction brought on by aging (Sawaki D et al., 2018). These results confirmed that OPN is primarily produced by VAT, compromising cardiac shape and function through a profibrotic secretome as we age. OPN is widely recognized for promoting autophagy (Zheng YH et al., 2012), although the role of autophagy in this situation has not yet been determined. This raises the possibility that inhibiting autophagy may have positive effects on aging. The possibility of induced fibroblast senescence as a cause of cardiac aging makes OPN-targeting interventions such as OPN inhibition and VAT elimination (Sawaki D et al., 2018).

Fibroblast Growth Factor 21 (FGF21)

FGF21 is a constituent of the hormone-like FGF group and controls stress responses, energy expenditure, and glucose and lipid metabolism as a longevity factor. FGF21 regulates ER stress, damaged mitochondria, and autophagy via the somatotropic and hypothalamic-pituitary-adrenal (HPA) axis. Type 2 diabetes, obesity, atherosclerosis, and cardiovascular diseases are age-related metabolic pathologies demonstrated to be decreased by FGF21. FGF21-induced healthy aging is thought to be mediated in the integrated stress response by several pathways, including AMPK activation, autophagy induction, and anti-inflammation. Like insulin resistance, FGF21 resistance can impair human health and hasten aging by disrupting autophagy (Salminen et al., 2017).

Urolithin A and Mitophagy Regulators

Urolithin A, a naturally occurring substance that triggers mitophagy, increases lifespan, and favors healthy aging, perhaps by promoting mitophagy and eliminating damaged mitochondria. In aged C. elegans and rats, urolithin A retained mitochondrial respiration and boosted a range of activities (including mobility, pharyngeal pumping, and exercise capacity) (Ryu D et al., 2016). Urolithin A and other nutrients contained in food, including tomatidine and nicotinamide riboside, may prevent sarcopenia with aging (Ryu D et al., 2016). These findings highlight the benefits of urolithin A and mitophagy enhancers for preserving cardiac and mitochondrial function as people age.

Additional Anti-Aging Medications with the Potential to Regulate Autophagy

Other anti-aging drugs that show promise in regulating autophagy include a number of them. By encouraging autophagy, b-guanidinopropionic acid and N-acetylglucosamine may lengthen life (Yang S et al., 2015). Although concrete evidence for an autophagy-permissive role in this context is still missing, natural curcumin and melatonin have also demonstrated protective effects against cardiovascular aging (Cai B et al., 2016). Although kallistatin is an endogenous protein, new research indicates that it may hold promise for lifetime extension and protection from senescence through Akt-Bcl2 inhibition, activation of Sirt1, and reduction of oxidative stress and inflammation (Chao J et al., 2017).

Dietary Restrictions and Lifestyle Changes

In addition to pharmaceutical treatments, calorie restriction is a lifestyle adjustment that prolongs the lifespan in several species and delays the onset of age-related illnesses (Zhang Y et al., 2018; Seals DR et al., 2018). The benefits of calorie restriction in extending longevity and health span and benefiting cardiac aging through autophagy activation have been demonstrated by epidemiological, clinical, and experimental research (Seals DR et al., 2018). In calorically restricted hearts, mTOR was suppressed, and autophagy was increased, highlighting the contribution of mTOR-dependent autophagy to the protective effects of calorie restriction on heart aging (Shinmura K et al., 2011). Autophagy also enhances the neuroendocrine and metabolic profiles of obesity, hyperglycemia, and lipid metabolism in aging (Zhang Y et al., 2018). Serum leptin levels are decreased by pharmacologically inducing autophagy, but leptin is removed genetically to cause autophagy. The accumulation of lipids and metabolic abnormalities associated with aging may contribute to decreased autophagy (Lapierre LR et al., 2011). In the heart, lipid droplets are formed under the regulation of autophagy; when autophagy is impaired, triglycerides and lipid droplets build up (Lapierre LR et al., 2011).

Mitochondrial dynamics are crucial for preserving the circulatory system's homeostasis. This procedure is connected to crucial biological processes such as metabolism and quality control. In particular, the ratio of mitochondrial fusion to fission relates to mitochondrial energy output. Because the circulatory system is constantly mechanically demanding, energy needs to be continuously provided, particularly in the heart. Metabolic remodeling occurs in several cardiovascular diseases. Changes in mitochondrial dynamics can be directly linked to a loss in energy efficiency because higher mitochondrial fission is associated with lower ATP production. For instance, mitochondrial dynamics are frequently altered under situations of metabolic excess, such as in diabetes, or dietary limitation, such as in prolonged ischemia—an imbalance in energy generation results from these modifications to mitochondrial structure and function. Future studies must fully identify the processes that start and maintain pathogenic and adaptive changes in mitochondrial dynamics.

Conclusion

The mitochondria-specific autophagic clearance process is conserved and is called mitophagy. A sophisticated regulatory mechanism regulates the harmony between mitochondrial biogenesis and mitophagy. These include mitochondrial biogenesis, fusion, and fission, which restore mitochondrial function, and mitophagy, which eliminates damaged mitochondria. Specifically, in mitophagy, the autophagosomes detect the target mitochondria and send them to the lysosome for destruction. Mitophagy has a dual function in pathologic circumstances of human disease, which may explain the substantial crosstalk between mitophagy and apoptosis, as they share some regulators. For lifespan and stress resistance, these conflicting processes must be appropriately coordinated.

Synthetic and natural substances that improve mitophagy and restore mitochondrial dynamics. Less adverse effects than conventional pharmaceuticals throughout an adequate duration is one of the significant advantages of using natural goods. When used in clinical settings, the stability, bioavailability, therapeutic window, and adverse effects of natural substances are yet unknown. Further limitations on clinical trials include the possibility of interactions between conventional drugs and herbal therapies. This technique should be considered in animal research, especially in the pharmacokinetics and pharmacodynamics methods, to maximize the utilization of natural products for sickness management. As a result, this chapter might serve as a helpful summary to aid researchers as they continue to study novel mitophagy-modulating medicines in hopes of one day curing various disorders.

References

Ak T, Gülçin I. (2008). Antioxidant and radical scavenging properties of curcumin. *Chem BiolInteract.* 174, 27–37.

Alfaras I, Di Germanio C, Bernier M, Csiszar A, Ungvari Z, Lakatta EG, de Cabo R. (2016). Pharmacological Strategies to Retard Cardiovascular Aging. *Circ Res* 118 (10), 1626–42.

Aman Y, Schmauck-Medina T, Hansen M, Morimoto RI, Simon AK, Bjedov I, Palikaras K, Simonsen A, Johansen T, Tavernarakis N, Rubinsztein DC, Partridge L, Kroemer G, Labbadia J, Fang EF (2021). Autophagy in healthy aging and disease. *Nat. Aging 1*. 634–650.

Arafa MH, Mohammad NS, Atteia HH, Abd-Elaziz HR. (2014) Protective effect of resveratrol against doxorubicin-induced cardiac toxicity and fibrosis in male experimental rats. *J PhysiolBiochem.* 70 (3), 701–711.

Ashrafi G, Schlehe JS, LaVoie MJ, Schwarz TL. (2014). Mitophagy of damaged mitochondria occurs locally in distal neuronal axons and requires PINK1 and Parkin. *J. Cell Biol.* 206, 655-670.

Ashrafizadeh M, Ahmadi Z, Farkhondeh T, Samarghandian S. (2020). Modulatory effects of statins on the autophagy: A therapeutic perspective. *J. Cell. Physiol.* 235, 3157–3168.

Bailes JE, Borlongan CV. (2020) Traumatic brain injury. *CNS Neuroscience & Therapeutics.* 26(6), 593-594.

Ban T, Ishihara T, Kohno H, Saita S, Ichimura A, Maenaka K, Oka T, Mihara K, Ishihara N. (2017). Molecular basis of selective mitochondrial fusion by heterotypic action between OPA1 and cardiolipin. Nat. *Cell Biol.* 19, 856-863.

Barnes PJ, Baker J, Donnelly LE. (2019). Cellular senescence as a mechanism and target in chronic lung diseases. *Am. J. Respir. Crit. Care Med.* 200, 556–564.

Bergsbaken T, Fink SL, Cookson BT. (2009). Pyroptosis: host cell death and inflammation. *Nature Reviews. Microbiology.* 7(2), 99–109.

Boyle K, Van Wickle J, Hill R, Marchese A, Kalyanaraman B, Dwinell MB. (2018). Mitochondria-targeted drugs stimulate mitophagy and abrogate colon cancer cell proliferation. *J Biol Chem.* 293(38):14891–14904.

Bravo-San Pedro JM, Kroemer G, Galluzzi L. (2017). Autophagy and mitophagy in cardiovascular disease. *Circ. Res.* 120, 1812-1824.

Bril F, Biernacki DM, Kalavalapalli S, Lomonaco R, Subbarayan SK, Lai J, Tio F, Suman A, Orsak BK, Hecht J, Cusi K. (2019). Role of Vitamin E for Nonalcoholic Steatohepatitis in Patients With Type 2 Diabetes: A Randomized Controlled Trial. *Diabetes Care.* 42 1481–1488.

Cai B, Ma W, Bi C, Yang F, Zhang L, Han Z, Huang Q, Ding F, Li Y, Yan G, Pan Z, Yang B, Lu Y. (2016). Long noncoding RNA H19 mediates melatonin inhibition of premature senescence of c-kit(+) cardiac progenitor cells by promoting miR-675. *J Pineal Res.* 61 (1), 82–95.

Carew J, Zhou Y, Albitar M, Carew JD, Keating MJ, Huang P. (2003). Mitochondrial DNA mutations in primary leukemia cells after chemotherapy: clinical significance and therapeutic implications. *Leukemia.* 17(8), 1437–1447.

Cassidy-Stone A, Chipuk JE, Ingerman E, Song C, Yoo C, Kuwana T, Kurth MJ, Shaw JT, Hinshaw JE, Green DR, Nunnari J. (2008). Chemical inhibition of the mitochondrial division dynamin reveals its role in Bax/Bak-dependent mitochondrial outer membrane permeabilization. *Dev Cell.* 14(2), 193–204.

Castro JP, Wardelmann K, Grune T, Kleinridders A. (2018). Mitochondrial chaperones in the brain: safeguarding brain health and metabolism. *Frontiers in Endocrinology.* 9.

Cen X, Chen Y, Xu X, Wu R, He F, Zhao Q, Sun Q, Yi C, Wu J, Najafov A, Xia H. (2020). Pharmacological targeting of MCL-1 promotes mitophagy and improves disease pathologies in an Alzheimer's disease mouse model. *Nat Commun.* 11(1):5731.

Cesari R, Martin E, Calin G, Pentimalli F, Bichi R, McAdams H, Trapasso F, Drusco A, Shimizu M, Masciullo V, D'Andrilli G, Scambia G, Picchio MC, Alder H, Godwin AK, Croce CM. (2003). Parkin, a gene implicated in autosomal recessive juvenile parkinsonism, is a candidate tumor suppressor gene on chromosome 6q25-q27. *Proc Natl Acad Sci USA.* 100(10), 5956–5961.

Chao H, Lin C, Zuo Q, Liu Y, Xiao M, Xu X, Li Z, Bao Z, Chen H, You Y, Kochanek PM, Yin H, Liu N, Kagan VE, Bayır H, Ji J. (2019) Cardiolipin-dependent mitophagy guides outcome after traumatic brain injury. *The Journal of Neuroscience.* 39(10), 1930–1943.

Chao J, Guo Y, Li P, Chao L. (2017). Role of Kallistatin Treatment in Aging and Cancer by Modulating miR-34a and miR-21 Expression. *Oxid Med Cell Longev*. 5025610.

Chen L, Liu L, Li Y, et al. (2018). Melatonin increases human cervical cancer HeLa cells apoptosis induced by cisplatin via inhibition of JNK/Parkin/mitophagy axis. *Vitro Cell Dev Biol Anim*. 54(1), 1–10.

Chen L, Yang S, Zumbrun EE, Guan H, Nagarkatti PS, Nagarkatti M. (2015). Resveratrol attenuates lipopolysaccharide-induced acute kidney injury by suppressing inflammation driven by macrophages. *Mol Nutr Food Res*. 59(5), 853–864.

Chen T, Li J, Liu J, Li N, Wang S, Liu H, Zeng M, Zhang Y, Bu P. (2015). Activation of SIRT3 by resveratrol ameliorates cardiac fibrosis and improves cardiac function via the TGF-beta/Smad3 pathway. *Am J Physiol Heart Circ Physiol*. 308 (5), H424–H434.

Chen Y, Chen H-N, Wang K, Zhang L, Huang Z, Liu J, Zhang Z, Luo M, Lei Y, Peng Y, Zhou ZG, Wei Y, Huang C. (2019). Ketoconazole exacerbates mitophagy to induce apoptosis by downregulating cyclooxygenase-2 in hepatocellular carcinoma. *J Hepatol*. 70(1), 66–77.

Chen Y, Meng J, Xu Q, Long T, Bi F, Chang C, Liu W. (2019). Rapamycin improves the neuroprotection effect of inhibition of NLRP3 inflammasome activation after TBI. *Brain Research*. 1710, 163–172.

Chen ZH, Lam HC, Jin Y, Kim HP, Cao J, Lee SJ, Ifedigbo E, Parameswaran H, Ryter SW, Choi AM. (2010). Autophagy protein microtubule-associated protein 1 light chain-3B (LC3B) activates extrinsic apoptosis during cigarette smoke-induced emphysema. *Proc. Natl. Acad. Sci. U. S. A*. 107, 18880–18885.

Chi H, Chang HY, Sang TK. (2018). Neuronal cell death mechanisms in major neurodegenerative diseases. *International Journal of Molecular Sciences*. 19, 10.

Chiao YA, Rabinovitch PS. (2015). The Aging Heart. *Cold Spring Harb Perspect Med*. 5 (9), a025148.

Chicote J, Yuste VJ, Boix J, Ribas J. (2020). Cell death triggered by the autophagy inhibitory drug 3-methyladenine in growing conditions proceeds with DNA damage. *Front. Pharmacol*. 11, 580343.

Chu CT, Bayir H, Kagan VE. (2014). LC3 binds externalized cardiolipin on injured mitochondria to signal mitophagy in neurons implications for Parkinson disease. *Autophagy*. 10(2), 376–378.

Cloonan SM, Choi AM. (2016). Mitochondria in lung disease. *J. Clin. Invest*. 126, 809-820.

Cojic M, Kocic R, Klisic A, Cvejanov-Kezunovic L, Kavaric N, Kocic G. (2020). A novel mechanism of vitamin D anti-inflammatory/antioxidative potential in type 2 diabetic patients on metformin therapy. *Arch Med Sci*. 16, 1004–1012.

Cornelissen, T., Haddad, D., Wauters, F., Van Humbeeck, C., Mandemakers, W., Koentjoro, B., Sue, C., Gevaert, K., De Strooper, B., Verstreken, P., & Vandenberghe, W. (2014). The deubiquitinase USP15 antagonizes Parkin-mediated mitochondrial ubiquitination and mitophagy. *Human molecular genetics*, *23*(19), 5227–5242.

Cornelissen T, Vilain S, Vints K, Gounko N, Verstreken P, Vandenberghe W. (2018) Deficiency of Parkin and PINK1 impairs age-dependent mitophagy in Drosophila. *eLife*. 7, e35878.

Dawson-Hughes B, Staten MA, Knowler WC, Nelson J, Vickery EM, LeBlanc ES, Neff LM, Park J, Pittas AG, and D2d Research Group. (2020). Interatrial Exposure to Vitamin D and New-Onset Diabetes Among Adults With Prediabetes: A Secondary Analysis From the Vitamin D and Type 2 Diabetes (D2d) Study. *Diabetes Care*. 43, 2916–2922.

De Cabo R, Navas P. (2016). Spermidine to the rescue for an aging heart. *Nat Med*. 22 (12), 1389–1390.

De Haes W, Frooninckx L, Van Assche R, Smolders A, Depuydt G, Billen J, Braeckman BP, Schoofs L, Temmerman L.(2014). Metformin promotes lifespan through mitohormesis via the peroxiredoxin PRDX-2. *Proc Natl Acad Sci U S A*. 111 (24), E2501–E2509.

Demidenko ZN, Zubova SG, Bukreeva EI, Pospelov VA, Pospelova TV, Blagosklonny MV.(2009). Rapamycin decelerates cellular senescence. *Cell Cycle*. 8 (12), 1888–1895.

Ding H, Li Y, Chen S, Wen Y, Zhang S, Luo E, Li X, Zhong W, Zeng H. (2022). Fisetin ameliorates cognitive impairment by activating mitophagy and suppressing neuroinflammation in rats with sepsis-associated encephalopathy. *CNS Neuroscience & Therapeutics*. 28, 2.

Ding HG, Li Y, Li XS, Liu XQ, Wang KR, Wen MY, Jiang WQ, Zeng HK. (2020). Hypercapnia promotes microglial pyroptosis via inhibiting mitophagy in hypoxemic adult rats. *CNS Neuroscience & Therapeutics*. 26(11), 1134–1146.

Dowdle WE, Nyfeler B, Nagel J, Elling RA, Liu S, Triantafellow E, Menon S, Wang Z, Honda A, Pardee G, Cantwell J, Luu C, Cornella-Taracido I, Harrington E, Fekkes P, Lei H, Fang Q, Digan ME, Burdick D,

Powers AF, Helliwell SB, D'Aquin S, Bastien J, Wang H, Wiederschain D, Kuerth J, Bergman P, Schwalb D, Thomas J, Ugwonali S, Harbinski F, Tallarico J, Wilson CJ, Myer VE, Porter JA, Bussiere DE, Finan PM, Labow MA, Mao X, Hamann LG, Manning BD, Valdez RA, Nicholson T, Schirle M, Knapp MS, Keaney EP, Murphy LO. (2014). Selective VPS34 inhibitor blocks autophagy and uncovers a role for NCOA4 in ferritin degradation and iron homeostasis in vivo. *Nat. Cell Biol.* 16, 1069–1079.

Drake LE, Springer MZ, Poole LP, Kim CJ, Macleod KF. (2017). Expanding perspectives on the significance of mitophagy in cancer. *Semin Cancer Biol*. 47, 110-124.

Duann P, Lianos EA, Ma J, Lin PH. (2016). Autophagy, innate immunity and tissue repair in acute kidney injury. *Int J Mol Sci.* 17(5), E662, 1–19.

Durcan TM, Tang MY, Perusse JR, Dashti EA, Aguileta MA, McLelland G L, Gros P, Shaler TA, Faubert D, Coulombe B, Fon EA. (2014). USP8 regulates mitophagy by removing K6-linked ubiquitin conjugates from parkin. *EMBO J*. 33, 2473–2491.

Egan DF, Chun MG, Vamos M, Zou H, Rong J, Miller CJ, Lou HJ, Raveendra-Panickar D, Yang CC, Sheffler DJ, Teriete P, Asara JM, Turk BE, Cosford ND, Shaw RJ. (2015). Small molecule inhibition of the autophagy kinase ULK1 and identification of ULK1 substrates. *Mol. Cell*. 59, 285–297.

Eirin A, Ebrahimi B, Zhang X, Zhu XY, Woollard JR, He Q, Textor SC, Lerman A, Lerman LO. (2014). Mitochondrial protection restores renal function in swine atherosclerotic renovascular disease. *Cardiovasc Res*. 103(4), 461–672.

Enyart DS, Crocker CL, Stansell JR, Cutrone M, Dintino MM, Kinsey ST, Brown SL, Baumgarner BL. (2020). Low-dose caffeine administration increases fatty acid utilization and mitochondrial turnover in C2C12 skeletal myotubes. *Physiol Rep.* 8, e14340.

Fang EF, Lautrup S, Hou Y, Demarest TG, Croteau DL, Mattson MP, Bohr VA. (2017). NAD(+) in Aging: Molecular Mechanisms and Translational Implications. *Trends Mol Med.* 23 (10), 899–916.

Fang EF, Hou Y, Palikaras K, Adriaanse BA, Kerr JS, Yang B, Lautrup S, Hasan-Olive MM, Caponio D, Dan X, Rocktäschel P, Croteau DL, Akbari M, Greig NH, Fladby T, Nilsen H, Cader MZ, Mattson MP, Tavernarakis N, Bohr VA. (2019). Mitophagy inhibits amyloid-β and tau pathology and reverses cognitive deficits in models of Alzheimer's disease. *Nature Neuroscience*. 22(3), 401–412.

Fei P, Wang W, Kim S, Wang S, Burns TF, Sax JK, Buzzai M, Dicker DT, McKenna WG, Bernhard EJ, El-Deiry WS. (2004). Bnip3L is induced by p53 under hypoxia, and its knockdown promotes tumor growth. *Cancer Cell*. 6(6), 597–609.

Filev AD, Silachev DN, Ryzhkov IA, Lapin KN, Babkina AS, Grebenchikov OA, Pisarev VM. (2021). Effect of xenon treatment on gene expression in brain tissue after traumatic brain injury in rats. *Brain Sciences*. 11, 7.

Fleury C, Mignotte B, Vayssiere JL. (2002). Mitochondrial reactive oxygen species in cell death signaling. *Biochimie*. 84(2-3), 131–141.

Flynn JM, O'Le. ary MN, Zambataro CA, Academia EC, Presley MP, Garrett BJ, Zykovich A, Mooney SD, Strong R, Rosen CJ, Kapahi P, Nelson MD, Kennedy BK, Melov S. (2013). Late-life rapamycin treatment reverses age-related heart dysfunction. *Aging Cell.* 12 (5), 851–62.

Friedman JR, Nunnari J. (2014). Mitochondrial form and function. *Nature*. 505(7483), 335–343.

Frojdo S, Cozzone D, Vidal H, Pirola L. (2007). Resveratrol is a class IA phosphoinositide 3-kinase inhibitor. *Biochem J.* 406 (3), 511–518.

Galluzzi L, Bravo-San Pedro JM, Levine B, Green DR, Kroemer G. (2017). Pharmacological modulation of autophagy: therapeutic potential and persisting obstacles. *Nat. Rev. Drug Discov*. 6, 487–511.

Gu W, Cui R, Ding T, Li X, Peng J, Xu W, Han F, Guo X. (2017). Simvastatin alleviates airway inflammation and remodelling through up-regulation of autophagy in mouse models of asthma. *Respirology (Carlton, Vic.)*. 22, 533–541.

Guan N, Ren YL, Liu XY, Zhang Y, Pei P, Zhu SN, Fan Q. (2015). Protective role of cyclosporine A and minocycline on mitochondrial disequilibrium-related podocyte injury and proteinuria occurrence induced by adriamycin. *Nephrol Dial Transplant*. 30(6), 957–969.

Han X, Tai H, Wang X, Wang Z, Zhou J, Wei X, Ding Y, Gong H, Mo C, Zhang J, Qin J, Ma Y, Huang N, Xiang R, Xiao H. (2016). AMPK activation protects cells from oxidative stress-induced senescence via autophagic flux restoration and intracellular NAD(+) elevation. *Aging Cell.* 15 (3), 416–427.

Han X, Xu T, Fang Q, Zhang H, Yue L, Hu G, Sun L. (2021). Quercetin hinders microglial activation to alleviate neurotoxicity via the interplay between NLRP3 inflammasome and mitophagy. *Redox Biology*. 44.

Hang W, He B, Chen J, Xia L, Wen B, Liang T, Wang X, Zhang Q, Wu Y, Chen Q, Chen J. (2018). Berberine Ameliorates High Glucose-Induced Cardiomyocyte Injury via AMPK Signaling Activation to Stimulate Mitochondrial Biogenesis and Restore Autophagic Flux. *Front Pharmacol*. 9, 1121.

Herranz-López M, Olivares-Vicente M, Rodríguez Gallego E, Encinar JA, Pérez-Sánchez A, Ruiz-Torres V, Joven J, Roche E, Micol V. (2020). Quercetin metabolites from Hibiscus sabdariffa contribute to alleviate glucolipotoxicity-induced metabolic stress in vitro. *Food Chem Toxicol*. 144, 111606.

Houssaini A, Breau M, Kebe K, Abid S, Marcos E, Lipskaia L, Rideau D, Parpaleix A, Huang J, Amsellem V, Vienney N, Validire P, Maitre B, Attwe A, Lukas C, Vindrieux D, Boczkowski J, Derumeaux G, Pende M, Bernard D, Meiners S, Adnot S. (2018). mTOR pathway activation drives lung cell senescence and emphysema. *JCI Insight*. 3, e93203.

Houtkooper RH, Mouchiroud L, Ryu D, Moullan N, Katsyuba E, Knott G, Williams RW, Auwerx J. (2013). Mitonuclear protein imbalance as a conserved longevity mechanism. *Nature*. 497 (7450), 451–457.

Hu Z, Yuan Y, Zhang X, Lu Y, Dong N, Jiang X, Xu J, Zheng D. (2021). Human umbilical cord mesenchymal stem cell-derived exosomes attenuate oxygenglucose deprivation/reperfusion-induced microglial pyroptosis by promoting FOXO3a-dependent mitophagy. *Oxidative Medicine and Cellular Longevity*. 2021, 19715.

Hua W, Liu H, Xia LX, Tian BP, Huang HQ, Chen ZY, Ju ZY, Li W, Chen ZH, Shen HH. (2015). Rapamycin inhibition of eosinophil differentiation attenuates allergic airway inflammation in mice. *Respirology*. 20, 1055–1065.

Im E, Yoo L, Hyun M, Shin WH, Chung KC. (2016). Covalent ISG15 conjugation positively regulates the ubiquitin E3 ligase activity of parkin. *Open Biol*. 6(8), 160193.

Iorio R, Celenza G, Petricca S. (2021). Mitophagy: Molecular mechanisms, new concepts on Parkin activation and the emerging role of AMPK/ULK1 axis. *Cells*. 11.

Jesinkey SR, Funk JA, Stallons LJ, Wills LP, Megyesi JK, Beeson CC, Schnellmann RG. (2014). Formoterol restores mitochondrial and renal function after ischemia-reperfusion injury. *J Am Soc Nephrol*. 25(6), 1157–1162.

Karaa A, Haas R, Goldstein A, Vockley J, Cohen BH. (2020). A randomized crossover trial of elamipretide in adults with primary mitochondrial myopathy. *J. Cachexia Sarcopenia Muscle*. 11, 909–918.

Karandish M, Mozaffari-Khosravi H, Mohammadi SM, Cheraghian B, Azhdari M. (2021). The effect of curcumin and zinc co-supplementation on glycemic parameters in overweight or obese prediabetic subjects: A phase 2 randomized, placebo-controlled trial with a multi-arm, parallel-group design. *Phytother Res*. 35, 4377–4387.

Karonova T, Stepanova A, Bystrova A, Jude EB. (2020). High-Dose Vitamin D Supplementation Improves Microcirculation and Reduces Inflammation in Diabetic Neuropathy Patients. *Nutrients*. 12, 2518.

Kennedy BK and Lamming DW. (2016). The Mechanistic Target of Rapamycin: The Grand ConducTOR of Metabolism and Aging. *Cell Metab*. 23 (6), 990–1003.

Kezic A, Spasojevic I, Lezaic V, Bajcetic M. (2016). Mitochondria-targeted antioxidants: future perspectives in kidney ischemia reperfusion injury. *Oxidative Med Cell Longev*. 2016, 2950503.

Kim I, Rodriguez-Enriquez S, Lemasters JJ. (2007). Selective degradation of mitochondria by mitophagy. *Archives of Biochemistry and Biophysics*. 462(2), 245–253.

Kim J, Yang G, Kim Y, Kim J, Ha J. (2016). AMPK activators: mechanisms of action and physiological activities. *Exp. Mol. Med*. 48, e224.

Ko YS, Jin H, Park SW, Kim HJ. (2020). Salvianolic acid B protects against oxLDL-induced endothelial dysfunction under high-glucose conditions by downregulating ROCK1-mediated mitophagy and apoptosis. *Biochem Pharmacol*. 174, 113815.

Korwitz A, Merkwirth C, Richter-Dennerlein R, Troder SE, Sprenger H-G, Quiros PM, Lopez-Otín C, Rugarli EI, Langer T. (2016). Loss of OMA1 delays neurodegeneration by preventing stress-induced OPA1 processing in mitochondria. *J. Cell Biol*. 212, 157–166.

Kuehlbrandt W. (2015). Structure and function of mitochondrial membrane protein complexes. *BMC Biology*. 13(1), 89.

Kulikov A, Luchkina E, Gogvadze V, Zhivotovsky B. (2017). Mitophagy: Link to cancer development and therapy. *Biochem Biophys Res Commun*. 482(3),432–439.

Kuma A, Komatsu M, Mizushima N. (2017). Autophagy-monitoring and autophagy-deficient mice. *Autophagy*. 13, 1619–1628.

Lameire NH, Bagga A, Cruz D, De Maeseneer J, Endre Z, Kellum JA, Liu KD, Mehta RL, Pannu N, Van Biesen W, Vanholder R. (2013). Acute kidney injury: an increasing global concern. *Lancet*. 382(9887), 170–179.

Landes T, Emorine LJ, Courilleau D, Rojo M, Belenguer P, Arnaune-Pelloquin L. (2010). The BH3-only Bnip3 binds to the dynamin Opa1 to promote mitochondrial fragmentation and apoptosis by distinct mechanisms. *EMBO Rep*. 11, 459-465.

Lapierre LR, Gelino S, Meléndez A, Hansen M. (2011). Autophagy and lipid metabolism coordinately modulate life span in germline-less C. elegans. *Curr Biol*. 21 (18), 1507–1514.

Lee Y, Lee HY, Hanna RA, Gustafsson AB. (2011). Mitochondrial autophagy by Bnip3 involves Drp1-mediated mitochondrial fission and recruitment of Parkin in cardiac myocytes. *Am. J. Physiol-Heart C*. 301, H1924-H1931.

Lee YS, Kim WS, Kim KH, Yoon MJ, Cho HJ, Shen Y, Ye JM, Lee CH, Oh WK, Kim CT, Hohnen-Behrens C, Gosby A, Kraegen EW, James DE, Kim JB. (2006). Berberine, a natural plant product, activates AMP-activated protein kinase with beneficial metabolic effects in diabetic and insulin-resistant states. *Diabetes*. 55 (8), 2256–2264.

Lempiainen J, Finckenberg P, Levijoki J, Mervaala E. (2012). AMPK activator AICAR ameliorates ischaemia reperfusion injury in the rat kidney. *Br J Pharmacol*. 166(6), 1905–1915.

Lenzlinger PM, Morganti-Kossmann MC, Laurer HK, TK MI. (2001). The duality of the inflammatory response to traumatic brain injury. *Molecular Neurobiology*. 24(1-3), 169–181.

Li J, Cao F, Yin HL, Huang ZJ, Lin ZT, Mao N, Sun B, Wang G. (2020) Ferroptosis: past, present and future. *Cell Death & Disease*. 11(2), 88.

Li MH, Zhang YJ, Yu YH, Yang SH, Iqbal J, Mi QY, Li B, Wang ZM, Mao WX, Xie HG, Chen SL (2014). Berberine improves pressure overload-induced cardiac hypertrophy and dysfunction through enhanced autophagy. *Eur J Pharmacol*. 728, 67–76.

Li W, Li Y, Siraj S, Jin H, Fan Y, Yang X, Huang X, Wang X, Wang J, Liu L, Du L, Chen Q. (2019). FUN14 domain-containing 1-mediated mitophagy suppresses hepatocarcinogenesis by inhibition of inflammasome activation in mice. *Hepatology*. 69(2), 604–621.

Li Z, Wang Y, Wu L, Dong Y, Zhang J, Chen F, Xie W, Huang J, Lu N. (2019). Apurinic endonuclease 1 promotes the cisplatin resistance of lung cancer cells by inducing Parkin-mediated mitophagy. *Oncol Rep*. 42(6), 2245–2254.

Lin C, Chao H, Li Z, Xu X, Liu Y, Hou L, Liu N, Ji J. (2016). Melatonin attenuates traumatic brain injury-induced inflammation: a possible role for mitophagy. *Journal of Pineal Research*. 61(2), 177–186.

Lin C, Li N, Chang H, Shen Y, Li Z, Wei W, Chen H, Lu H, Ji J, Liu N. (2020). Dual effects of thyroid hormone on neurons and neurogenesis in traumatic brain injury. *Cell Death & Disease*. 11(8), 671.

Lin CH, Lin CC, Ting WJ, Pai PY, Kuo CH, Ho TJ, Kuo WW, Chang CH, Huang CY, Lin WT. (2014). Resveratrol enhanced FOXO3 phosphorylation via synergetic activation of SIRT1 and PI3K/Akt signaling to improve the effects of exercise in elderly rat hearts. *Age (Dordr)*. 36 (5), 9705.

Liu Y, Fu Y, Hu X, Chen S, Miao J, Wang Y, Zhou Y, Zhang Y. (2020). Caveolin-1 knockdown increases the therapeutic sensitivity of lung cancer to cisplatin-induced apoptosis by repressing Parkin-related mitophagy and activating the ROCK1 pathway. *J Cell Physiol*. 235(2), 1197–1208.

Lone J, Choi JH, Kim SW, Yun JW. (2016). Curcumin induces brown fat-like phenotype in 3T3-L1 and primary white adipocytes. *J Nutr Biochem*. 27, 193–202.

Ma M, Lin X, Liu H, Zhang R, Chen RX. (2020). Suppression of DRP1-mediated mitophagy increases the apoptosis of hepatocellular carcinoma cells in the setting of chemotherapy. *Oncol Rep*. 43(3), 1010–1018.

Ma W, Jin Q, Guo H, Han X, Xu L, Lu S, Wu C. (2022). Metformin ameliorates inflammation and airway remodeling of experimental allergic asthma in mice by restoring AMPKα activity. *Front. Pharmacol*. 13, 780148.

Madeo F, Eisenberg T, Pietrocola F, Kroemer G. (2018). Spermidine in health and disease. *Science*. 359 (6374).

Magrane J, Cortez C, Gan WB, Manfredi G. (2014). Abnormal mitochondrial transport and morphology are common pathological denominators in SOD1 and TDP43 ALS mouse models. *Hum. Mol. Genet.* 23, 1413-1424.

Maiti P, Peruzzaro S, Kolli N, Andrews M, Al-Gharaibeh A, Rossignol J, Dunbar GL. (2019). Transplantation of mesenchymal stem cells overexpressing interleukin-10 induces autophagy response and promotes neuroprotection in a rat model of TBI. *Journal of Cellular and Molecular Medicine*. 23(8), 5211–5224.

Maneechotesuwan K, Kasetsinsombat K, Wongkajornsilp A, Barnes PJ. (2021). Role of autophagy in regulating interleukin-10 and the responses to corticosteroids and statins in asthma. *Clin. Exp. Allergy*. 51, 1553–1565.

Maneechotesuwan K, Wongkajornsil A, Adcock IM, Barnes PJ. (2015). Simvastatin suppressses airway IL-17 and upregulates IL-10 in patients with stable *COPD*. 148, 1164–1176.

Martinez-Vicente M, Talloczy Z, Wong E, Tang GM, Koga H, Kaushik S, de Vries R, Arias E, Harris S, Sulzer D, Cuervo AM. (2010). Cargo recognition failure is responsible for inefficient autophagy in Huntington's disease. *Nat. Neurosci*. 13, 567-U574.

Martin-Maestro P, Gargini R, Perry G, Avila J, Garcia-Escudero V. (2016). PARK2 enhancement is able to compensate mitophagy alterations found in sporadic Alzheimer's disease. *Hum. Mol. Genet*. 25, 792-806.

Martin-Montalvo A, Mercken EM, Mitchell SJ, Palacios HH, Mote PL, Scheibye-Knudsen M, Gomes AP, Ward TM, Minor RK, Blouin MJ, Schwab M, Pollak M, Zhang Y, Yu Y, Becker KG, Bohr VA, Ingram DK, Sinclair DA, Wolf NS, Spindler SR, Bernier M, de Cabo R. (2013). Metformin improves healthspan and lifespan in mice. *Nat Commun*. 4, 2192.

Martins WK, Silva M, Pandey K, Maejima I, Ramalho E, Olivon VC, Diniz SN, Grasso D. (2021). Autophagy-targeted therapy to modulate age-related diseases: Success, pitfalls, and new directions. *Curr. Res. Pharmacol. Drug Discov.* 2, 100033.

Mercado N, Colley T, Baker JR, Vuppussetty C, Kono Y, Clarke C, Tooze S, Johansen T, Barnes PJ. (2019). Bicaudal D1 impairs autophagosome maturation in chronic obstructive pulmonary disease. *FASEB Bioadv*. 1, 688–705.

Merkwirth C, Dargazanli S, Tatsuta T, Geimer S, Lower B, Wunderlich FT, von Kleist-Retzow JC, Waisman A, Westermann B, Langer T. (2008). Prohibitins control cell proliferation and apoptosis by regulating OPA1-dependent cristae morphogenesis in mitochondria. *Gene Dev*. 22, 476-488.

Mitani A, Ito K, Vuppusetty C, Barnes PJ, Mercado N. (2016). Restoration of corticosteroid sensitivity in chronic obstructive pulmonary disease by inhibition of mammalian target of rapamycin. *Am. J. Respir. Crit. Care Med.* 193, 143–153.

Mitchell SJ, Bernier M, Aon MA, Cortassa S, Kim EY, Fang EF, Palacios HH, Ali A, Navas-Enamorado I, Di Francesco A, Kaiser TA, Waltz TB, Zhang N, Ellis JL, Elliott PJ, Frederick DW, Bohr VA, Schmidt MS, Brenner C, Sinclair DA, Sauve AA, Baur JA, de Cabo R. (2018). Nicotinamide Improves Aspects of Healthspan, but Not Lifespan, in Mice. *Cell Metab.* 27 (3), 667–676 e4.

Mizumura K, Cloonan SM, Nakahira K, Bhashyam AR, Cervo M, Kitada T, Glass K, Owen CA, Mahmood A, Washko GR, Hashimoto S, Ryter SW, Choi AM. (2014). Mitophagy-dependent necroptosis contributes to the pathogenesis of COPD. *J. Clin. Invest.* 124, 3987-4003.

Mokhtari M, Razzaghi R, Momen-Heravi M. (2021). The effects of curcumin intake on wound healing and metabolic status in patients with diabetic foot ulcer: A randomized, double-blind, placebo-controlled trial. *Phytother Res*. 35, 2099–2107.

Mukherjee S, Ray D, Lekli I, Bak I, Tosaki A, Das DK. (2010). Effects of Longevinex (modified resveratrol) on cardioprotection and its mechanisms of action. *Can J Physiol Pharmacol.* 88 (11), 1017–1025.

Murata H, Takamatsu H, Liu S, Kataoka K, Huh NH, Sakaguchi M. (2015). NRF2 regulates PINK1 expression under oxidative stress conditions. *PLoS One*. 10(11).

Nair S, Ren J. (2012). Autophagy and cardiovascular aging: lesson learned from rapamycin. *Cell Cycle*. 11 (11), 2092–2099.

Neff F, Flores-Dominguez D, Ryan DP, Horsch M, Schröder S, Adler T, Afonso LC, Aguilar-Pimentel JA, Becker L, Garrett L, Hans W, Hettich MM, Holtmeier R, Hölter SM, Moreth K, Prehn C, Puk O, Rácz I, Rathkolb B, Rozman J, Naton B, Ordemann R, Adamski J, Beckers J, Bekeredjian R, Busch DH, Ehninger G, Graw J, Höfler H, Klingenspor M, Klopstock T, Ollert M, Stypmann J, Wolf E, Wurst W, Zimmer A,

Fuchs H, Gailus-Durner V, Hrabe de Angelis M, Ehninger D. (2013). Rapamycin extends murine lifespan but has limited effects on aging. *J Clin Invest.* 123 (8), 3272–3291.

Niu F, Dong J, Xu X, Zhang B, Liu B. (2019). Mitochondrial division inhibitor 1 prevents early-stage induction of mitophagy and accelerated cell death in a rat model of moderate controlled cortical impact brain injury. *World Neurosurgery.* 122, E1090–E1101.

Niu K, Fang H, Chen Z, Zhu Y, Tan Q, Wei D, Li Y, Balajee AS, Zhao Y. (2020). USP33 deubiquitinates PRKN/parkin and antagonizes its role in mitophagy. *Autophagy.* 16, 724–734.

Okamoto K. (2014). Organellophagy: eliminating cellular building blocks via selective autophagy. *Journal of Cell Biology.* 205(4), 435–445.

Papp Z, Edes I, Fruhwald S, De Hert SG, Salmenpera M, Leppikangas H, Mebazaa A, Landoni G, Grossini E, Caimmi P, Morelli A, Guarracino F, Schwinger RH, Meyer S, Algotsson L, Wikström BG, Jörgensen K, Filippatos G, Parissis JT, González MJ, Parkhomenko A, Yilmaz MB, Kivikko M, Pollesello P, Follath F (2012). Levosimendan: molecular mechanisms and clinical implications: consensus of experts on the mechanisms of action of levosimendan. *Int J Cardiol.* 159(2), 82–87.

Parikh SM, Yang Y, He LY, Tang CY, Zhan M, Dong Z. (2015). Mitochondrial Function and Disturbances in the Septic Kidney. *Semin. Nephrol.* 35, 108-119.

Pittas AG, Dawson-Hughes B, Sheehan P, Ware JH, Knowler WC, Aroda VR, Brodsky I, Ceglia L, Chadha C, Chatterjee R, Desouza C, Dolor R, Foreyt J, Fuss P, Ghazi A, Hsia DS, Johnson KC, Kashyap SR, Kim S, LeBlanc ES, Lewis MR, Liao E, Neff LM, Nelson J, O'Neil P, Park J, Peters A, Phillips LS, Pratley R, Raskin P, Rasouli N, Robbins D, Rosen C, Vickery EM, Staten M; D2d Research Group. (2019). Vitamin D Supplementation and Prevention of Type 2 Diabetes. *N Engl J Med.* 381, 520–530.

Poulogiannis G, McIntyre R, Dimitriadi M, Apps JR, Wilson CH, Ichimura K, Luo F, Cantley LC, Wyllie AH, Adams DJ, Arends MJ. (2010). PARK2 deletions occur frequently in sporadic colorectal cancer and accelerate adenoma development in Apc mutant mice. *Proc Natl Acad Sci U S A.* 107(34), 15145–15150.

Qin X, Jiang M, Zhao Y, Gong J, Su H, Yuan F, Fang K, Yuan X, Yu X, Dong H, Lu F. (2020). Berberine protects against diabetic kidney disease via promoting PGC-1α-regulated mitochondrial energy homeostasis. *Br J Pharmacol.* 177, 3646–3661.

Raichle ME, Gusnard DA. (2002). Appraising the brain's energy budget. *Proceedings of the National Academy of Sciences of the United States of America.* 99(16), 10237–10239.

Ramirez A, Old W, Selwood DL, Liu X. (2022). Cannabidiol activates PINK1-Parkin-dependent mitophagy and mitochondrial-derived vesicles. *Eur. J. Cell Biol.* 101, 151185.

Ren YZ, Zhang BZ, Zhao XJ, Zhang XY. (2020). Resolvin D1 ameliorates cognitive impairment following traumatic brain injury via protecting astrocytic mitochondria. *Journal of Neurochemistry.* 154(5), 530–546.

Rodolfo C, Campello S, Cecconi F. (2017). Mitophagy in neurodegenerative diseases. *Neurochem. Int.* 117, 156–166.

Rosdah AA, Holien JK, Delbridge LM, Dusting GJ, Lim SY. (2016). Mitochondrial fission – a drug target for cytoprotection or cytodestruction? *Pharmacol Res Perspect.* 4(3), e00235.

Ryu D, Mouchiroud L, Andreux PA, Katsyuba E, Moullan N, Nicolet-Dit-Félix AA, Williams EG, Jha P, Lo Sasso G, Huzard D, Aebischer P, Sandi C, Rinsch C, Auwerx J. (2016). Urolithin A induces mitophagy and prolongs lifespan in C. elegans and increases muscle function in rodents. *Nat Med.* 22 (8), 879–888.

Saito T, Sadoshima J. (2015). Molecular mechanisms of mitochondrial autophagy/mitophagy in the heart. *Circulation Research.* 116(8), 1477–1490.

Salminen A, Kaarniranta K, Kauppinen A. (2017). Regulation of longevity by FGF21: Interaction between energy metabolism and stress responses. *Ageing Res Rev.* 37, 79–93.

Saraswat K, Rizvi SI. (2017). Novel strategies for anti-aging drug discovery. *Expert Opin Drug Discov.* 12 (9), 955–966.

Sawaki D, Czibik G, Pini M, Ternacle J, Suffee N, Mercedes R, Marcelin G, Surenaud M, Marcos E, Gual P, Clément K, Hue S, Adnot S, Hatem SN, Tsuchimochi I, Yoshimitsu T, Hénégar C, Derumeaux G (2018). Visceral Adipose Tissue Drives Cardiac Aging Through Modulation of Fibroblast Senescence by Osteopontin Production. *Circulation.* 138(8), 809–822.

Schwarz C, Stekovic S, Wirth M, Benson G, Royer P, Sigrist SJ, Pieber T, Dammbrueck C, Magnes C, Eisenberg T, Pendl T, Bohlken J, Köbe T, Madeo F, Flöel A. (2018). Safety and tolerability of spermidine supplementation in mice and older adults with subjective cognitive decline. *Aging (Albany NY).* 10 (1), 19–33.

Seals DR, Brunt VE, Rossman MJ. (2018). Strategies for Optimal Cardiovascular Aging. *Am J Physiol Heart Circ Physiol.* 315(2), H183–H188.

Sekine S, Wang C, Sideris DP, Bunker E, Zhang Z, Youle RJ. (2019). Reciprocal roles of Tom7 and OMA1 during mitochondrial import and activation of PINK1. *Mol. Cell.* 73, 1028–1043.e1025.

Shi Y, Williamson G. (2016). Quercetin lowers plasma uric acid in pre-hyperuricaemic males: a randomised, double-blinded, placebo-controlled, cross-over trial. *Br J Nutr.* 115, 800–806.

Shinmura K, Tamaki K, Sano M, Murata M, Yamakawa H, Ishida H, Fukuda K. (2011). Impact of long-term caloric restriction on cardiac senescence: caloric restriction ameliorates cardiac diastolic dysfunction associated with aging. *J Mol Cell Cardiol.* 50 (1), 117–127.

Shirakabe A, Ikeda Y, Sciarretta S, Zablocki DK, Sadoshima J. (2016). Aging and Autophagy in the Heart. *Circ Res.* 118 (10), 1563–1576.

Shirihai OS, Song MS, Dorn GW. (2015). How mitochondrial dynamism orchestrates mitophagy. *Circ. Res.* 116, 1835-1849.

Shivaprasad C, Gautham K, Ramdas B, Gopaldatta KS, Nishchitha K. (2020). Metformin Usage Index and assessment of vitamin B12 deficiency among metformin and non-metformin users with type 2 diabetes mellitus. *Acta Diabetol.* 57, 1073–1080.

Sinha RA, Yen PM. (2016). Thyroid hormone-mediated autophagy and mitochondrial turnover in NAFLD. *Cell Biosci.* 6.

Smith BK, Marcinko K, Desjardins EM, Lally JS, Ford RJ, Steinberg GR. (2016). Treatment of nonalcoholic fatty liver disease: role of AMPK. *Am. J. Physiol. Endocrinol. Metab.* 311, E730-E740.

Sumida M, Doi K, Ogasawara E, Yamashita T, Hamasaki Y, Kariya T, Takimoto E, Yahagi N, Nangaku M, Noiri E. (2015). Regulation of mitochondrial dynamics by dynamin-related protein-1 in acute cardiorenal syndrome. *J Am Soc Nephrol.* 26(10), 2378–2387.

Sunata K, Kabata H, Kuno T, Takagi H, So M, Masaki K, Fukunaga K. (2022). The effect of statins for asthma. A systematic review and meta-analysis. *J. Asthma.* 59, 801–810.

Suzuki T, Yamaguchi H, Kikusato M, Hashizume O, Nagatoishi S, Matsuo A, et al. (2016). Mitochonic acid 5 binds mitochondria and ameliorates renal tubular and cardiac myocyte damage. *J Am Soc Nephrol.* 27(7), 1925–1932.

Suzuki T, Yamaguchi H, Kikusato M, Matsuhashi T, Matsuo A, Sato T, Oba Y, Watanabe S, Minaki D, Saigusa D, Shimbo H, Mori N, Mishima E, Shima H, Akiyama Y, Takeuchi Y, Yuri A, Kikuchi K, Toyohara T, Suzuki C, Kohzuki M, Anzai J, Mano N, Kure S, Yanagisawa T, Tomioka Y, Toyomizu M, Ito S, Osaka H, Hayashi K, Abe T. (2015). Mitochonic Acid 5 (MA-5), a derivative of the plant hormone Indole-3-acetic acid, improves survival of fibroblasts from patients with mitochondrial diseases. *Tohoku J Exp Med.* 236(3), 225–232.

Swerdlow NS, Wilkins HM. (2020). Mitophagy and the brain. *International Journal of Molecular Sciences.* 21(24).

Szeto HH, Birk AV. (2014). Serendipity and the discovery of novel compounds that restore mitochondrial plasticity. *Clin Pharmacol Ther.* 96(6), 672–683.

Szeto HH. (2014). First-in-class cardiolipin-protective compound as a therapeutic agent to restore mitochondrial bioenergetics. *Br J Pharmacol.* 171(8), 2029–2050.

Tabara LC, Poveda J, Martin-Cleary C, Selgas R, Ortiz A, Sanchez-Nino MD. (2014). Mitochondria-targeted therapies for acute kidney injury. *Expert Rev Mol Med.* 16, e13.

Tang WX, Wu WH, Qiu HY, Bo H, Huang SM. (2013). Amelioration of rhabdomyolysis-induced renal mitochondrial injury and apoptosis through suppression of Drp-1 translocation. *J Nephrol.* 26(6), 1073–1082.

Tran I, Ji C, Ni I, Min T, Tang D, Vij N. (2015). Role of cigarette smoke-induced aggresome formation in chronic obstructive pulmonary disease-emphysema pathogenesis. *Am. J. Respir. Cell Mol. Biol.* 53, 159–173.

Tran MT, Zsengeller ZK, Berg AH, Khankin EV, Bhasin MK, Kim W, Clish CB, Stillman IE, Karumanchi SA, Rhee EP, Parikh SM. (2016). PGC1alpha drives NAD biosynthesis linking oxidative metabolism to renal protection. *Nature.* 531(7595), 528–532.

Trotta AP, Chipuk JE. (2017). Mitochondrial dynamics as regulators of cancer biology. *Cell Mol. Life Sci.* 74, 1999-2017.

Turdi S, Fan X, Li J, Zhao J, Huff AF, Du M, Ren J. (2010). AMP-activated protein kinase deficiency exacerbates aging-induced myocardial contractile dysfunction. *Aging Cell*. 9 (4), 592–606.

Tzoulaki I, Molokhia M, Curcin V, Little MP, Millett CJ, Ng A, Hughes RI, Khunti K, Wilkins MR, Majeed A, Elliott P.(2009). Risk of cardiovascular disease and all cause mortality among patients with type 2 diabetes prescribed oral antidiabetes drugs: retrospective cohort study using UK general practice research database. *BMJ*. 339, b4731.

Vazquez F, Lim JH, Chim H, Bhalla K, Girnun G, Pierce K, Clish CB, Granter SR, Widlund HR, Spiegelman BM, Puigserver P (2013). PGC1 alpha expression defines a subset of human melanoma tumors with increased mitochondrial capacity and resistance to oxidative stress. *Cancer Cell*. 23, 287-301.

Villa E, Proics E, Rubio-Patino C, Obba S, Zunino B, Bossowski JP, Rozier RM, Chiche J, Mondragón L, Riley JS, Marchetti S, Verhoeyen E, Tait SWG, Ricci JE. (2017). Parkin-independent mitophagy controls chemotherapeutic response in cancer cells. *Cell Rep*. 20(12), 2846–2859.

Wall CE, Rose CM, Adrian M, Zeng YJ, Kirkpatrick DS, Bingol B. (2019). PPEF2 opposes PINK1-mediated mitochondrial quality control by dephosphorylating ubiquitin. *Cell Rep*. 29, 3280–3292.e3287.

Wang CW, Klionsky DJ. (2003). The molecular mechanism of autophagy. *Mol. Med*. 9, 65-76.

Wang H, Guan Y, Karamercan MA, Ye L, Bhatti T, Becker LB, Baur JA, Sims CA. (2015). Resveratrol rescues kidney mitochondrial function following hemorrhagic shock. *Shock*. 44(2), 173–180.

Wang X, Ma H, Sun J, et al. (2022). Mitochondrial ferritin deficiency promotes osteoblastic ferroptosis via mitophagy in type 2 diabetic osteoporosis. *Biological Trace Element Research*. 200(1), 298–307.

Wang X, Zhang JQ, Xiu CK, Yang J, Fang JY, Lei Y. (2020). Ginseng-Sanqi-Chuanxiong (GSC) Extracts Ameliorate Diabetes-Induced Endothelial Cell Senescence through Regulating Mitophagy via the AMPK Pathway. *Oxid Med Cell Longev*. 2020, 7151946.

Wang Y, Liu N, Lu B. (2019). Mechanisms and roles of mitophagy in neurodegenerative diseases. *CNS Neuroscience & Therapeutics*. 25(7), 859–875.

Wang Y, Serricchio M, Jauregui M, Shanbhag R, Stoltz T, Di Paolo CT, Kim PK, McQuibban GA. (2015). Deubiquitinating enzymes regulate PARK2-mediated mitophagy. *Autophagy*. 11, 595–606.

Wang Z, Bao H, Ge Y, Zhuang S, Peng A, Gong R. (2015). Pharmacological targeting of GSK3beta confers protection against podocytopathy and proteinuria by desensitizing mitochondrial permeability transition. *Br J Pharmacol*. 172(3), 895–909.

Wang Z, Ge Y, Bao H, Dworkin L, Peng A, Gong R. (2013). Redox-sensitive glycogen synthase kinase 3beta-directed control of mitochondrial permeability transition: rheostatic regulation of acute kidney injury. *Free Radic Biol Med*. 65, 849–858.

Waris G, Ahsan H. (2006). Reactive oxygen species: role in the development of cancer and various chronic conditions. *J Carcinog*. 5, 14.

Wauer T, Swatek KN, Wagstaff JL, Gladkova C, Pruneda JN, Michel MA, Gersch M, Johnson CM, Freund SM, Komander D. (2015). Ubiquitin Ser65 phosphorylation affects ubiquitin structure, chain assembly and hydrolysis. *EMBO J*. 34, 307–325.

Weichhart T. (2018). mTOR as Regulator of Lifespan, Aging, and Cellular Senescence: A Mini-Review. *Gerontology*. 64 (2), 127–134.

Wong YC, Holzbaur ELF. (2014). The regulation of autophagosome dynamics by huntingtin and HAP1 is disrupted by expression of mutant huntingtin, leading to defective cargo degradation. *J. Neurosci*. 34, 1293-1305.

Wu J, Zhong W, Zhang H, Yin Y. (2020). Mammalian target of rapamycin signaling enhances ovalbumin-induced neutrophilic airway inflammation by promoting Th17 cell polarization in murine noneosinophilic asthma model. *Pediatr. Allergy Immunol. Pulmonol*. 33, 25–32.

Wu Q, Gao C, Wang H, Zhang X, Li Q, Gu Z, Shi X, Cui Y, Wang T, Chen X, Wang X, Luo C, Tao L. (2018). Mdivi-1 alleviates blood-brain barrier disruption and cell death in experimental traumatic brain injury by mitigating autophagy dysfunction and mitophagy activation. *The International Journal of Biochemistry & Cell Biology*. 94, 44–55.

Wu TD, Keet CA, Fawzy A, Segal JB, Brigham EP, McCormack MC. (2019). Association of metformin initiation and risk of asthma exacerbation. A claims-based cohort study. *Ann Am. Thorac. Soc.* 16, 1527–1533.

Wu WX, Lin CX, Wu K, Jiang L, Wang XJ, Li W, Zhuang HX, Zhang XL, Chen H, Li SP, et al. (2016). FUNDC1 regulates mitochondrial dynamics at the ER-mitochondrial contact site under hypoxic conditions. *EMBO J.* 35, 1368-1384.

Xufeng Cen, Xiaoyan Xu &Hongguang Xia (2021). Targeting MCL1 to induce mitophagy is a potential therapeutic strategy for Alzheimer disease. *Autophagy.* 17:3, 818-819.

Yamano K, Fogel AI, Wang C, van der Bliek AM, Youle RJ. (2014). Mitochondrial Rab GAPs govern autophagosome biogenesis during mitophagy. *Elife.* 3, e01612.

Yamashita S, Kanki T. (2017). How autophagy eats large mitochondria: Autophagosome formation coupled with mitochondrial fragmentation. *Autophagy.* 13, 980-981.

Yan C, Luo L, Guo C, Goto S, Urata Y, Shao JH, Li TS. (2017). Doxorubicin-induced mitophagy contributes to drug resistance in cancer stem cells from HCT8 human colorectal cancer cells. *Cancer Lett.* 388, 34–42.

Yang LY, Greig NH, Tweedie D, Jung YJ, Chiang YH, Hoffer BJ, Miller JP, Chang KH, Wang JY. (2020). The p53 inactivators pifithrin-μ and pifithrin-α mitigate TBI-induced neuronal damage through regulation of oxidative stress, neuroinflammation, autophagy and mitophagy. *Experimental Neurology.* 324, 113135.

Yang S, Long LH, Li D, Zhang JK, Jin S, Wang F, Chen JG. (2015). beta-Guanidinopropionic acid extends the lifespan of Drosophila melanogaster via an AMP-activated protein kinase-dependent increase in autophagy. *Aging Cell.* 14 (6), 1024–1033.

Yao N, Wang C, Hu N, Li Y, Liu M, Lei Y, Chen M, Chen L, Chen C, Lan P, Chen W, Chen Z, Fu D, Ye W, Zhang D. (2019). Inhibition of PINK1/Parkin-dependent mitophagy sensitizes multidrug-resistant cancer cells to B5G1, a new betulinic acid analog. *Cell Death Dis.* 10(3), 232.

Ye X, Sun X, Starovoytov V, Cai Q. (2015). Parkin-mediated mitophagy in mutant hAPP neurons and Alzheimer's disease patient brains. *Hum. Mol. Genet.* 24, 2938-2951.

Youle RJ and Narendra DP. (2011). Mechanisms of mitophagy. *Nature Reviews Molecular Cell Biology.* 12(1), 9–14.

Youle RJ, van der Bliek AM. (2012). Mitochondrial fission, fusion, and stress. *Science.* 337(6098), 1062–1065.

Yu Y, Zhao Y, Teng F, Li J, Guan Y, Xu J, Lv X, Guan F, Zhang M, Chen L. (2018). Berberine Improves Cognitive Deficiency and Muscular Dysfunction via Activation of the AMPK/SIRT1/PGC-1a Pathway in Skeletal Muscle from Naturally Aging Rats. *J Nutr Health Aging.* 22 (6), 710–717.

Zakarya R, Sapkota A, Chan YL, Shah J, Saad S, Bottle SE, Oliver BG, Gorrie CA, Chen H. (2020). Nitroxides affect neurological deficits and lesion size induced by a rat model of traumatic brain injury. *Nitric Oxide-Biology and Chemistry.* 97, 57–65.

Zeng XH, Zeng XJ, Li YY. (2003). Efficacy and safety of berberine for congestive heart failure secondary to ischemic or idiopathic dilated cardiomyopathy. *Am J Cardiol.* 92 (2), 173–176.

Zhang C, Liu Z, Bunker E, Ramirez A, Lee S, Peng Y, Tan AC, Eckhardt SG, Chapnick DA, Liu X. (2017). Sorafenib targets the mitochondrial electron transport chain complexes and ATP synthase to activate the PINK1-Parkin pathway and modulate cellular drug response. *J Biol Chem.* 292(36), 15105–15120.

Zhang D, Contu R, Latronico MV, Zhang J, Rizzi R, Catalucci D, Miyamoto S, Huang K, Ceci M, Gu Y, Dalton ND, Peterson KL, Guan KL, Brown JH, Chen J, Sonenberg N, Condorelli G. (2010). MTORC1 regulates cardiac function and myocyte survival through 4E-BP1 inhibition in mice. *J Clin Invest.* 120 (8), 2805–2816.

Zhang Q, Song W, Zhao B, Xie J, Sun Q, Shi X, Yan B, Tian G, Liang X. (2021). Quercetin Attenuates Diabetic Peripheral Neuropathy by Correcting Mitochondrial Abnormality via Activation of AMPK/PGC-1α Pathway in vivo and in vitro. *Front Neurosci.* 15, 636172.

Zhang X, Chen Y, Jenkins LW, Kochanek PM, Clark RS. (2005). Bench-to-bedside review: apoptosis/programmed cell death triggered by traumatic brain injury. *Critical Care.* 9(1), 66–75.

Zhang Y, Mi SL, Hu N, Doser TA, Sun A, Ge J, Ren J. (2014). Mitochondrial aldehyde dehydrogenase 2 accentuates aging-induced cardiac remodeling and contractile dysfunction: role of AMPK, Sirt1, and mitochondrial function. *Free Radic Biol Med.* 71, 208–220.

Zhang Y, Sowers JR, Ren J. (2018). Targeting autophagy in obesity: from pathophysiology to management. *Nat Rev Endocrinol*. 14 (6), 356–376.

Zhang Y, Wang X, Zhang H, Tang H, Hu H, Wang S, Wong VKW, Li Y, Deng J. (2021). Autophagy modulators from Chinese herbal medicines: Mechanisms and therapeutic potentials for asthma. *Front. Pharmacol.* 12, 710679.

Zhang Z, Li X, Li F, An L. (2016). Berberine alleviates postoperative cognitive dysfunction by suppressing neuroinflammation in aged mice. *Int Immunopharmacol.* 38, 426–33.

Zheng S, Jian D, Gan H, Wang L, Zhao J, Zhai X. (2021). FUNDC1 inhibits NLRP3-mediated inflammation after intracerebral hemorrhage by promoting mitophagy in mice. *Neuroscience Letters*. 756.

Zheng YH, Tian C, Meng Y, Qin YW, Du YH, Du J, Li HH. (2012). Osteopontin stimulates autophagy via integrin/CD44 and p38 MAPK signaling pathways in vascular smooth muscle cells. *J Cell Physiol*. 227 (1), 127–135.

Zhong Z, Umemura A, Sanchez-Lopez E, Liang S, Shalapour S, Wong J, He F, Boassa D, Perkins G, Ali SR, McGeough MD, Ellisman MH, Seki E, Gustafsson AB, Hoffman HM, Diaz-Meco MT, Moscat J, Karin M. (2016). NFkappaB restricts inflammasome activation via elimination of damaged mitochondria. *Cell*. 164, 896-910.

Zhou J, Li G, Zheng Y, Shen HM, Hu X, Ming QL, Huang C, Li P, Gao N (2015). A novel autophagy/mitophagy inhibitor liensinine sensitizes breast cancer cells to chemotherapy through DNM1Lmediated mitochondrial fission. *Autophagy*. 11(8), 1259–1279.

Zhu X, Yang L, Xu F, Lin L, Zheng G. (2017). Combination therapy with catechins and caffeine inhibits fat accumulation in 3T3-L1 cells. *Exp Ther Med*. 13, 688–694.

Chapter 15

Mitophagy and Stem Cells

Annu Devi[1]
Madhu Rani[2]
Shashi Prakash Singh[3]
Rashmi Kumari[2]
Asmita Bharti[4]
Moshahid Alam Rizvi[2]
and Anil Kumar[5,*]

[1]Special Centre for Molecular Medicine, Jawaharlal Nehru University, New Delhi, India
[2]Genome Biology Laboratory, Department of Biosciences, Jamia Millia Islamia, New Delhi, India
[3]Pharmacology and Therapeutics, Centre for Genetics and Pharmacology, Roswell Park Comprehensive Cancer Care, Buffalo, NY, USA
[4]Department of Biochemistry, Indira Gandhi National Open University, Department of Biochemistry, New Delhi, India
[5]Directorate of Education, Rani Lakshmi Bai Central Agricultural University, Jhansi, Uttar Pradesh India

Abstract

Mitophagy is a chief autophagic process in which autophagic lysosomes selectively remove damaged mitochondria. A decline in the mitophagy phenomenon can trigger the process of aging. Mitophagy is associated with longevity and health and is critical in protecting stem cells against metabolic stress damage. Mitophagy reduces the metabolic capacity of stem cells via the elimination of active mitochondria; therefore, it is crucial for maintaining the regenerative power of old stem cells. Further, mitophagy also influences the self-renewal potential of stem cells along with their regeneration potential. Mitophagy is a crucial survival approach to overcome nutrient deficiency and regulate mitochondrial function and integrity that influences proliferation, differentiation potential, and survival of stem cells under normal conditions. Several shreds of evidence show the involvement of mitophagy in the health and longevity of the human body, but the participation of mitophagy in stem cell development is less explored and not entirely understood. In this book chapter, we have described the potential significance of mitophagy in maintaining the various functions of several types of stem cells. We have further addressed the role of mitophagy in stem

* Corresponding Author's Email: anilkumar.mbge@gmail.com.

In: Mitophagy and Its Role in Health and Disease
Editors: Shalini Mani and Samiksha Wasnik
ISBN: 979-8-89113-470-6

cell-related diseases. Our book chapter might provide new insights into mitophagy in stem cell biology.

Keywords: mitophagy, PINK1-parkin signaling pathway, stem cell

Introduction

Stem cells are unique and undifferentiated cells that can give rise to various cell types in the body. They have a self-renewal property and therefore can maintain their population. These cells are found in the embryo and the adult organism (Zakrzewski et al., 2019).

Stem cells undergo differentiation through various stages to form multiple body cells; the differentiation capacity is reduced with each step. Based on their potential to differentiate, these can be classified as totipotent, pluripotent, multipotent, oligopotent, and unipotent. Totipotent stem cells can form any cell of the organism and embryonic and extra-embryonic structures like the zygote. Pluripotent stem cells (PSCs) can form any cell of the three germ layers but not extra-embryonic structures. These include embryonic stem cells (ESCs) and induced pluripotent stem cells (iPSCs). Multipotent stem cells can differentiate into multiple lineages but cannot form any cell of the three germ layers. These include adult stem cells such as hematopoietic stem cells (HSCs), Mesenchymal stem cells (MSCs), and neural stem cells (NSCs). Oligopotent stem cells can divide into limited cell types, whereas unipotent cells have very narrow differentiation potential (Foster et al., 2002). Therefore, stem cells are crucial for the development of an organism, and in the adult organism, these cells are significant for replacing the dead or injured cells to maintain physiological homeostasis. Also, stem cells have become an attractive tool for regenerative and tissue engineering studies (Bianco, 2001). Therefore, it is crucial to understand the mechanisms crucial for stem cell homeostasis and maintenance of differentiation potential.

Recently, mitochondria, an organelle involved in regulating the bioenergetics and survival of the cell, has emerged to be crucial in controlling multiple facets of stem cell biology (Cairns et al., 2020b). Although the number of mitochondria in various stem cells is reported to be low, and they are majorly dependent on glycolysis, mitochondria have been linked to the maintenance of homeostasis and properties such as differentiation and self-renewal of stem cells (Lisowski et al., 2018), (H. Zhang et al., 2018). Therefore, the regulation of mitochondrial structure and function is critical in stem cells. There are various mechanisms by which the structure and function of mitochondria are regulated, such as fission, fusion, and mitophagy. Mitophagy is increasingly associated with multiple stem cell properties (Lin et al., 2021), (Cairns et al., 2020a).

Mitophagy Signaling Pathways

Mitophagy is the specific form of autophagy for removing mitochondria through the formation of autophagosomes and subsequent degradation by fusion with lysosomes (Killackey et al., 2020). Mitophagy is crucial for eliminating non-functional or damaged mitochondria or maintaining the number of mitochondria for metabolic purposes. It is also critical because defective mitochondria produce large amounts of reactive oxygen species, creating a state of oxidative stress that could be detrimental to the cell. There are different pathways for the activation of mitophagy that involve various proteins (Figure 1).

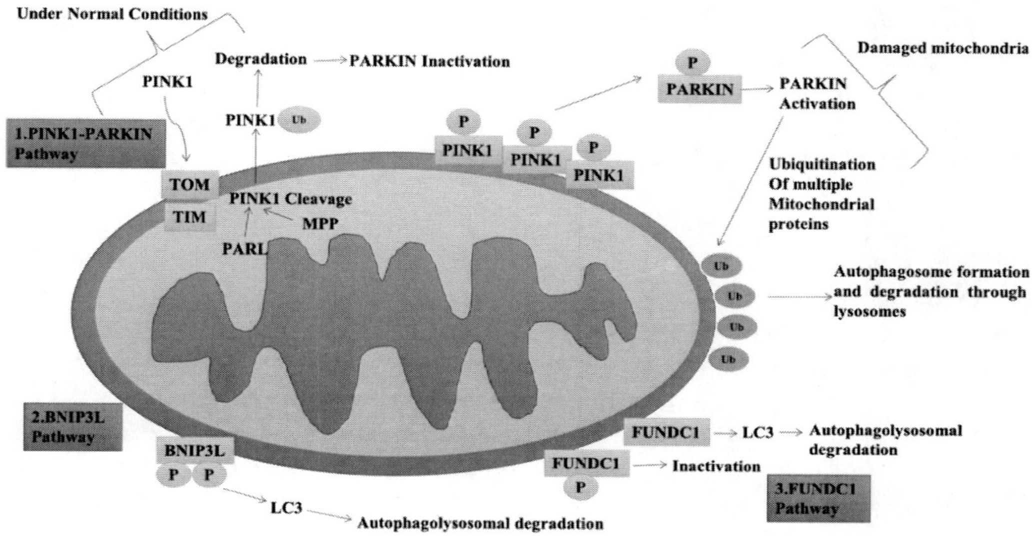

Figure 1. Three different pathways of mitophagy activation PINK1-PARKIN pathway, the BNIP3L pathway, and the FUNDC1 pathway.

PINK1-Parkin Pathway

The major pathway of mitophagy in response to various stressors includes the serine/threonine kinase PINK1 (PTEN-induced putative kinase 1) and the E3 ubiquitin ligase Parkin. PINK1 is continuously imported and subsequently degraded in mitochondria under normal conditions. The transmembrane potential of the mitochondria causes the import of PINK1 into the inner mitochondrial membrane (IMM) from the cytosol through two translocases (TOM and TIM) in the inner and outer mitochondrial membranes. In the IMM, PINK1 is subjected to two cleavages by matrix processing peptidases (MPP) that remove the N-terminal mitochondrial targeting sequences (MTS) and the rhomboid-like serine protease PARL that cleaves in the hydrophobic domain in the IMM. The second cleavage by PARL exposes amino acid F104 at the N terminal, due to which the PINK1 protein is released from the mitochondria and is degraded by the ubiquitin proteases in the cytosol. As PINK1 is degraded, the PARKIN remains inactivated in the cytosol, and consequently, the mitophagy pathway is not activated in healthy mitochondria (Nguyen et al., 2016).

On the other hand, when any damage occurs to the mitochondria, the PINK1 protein is not translocated to the IMM and is retained in the OMM, its degradation by PARL protein is inhibited. As a result, the PINK1 protein accumulates on the outer mitochondrial membrane, where it homodimerizes and undergoes auto-phosphorylation, which is necessary for its activation. An E3 ubiquitin ligase Parkin is then recruited and activated by PINK1 through phosphorylation. Under normal circumstances, Parkin is kept inactivated in an auto-regulatory manner. Parkin's N terminal Ubiquitin-like domain (UBL) negatively regulates its E3 ubiquitin ligase activity. Phosphorylation by PINK1 at S65 residue in the UBL domain results in its activation (Eiyama & Okamoto, 2015). Upon activation, multiple proteins on the outer mitochondrial membrane surface get ubiquitinated by Parkin. Consequently, the damaged mitochondria are engulfed by autophagosomes, followed by lysosome degradation (Lin et al., 2021).

BNIP3L/NIX Signaling Pathway

Although the PINK1/PARKIN pathway is critical in responding to different stress conditions, cells employ other programmed mitophagy pathways such as the BNIP3L/NIX pathway crucial for developing and differentiating various cells. Mitochondrial elimination through this pathway is crucial in developing erythrocytes, neurogenesis of retinal ganglion cells, and differentiation to cardiomyocytes (Marinković & Novak, 2021). BNIP3L is a member of the BCL2 family of pro-apoptotic proteins. It acts as a receptor for mitophagy on the outer membrane of the mitochondria. There are different mechanisms by which BNIP3L can induce autophagy. The mitophagy pathway activated by BNIP3L is not dependent on ubiquitination. In the developing reticulocytes, BNIP3L was shown to directly recruit the autophagosomes through the LIR (LC3-interacting region) domain located at the N terminal end. Although the detailed mechanisms by which BNIP3L recruits the autophagosomes are unclear, the LIR domain interacts with LC3/GABARAP proteins (Li et al., 2022). The phosphorylation of BNIP3L at Ser34 and Ser35 is crucial for its interaction with the LC3/GABARAP proteins (Rogov et al., 2017). Additionally, the homodimerization of BNIP3L is shown to recruit LC3 proteins more efficiently (Marinković et al., 2020).

FUNDC1 Signaling Pathway

Another pathway for mitophagy that includes a receptor on the outer membrane of mitochondria is the protein FUNDC1 (Fun14 domain-containing protein 1). Like BNIP3L/Nix, FUNDC1 can induce mitophagy by directly interacting with LC3 proteins (Onishi et al., 2021). FUNDC1 induces mitophagy under hypoxic conditions. Further, it was revealed that FUNDC1 consists of an LC3-binding domain through which it interacts with LC3. Contrary to BNIP3L, phosphorylation negatively regulates the activity of FUDC1, and interaction with LC3 is induced by dephosphorylation, thus activating mitophagy (L. Liu et al., 2012). The dephosphorylation is done by a phosphatase PGAM5 located in the mitochondria at Serine 13. Inactivation of FUNDC1 through phosphorylation is achieved by CK2 (Chen et al., 2014). Reports have also shown the role of phosphorylation in activating FUNDC1. It has been shown that under hypoxic conditions, a kinase ULK1 translocates to the mitochondria and phosphorylates FUNDC1 at Serine 17, thus enhancing its binding efficiency with LC3 and activating mitophagy (Wu et al., 2014). Thus, phosphorylation at different sites on FUNDC1 results in differential receptor regulation and mitophagy activation. In the following sections, we have discussed the role of mitophagy in various aspects of different stem cells (Figure 2, Table 1).

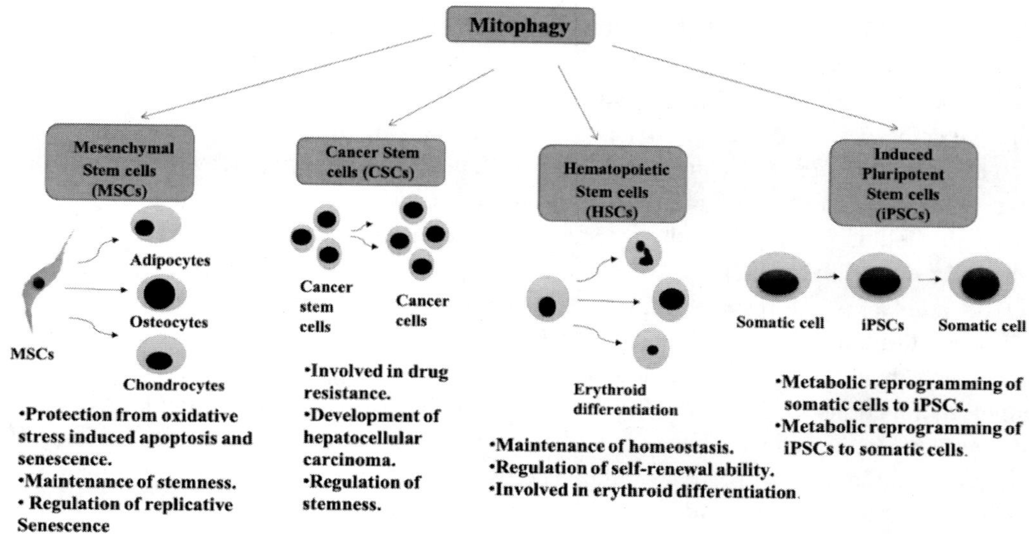

Figure 2. Role of mitophagy in various properties and functions of different types of stem cells.

Table 1. The functions of mitophagy in different types of stem cells

Type of stem cell	Function of mitophagy	References
Mesenchymal stem cell	Protection from oxidative stress	(Fan et al., 2019)
Mesenchymal stem cell	Maintenance of the stemness property	(Feng et al., 2021)
Mesenchymal stem cell	nvolved in replicative senescence	(Lee et al., 2020)
Colorectal Cancer stem cell	Involved in doxorubicin resistance	(Yan et al., 2017a)
Hepatic Cancer stem cell	Involved in the development of hepatocellular carcinoma	(K. Liu et al., 2017a)
Hepatic Cancer stem cell	Involved in hepatitis B-induced hepatocellular carcinoma	(Cells et al., 2020)
Pancreatic Cancer stem cell	Involved in the maintenance of stemness	(Alcalá et al., 2020)
Leukemia stem cell	Involved in self-renewal property	(Pei et al., 2018b)
Hematopoietic stem cell	Involved in maintaining homeostasis	(Murakami et al., 2021)
Hematopoietic stem cell	Involved in self-renewal property	(Ito et al., 2016)
Hematopoietic stem cell	Involved in erythroid differentiation	(Kuhikar et al., 2020)
Induced Pluripotent stem cell	Involved in metabolic reprogramming of iPSCs into vascular endothelial cells	(Krantz et al., 2021)
Induced Pluripotent stem cell	Involved in metabolic reprogramming of somatic cells into iPSCs	(Xiang et al., 2017), (Vazquez-Martin et al., 2016)

The Role of Mitophagy in Various Stem Cells

Various studies have been conducted on the effect of mitophagy on different types of stem cells, which are discussed below.

The Effect of Mitophagy on Mesenchymal Stem Cells (MSCs)

Mesenchymal stem cells (MSCs) are adult stem cells that can self-renew and differentiate into multiple cell types. MSCs are mostly isolated from the bone marrow, but other tissue sources such as adipose tissue, dental pulp, umbilical cord, umbilical cord blood, and peripheral blood have also been reported for their isolation (Andrzejewska et al., 2019). The International Society of Cellular Therapy has characterized MSCs based on specific properties such as plastic adherence during in vitro culture, the presence of surface molecules like CD73, CD90, and CD105, and lack of expression of CD34, CD45, CD14, CD11b, CD19, and MHC—II and the ability to differentiate into multiple mesodermal lineages like adipocytes, osteoblasts, and chondroblasts (Lv et al., 2014). MSCs are critical in various physiological processes, showing numerous functions such as tissue repair, regeneration, and immunomodulation. MSCs have different crucial functions; therefore, they have been used in therapies for diseases like myocardial infarction, acute renal failure, hepatic failure, lung injury, diabetes, rheumatoid arthritis, and many bacterial and viral infections (Uccelli et al., 2008). Therefore, understanding the mechanisms underlying the self-renewal, multi-differentiation, and senescence of MSCs is urgently required to improve the therapeutic efficacy of these cells. Various mitochondrial functions and mitophagy are reported to be critical in determining many of these properties.

One of the crucial functions of MSCs is tissue repair and regeneration in the injured sites. However, the oxidative stress levels in the injured tissues could lead to the damage and death of these MSCs, thereby stalling the repair process. Mitophagy is shown to protect bone-marrow MSCs (BMSCs) from oxidative stress. Fan et al. showed that the induction of apoptosis is through the mitochondrial pathway in H2O2-induced oxidative stress models of BMSCs. The induced apoptosis was evident through the decrease in the mitochondrial membrane potential, increased expression of cytosolic Cytochrome C and pro-apoptotic Bcl2, along with low levels of anti-apoptotic protein Bax. During the early stages of oxidative stress, mitophagy was induced, as shown by high mitophagy markers LC3 and p62. Mitophagy was initiated through the activation and phosphorylation of JNK. Further, it was shown that mitophagy could protect these cells from apoptosis by removing the damaged mitochondria; however, prolonged oxidative stress reduces mitophagy leading to apoptosis of these cells (Fan et al., 2019).

Mitophagy is also vital for maintaining the stemness of MSCs. Feng et al. showed that mitochondrial fission is crucial for maintaining the stemness of MSCs, which is mediated through mitophagy induction. It was shown that when treating MSCs with the mitochondrial fission inducer tryphostin A9 (TA9), there are increased levels of fission in the cells, accompanied by enhanced levels of mitophagy. Mitochondrial fission followed by mitophagy was shown to be critical in removing damaged mitochondria and retaining the stemness by MSCs. When carbonyl cyanide 4-(trifluoromethoxy) phenylhydrazone (FCCP) was used to induce mitophagy without any effect on the fission process selectively, the stemness of MSCs could still be maintained, which was shown by upregulated levels of Oct4 and Sox2 genes. On the contrary, when mitophagy was inhibited by silencing the essential mitophagy genes Pink1 or Parkinalong with the induction of fission, the levels of Oct4 and sox2 (stemness marker) reduced. This indicates the vital role of mitophagy in the fission-induced stemness of MSCs (Feng et al., 2021).

Studies have also associated MSC's replicative senescence with a reduction in mitophagy. It is reported that mitophagy reduction in senescent MSCs is due to the inhibition of the process of mitofission. As a result, mitochondrial dysfunction is enhanced in these cells. Lee et al.

showed that when MSCs are treated with melatonin, it upregulates mitophagy and mitochondrial function, thus rescuing these cells from replicative senescence. It was shown that Melatonin increases the expression of heat shock 70 KDa protein 1L (HSPA1L), which interacts with cellular prion protein (PrP^C), recruiting it to the mitochondrial membrane. The protein complex then binds to a mitochondrial membrane protein COX4IA, enhancing the mitochondrial membrane potential, the anti-oxidant activity, and mitophagy (Lee et al., 2020).

Another report highlighted mitophagy's crucial role in protecting MSCs from oxidative stress-induced apoptosis and senescence. It was shown that reduction in mitophagy due to oxidative stress induces apoptosis and senescence in MSCs. Zhang et al. showed that transplantation of BMSCst to treat early steroid-induced osteonecrosis of the femoral head (ONFH) is highly effective, but due to the oxidative stress environment in the implanted tissues, MSCs undergo apoptosis. It was revealed that oxidative stress induces the up-regulation of p53, inhibiting the Parkin-mediated mitophagy and accumulating damaged mitochondria. High levels of ROS and increased DNA damage also result in the apoptosis of these cells. However, when Parkin was over-expressed in BMSCs using a lentiviral vector with simultaneousp53 down-regulation using shRNA, the apoptosis and senescence in these cells decreased, indicating the significance of mitophagy (F. Zhang et al., 2020a).

As mitophagy is crucial for the various properties of MSCs, understanding the underlying mechanisms by which mitophagy is regulated in these cells is vital for developing therapeutic interventions. Tsujimoto et al. reported miR-155 as a regulator of mitophagy in MSCs. It was shown that miR-155 was upregulated in aged bone marrow tissues. On transfection of MSCs with miR-155 mimics, a decrease in the proteins associated with mitophagy was observed, such as PINK1, ATG5, and ATG12, along with a reduction in mitochondrial membrane potential. These changes were associated with low levels of mitophagy. Bcl-2-associated athanogene 5 (BAG5), a stabilizer of PINK1, was shown to be the direct target of miR-155. When the levels of BAG5 reduce due to miR-155, it causes the ubiquitination and destabilization of PINK1, a critical protein of mitophagy. Hence, miR-155 disrupts mitophagy in MSCs (Tsujimoto et al., 2020).

The Effect of Mitophagy on Cancer Stem Cells (CSCs)

Cancer stem cells (CSCs) are a small subset of specialized cells within the tumor that have the property of self-renewal and are responsible for the long-term sustenance of the tumor. CSCs resist the various chemotherapy regimens and radiotherapies employed to treat tumors. Therefore, even after complete treatment, CSCs may remain in the body, causing a reoccurrence of the tumor. These may also be responsible for tumor metastasis to other sites (Clevers, 2011). Therefore, understanding the various mechanisms that may contribute to these cells' multiple functions is vital for treating various tumors.

Mitophagy contributes to the drug resistance of cancer stem cells (CSC) in various cancers. In colorectal cancer stem cells, mitophagy was involved in doxorubicin resistance. From the parental cancer cell line HCT8, CSCs were isolated using the markers $CD133^+$/$CD44^+$ and treated with doxorubicin. Cell survival after doxorubicin treatment was greater in CSCs than in parental cells. Also, the apoptosis levels were low in CSCs, as observed by the lower expression of various apoptotic markers such as cleaved PARP, cleaved caspase-3, and caspase-7. Mitophagy was seen to be enhanced in CSCs than in the parental cells as observed by lower levels of mitochondrial membrane protein TOMM20, lower superoxide levels, increased colocalization of mitochondria and autophagosomes, and higher levels of mitophagy regulator protein BNIP3L. Therefore, to check the significance of mitophagy in CSCs, the expression of BNIP3L was reduced using specific siRNA. After silencing BNIP3L in CSCs, the levels of cleaved PARP increased, along with decreased mitochondrial localization with LC3B upon doxorubicin treatment. Thus, CSC sensitivity was enhanced upon mitophagy inhibition (Yan et al., 2017a).

Liu et al. reported a significant role of mitophagy in hepatic CSCs and the development of hepatocellular carcinoma. They showed that on treating HepG2 cells with the mitophagy inhibitor Mdivi-1, the CSC population among HepG2 ($CD133^+$) reduced along with its self-renewal ability. The effects were reversed when the cells were treated with inducer carbonyl cyanide m-chlorophenylhydrazone (CCCP). Mitophagy controls the activities of a crucial tumor-suppressor protein, p53. The association of p53 with the mitochondrial membrane and its subsequent degradation through the CCCP-mediated mitophagy activation was observed. However, upon inhibition of mitophagy by Mdivi-1, p53 gets phosphorylated (pS392) by PTEN-induced putative kinase 1 (PINK) and translocates to the nucleus, where it binds to the promoter of NANOG and blocks the binding with the transcription factors OCT4 and SOX2. The transcription of NANOG is thus inhibited. NANOG is crucial for maintaining the stemness properties of CSCs (K. Liu et al., 2017a).

Chen et al. reported the underlying mechanism of hepatitis B virus-induced hepatocellular carcinoma based on mitophagy. They showed that the hepatitis protein x (HBx) could induce hepatocellular carcinoma (HCC). A mouse subcutaneous xenograft model revealed a rapidly growing tumor with increased volume and weight observed on implanting xenograft cells from HBx-expressing Huh7 and MHCC-97H. The expression level of stemness-related genes such as ABCG2, NANOG, BMI1, KLF4, and OCT4 was higher when compared with the control vector. On analyzing the metabolic status of these cells, there were higher glycolysis rates. These experiments indicated that HBx promotes the stemness of liver CSCs (LCSCs) through the metabolic reprogramming of these cells. Glycolytic reprogramming was also shown by using inhibitors such as STF31 (inhibits glucose transport) and 2-deoxy-D-glucose (2-DG) (inhibits the formation of glucose-6-phosphate from glucose). Upon treatment with these inhibitors, the downregulation of stemness-related genes was observed. However, on further exploration, BNIP3L-dependent mitophagy was found to be responsible for this metabolic reprogramming of LCSC. Upregulated levels of proteins associated with BNIP3L-dependent mitophagy were demonstrated, such as HIF-1alpha, Beclin1, BNIP3L, and LC3B. Mitophagy-dependent metabolic reprogramming of LCSCs by HBx protein to increase stemness was established (Cells et al., 2020).

Alcala *et al*. evaluated the role of mitophagy in pancreatic cancer. It was shown that the improper functioning of mitophagy resulted in reduced stemness of pancreatic CSC. The interferon-stimulated gene 15 (ISG 15) increased expression at RNA and protein levels, and ISGylation of cellular proteins was observed in pancreatic CSCs. Higher levels of ISG15 are also detected in the plasma of pancreatic ductal adenocarcinoma (PDAC) patients. Even a clinical correlation is made between patient survival and ISG15 expression. There was decreased survival in patients expressing higher levels of ISG15. When the ISG15 expression was reduced using CRISPR-mediated gene editing, the stemness of pancreatic cancer stem cells was reduced. This was due to the decreased ISGylation of mitochondrial proteins, which inhibited mitophagy, accumulation of damaged mitochondria, and low oxidative phosphorylation. Thus the importance of mitophagy in maintaining the stemness of pancreatic CSCs was elucidated (Alcalá et al., 2020).

A higher expression of mitochondrial dynamics regulator FIS1 was observed in acute myeloid leukemia stem cells. This was accompanied by a distinct mitochondrial morphology in leukemia stem cells (LSC) compared with non-stem cells. The number and area of mitochondria in LSC were low, and they were not uniformly distributed but confined to a small localized area, unlike in non-stem cells. Furthermore, the number of mitochondria increased after the depletion of FIS1 with shRNA knockdown. This was linked with decreased mitophagy in these FIS1-depleted cells. When the mitochondrial stress inducer Valinomycin was used, mitochondria accumulated in FIS1-depleted cells, indicating a loss of mitophagy activation in these cells. Also, the loss of FIS1 reduced the stemness potential of LSC. It was observed that the inactivation of GSK and cell cycle arrest follows the loss of FIS1. They also demonstrated that AMPK signaling was responsible for FIS1 activation (Pei et al., 2018a).

The Effect of Mitophagy on Hematopoietic Stem Cells (HSCs)

HSCs are the multipotent cells that give rise to all types of blood cells in the body. They are capable of self-renewal and are generally present in hypoxic bone marrow niches.

Several reports have shown the critical role of mitophagy in maintaining the homeostasis of HSC. Murakami et al. showed the vital role of the maintenance of mitochondria through mitophagy in the survival and stress response of HSCs. They showed that in HSC, mitophagy is regulated through an important enzyme O-linked N-acetylglucosamine transferase (OGT) that adds N-acetylglucosamine moiety to serine or threonine amino acids of multiple proteins involved in various pathways such as metabolism and stress response. OGT interacts with a crucial mitophagy protein, PINK1, and regulates mitophagy. When OGT was downregulated through knockout, the levels of Pink1 also got reduced, impairing mitophagy. This resulted in the accumulation of defective mitochondria, thus affecting the maintenance and stress response of HSC (Murakami et al., 2021).

Another study by Jin et al. showed that an ATPase Atad3a, located in the inner membrane of the mitochondria, is involved in the transportation of Pink1 into the mitochondria for subsequent processing. Thus, Atad3a is involved in suppressing mitophagy. They showed that HSCs have higher expression of Atad3a and, upon its knockdown in mice, resulted in lower survival along with abnormal hematopoiesis and a low number of cells in the bone marrow. The number of erythrocytes, leukocytes, and platelets also decreased (Jin et al., 2018).

Ito et al. demonstrated the significant role of mitophagy in regulating the self-renewal ability of HSCs. They identified a subset of the HSC population that was positive for the marker *Tie2* and exhibited minimum heterogeneity. Further, they revealed the critical involvement of fatty acid oxidation pathways and PPAR in controlling the self-renewal ability of HSC. Induction of PPAR-fatty acid oxidation pathways by using a specific agonist upregulates the transcription of Pink1, which is a critical regulator of mitophagy through FOXO3a, increased localization of Tom20 with LAMP1, and Parkin localization with pyruvate dehydrogenase. Furthermore, these findings were proved by using *Ppard* knockout mice (Ito et al., 2016).

Mitophagy is also crucial for the erythroid differentiation of HSCs, as shown in a study by Kuhikar et al. They used TGF-β1 at a low concentration (10ng/ml) in HSCs cultured for differentiation to RBCs. Increased maturation of RBCs was observed as monitored by the loss of the transferrin receptor, increased expression of glycophorin A, and increased enucleation of RBCs. It was observed that TGF-β1 arrests the cells in the G0/G1 phase of the cell cycle by enhancing the expression of CDKI p27. Mitophagy was observed to be upregulated, as evidenced by a loss in mitochondrial mass and low levels of ROS. Also, upregulation was observed in various genes associated with mitophagy, like BNIP3L/NIX (Kuhikar et al., 2020).

The Effect of Mitophagy on Induced Pluripotent Stem Cells (IPSs)

iPSCs are pluripotent stem cells derived from various somatic cells by reprogramming them by introducing transcription factors abundantly in ES cells. Four major transcription factors cause this reprogramming of Oct3/4, Sox2, Klf4, and c-Myc (Okita & Yamanaka, n.d.). These are extensively studied as these cells play a significant role in studying various disease pathogenesis and drug testing. Several reports have shown the critical role of mitophagy in metabolic reprogramming during the differentiation of these iPSCs. Krantz et al. described the crucial role of mitophagy in the differentiation of human iPSCs into vascular endothelial cells. He showed extensive mitophagy during the initial phase of differentiation, which activates mitochondrial biogenesis by the metabolic requirements of the differentiated cell. It was reported that during mitophagy PGAM which is a mitochondrial phosphatase, is cleaved, activating the PGC1-α transcription co-activator through β catenin. PGC1-α upregulates mitochondrial biogenesis (Krantz et al., 2021). Another report also showed that mitophagy is critical for iPSC reprogramming from somatic cells. iPSCs derived from the fibroblasts of PINK1 knockout mice were highly unstable and showed nonspecific differentiation, resulting in heterogeneous populations (Vazquez-Martin et al., 2016). A study by Xiang et al. also highlighted the involvement of BNIP3L/NIX-mediated mitophagy in reprogramming somatic cells to iPSCs (Xiang et al., 2017). Mitophagy is crucial for metabolic reprogramming from somatic to iPSCs and later in the differentiation of iPSCs. Therefore, understanding the key players involved in the control of mitophagy is critical.

Mitophagy in Stem Cell-Related Diseases

Numerous studies have demonstrated the link between stem cells and mitophagy. When mitochondrial dysfunction arises in stem cells, this can result in the onset of several diseases, including cancer, neurological disorders, diseases of the bones, and age-related diseases.

Mitophagy in Stem Cell-Related Cancer Disease

The association of mitophagy with cancer is multifaceted, which may lead to changes in the upregulation and downregulation of various gene expressions (Cairns et al., 2020b). The discovery of cancer stem cells (CSCs) has motivated researchers to examine how mitophagy functions in newly discovered cancer subtypes. CSCs influence invasion, differentiation, stemness, drug resistance, and cancer progression and recurrence. Various CSC subtypes exhibit distinct mitochondrial properties such as metabolism, reactive oxygen species levels, and dynamics. It has been investigated lung and nasopharyngeal CSC models where immature mitochondria have been observed, similar to in iPSCs and embryonic stem cells. On the other side of the pancreas, CSCs, called gliomaspheres, have been found to use mitochondrial OXPHOS to produce energy. It often seems that the dependence on mitophagy in CSCs differs depending on the model type due to the diverse mitochondrial fates among the multiple CSC subtypes and the varying characteristics of malignancies. However, inhibition of p53 expression and induction of the NANOG pathway and PINK1-dependent mitophagy signaling appears to be essential for preserving hepatic CSCs (K. Liu et al., 2017b). Moreover, in the case of AML stem cells (LSCs), an upsurge in the expression level of the mitochondrial DRP1 receptor FIS-1 has been analyzed, resulting in the enhancement of mitophagy and modification of mitochondrial structure along with the generation of ROS.

Drug resistance in some tumors has been significantly attributed to mitophagy in some CSCs, as discussed earlier in the chapter. For example, in the case of colorectal CSCs, treatment with the chemotherapeutic drug doxorubicin induces mitophagy, which has been found by an augmented co-localization of TOMM20/LC3B, resulting in an improvement of drug resistance in CSCs. In addition, no significant changes were observed in PINK1 and PARKIN expression with doxorubicin, whereas NIX upregulation has been found in CSCs (Yan et al., 2017b).

Mitophagy in Stem Cell-Related Neurodegenerative Disease

The role of mitochondria in regulating metabolism, ROS signaling, and stem cell fate decisions in brain stem cell populations has been well recognized in several studies. Evidence has shown that mitophagy is linked to the depletion of neural stem cells, leading to impaired neurogenesis. Recent investigations have indicated that certain neurodegenerative disorders, including Parkinson's, Alzheimer's, Huntington's, and autism spectrum disorders, have reduced neurogenesis. In the case of Alzheimer's, it has been observed that any mutation in the Presenilin 1 (Ps1) gene results in a reduction in the proliferation of stem cells, neuronal differentiation, and survival in knock-in mice. Therefore, the use of bexarotene to increase mitophagy in human iPSCs, which are produced by the PS1 M146L mutation, encourages the functional ability of these cells.

Various studies have determined that mitophagy may be induced by bexarotene treatment by increasing the co-localization of TOM20-LAMP1. Similarly, ubiquitin phosphorylation is augmented at Ser65, triggering the activation of Parkin and enhancing its co-localization with the mitochondria. Therefore, it is conceivable that the reported rescue effects of bexarotene treatment may be due to increased mitophagy.

Various animal model experiments have proved that Parkinson's disorder disrupts Neurogenesis. In this disease, the aberrant dendritic structure of the double cortin-positive neonatal neurons hampered their maturation. The neural stem cell (NSC) number in vivo was not affected, but in cultured adult hippocampal PINK1-deficient NSCs, it has shown increased apoptosis, decreased growth rate, and impaired mitochondria by analyzing the improvement of glycolysis and loss of membrane potential. Hence, all these factors suggest that abnormal mitophagy plays a pivotal role in Parkinson's disease-related regulation of NSCs and neurogenesis.

It has been discovered that the autosomal dominant Huntington's disease caused by CAG expansion in the HD gene occurs in various neurological disorders (Kim et al., 2021). Children at risk for Huntington's disease showed impairments in brain development, and multiple studies on animals have revealed the significance of Huntington's disease in improving the brain. Several studies have determined the involvement of defective autophagy in this disorder, specifically the effect of mutant HTT on the trafficking of the autophagosome, which is of particular relevance in this case. Furthermore, the essential role of HTT in the interaction between mitophagy receptors, injured mitochondria, and developing autophagosomes has recently been found.

An association between aberrant mitophagy and abnormal neural stem cells was found in studies on autism spectrum disorder (ASD). ASD is a neurodevelopmental condition caused by genetically inherited or non-genetic traits (Park et al., 2016). A Wdfy3 gene deficiency in mice controls neural progenitor divisions, resulting in increased early progenitor cell proliferation and decreased intermediate progenitor numbers. The cell cycle length was reduced by around 30%, and the migration of neurons was substantially hampered in the developing brain. It was later proposed that the Wdfy gene plays a crucial role in mitophagy and the regulation of mitochondrial quality.

ASD progression may merit further study because of the peculiar similarities between alterations in the early progenitor cells and other potential problems related to mitophagy. Impaired mitophagy may also impact the development of schizophrenia. The protein disrupted-in-schizophrenia-1 (DISC1) function in various physiological illnesses has been well investigated. Through the GSK3/catenin pathway, DISC1 loss is linked to poor neurogenesis and proliferation of embryonic and adult brain progenitor cells. In addition, it has been demonstrated that the deletion of DISC1 affects dendritic growth during the maturation of neurons from newborn dentate granule neurons. DISC1 has recently been identified as a receptor for mitophagy as it can localize itself to the mitochondrial membrane and bind to LC3 directly.

Mitophagy in Stem Cell-Related Bone Marrow Transplantation and Bone Diseases

According to a recent study, it has also been revealed that during ex vivo development for therapeutic purposes, MSCs were protected against senescence via mitophagy. In addition, inducing mitophagy also plays an acritical role in the regulation of activation of the inflammasome, which helps protect BMSCs. This increased the possibility that mitophagy mechanisms regulate mitochondrial integrity by removing damaged and malfunctioning mitochondria and protecting the transplanted BMSCs. For example, inducing the expression level of Parkin and reducing the expression level of the P53 gene after BMSC transplantation restored the damaged area in osteonecrotic mice, indicating that prompting mitophagy may be a successful treatment for osteonecrosis and other bone-related diseases (F. Zhang et al., 2020b).

Mitophagy in Stem Cell-Related Aging

Multiple studies have indicated that the mechanism of mitophagy plays a pivotal role in stem cell-linked aging, an emerging subject that remains to be explored. Basal autophagy is necessary for quiescent muscle satellite cells (García-Prat et al., 2016). In aged stem cells, the mitophagy-promoting drug rapamycin, which is employed to boost autophagy in geriatric satellite cells, was found to have the potential to regain the composition of mitochondria and reduce the extent of ROS, indicating that the mitophagy mechanism is crucial during the aging process. Moreover, HSCs have also shown the same dynamics as older stem cells, revealing high proportions of metabolically active mitochondria containing more significant fused networks. According to bioenergetic flux analysis, aged stem cells require OXPHOS and mitochondria for metabolism, which has also been observed in naive autophagy lacking HSCs that possess Atg12 mutations. The investigations suggest that mitophagy, impaired in old HSCs, may be essential for ridding naive stem cells of active mitochondria. Correspondingly, Drosophila intestinal stem cells exhibit higher senescence when expression of PINK1 and Parkin levels are reduced.

According to recently investigated data, damaged mitophagy and NAD+ exhaustion are linked to the development of Werner syndrome. This disease causes premature aging, which is very similar to general aging, such as the loosening of the skin, gray skin, cataracts, and gray hair with specific metabolic characteristics (Gudmundsrud et al., 2021). A study demonstrated that NAD^+ replenishment increased the quantity and ability of cell division in mitotic cells in a *C. elegans* model of Werner disease, extending the lifespan of the worms and suggesting better stem cell activity. This was also analyzed in a drosophila illness model for ISCs, where precursors to NAD^+ supplementation increased the longevity and cell division capacity of ISCs. Significantly, the NIX homolog pathway DCT-1 was necessary for this NAD^+-mediated effect. Urolithin A, a specific inducer of mitophagy, produces similar results to NAD^+ supplementation and is demonstrated to extend life; it was subsequently used to further solidify the role of mitophagy.

Conclusion

Stem cells are undifferentiated cells of the body that can self-renew and differentiate into multiple cell types. There are multiple types of stem cells, including embryonic stem cells, mesenchymal stem cells, hematopoietic stem cells, cancer stem cells, and induced pluripotent stem cells. Stem cells are critical for forming different types of tissues and play a vital role in the repair and regeneration of cells in an adult organism. Therefore, it is critical to understand the various components involved in the regulation of stem cell homeostasis and its properties. One such factor is mitochondria, which play a significant role in various aspects of stem cell biology, such as the maintenance of homeostasis, self-renewal property, and differentiation potential. Multiple processes regulate the various functions of mitochondria in cells, such as mitophagy, a specific form of autophagy that removes damaged or non-functional mitochondria from the cell and maintains the normal functioning of stem cells. Mitophagy regulates various properties of stem cells, such as self-renewal and multi-differentiation, and protection from oxidative stress. Mitophagy is also linked to the development of drug resistance in cancer stem cells. Therefore, understanding mitophagy's role is critical for developing novel therapeutics for various cancers and other stem cell-related diseases.

References

Alcalá, S., Sancho, P., Martinelli, P., Navarro, D., Pedrero, C., Martín-Hijano, L., Valle, S., Earl, J., Rodríguez-Serrano, M., Ruiz-Cañas, L., Rojas, K., Carrato, A., García-Bermejo, L., Fernández-Moreno, M. Á., Hermann, P. C., & Sainz, B. (2020). ISG15 and ISGylation are required for pancreatic cancer stem cell mitophagy and metabolic plasticity. *Nature Communications*, *11*(1), 1–17.

Andrzejewska, A., Lukomska, B., & Janowski, M. (2019). Concise Review: Mesenchymal Stem Cells: From Roots to Boost. *Stem Cells*, *37*(7), 855–864.

Bianco. (2001). www.nature.com.

Cairns, G., Thumiah-Mootoo, M., Burelle, Y., & Khacho, M. (2020a). Mitophagy: A new player in stem cell biology. *Biology*, *9*(12), 1–24.

Cairns, G., Thumiah-Mootoo, M., Burelle, Y., & Khacho, M. (2020b). Mitophagy: A New Player in Stem Cell Biology. *Biology 2020, Vol. 9, Page 481*, *9*(12), 481.

Cells, C., Huang, Z., Lin, Z., & Lin, Y. (2020). *BNIP3L-Dependent Mitophagy Promotes Metabolism Reprogramming*. 1–24.

Chen, G., Han, Z., Feng, D., Chen, Y., Chen, L., Wu, H., Huang, L., Zhou, C., Cai, X., Fu, C., Duan, L., Wang, X., Liu, L., Liu, X., Shen, Y., Zhu, Y., & Chen, Q. (2014). A regulatory signaling loop comprising the PGAM5 phosphatase and CK2 controls receptor-mediated mitophagy. *Molecular Cell*, *54*(3), 362–377.

Clevers, H. (2011). *The cancer stem cell: premises, promises and challenges*. 17.

Eiyama, A., & Okamoto, K. (2015). PINK1/Parkin-mediated mitophagy in mammalian cells. *Current Opinion in Cell Biology*, *33*, 95–101.

Fan, P., Yu, X. Y., Xie, X. H., Chen, C. H., Zhang, P., Yang, C., Peng, X., & Wang, Y. T. (2019). Mitophagy is a protective response against oxidative damage in bone marrow mesenchymal stem cells. *Life Sciences*, *229*(March), 36–45.

Feng, X., Yin, W., Wang, J., Feng, L., & Kang, Y. J. (2021). Mitophagy promotes the stemness of bone marrow-derived mesenchymal stem cells. *Experimental Biology and Medicine*, *246*(1), 97–105.

Foster, C. S., Dodson, A., Karavana, V., Smith, P. H., & Ke, Y. (2002). An introduction to stem cells. *Journal of Pathology*, *197*(4), 419–423.

García-Prat, L., Martínez-Vicente, M., Perdiguero, E., Ortet, L., Rodríguez-Ubreva, J., Rebollo, E., Ruiz-Bonilla, V., Gutarra, S., Ballestar, E., Serrano, A. L., Sandri, M., & Muñoz-Cánoves, P. (2016). Autophagy maintains stemness by preventing senescence. *Nature 2016 529:7584*, *529*(7584), 37–42.

Gudmundsrud, R., Skjånes, T. H., Gilmour, B. C., Caponio, D., Lautrup, S., & Fang, E. F. (2021). Crosstalk among DNA Damage, Mitochondrial Dysfunction, Impaired Mitophagy, Stem Cell Attrition, and Senescence in the Accelerated Ageing Disorder Werner Syndrome. *Cytogenetic and Genome Research*, *161*(6–7), 297–304.

Ito, K., Turcotte, R., Cui, J., Zimmerman, S. E., Pinho, S., Mizoguchi, T., Arai, F., Runnels, J. M., Alt, C., Teruya-Feldstein, J., Mar, J. C., Singh, R., Suda, T., Lin, C. P., Frenette, P. S., & Ito, K. (2016). Self-renewal of a purified Tie2+ hematopoietic stem cell population relies on mitochondrial clearance. *Science*, *354*(6316), 1156–1160.

Jin, G., Xu, C., Zhang, X., Long, J., Rezaeian, A. H., Liu, C., Furth, M. E., Kridel, S., Pasche, B., Bian, X. W., & Lin, H. K. (2018). Atad3a suppresses Pink1-dependent mitophagy to maintain homeostasis of hematopoietic progenitor cells article. *Nature Immunology*, *19*(1), 29–40.

Killackey, S. A., Philpott, D. J., & Girardin, S. E. (2020). *Mitophagy pathways in health and disease*.

Kim, A., Lalonde, K., Truesdell, A., Welter, P. G., Brocardo, P. S., Rosenstock, T. R., & Gil-mohapel, J. (2021). New Avenues for the Treatment of Huntington's Disease. *International Journal of Molecular Sciences 2021, Vol. 22, Page 8363*, *22*(16), 8363.

Krantz, S., Kim, Y. M., Srivastava, S., Leasure, J. W., Toth, P. T., Marsboom, G., & Rehman, J. (2021). Mitophagy mediates the metabolic reprogramming of induced pluripotent stem cells undergoing endothelial differentiation. *Journal of Biological Chemistry*, *297*(6).

Kuhikar, R., Khan, N., Philip, J., Melinkeri, S., Kale, V., & Limaye, L. (2020). Transforming growth factor β1 accelerates and enhances in vitro red blood cell formation from hematopoietic stem cells by stimulating mitophagy. *Stem Cell Research and Therapy*, *11*(1), 1–15.

Lee, J. H., Yoon, Y. M., Song, K. H., Noh, H., & Lee, S. H. (2020). Melatonin suppresses senescence-derived mitochondrial dysfunction in mesenchymal stem cells via the HSPA1L–mitophagy pathway. *Aging Cell*, *19*(3), 1–14.

Li, Y., Zheng, W., Lu, Y., Zheng, Y., Pan, L., Wu, X., Yuan, Y., Shen, Z., Ma, S., Zhang, X., Wu, J., Chen, Z., & Zhang, X. (2022). BNIP3L/NIX-mediated mitophagy: molecular mechanisms and implications for human disease. *Cell Death and Disease*, *13*(1).

Lin, Q., Chen, J., Gu, L., Dan, X., Zhang, C., & Yang, Y. (2021). New insights into mitophagy and stem cells. *Stem Cell Research and Therapy*, *12*(1), 1–14.

Lisowski, P., Kannan, P., Mlody, B., & Prigione, A. (2018). Mitochondria and the dynamic control of stem cell homeostasis. *EMBO Reports*, *19*(5), 1–12.

Liu, K., Lee, J., Kim, J. Y., Wang, L., Tian, Y., Chan, S. T., Cho, C., Machida, K., Chen, D., & Ou, J. H. J. (2017a). Mitophagy Controls the Activities of Tumor Suppressor p53 to Regulate Hepatic Cancer Stem Cells. *Molecular Cell*, *68*(2), 281-292.e5.

Liu, K., Lee, J., Kim, J. Y., Wang, L., Tian, Y., Chan, S. T., Cho, C., Machida, K., Chen, D., & Ou, J. H. J. (2017b). Mitophagy Controls the Activities of Tumor Suppressor p53 to Regulate Hepatic Cancer Stem Cells. *Molecular Cell*, *68*(2), 281-292.e5.

Liu, L., Feng, D., Chen, G., Chen, M., Zheng, Q., Song, P., Ma, Q., Zhu, C., Wang, R., Qi, W., Huang, L., Xue, P., Li, B., Wang, X., Jin, H., Wang, J., Yang, F., Liu, P., Zhu, Y., … Chen, Q. (2012). Mitochondrial outer-membrane protein FUNDC1 mediates hypoxia-induced mitophagy in mammalian cells. *Nature Publishing Group*.

Lv, F. J., Tuan, R. S., Cheung, K. M. C., & Leung, V. Y. L. (2014). Concise Review: The Surface Markers and Identity of Human Mesenchymal Stem Cells. *Stem Cells*, *32*(6), 1408–1419.

Marinković, M., & Novak, I. (2021). A brief overview of BNIP3L/NIX receptor-mediated mitophagy. *FEBS Open Bio*, *11*(12), 3230–3236.

Marinković, M., Šprung, M., & Novak, I. (2020). *Dimerization of mitophagy receptor BNIP3L/NIX is essential for the recruitment of autophagic machinery*.

Murakami, K., Kurotaki, D., Kawase, W., Soma, S., Fukuchi, Y., Kunimoto, H., Yoshimi, R., Koide, S., Oshima, M., Hishiki, T., Hayakawa, N., Matsuura, T., Oda, M., Yanagisawa, K., Kobayashi, H., Haraguchi, M., Atobe, Y., Funakoshi, K., Iwama, A., Nakajima, H. (2021). OGT Regulates Hematopoietic Stem Cell Maintenance via PINK1-Dependent Mitophagy. *Cell Reports*, *34*(1), 108579.

Nguyen, T. N., Padman, B. S., & Lazarou, M. (2016). Deciphering the Molecular Signals of PINK1/Parkin Mitophagy. *Trends in Cell Biology*, *26*(10), 733–744.

Okita, K., & Yamanaka, S. (n.d.). *Induced Pluripotent Stem Cells*.

Onishi, M., Yamano, K., Sato, M., Matsuda, N., & Okamoto, K. (2021). Molecular mechanisms and physiological functions of mitophagy. *The EMBO Journal*, *40*(3), 1–27.

Park, H. R., Lee, J. M., Moon, H. E., Lee, D. S., Kim, B. N., Kim, J., Kim, D. G., & Paek, S. H. (2016). A Short Review on the Current Understanding of Autism Spectrum Disorders. *Experimental Neurobiology*, *25*(1), 1–13.

Pei, S., Minhajuddin, M., Adane, B., Khan, N., Stevens, B. M., Mack, S. C., Lai, S., Rich, J. N., Inguva, A., Shannon, K. M., Kim, H., Tan, A. C., Myers, J. R., Ashton, J. M., Neff, T., Pollyea, D. A., Smith, C. A., & Jordan, C. T. (2018a). AMPK/FIS1-Mediated Mitophagy Is Required for Self-Renewal of Human AML Stem Cells. *Cell Stem Cell*, *23*(1), 86-100.e6.

Pei, S., Minhajuddin, M., Adane, B., Khan, N., Stevens, B. M., Mack, S. C., Lai, S., Rich, J. N., Inguva, A., Shannon, K. M., Kim, H., Tan, A. C., Myers, J. R., Ashton, J. M., Neff, T., Pollyea, D. A., Smith, C. A., & Jordan, C. T. (2018b). AMPK/FIS1-Mediated Mitophagy Is Required for Self-Renewal of Human AML Stem Cells. *Cell Stem Cell*, *23*(1), 86-100.e6.

Rogov, V. V., Suzuki, H., Marinković, M., Lang, V., Kato, R., Kawasaki, M., Buljubašić, M., Šprung, M., Rogova, N., Wakatsuki, S., Hamacher-Brady, A., Dötsch, V., Dikic, I., Brady, N. R., & Novak, I. (2017). Phosphorylation of the mitochondrial autophagy receptor Nix enhances its interaction with LC3 proteins. *Scientific Reports*, *7*(1), 1–12.

Tsujimoto, T., Mori, T., Houri, K., Onodera, Y., Takehara, T., Shigi, K., Nakao, S., Teramura, T., & Fukuda, K. (2020). miR-155 inhibits mitophagy through suppression of BAG5, a partner protein of PINK1. *Biochemical and Biophysical Research Communications*, *523*(3), 707–712.

Uccelli, A., Moretta, L., & Pistoia, V. (2008). Mesenchymal stem cells in health and disease. *Nature Reviews Immunology*, *8*(9), 726–736.

Vazquez-Martin, A., Van den Haute, C., Cufí, S., Corominas-Faja, B., Cuyàs, E., Lopez-Bonet, E., Rodriguez-Gallego, E., Fernández-Arroyo, S., Joven, J., Baekelandt, V., & Menendez, J. A. (2016). Mitophagy-driven mitochondrial rejuvenation regulates stem cell fate. *Aging*, *8*(7), 1330–1352.

Wu, W., Tian, W., Hu, Z., Chen, G., Huang, L., Li, W., Zhang, X., Xue, P., Zhou, C., Liu, L., Zhu, Y., Zhang, X., Li, L., Zhang, L., Sui, S., Zhao, B., & Feng, D. (2014). ULK1 translocates to mitochondria and phosphorylates FUNDC1 to regulate mitophagy. *EMBO Reports*, *15*(5), 566–575.

Xiang, G., Yang, L., Long, Q., Chen, K., Tang, H., Wu, Y., Liu, Z., Zhou, Y., Qi, J., Zheng, L., Liu, W., Ying, Z., Fan, W., Shi, H., Li, H., Lin, X., Gao, M., Liu, J., Bao, F., Liu, X. (2017). BNIP3L-dependent mitophagy accounts for mitochondrial clearance during 3 factors-induced somatic cell reprogramming. *Autophagy*, *13*(9), 1543–1555.

Yan, C., Luo, L., Guo, C. Y., Goto, S., Urata, Y., Shao, J. H., & Li, T. S. (2017a). Doxorubicin-induced mitophagy contributes to drug resistance in cancer stem cells from HCT8 human colorectal cancer cells. *Cancer Letters*, *388*(December), 34–42.

Yan, C., Luo, L., Guo, C. Y., Goto, S., Urata, Y., Shao, J. H., & Li, T. S. (2017b). Doxorubicin-induced mitophagy contributes to drug resistance in cancer stem cells from HCT8 human colorectal cancer cells. *Cancer Letters*, *388*, 34–42.

Zakrzewski, W., Dobrzyński, M., Szymonowicz, M., & Rybak, Z. (2019.). *Stem cells: past, present, and future.*

Zhang, F., Peng, W., Zhang, J., Dong, W., Wu, J., Wang, T., & Xie, Z. (2020a). P53 and Parkin co-regulate mitophagy in bone marrow mesenchymal stem cells to promote the repair of early steroid-induced osteonecrosis of the femoral head. *Cell Death and Disease*, *11*(1), 1–16.

Zhang, F., Peng, W., Zhang, J., Dong, W., Wu, J., Wang, T., & Xie, Z. (2020b). P53 and Parkin co-regulate mitophagy in bone marrow mesenchymal stem cells to promote the repair of early steroid-induced osteonecrosis of the femoral head. *Cell Death & Disease 2020 11:1*, *11*(1), 1–16.

Zhang, H., Menzies, K. J., & Auwerx, J. (2018). *The role of mitochondria in stem cell fate and aging.*

Chapter 16

Mitophagy and Lung Cancer

Ashok Kumar Tiwari[1]
Devatman Jauhari[1]
Debdarshan Dutta[1]
Dhruv Sharma[1]
Priyanka[1]
Harshit Devtalla[1]
Nilay Solanki[2]
Megha Chouhan[1]
and Rachana R[1,*]
[1]Department of Biotechnology, JIIT, Noida, Uttar Pradesh, India
[2]Department of Pharmacology, Ramanbhai Patel College of Pharmacy, Charotar University of Science and Technology, Changa, Gujarat, India

Abstract

Llung cancer is a terrible disease with high morbidity and mortality worldwide. Cancer of lung cells is linked with mitophagy, i.e., autophagy's selective elimination of damaged or malfunctioning mitochondria. Because defective mitophagy can contribute to the development and progression of lung cancer, this process is now being suggested as a target for a feasible treatment strategy to cope with lung cancer. This review summarizes the current understanding of the function of mitophagy in lung cancer and examines the potential of mitophagy-targeted treatments as novel treatment options.

Keywords: lung cancer, mitochondria, autophagy, mitophagy

* Corresponding Author's Email: rachana.dr@iitbombay.org.

In: Mitophagy and Its Role in Health and Disease
Editors: Shalini Mani and Samiksha Wasnik
ISBN: 979-8-89113-470-6

Introduction

Cancer

Our tissues comprise billions of individual cells. Cellular death is normal, and controlled cellular division is required to keep cell count stable. Abnormalities in cell cycle regulation, genetic mutations, and many other factors may result in unregulated cellular division (Shostak, 2017). This aberrant cell division can result in the development of a tumor, which can be benign or malignant. A benign tumor is a noncancerous tumor that does not spread and usually does not return once removed; on the other hand, a malignant tumor, also known as a cancerous tumor, can recur and spread to other tissues within our body, resulting in the formation of additional tumors (Geiger and Peeper, 2019). Cancer can manipulate the immune system to enhance cancer cells' growth and survival chances (Gonzalez, Hagerling, and ZWerb 2018). Cancer cells can also change how they take nutrients, eventually encouraging rapid cellular development and cancer progression (Papalazarou and Maddocks, 2021).

Overall, there are eight main characteristics of cancer cells: uncontrolled cell proliferation, evasion of tumor suppressors, resistance to apoptosis, hyperactive angiogenesis, activation of cellular invasion and metastasis, abnormal cellular metabolism, and escape from the immune system (Hanahan, 2022). Most of these hallmarks of cancer are associated with mitochondrial functioning or dynamics because mitochondria are the primary regulators of various cellular processes such as energy production, oxidative environment, inflammation, immunity, cell death, and migration (Giampazolias & Tait, 2016).

Mitochondria in Lung Cells

Mitochondria are double-membraned cell organelles responsible for oxidative phosphorylation, iron homeostasis, apoptosis, and several other metabolic functions; they are often called the cellular dynamo. Mitochondria are endosymbiotic because they possess their DNA, proteome, and transcriptional machinery (Chinnery & Schon, 2003; Ernster & Schatz, 1981). Human mitochondrial DNA (mtDNA) encodes 37 genes responsible for translating and coding respiratory complexes I, II, and IV. mtDNA is passed directly from mother to offspring and replicates independently of nuclear DNA (nDNA). Unlike nDNA, mtDNA is highly prone to damage prompted by reactive oxygen species (ROS) due to the absence of histones in mitochondria, which protect nDNA from ROS (Nam et al., 2017).

Over 40 distinct cell types comprise the mammalian lungs, which are precisely distributed with varying levels of mitochondria (Cloonan & Choi, 2016). Alveolar epithelium Type II cells (AECs) have more mitochondria than other lung cells because type II AECs continuously pour the surfactant onto the epithelial cell surface via exocytosis (Massaro et al., 1975). This increase in mitochondrial number and other functional changes in cells such as ciliated bronchial epithelial cells, alveolar macrophages, and vascular smooth muscle cells enhances the lungs' capability to adapt to stress and other special physiological requirements (Carraway et al., 2008). However, mitochondria are also responsible for ROS production, and an imbalance in this function can result in diseases such as chronic airway disease, lung cancer, and cystic fibrosis (Liu et al., 2017). The imbalanced production of ROS causes mutations in mtDNA, which are directly and indirectly associated with lung cancer (Yang et al., 2016).

Lung Cancer

Lung cancers are histologically classified into small-cell lung cancer (SCLC) and non-SLC). Nearly 15% of total lung cancer cases are attributed to SCLC, which has neuroendocrine characteristics, whereas the remaining 85% are attributed to NSCLC. NSCLCs are further subclassified as adenocarcinoma, squamous, and large-cell carcinoma (Dela et al., 2011).

Numerous oncogenes and tumor suppressor genes, including RAS, MYC, and TP53, alter mitochondrial function, resulting in an electron transport surge, increased oxygen absorption, and hyperROS production (Nam et al., 2017). Increased ROS production causes damage and mutations in mtDNA, which can ultimately lead to cancer. In addition, increased copies of mitochondrial DNA are an associated risk factor for lung cancer in regular smokers (Hosgood et al., 2010). Therefore, new strategies concentrating on mtDNA alterations to detect and treat lung malignancies cannot be ruled out (Hosgood et al., 2010). In addition, when mitochondria become defective, the mechanisms of autophagy and mitophagy remove the defective mitochondria. This mechanism is beneficial for normal cells to stay away from various diseased conditions. However, during the cancer progression, cancer cells use these strategies to survive; hence this mechanism remains exacerbated and favors cancer development (Yun and Lee, 2018). The following section describes the basic features of autophagy and related events and mechanisms.

Autophagy

Autophagy is an intracellular physiological process employed in the lysosomal degradation of cellular components, including aggregated proteins, macromolecules, dysfunctional cell organelles, and pathogens. All these components are transported to lysosomes by autophagosomes, where they are catabolized to simpler molecules such as amino acids, nucleotides, sugars, fatty acids, and ATP by the action of lysosomal hydrolases (Li et al., 2020). Duve first identified autophagy in 1963, and according to his definition, it occurs when some special cellular protein-containing vesicles fuse with lysosomes (Benbrook & Long, 2012). Autophagosomes are protein-containing vesicles that transport substrates to the lysosome for breakdown (Aman et al., 2021). Two other essential functions of autophagy are improving antigen presentation on the cell surface and preventing cellular senescence. Many illnesses, including cancer, diabetes, autoimmune disorders, and cardiomyopathy, can be avoided if there is a balance in autophagy function as it breaks down damaged or dysfunctional organelles and intracellular aggregates before they become hazardous for the cells (Glick et al., 2010).

Macroautophagy, microautophagy, and chaperone-mediated autophagy (CMA) are the identified subcategories of autophagy (Saftig et al., 2008). Macroautophagy is essential for cellular homeostasis because it degrades and recycles macromolecules. In this case, a double-membrane vesicle (autophagosome) amalgamates with lysosomes to form an autolysosome. Cargo is unloaded and digested by acidic hydrolases inside the autolysosome.

Micro-autophagy is characterized by the direct uptake of cytosolic components by lysosomes (Saftig et al., 2008). Upon recognition by the lysosomal membrane receptor lysosomal-associated membrane protein 2A (LAMP-2A), chaperone proteins (such as Hsc-70) facilitate the transportation of targeted proteins to the lysosome, where they undergo unfolding and degradation (Glick et al., 2010).

Chaperone-mediated autophagy is a lysosome-dependent proteolysis pathway (Dice, 2007). This begins by binding the heat shock cognate 70-kDa protein (hsc-70) with the KFERQ motif of the target protein. This now acts as a chaperone, further guiding the targeted proteins to LAMP-2a. After docking the targeted proteins outside the lysosomal membrane, they unfold and are internalized into the lysosomal lumen with the help of other chaperone and core chaperone proteins (such as Hsc-90). Targeted proteins are further degraded due to the proteases and hydrolysis that works at lower pH in the lysosome (Kaushik & Cuervo, 2018).

Mitophagy

Mitophagy is a subclassification of macroautophagy, in which dysfunctional or excessive numbers of mitochondria are degraded by the action of lysosomal hydrolyzes (Chourasia et al., 2015). Hypoxic conditions, starvation, and oxidative stress induce mitophagy, which helps in reducing the number of mitochondria to create a balance between reactive oxygen species (ROS) and antioxidants, to promote efficient use of oxygen and to conserve nutrients for cell survival during stress (Chourasia et al., 2015) (Figure 1). In addition, the improper modulation of mitochondria during oncogenic stress and the accumulation of defective mitochondria leads to tumorigenicity and tumor progression (Strohecker et al., 2013).

In various studies on mouse models, autophagy-deficient tumors are characterized by nonfunctional mitochondria along with alteration in mitochondrial morphology, which ultimately leads to the reduction in oxygen consumption during aerobic respiration and increased glucose uptake by these tumor cells (Chourasia et al. 2015). Several mitophagy modulators like Parkin, BNIP3, PARK2, BNIP3L, OPTN, and p62/SQSTM1 are often deregulated in various cancer types (Drake et al., 2017). The PARK2 gene encodes for Parkin, which recruits the mitophagy cargo adaptors like p62/Sqstm1, OPTN, and NDP52 (Cesari et al., 2003). Hence, the inactivation of PARK2 is directly associated with inhibiting mitophagy in cancer.

Mitochondria and Lung Cancer

The development of lung cancer is a complex process involving hereditary and environmental variables. Recent research has demonstrated that deregulation of mitochondrial activities is interlinked with lung cancer growth and progression (Qi et al., 2020). These "powerhouses" are necessary for various biological functions such as ATP production, cell signaling, and programmed cell death. Under certain conditions, ROS generation, increased glycolysis, and reduced oxidative phosphorylation are the key indicators of mitochondrial dysfunction in lung cancer cells. According to Cheng et al. (2019), these alterations provide a pro-tumorigenic milieu that promotes the survival and growth of cancer cells. This metabolic change is essential for the survival of cancer cells. Following are some mitochondrial dysfunctions that are related to lung cancer.

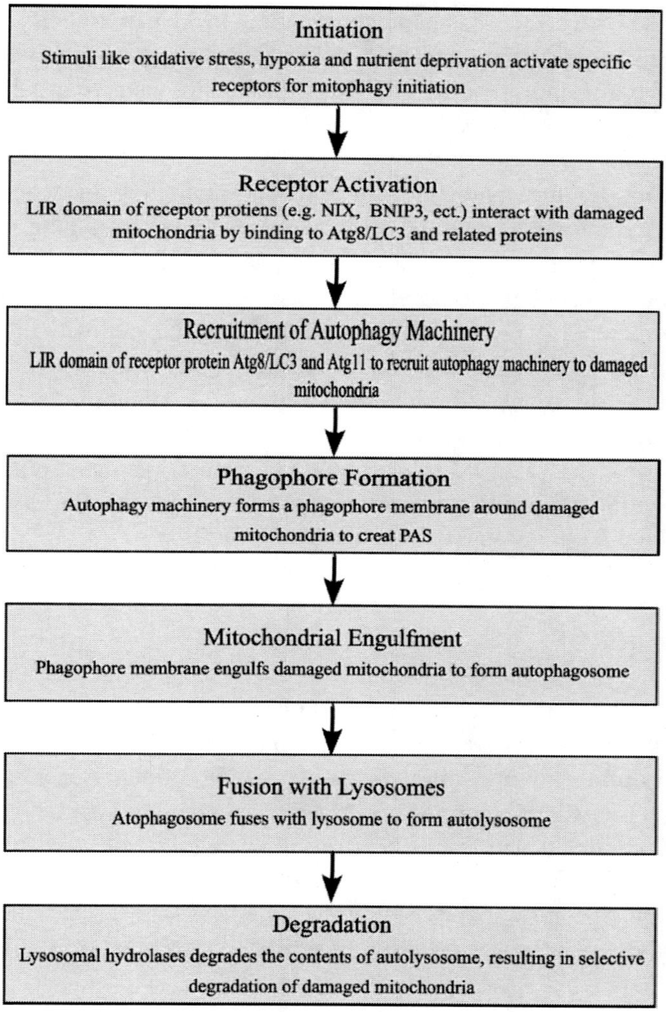

Figure 1. Simplified Pathway for Mitophagy.

Altered Respiration

More often, tumor cells exhibit higher rates of aerobic glycolysis and, simultaneously, reduced mitochondrial respiration. This phenomenon is known as the Warburg effect (Lu et al., 2015). This renders the mitochondrial dysfunctional and makes tumor cells dependent on glycolysis for ATP production, despite oxygen (Vander et al., 2009). Although the production of ATPs during glycolysis is 16 times lower than that during mitochondrial respiration, it escalates cellular proliferation and growth (Plas and Thompson 2005).

An imbalance between glycolysis and mitochondrial respiration results in the accretion of mitochondrial metabolites in the cytoplasm, eventually altering cellular metabolism. This created a hypoxic condition in the cell due to highly increased cell proliferation and lowered blood supply in the tumor (Nelson et al., 2004). Hypoxia is the primary driving force for cancer progression through hypoxia-induced genomic instability (Nelson et al. 2004). Hypoxia has also been associated with decreased apoptosis, resulting in an increased life span of cancer cells and thus allowing cancer cells to proliferate. Furthermore, hypoxic conditions stabilize the hypoxia-inducible transcription factor (HIF-1α), which helps cancerous cells adapt to stressful environments by transcribing and synthesizing almost 70 hypoxia-related factors (Roberts and Thomas 2013).

Along with HIF-1α, other tumor-inducing elements, including oncogenic mutations, tumor suppressor gene mutations, and mutations of signaling kinases, help regulate glycolytic conversion and compensate for low-ATP yields in cancer. During such stressful conditions, 5' adenosine monophosphate-activated protein kinase (AMPK) signaling is activated to inhibit cell proliferation and energy production through the oxidative pathway initiates (Roberts and Thomas 2013). In contrast, tumor cells try to suppress AMPK pathways to continue cellular growth under low nutrients and stressful conditions. In this situation, tumor cells continue to proliferate by surpassing cellular checkpoints through oncogene activation or loss of function in tumor suppressor genes (e.g., PTEN and p53). Therefore, the metabolic imbalance created by mitochondria and impaired mitophagy lead to the loss of the AMPK cascade and ultimately to cancer growth and progression (Roberts and Thomas 2013).

Overproduction of ROS

The Complex I subunit of the electron transport chain (ETC) plays a significant role in tumor growth and metastasis (Solaini et al., 2011). The mutations in Complex-I hinder its integration with Complex III during supercomplex formation, which ultimately leads to the alteration in the production and transfer of electrons in ETC, and eventually, these cells cannot detect ROS overproduction. This imbalance finally leads to escalated energy loss, ROS overproduction, and oxidative stress (Solaini et al., 2011). Complex I has often been altered in lung cancer and, more occasionally, in nonsmokers. Forcible overexpression of mutated mitochondrial Complex-I resulted in increased superoxide particles, invasion, and proliferation *in vitro* (Dasgupta et al., 2012).

Mutation in Mitochondrial DNA

Interestingly, the oxidative stress developed due to an imbalance in ROS and antioxidant levels in tumor cells leads to alterations in the mitochondrial DNA (mtDNA), mitochondrial uncoupling, and proton leakage (Servais et al., 2003). An accumulation of mutations in mtDNA promotes cancer growth and metastasis via deregulated mitophagy. As per Fliss et al. (2000), mutations in mtDNA's non-coding displacement (D) loop are often associated with lung cancer. These alterations in D-loop can be used as the biomarkers of carcinogenesis in lung cancer (Fliss et al., 2000).

Furthermore, lung cancer has been linked to copy number changes and mutations in mitochondrial DNA (mtDNA). These changes may result in increased ROS generation and mitochondrial malfunction, which may favor the survival and growth of cancer cells. It has also been noted that lung cancer patients with mitochondrial DNA mutations have poor prognoses and are unsusceptible to treatment (Dai et al., 2013).

Dysregulated Mitochondrial Dynamics

Llung cancer has been shown to develop and progress because of dysregulated mitochondrial dynamics, which include mitochondrial fission and fusion, in addition to Warburg's effect (Dai & Jiang 2019). An increase in ROS generation, DNA damage, and abnormal mitophagy can result from the dysregulation of mitochondrial fission and fusion, encouraging the survival and growth of cancer cells. Specific proteins such as Drp1, which are primarily responsible for controlling mitochondrial fusion, have been linked to poor prognosis and therapeutic resistance for lung cancer (Dai & Jiang, 2019).

Furthermore, to understand the role of mitophagy to a deeper extent in lung cancer, it was blocked by Chourasia et al. (2015) by removing the genes responsible for mitophagy. This experiment demonstrated that mitophagy inhibited tumor growth in its early phases and encouraged mitophagy in later stages (Wang et al., 2020). The following section further elaborates on the role of mitophagy in lung cancer.

Alteration in Mitophagy and Lung Cancer

Several studies have identified mitophagy regulators that play crucial roles in the development and progression of lung cancer (Lu et al., 2023). Hypoxia, a benchmark characteristic of solid tumors, has been illustrated that it promotes mitophagy and helps cancer cell survival (Muz et al., 2015). Hypoxia-induced mitophagy in NSCLC cells increases cell survival by removing dysfunctional mitochondria, thereby reducing oxidative stress and DNA damage (Sureshbabu and Bhandari, 2013). In addition, mitophagy induction has been ruled out as being associated with chemotherapy resistance in lung cancer cells, indicating that mitophagy inhibition may be a viable therapeutic strategy for treating chemotherapy-sensitized cancer (Guan et al., 2021).

Mitophagy is also involved in the modulation of immune responses, which play crucial roles in lung cancer progression (Wang et al., 2020). Inhibition of autophagy and mitophagy in dendritic cells impairs antigen presentation and promotes immune tolerance, which may lead to immune evasion in cancer cells (Xia, Green, and Zou, 2021). In addition, the regulation of mitophagy in T cells may be essential for the immune surveillance of cancer cells.

As mentioned elsewhere, hypoxia induces mitophagy, which in turn promotes lung cancer survival and chemotherapy resistance (Wang et al., 2022a). Following is a description of various proteins that are directly or indirectly involved in the mitophagy pathway and are found to be affected in the case of lung cancer.

Role of PINK1 in Lung Cancer

PINK1 levels are usually low in healthy mitochondria; however, when mitochondrial damage occurs, PINK1 builds up on the OMM (Pickrel and Youle, 2015). This buildup of PINK1 causes the E3 ubiquitin ligase Parkin to be recruited and turned on. Parkin then ubiquitinates OMM proteins, which allows autophagosomes to find and remove damaged mitochondria (Figure 2) (Fritsch et al., 2020). It has been shown that PINK1 is essential for protecting cells from damage caused by oxidative stress, and its misregulation has been linked to Parkinson's disease (PD) and cancer (Dorszewska et al., 2021). Mutations in PINK1 cause damaged mitochondria to build up in patients with PD, which makes neurodegeneration more likely. Recent research has shown that PINK1 instability may contribute to the growth and spread of several types of cancer (Li et al., 2021). In some cases, the expression of PINK1 is decreased, whereas in others, it is turned up (Rojas-Charry et al., 2014). On the other hand, it has been shown that the PINK1 protein goes up in lung cancer (Liu et al., 2018). Zhu and colleagues (2020) established that lung cancer tissues had more PINK1 than normal ones. Overexpression of PINK1 was also linked to a higher chance of apoptosis and slower growth of tumors (Liu et al., 2018). This shows that PINK1 may help cancer grow in lung cancer. Other types of cancer, such as ovarian and colon cancer, have also been linked to PINK1 instability as PINK1 expression was found to be downregulated in ovarian cancer, and its overexpression was linked to less cell growth and more cell death (Wang et al., 2018). PINK1 expression was demonstrated to be raised in colon cancer, and its overexpression was linked to hypercell growth and tumor growth (Yin et al., 2021). This shows that PINK1's role in cancer is probably complicated and may depend on the type of cancer. In some cases, PINK1 may help to halt tumor growth, whereas in other cases, it may help the growth of a tumor. More studies are required to fully understand what role PINK1 plays in cancer and find possible treatment targets.

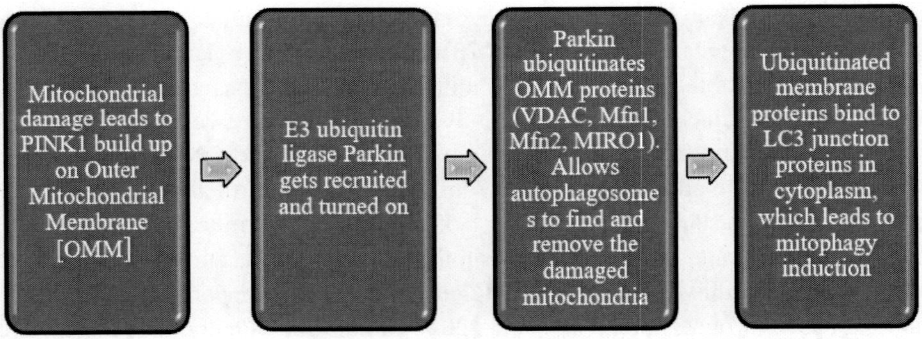

Figure 2. Role of PINK1 and mitophagy.

Role of PRKN in Lung Cancer

Parkin RBR E3 ubiquitin-protein ligase (PRKN), also called Parkin, is a cytoplasmic protein that regulates mitophagy (Sun et al., 2023). Parkin belongs to the E3 ubiquitin family of ligases that ubiquitinate mitochondrial proteins, allowing autophagy to selectively degrade damaged or undesired mitochondria (Glauser et al., 2011). Parkin is inducted into the OMM in response to mitochondrial injury by the PTEN-induced putative kinase 1 (PINK1) of mitochondria. It ubiquitinates mitochondrial proteins after being recruited to the OMM, which causes autophagosomes to recognize and sequester damaged mitochondria for destruction. This mechanism is essential for cellular homeostasis and prevents the accumulation of damaged mitochondria, which can result in cellular malfunction and illness (Ma et al., 2020). Parkin also interacts with other mitophagy proteins, such as p62 and LC3, which are essential for autophagosome production and fusion with lysosomes for destruction. This shows that Parkin is crucial in organizing the entire mitophagy process. According to recent research, Parkin dysregulation may aid in the genesis and development of various forms of cancer. Parkin's involvement in cancer appears to be complicated and may vary based on the kind of cancer (Liu J et al., 2018). In one investigation, parkin expression was overexpressed in lung cancer tissues relative to normal lung tissues (Wang et al., 2020). Parkin overexpression was also associated with an increased likelihood of apoptosis and reduced tumor development.

Parkin dysregulation has also been linked to other cancers, such as ovarian and stomach cancer. Parkin expression was considerably reduced in breast cancer tissues compared with normal breast tissues in one research (Tay et al., 2010). Furthermore, reduced Parkin expression was shown to be linked to a more advanced tumor stage and a worse prognosis. Parkin may have tumor-suppressive properties in breast cancer (Liu J et al., 2018).

Parkin expression was discovered to be downregulated in ovarian cancer, and its overexpression was related to lower cell proliferation and higher apoptosis. Parkin expression was raised in gastric cancer, and its overexpression was linked to increased cell proliferation and tumor formation (Liu J et al., 2018).

Parkin may serve a tumor-suppressive function in some circumstances, whereas it may also play a tumor-promoting role in others (Sun et al., 2023). More studies are needed to understand Parkin's involvement in cancer and find possible therapeutic targets for cancer therapy.

Role of BNIP3

A transmembrane protein BNIP3, localized on the outer mitochondrial membrane (OMM), is a well-established mitophagy regulatory protein. Two mitochondrial outer membrane proteins, BNIP3 and BNIP3L/NIX, have been implicated in hypoxia-induced mitophagy (Li et al., 2021). NIX-mediated mitophagy was first discovered during the maturation of red blood cells, whereas BNIP3 is broadly distributed in numerous tissues, including the heart, liver, and muscle. NIX is predominantly expressed in hematopoietic tissue (Zhang and Ney, 2009). It has been demonstrated that hypoxia-inducible factor 1 (HIF1) upregulates the expression of BNIP3, which promotes mitophagy. In addition, it has been demonstrated that BNIP3 knockout upregulates NIX expression, indicating a compensatory role for NIX in mitophagy regulation (Sowter et al., 2001; Li et al., 2021).

In stroke, during delayed neuronal death, BNIP3 in association with LC3 triggers mitophagy excessively (Shi et al., 2014). BNIP3 modulates mitochondrial membrane permeability, which is essential for mitophagy initiation (Quinsay et al., 2010). BNIP3 can interact with the mitochondrial fission protein Drp1, resulting in mitochondrial fragmentation and the formation of autophagosomes for degradation (Lee et al., 2011).

Like the above-mentioned other proteins, BNIP3 dysregulation has been associated with the development and progression of various cancer types. BNIP3's function in cancer appears complex and may vary depending on the form of cancer. BNIP3 has been suggested to be an essential link between cell death and cell survival. It has to impact cancer development and progression; its action appears to be tissue-specific as it is expressed differently in normal cells and cancer types (Gorbunova et al., 2020).

Lung cancer upregulates BNIP3 expression (Gorbunova et al., 2020). In addition, the dysregulation of BNIP3 has been linked to other forms of cancer, including colon and prostate cancer. The overexpression of BNIP3 was associated with decreased cell proliferation and increased apoptosis in colon cancer (Vijayalingam et al., 2010). BNIP3 expression was elevated in prostate cancer, and its overexpression was linked to increased cell proliferation and tumor growth (Chourasia et al., 2015).

Role of PARK2

The tumor suppressor gene PARK2, also known as Parkin, is often altered (somatically and genetically) in many different forms of cancer, including lung cancer (Zhang et al., 2020). The protein encoded by PARK2 functions as an E3 ubiquitin ligase, which directs proteins to be degraded by the proteasome or autophagy processes. Damaged mitochondria and compromised cellular function result from these Parkin mutations, which reduce its capacity to ubiquitinate outer membrane proteins and trigger mitophagy (Ge et al., 2020). Lung cancer cells with PARK2 tumor suppressor gene mutations, including somatic and germline, impair PINK1/Parkin-mediated mitophagy. As discussed, mitophagy, the process of eliminating damaged mitochondria, relies heavily on the Parkin/PINK1 pathway (Durcan and Fon, 2015). Mitophagy is hindered, and damaged mitochondria accumulate when mutations in the tumor suppressor gene PARK2 disrupt the PINK1/Parkin pathway (Zhang et al., 2020). Parkin has an essential role in the initiation, development, and progression of cancer, as evidenced by studies showing that the reexpression of Parkin in cancer cells, could restore mitophagy and decrease tumor growth (Denisenko et al., 2021).

Impaired mitochondrial function and the buildup of damaged mitochondria are linked to these germ-line alterations. Cancer risk is increased when the PARK2 gene is mutated at birth (Iwakawa et al., 2012), (Xiong et al., 2015). Reactivating Parkin in cancer cells can restore mitophagy and reduce tumor development, suggesting that medicines based on mitophagy may be helpful in the fight against the disease (Denisenko et al., 2021). Understanding the significance of the PINK1/Parkin pathway in cancer and creating effective mitophagy-based treatments for cancer requires a more profound understanding of the subject.

Role of TP53

TP53 is responsible for regulating cell cycle checkpoints, apoptosis, and nucleic acid repair. It has also been reported to be an essential tumor suppressor gene and is majorly expressed in the lung epithelium (Borrero and El-Deiry, 2021). It is a crucial regulator of NSCLC, and its suppression has been reported to cause increased tumor growth, higher mortality rate, and poor response to anti-cancer treatments (Sosa et al., 2018).

TP53-mediated mitochondrial dysfunction is known to be centered around the PINK1-Parkin-controlled mitophagy initiation (Goiran et al., 2018). Jin and Youle (2012), in their study of nuclear TP53-mediated autophagy, reported that TP53 is associated using a selective downregulation of PINK1 and provided *in-vivo* and *ex-vivo* evidence to support their hypothesis. They reported that PINK1 repression resulted from deleting the TP53 responsive element on the PINK1 promoter sequence and experimentally validated the phenomenon using a chromatin immunoprecipitation assay.

Another recent study involving the Next-generation sequencing (NGS) of tumor tissue data from 314 Chinese patients with NSCLC found that mutation in TP53 is a driver mutation in over 600 mutations across patient data. Expression of both wild-type and mutated TP53 was studied using H2172 cells. Moreover, the results showcased the increased cell death in the cells expressing mutated TP53. These observations pointed out the role of TP53 mutation promoted mitophagy and support tumor cell proliferation (Wang et al., 2022b).

Role of FUNDC1 in Cancer

FUNDC1, an additional mitochondrial outer membrane protein, induces mitophagy independent of Parkin, an E3 ubiquitin ligase associated with PD. FUN14 domain containing 1 (FUNDC1) is an essential molecule in receptor-dependent mitophagy. FUNDC1 has been suggested to be used as a prognostic biomarker for cervical cancer (Hou et al., 2017). A new therapeutic target is also suggested to improve chemo-radiotherapy's antitumor effects (Hou et al., 2017). Studying the expression level of FUNDC1 mRNA, it remains overexpressed in various tumor tissues, including lung cancers, compared with those of the respective controls. (Zhang, 2021). FUNDC1 mediates hypoxia-induced mitophagy through its LIR motif-mediated interaction with LC3 (Di Rita et al., 2018). FUNDC1 phosphorylation and ubiquitin modification can also influence mitophagy (Liu et al., 2012). Chen et al. (2014) discovered that ULK1 and PGAM5 catalyze the dephosphorylation and phosphorylation of Ser13 and Ser17 present on FUNDC1, respectively, thereby transducing hypoxia signals and inducing FUNDC1-mediated mitophagy. In addition, it has been demonstrated that MARCH 5 promotes the degradation of FUNDC1 through ubiquitin modification, thereby inhibiting mitophagy (Chen et al., 2017).

Targeting Mitophagy in Cancer Treatment

Mitophagy-inducing medicines are effective anticancer treatments in several studies. Metformin, a medicine used to treat type 2 diabetes, has also been shown to inhibit the development of cancer cells by inducing mitophagy. Metformin stimulates the cellular energy sensor AMP-activated protein kinase (AMPK), which controls metabolism and mitophagy. Phosphorylation of ULK1 by AMPK initiates mitophagy, resulting in the removal of dysfunctional mitochondria (Laker et al., 2017).

The antihelminthic medication niclosamide has been demonstrated to stimulate mitophagy and suppress cancer cell proliferation and is licensed for use in the United States (Chen et al., 2018). Niclosamide triggers mitophagy, which activates the PINK1/Parkin pathway by blocking mitochondrial complex I. Damaged mitochondria are removed, and cancer cell proliferation is halted once mitophagy is triggered (Masaldan et al., 2022; Vara-Perez et al., 2019).

Tartrated Dihydroergotamine (DHE) efficiently suppresses the proliferation of several lung cancer cell lines. The cancer cell viability of those treated with DHE was lower than that of the control cells. DHE inhibited ATP synthesis triggered mitophagy for dysfunctional mitochondria, and enhanced ROS production. These data support further investigation into the therapeutic potential of DHE for treating lung cancer (Chang et al., 2016).

Sorafenib (Nexavar), a Raf serine/threonine kinase inhibitor, is being used to treat late-stage kidney cancer (advanced renal cell carcinoma), liver cancer (hepatocellular carcinoma) that cannot be treated by surgery, and differentiated thyroid cancer that has come back or spread to other parts of your body (Xie et al., 2012; Escudier et al., 2009). Sorafenib was shown to stabilize PINK1 on the OMM and recruit Parkin to damage mitochondria, inhibiting complexes II and III of the electron transport chain and triggering mitophagy (Zhang et al., 2017).

Ceramide is a sphingolipid molecule that regulates cellular signaling pathways involved in cell proliferation, differentiation, and death. According to many studies, ceramide buildup has been linked to mitophagy-related cell death in cancer cells (Dany & Ogretmen, 2015). It can induce stress in certain malignancies to cause mitophagy-related cell death. Thus, excessive synthesis of the ceramide CerS1 results in mitophagy and caspase-independent cell death. Furthermore, the ceramide analog LCL85 promotes mitophagy-related cell death in pancreatic cancer cells by altering the mitochondrial membrane potential and activating the PINK1/PARKIN pathway. Similarly, fumonisin B1, a ceramide synthase inhibitor, produces ceramide accumulation and causes mitophagy-related cell death in lung cancer cells by activating the PINK1/PARKIN pathway and boosting ROS generation (Denisenko et al., 2021).

A broad-spectrum antifungal drug called ketoconazole has been identified as an anticancer agent in several cancer types (Doheny et al., 2022). Ketoconazole has an anticancer effect though the molecular mechanism underlying this activity is unclear. Recent research has shown that ketoconazole increases mitophagy both *in vitro* and *in vivo*, which inhibits the proliferation of hepatocellular carcinoma (HCC) cells. Prostaglandin-endoperoxide synthase 2 (PTGS2) is suppressed by ketoconazole, leading to mitochondrial malfunction and death (Chen et al., 2019). These findings connect PTGS2 to the mitophagy system and suggest that ketoconazole may be an effective therapy for HCC. In addition, sodium selenite and a new BH3-mimetic compound called AT101 caused increased mitophagic cell death in glioma cells (Denisenko et al., 2021).

Conclusion

Mitophagy is essential for a cell to survive under certain stress conditions. In contrast, cancer cells use mitophagy for their survival; therefore, this process can be targeted as a potential process to treat cancer. There has been a special relationship between the lungs and mitochondria as the number and dynamics differ in the lungs and other cells. This further makes the process of mitophagy essential to study as it can have implications for treating cancers, including lung cancer. It is essential to mention that various mitophagy proteins behave differently in different types of cancers compared with lung cancer; therefore, studying the role of these unique proteins becomes more relevant in the case of lung cancer.

The possibility of targeting mitophagy as a novel treatment method has been further highlighted by the discovery of abnormalities in mitophagy involved in the etiology of lung cancer. It has emerged as a potential therapeutic strategy for cancer treatment because it can sensitize cancer cells to shift toward apoptosis and reduce tumor growth by eliminating damaged mitochondria, reducing ROS production, and inhibiting cancer growth.

Metformin and niclosamide are two medications that induce mitophagy and have shown potential as anticancer therapies. New mitophagy-inducing medications that can be employed in cancer therapy require further research to understand the specific mechanism of mitophagy regulation for cancer.

In conclusion, mitochondria are essential in initiating and developing lung cancer. The survival and growth of cancer cells can be aided by dysregulation of mitochondrial function, such as the "Warburg effect," dysregulated mitochondrial dynamics, mtDNA mutations, and consequent changes in the process of mitophagy. A possible treatment approach for lung cancer has been identified as focusing on mitochondrial activity. More research is necessary to completely comprehend the molecular processes underlying the deregulation of mitochondrial activity in lung cancer and to create efficient treatment approaches that target mitochondrial function.

References

Aman, Yahyah, Tomas Schmauck-Medina, Malene Hansen, Richard I. Morimoto, Anna Katharina Simon, Ivana Bjedov, Konstantinos Palikaras, Anne Simonsen, Terje Johansen, Nektarios Tavernarakis, David C. Rubinsztein, Linda Partridge, Guido Kroemer, John Labbadia, and Evandro F. Fang. 2021. "Autophagy in Healthy Aging and Disease." *Nature Aging* 1, no. 8: 634–50.https://doi.org/10.1038/s43587-021-00098-4

Benbrook, D. M., and A. Long. 2012. "Integration of Autography, Proteasomal Degradation, Unfolded Protein Response and Apoptosis." *Experimental Oncology* 34, no.3 : 286–297.

Borrero, Liz J., Hernández, and Wafik S. El-Deiry. 2021 "Tumor suppressor p53: Biology, signaling pathways, and therapeutic targeting." *Biochimica et & Biophysica Acta (BBA) – Reviews on Cancer* 1876 1: 188556.https://doi.org/10.1016/j.bbcan.2021.188556

Carraway, Martha S., Hagir B. Suliman, Corrine Kliment, Karen E. Welty-Wolf, Tim D. Oury, and Claude A. Piantadosi. 2008. "Mitochondrial Biogenesis in the Pulmonary Vasculature During Inhalational Lung Injury and Fibrosis." *Antioxidants & Redox Signaling* 10, no. 2: 269–75.https://doi.org/10.1089/ars.2007.1910

Cesari, Rossano, Eric S. Martin, George A. Calin, Francesca Pentimalli, Roberta Bichi, Holly McAdams, Francesco Trapasso, Alessandra Drusco, Masayoshi Shimizu, Valeria Masciullo, Giuseppina D'Andrilli, Giovanni Scambia, Maria Cristina Picchio, Hansjuerg Alder, Andrew K. Godwin, and Carlo M. Croce. 2003. "Parkin, a Gene Implicated in Autosomal Recessive Juvenile Parkinsonism, Is a Candidate Tumor Suppressor Gene on Chromosome 6q25–q27." *Proceedings of the National Academy of Sciences of the United States of America* 100, no. 10: 5956–61.https://doi.org/10.1073/pnas.0931262100

Chang, Seung-Hee, Ah Young Lee, Kyeong-Nam Yu, Jongsun Park, Kwang Pyo Kim, and Myung-Haing Cho. 2016. "Dihydroergotamine Tartrate Induces Lung Cancer Cell Death Through Apoptosis and Mitophagy." *Chemotherapy* 61, no. 6: 304–12.https://doi.org/10.1159/000445044

Chen, Guo, Zhe Han, D. U. Feng, Yanfang Chen, Linbo Chen, Hao Wu, Li Huang, Changqian Zhou, Xiangyu Cai, Changying Fu, Liangwei Duan, Xiaohui Wang, Lei Liu, Xinqi Liu, Yuequan Shen, Yushan Zhu, and Quan Chen. 2014. "A Regulatory Signaling Loop Comprising the PGAM5 Phosphatase and CK2 Controls Receptor-Mediated Mitophagy." *Molecular Cell* 54, no. 3: 362–77.https://doi.org/10.1016/j.molcel.2014.02.034

Chen, Wei, Wei, Robert A. Mook Jr., Richard T. Premont, and Jiangbo Wang. 2018. "Niclosamide: Beyond an Antihelminthic Drug." *Cellular Signalling* 41: 89–96.https://doi.org/10.1038/s41467-017-00520-9

Chen, Ziheng, Lei Liu, Qi Cheng, Yanjun Li, Hao Wu, Weilin Zhang, Yueying Wang, Sheikh Arslan Sehgal, Sami Siraj, Xiaohui Wang, Jun Wang, Yushan Zhu, and Quan Chen. 2017. "Mitochondrial E3 Ligase

MARCH 5 Regulates FUNDC 1 to Fine-Tune Hypoxic Mitophagy," *EMBO Reports* 18, no. 3: 495–509.https://doi.org/10.15252/embr.201643309

Cheng, Meng-Hsuan, Hung-Ling Huang Huang, Yen-You Lin, Kuan-Hao Tsui, Pei-Chin Chen, Shu-Yu Cheng, Inn-Wen Chong, Ping-Jyun Sung, Ming-Hong Tai, Zhi-Hong Wen, Nan-Fu Chen, and Hsiao-Mei Kuo. 2019. "BA6 Induces Apoptosis via Stimulation of Reactive Oxygen Species and Inhibition of Oxidative Phosphorylation in Human Lung Cancer Cells." *Oxidative Medicine & Cellular Longevity*: 6342104.https://doi.org/10.1155/2019/6342104

Chinnery, Patrick F., and Eric A. Schon. 2003. "Mitochondria." *Journal of Neurology, Neurosurgery, & Psychiatry* 74, no. 9: 1188–99.https://doi.org/10.1136/jnnp.74.9.1188

Chourasia, Aparajita H., H. Aparajita, Michelle L. Boland, and Kay F. Macleod. 2015. "Mitophagy and Cancer." *Cancer & Metabolism* 3, no. 1: 1–11.https://doi.org/10.1186/s40170-015-0130-8

Cloonan, Suzanne M., and Augustine M. K. Choi. 2016. "Mitochondria in Lung Disease." *Journal of Clinical Investigation* 126, no. 3: 809–20.https://doi.org/10.1172/JCI81113

Dai, Ji-Gang, Zai-Yong Zhang Zhang, Quan-Xing Liu, and Jia-Xin Min. 2013. "Mitochondrial Genome Microsatellite Instability and Copy Number Alteration in Lung Carcinomas." *Asian Pacific Journal of Cancer Prevention: APJCP* 14, no. 4: 2393–9.https://doi.org/10.7314/apjcp.2013.14.4.2393

Dai, Wenting, and Lei Jiang. 2019. "Dysregulated Mitochondrial Dynamics and Metabolism in Obesity, Diabetes, and Cancer." *Frontiers in Endocrinology* 10: 570.https://doi.org/10.3389/fendo.2019.00570

Dany, Mohammed, and Besim Ogretmen. 2015. "Ceramide Induced Mitophagy and Tumor Suppression." *Biochimica & Biophysica Acta* 1853, no. 10 Pt B: 2834–45.https://doi.org/10.1016/j.bbamcr.2014.12.039

Dasgupta, Santanu, Ethan Soudry, Nitai Mukhopadhyay, Chunbo Shao, John Yee, Stephan Lam, Wan Lam, Wei Zhang, Adi F. Gazdar, Paul B. Fisher, and David Sidransky. 2012. "Mitochondrial DNA Mutations in Respiratory Complex-I in Never-Smoker Lung Cancer Patients Contribute to Lung Cancer Progression and Associated with EGFR Gene Mutation." *Journal of Cellular Physiology* 227, no. 6: 2451–60.https://doi.org/10.1002/jcp.22980

Dela Cruz, Charles S., Lynn T. Tanoue, and Richard A. Matthay. 2011. "Lung Cancer: Epidemiology, Etiology, and Prevention." *Clinics in Chest Medicine* 32, no. 4: 605–44.https://doi.org/10.1016/j.ccm.2011.09.001

Denisenko, Tatiana V., Vladimir Gogvadze, and Boris Zhivotovsky. 2021. "Mitophagy in Carcinogenesis and Cancer Treatment." *Discover. Oncology* 12, no. 1: 58.https://doi.org/10.1007/s12672-021-00454-1

Di Rita, Anthea, Angelo Peschiaroli, Pasquale D Acunzo, Daniela Strobbe, Zehan Hu, Jens Gruber, Mads Nygaard, Matteo Lambrughi, Gerry Melino, Elena Papaleo, Jörn Dengjel, Said El Alaoui, Michelangelo Campanella, Volker Dötsch, Vladimir V. Rogov, Flavie Strappazzon, and Francesco Cecconi. 2018. "HUWE1 E3 Ligase Promotes PINK1/PARKIN-Independent Mitophagy by Regulating AMBRA1 Activation via IKKα." *Nature Communications* 9, no. 1: 3755.https://doi.org/10.1038/s41467-018-05722-3

Dice, J. Fred. 2007. "Chaperone-Mediated Autophagy." *Autophagy* 3, no. 4: 295–9.https://doi.org/10.4161/auto.4144

Doheny, Daniel, Sara Manore, Sherona R. Sirkisoon, Dongqin Zhu, Noah R. Aguayo, Alexandria Harrison, Mariana Najjar, Marlyn Anguelov, Anderson O. 'Brien Cox, Cristina M. Furdui, Kounosuke Watabe, Thomas Hollis, Alexandra Thomas, Roy Strowd, and Hui-Wen Lo. 2022. "An FDA-Approved Antifungal, Ketoconazole, and Its Novel Derivative Suppress tGLI1-Mediated Breast Cancer Brain Metastasis by Inhibiting the DNA-Binding Activity of Brain Metastasis-Promoting Transcription Factor tGLI1." *Cancers* 14, no. 17: 4256.https://doi.org/10.3390/cancers14174256

Dorszewska, Jolanta, Marta Kowalska, Michał Prendecki, Thomas Piekut, Joanna Kozłowska, and Wojciech Kozubski. 2021. "Oxidative Stress Factors in Parkinson's Disease." *Neural Regeneration Research* 16, no. 7: 1383–91.

Drake, Lauren E., Maya Z. Springer, Logan P. Poole, Casey J. Kim, and Kay F. Macleod. 2017. "Expanding Perspectives on the Significance of Mitophagy in Cancer." In *Seminars in Cancer Biology*. Academic Press 47: 110–24.https://doi.org/10.1016/j.semcancer.2017.04.008

Durcan, Thomas M., and Edward A. Fon. 2015. "The Three 'P's of Mitophagy: PARKIN, PINK1, and Post-translational Modifications.'" *Genes & Development* 29, no. 10: 989–99.https://doi.org/10.1101/gad.262758.115

Ernster, Lars, and Gottfried Schatz. 1981. "Mitochondria: a Historical Review." *Journal of Cell Biology* 91, no. 3 Pt 2: 227s–55s.https://doi.org/10.1083/jcb.91.3.227s

Escudier, Bernard, Tim Eisen, Walter M. Stadler, Cezary Szczylik, Stéphane Oudard, Michael Staehler, Sylvie Negrier, Christine Chevreau, Apurva A. Desai, Frédéric Rolland, Tomasz Demkow, Thomas E. Hutson, Martin Gore, Sibyl Anderson, Gloria Hofilena, Minghua Shan, Carol Pena, Chetan Lathia, and Ronald M. Bukowski. 2009. "Sorafenib for Treatment of Renal Cell Carcinoma: Final Efficacy and Safety Results of the Phase III Treatment Approaches in Renal Cancer Global Evaluation Trial." *Journal of Clinical Oncology* 27, no. 20: 3312–8.https://doi.org/10.1200/JCO.2008.19.5511

Fliss, Makiko S., Henning Usadel, Otávia L. Caballero, Li Wu, Martin R. Buta, Scott M. Eleff, Jin Jen, and David Sidransky. 2000. "Facile Detection of Mitochondrial DNA Mutations in Tumors and Bodily Fluids." *Science* 287, no. 5460: 2017–9.https://doi.org/10.1126/science.287.5460.2017

Fritsch, Lauren E., M. Elyse Moore, Shireen A. Sarraf, and Alicia M. Pickrell. 2020. "Ubiquitin and Receptor-Dependent Mitophagy Pathways and Their Implication in Neurodegeneration." *Journal of Molecular Biology* 432, no. 8: 2510–24.https://doi.org/10.1016/j.jmb.2019.10.015

https://doi.org/10.1038/s41401-019-0260-6Ge, Preston, Valina L. Dawson, and Ted M. Dawson. 2020. "PINK1 and Parkin Mitochondrial Quality Control: A Source of Regional Vulnerability in Parkinson's Disease." *Molecular Neurodegeneration* 15, no. 1: 20.https://doi.org/10.1186/s13024-020-00367-7

Geiger, Thomas R., and Daniel S. Peeper. 2009 Dec. "Metastasis Mechanisms." *Biochimica & Biophysica Acta* 1796, no. 2: 293–308.https://doi.org/10.1016/j.bbcan.2009.07.006

Giampazolias, Evangelos, and Stephen W. G. Tait. 2016. "Mitochondria and the Hallmarks of Cancer." *FEBS Journal* 283, no. 5: 803–14.https://doi.org/10.1111/febs.13603

Glauser, Liliane, Sarah Sonnay, Klodjan Stafa, and Darren J. Moore. 2011. "Parkin Promotes the Ubiquitination and Degradation of the Mitochondrial Fusion Factor Mitofusin 1." *Journal of Neurochemistry* 118, no. 4: 636–45.https://doi.org/10.1111/j.1471-4159.2011.07318.x

Glick, Danielle, Sandra Barth, and Kay F. Macleod. 2010. "Autophagy: Cellular and Molecular Mechanisms." *Journal of Pathology* 221, no. 1: 3–12.https://doi.org/10.1002/path.2697

Goiran, Thomas, Eric Duplan, Lila Rouland, Wejdane El Manaa, Inger Lauritzen, Julie Dunys, Han You, Frédéric Checler, and Cristine Alves da Costa. 2018 Jan. 19. "Nuclear p53-Mediated Repression of Autophagy Involves PINK1 Transcriptional Down-Regulation." *Cell Death & Differentiation* 25, no. 5: 873–84.https://doi.org/10.1038/s41418-017-0016-0

Gonzalez, Hugo, Catharina Hagerling, and Zena Werb. 2018. "Roles of the Immune System in Cancer: from Tumor Initiation to Metastatic Progression." *Genes & Development* 32, no. 19–20: 1267–84.

Gorbunova, Anna S., Maria A. Yapryntseva, Tatiana V. Denisenko, and Boris Zhivotovsky. 2020. "BNIP3 in Lung Cancer: to Kill or Rescue?." *Cancers* 12, no. 11: 3390.https://doi.org/10.3390/cancers12113390

Guan, Yanjie, Yifei Wang, Bo Li, Kai Shen, Quanfu Li, Yingyin Ni, and Lei Huang. 2021. "Mitophagy in Carcinogenesis, Drug Resistance and Anticancer Therapeutics." *Cancer Cell International* 21, no. 1: 350.https://doi.org/10.1186/s12935-021-02065-w

Hanahan, Douglas. 2022. "Hallmarks of Cancer: New Dimensions." *Cancer Discovery* 12, no. 1: 31–46.https://doi.org/10.1158/2159-8290.cd-21-1059

Hosgood, 3rd, H. Dean, Chin-San Liu, Nathaniel Rothman, Stephanie J. Weinstein, Matthew R. Bonner, Min Shen, Unhee Lim, Jarmo Virtamo, W. L. Cheng, Demetrius Albanes, and Qing Lan. 2010. "Mitochondrial DNA Copy Number and Lung Cancer Risk in a Prospective Cohort Study." *Carcinogenesis* 31, no. 5: 847–9.https://doi.org/10.1093/carcin/bgq045

Hou, Hailing, Puchun Er, Jingjing Cheng, Xiuli Chen, Xiaofeng Ding, Yuwen Wang, Xi Chen, Zhiyong Yuan, Qingsong Pang, Ping Wang, and Dong Qian. 2017. "High Expression of FUNDC1 Predicts Poor Prognostic Outcomes and Is a Promising Target to Improve Chemoradiotherapy Effects in Patients with Cervical Cancer." *Cancer Medicine* 6, no. 8: 1871–81.https://doi.org/10.1002/cam4.1112

Iwakawa, Reika, Hirokazu Okayama, Takashi Kohno, Aiko Sato-Otsubo, Seishi Ogawa, and Jun Yokota. 2012 May. "Contribution of Germline Mutations to PARK2 Gene Inactivation in Lung Adenocarcinoma." *Genes, Chromosomes & Cancer* 51, no. 5: 462–72.

Jin, Seok Min, and Richard J. Youle. 2012. "PINK1- And Parkin-Mediated Mitophagy at a Glance." *Journal of Cell Science* 125, no. 4: 795–9.https://doi.org/10.1242/jcs.093849

Kaushik, Susmita, and Ana Maria Cuervo. 2018. "The Coming of Age of Chaperone-Mediated Autophagy." *Nature Reviews. Molecular Cell Biology* 19, no. 6: 365–81.https://doi.org/10.1038/s41580-018-0001-6

Laker, Rhianna C., Joshua C. Drake, Rebecca J. Wilson, Vitor A. Lira, Bevan M. Lewellen, Karen A. Ryall, Carleigh C. Fisher, Mei Zhang, Jeffrey J. Saucerman, Laurie J. Goodyear, Mondira Kundu, and Zhen Yan. 2017. "AMPK Phosphorylation of Ulk1 Is Required for Targeting of Mitochondria to Lysosomes in Exercise-Induced Mitophagy." *Nature Communications* 8, no. 1: 548.https://doi.org/10.1038/s41467-017-00520-9

Lee, Youngil, Hwa-Youn Lee, Rita A. Hanna, and Åsa B. Gustafsson. 2011. "Mitochondrial Autophagy by Bnip3 Involves Drp1-Mediated Mitochondrial Fission and Recruitment of Parkin in Cardiac Myocytes." *American Journal of Physiology. Heart & Circulatory Physiology* 301, no. 5: H1924–31.https://doi.org/10.1152/ajpheart.00368.2011

Li, Xiaohua, Shikun He, and Binyun Ma. 2020. "Autophagy and Autophagy-Related Proteins in Cancer." *Molecular Cancer* 19, no. 1: 1–16.https://doi.org/10.1186/s12943-020-1138-4

Li, Yan-Qin, Fan Zhang, Li-Ping Yu, Jian-Kang Mu, Ya-Qin Yang, Jie Yu, and Xing-Xin Yang. 2021. "Targeting PINK1 Using Natural Products for the Treatment of Human Diseases." *BioMed Research International* 2021: 4045819.https://doi.org/10.1155/2021/4045819

Li, Yue, Wanqing Zheng, Yangyang Lu, Yanrong Zheng, Ling Pan, Xiaoli Wu, Yang Yuan, Zhe Shen, Shijia Ma, Xingxian Zhang, Jiaying Wu, Zhong Chen, and Xiangnan Zhang. 2021. "BNIP3L/NIX-Mediated Mitophagy: Molecular Mechanisms and Implications for Human Disease." *Cell Death & Disease* 13, no. 1: 14https://doi.org/10.1038/s41419-021-04469-y.

Liu, Juan, Cen Zhang, Wenwei Hu, and Zhaohui Feng. 2018. "Parkinson's Disease-Associated Protein Parkin: an Unusual Player in Cancer." *Cancer Communications* 38, no. 1: 40.https://doi.org/10.1186/s40880-018-0314-z

Liu, Lei, Du Feng, Guo Chen, Ming Chen, Qiaoxia Zheng, Pingping Song, Qi Ma, Chongzhuo Zhu, Rui Wang, Wanjun Qi, Lei Huang, Peng Xue, Baowei Li, Xiaohui Wang, Haijing Jin, Jun Wang, Fuquan Yang, Pingsheng Liu, Yushan Zhu, Senfang Sui, and Quan Chen. 2012. "Mitochondrial Outer-Membrane Protein FUNDC1 Mediates Hypoxia-Induced Mitophagy in Mammalian Cells." *Nature Cell Biology* 14, no. 2: 177–85.https://doi.org/10.1038/ncb2422

Liu, Lihua, Zhongfu Zuo, Sijing Lu, Lihua Wang, Aihua Liu, and Xuezheng Liu. 2018. "Silencing of PINK1 Represses Cell Growth, Migration and Induces Apoptosis of Lung Cancer Cells." *Biomedicine & Pharmacotherapy* 106: 333–41.

Liu, Xiaojing, and Zhihong Chen. 2017. "The Pathophysiological Role of Mitochondrial Oxidative Stress in Lung Diseases." *Journal of Translational Medicine* 15, no. 1: 207.https://doi.org/10.1186/s12967-017-1306-5

Lu, Jianrong, Ming Tan, and Qingsong Cai. 2015. "The Warburg Effect in Tumor Progression: Mitochondrial Oxidative Metabolism as an Anti-Metastasis Mechanism." *Cancer Letters* 356, no. 2 Pt A: 156–64.https://doi.org/10.1016/j.canlet.2014.04.001

Lu, Yingying, Zhijia Li, Shuangqian Zhang, Tongtong Zhang, Yanjun Liu, and Lan Zhang. 2023. "Cellular Mitophagy: Mechanism, Roles in Diseases and Small Molecule Pharmacological Regulation." *Theranostics* 13, no. 2: 736–66.https://doi.org/10.7150/thno.79876

Ma, Kaili, Guo Chen, Wenhui Li, Oliver Kepp, Yushan Zhu, and Quan Chen. 2020. "Mitophagy, Mitochondrial Homeostasis, and Cell Fate." *Frontiers in Cell & Developmental Biology* 8: 467.https://doi.org/10.3389/fcell.2020.00467

Masaldan, Shashank, Sylvie Callegari, and Grant Dewson. 2022. "Therapeutic Targeting of Mitophagy in Parkinson's Disease." *Biochemical Society Transactions* 50, no. 2: 783–97.https://doi.org/10.1042/bst20211107

Massaro, Gloria D., Dorothy Berlin Gail, and D. Massaro. 1975. "Lung Oxygen Consumption and Mitochondria of Alveolar Epithelial and Endothelial Cells." *Journal of Applied Physiology* 38, no. 4: 588–92.https://doi.org/10.1152/jappl.1975.38.4.588

Muz, Barbara, Pilar de la Puente, Feda Azab, and Abdel Kareem Azab. 2015. "The Role of Hypoxia in Cancer Progression, Angiogenesis, Metastasis, and Resistance to Therapy." *Hypoxia* 3: 83–92.https://doi.org/10.2147/hp.s93413

Nam, Hae-Seong, Evgeny Izumchenko, Santanu Dasgupta, and Mohammad O. Hoque. 2017. "Mitochondria in Chronic Obstructive Pulmonary Disease and Lung Cancer: Where Are We Now?." *Biomarkers in Medicine* 11, no. 6: 475–89.https://doi.org/10.2217/bmm-2016-0373

Nelson, Deirdre A., Ting-Ting Tan, Arnold B. Rabson, Diana Anderson, Kurt Degenhardt, and Eileen White. 2004. "Hypoxia and Defective Apoptosis Drive Genomic Instability and Tumorigenesis." *Genes & Development* 18, no. 17: 2095–107.https://doi.org/10.1101/gad.1204904

Papalazarou, Vasileios, and Oliver Dk Maddocks. 2021. "Supply and Demand: Cellular Nutrient Uptake and Exchange in Cancer." *Molecular Cell* 81, no. 18: 3731–48.https://doi.org/10.1016/j.molcel.2021.08.026

Pickrell, Alicia M., and Richard J. Youle. 2015. "The Roles of PINK1, Parkin, and Mitochondrial Fidelity in Parkinson's Disease." *Neuron* 85, no. 2: 257–73.https://doi.org/10.1016/j.neuron.2014.12.007

Plas, David R., and Craig B. Thompson. 2005. "Akt-Dependent Transformation: There Is More to Growth than Just Surviving." *Oncogene* 24, no. 50: 7435–42.https://doi.org/10.1038/sj.onc.1209097

Qi, Miao, Dan Dai, Jin Liu, Zhongqi Li, Panpan Liang, Yue Wang, Lu Cheng, Yihong Zhan, Zhifeng An, Yaoyao Song, Yana Yang, Xiaohui Yan, Hui Xiao, and Huanjie Shao. 2020. "AIM2 Promotes the Development of Non-small Cell Lung Cancer by Modulating Mitochondrial Dynamics." *Oncogene* 39, no. 13: 2707–23.https://doi.org/10.1038/s41388-020-1176-9

Quinsay, Melissa N., Robert L. Thomas, Youngil Lee, and Asa B. Gustafsson. 2010. "Bnip3-Mediated Mitochondrial Autophagy Is Independent of the Mitochondrial Permeability Transition Pore." *Autophagy* 6, no. 7: 855–62.https://doi.org/10.4161/auto.6.7.13005

Roberts, Emily R., and Kelly Jean Thomas. 2013. "The Role of Mitochondria in the Development and Progression of Lung Cancer." *Computational & Structural Biotechnology Journal* 6, no. 7: e201303019.

Rojas-Charry, Liliana, Mark R. Cookson, Andrea Niño, Humberto Arboleda, and Gonzalo Arboleda. 2014. "Downregulation of Pink1 Influences Mitochondrial Fusion–Fission Machinery and Sensitizes to Neurotoxins in Dopaminergic Cells." *Neurotoxicology* 44: 140–8.https://doi.org/10.1016/j.neuro.2014.04.007

Saftig, Paul, Wouter Beertsen, and Eeva-Liisa Eskelinen. 2008. "LAMP-2: a Control Step for Phagosome and Autophagosome Maturation." *Autophagy* 4, no. 4: 510–2.https://doi.org/10.4161/auto.5724

Servais, S., K. Couturier, H. Koubi, J. L. Rouanet, D. Desplanches, M. H. Sornay-Mayet, B. Sempore, J. M. Lavoie, and Roland Favier. 2003. "Effect of Voluntary Exercise on H2O2 Release by Subsarcolemmal and Intermyofibrillar Mitochondria." *Free Radical Biology & Medicine* 35, no. 1: 24–32.https://doi.org/10.1016/s0891-5849(03)00177-1

Shi, Ruo-Yang, Sheng-Hua Zhu Zhu, Victor Li, Spencer B. Gibson, Xing-Shun Xu, and Ji-Ming Kong. 2014. "BNIP3 Interacting with LC3 Triggers Excessive Mitophagy in Delayed Neuronal Death in Stroke." *CNS Neuroscience & Therapeutics* 20, no. 12: 1045–55.https://doi.org/10.1111/cns.12325

Shostak, Anton. 2017. "Circadian Clock, Cell Division, and Cancer: from Molecules to Organism." *International Journal of Molecular Sciences* 18, no. 4: 873.https://doi.org/10.3390/ijms18040873

Solaini, Giancarlo, Gianluca Sgarbi, and Alessandra Baracca. 2011. "Oxidative Phosphorylation in Cancer Cells." *Biochimica & Biophysica Acta* 1807, no. 6: 534–42.https://doi.org/10.1016/j.bbabio.2010.09.003

Sosa Iglesias, Venus, Lorena Giuranno, Ludwig J. Dubois, Jan Theys, and Marc Vooijs. 2018 Jul. 24. "Drug Resistance in Non-small Cell Lung Cancer: a Potential for NOTCH Targeting?." *Frontiers in Oncology* 8: 267.https://doi.org/10.3389/fonc.2018.00267

Sowter, Heidi M., Peter J. Ratcliffe, Peter Watson, Arnold H. Greenberg, and Adrian L. Harris. 2001. "HIF-1-dependent Regulation of Hypoxic Induction of the Cell Death Factors BNIP3 and NIX in Human Tumors." *Cancer Research* 61, no. 18: 6669–73.

Strohecker, Anne M., Jessie Yanxiang Guo, Gizem Karsli-Uzunbas, Sandy M. Price, Guanghua Jim Chen, Robin Mathew, Martin McMahon, and Eileen White. 2013. "Autophagy Sustains Mitochondrial Glutamine Metabolism and Growth of BrafV600E–Driven Lung TumorsAutophagy Promotes BrafV600E-Driven Lung Tumor Growth." *Cancer Discovery* 3, no. 11: 1272–85.https://doi.org/10.1158/2159-8290.cd-13-0397

Sun, Xin, Guiqin Ye, Yuanyuan Mai, Yuhan Shu, Lei Wang, and Jianbin Zhang. 2023. "Parkin Exerts the Tumor-Suppressive Effect Through Targeting Mitochondria." *Medicinal Research Reviews* 43, no. 4: 855–71.https://doi.org/10.1002/med.21938

Sureshbabu, Angara, and Vineet Bhandari. 2013. "Targeting Mitochondrial Dysfunction in Lung Diseases: Emphasis on Mitophagy." *Frontiers in Physiology* 4: 384.https://doi.org/10.3389/fphys.2013.00384

Tay, Shiam-Peng, Calvin W. S. Yeo, Chou Chai, Pei-Jou Chua, Hui-Mei Tan, Alex X. Y. Ang, Daniel L. H. Yip, Jian-Xiong Sung, Puay Hoon Tan, Boon-Huat Bay, Siew-Heng Wong, Carol Tang, Jeanne M. M. Tan, and Kah-Leong Lim. 2010. "Parkin Enhances the Expression of Cyclin-Dependent Kinase 6 and Negatively Regulates the Proliferation of Breast Cancer Cells." *Journal of Biological Chemistry* 285, no. 38: 29231–8https://doi.org/10.1074/jbc.m110.108241.

Vander Heiden, Matthew G., Lewis C. Cantley, and Craig B. Thompson. 2009. "Understanding the Warburg Effect: the Metabolic Requirements of Cell Proliferation." *Science* 5930, no. 5930: 1029–33.https://doi.org/10.1126/science.1160809

Vara-Perez, Monica, Blanca Felipe-Abrio, and Patrizia Agostinis. 2019. "Mitophagy in Cancer: a Tale of Adaptation." *Cells* 8, no. 5: 493.

Vijayalingam, S., Sreeraj G. Pillai, Ramachandran Rashmi, T. Subramanian, John E. Sagartz, and G. Chinnadurai. 2010. "Overexpression of BH3-Only Protein BNIP3 Leads to Enhanced Tumor Growth." *Genes & Cancer* 1, no. 9: 964–71.https://doi.org/10.1177/1947601910386110

Wang, Jiying, Zhiyi Dong, Zhaoying Sheng, and Yong Cai. 2022a. "Hypoxia-Induced PVT1 Promotes Lung Cancer Chemoresistance to Cisplatin by Autophagy via PVT1/miR-140-3p/ATG5 Axis." *Cell Death Discovery* 8, no. 1: 104.https://doi.org/10.1038/s41420-022-00886-w

Wang, Yigang, Hui-Hui Liu, Yu-Ting Cao, Lei-Lei Zhang, Fang Huang, and Cong Yi. 2020. "The Role of Mitochondrial Dynamics and Mitophagy in Carcinogenesis, Metastasis and Therapy." *Frontiers in Cell & Developmental Biology* 8: 413.https://doi.org/10.3389/fcell.2020.00413

Wang, Yiyang, Jieliang Shen, Yanyang Chen, Huzhe Liu, Hao Zhou, Zhibiao Bai, Zhenming Hu, and Xiuming Guo. 2018. "PINK1 Protects Against Oxidative Stress Induced Senescence of Human Nucleus Pulposus Cells via Regulating Mitophagy." *Biochemical & Biophysical Research Communications* 504, no. 2: 406–14.https://doi.org/10.1016/j.bbrc.2018.06.031

Wang, Yuanli, Kah Yong Goh, Zhencheng Chen, Wen Xing Lee, Sze Mun Choy, Jia Xin Fong, Yun Ka Wong, Dongxia Li, Fangrong Hu, and Hong-Wen Tang. 2022b. "A Novel TP53 Gene Mutation Sustains Non-small Cell Lung Cancer Through Mitophagy." *Cells* 11, no. 22: 3587.https://doi.org/10.3390/cells11223587

Xia, Houjun, Douglas R. Green, and Weiping Zou. 2021. "Autophagy in Tumour Immunity and Therapy." *Nature Reviews. Cancer* 21, no. 5: 281–97. https://doi.org/10.1038/s41568-021-00344-2https://doi.org/10.1038/s41568-021-00344-2.

Xie, Bingru, David H. Wang, and Stuart Jon Spechler. 2012. "Sorafenib for Treatment of Hepatocellular Carcinoma: a Systematic Review." *Digestive Diseases & Sciences* 57, no. 5: 1122–9.https://doi.org/10.1007/s10620-012-2136-1.

Xiong, Donghai, Yian Wang, Elena Kupert, Claire Simpson, Susan M. Pinney, Colette R. Gaba, Diptasri Mandal, Ann G. Schwartz, Ping Yang, Mariza de Andrade, Claudio Pikielny, Jinyoung Byun, Yafang Li, Dwight Stambolian, Margaret R. Spitz, Yanhong Liu, Christopher I. Amos, Joan E. Bailey-Wilson, Marshall Anderson, and Ming You. 2015. "A Recurrent Mutation in PARK2 Is Associated with Familial Lung Cancer." *American Journal of Human Genetics* 96, no. 2: 301–8.https://doi.org/10.1016/j.ajhg.2014.12.016

Yang, Yuhui, Svetlana Karakhanova, Werner Hartwig, Jan G. D'Haese, Pavel P. Philippov, Jens Werner, and Alexandr V. Bazhin. 2016. "Mitochondria and Mitochondrial ROS in Cancer: Novel Targets for Anticancer Therapy." *Journal of Cellular Physiology* 231, no. 12: 2570–81.https://doi.org/10.1002/jcp.25349

Yin, Kunlun, Jordan Lee, Zhaoli Liu, Hyeoncheol Kim, David R. Martin, Dandan Wu, Meilian Liu, and Xiang Xue. 2021. "Mitophagy Protein PINK1 Suppresses Colon Tumor Growth by Metabolic Reprogramming via p53 Activation and Reducing Acetyl-CoA Production." *Cell Death & Differentiation* 28, no. 8: 2421–35.https://doi.org/10.1038/s41418-021-00760-9

Yun, Chul Won, and Sang Hun Lee. 2018. "The Roles of Autophagy in Cancer." *International Journal of Molecular Sciences* 19, no. 11: 3466.https://doi.org/10.3390/ijms19113466

Zhang, Ji, and Paul A. Ney. 2009. "Role of BNIP3 and NIX in Cell Death, Autophagy, and Mitophagy." *Cell Death & Differentiation* 16, no. 7: 939–46.https://doi.org/10.1038/cdd.2009.16

Zhang, Weilin. 2021. "The Mitophagy Receptor FUN14 Domain-Containing 1 (FUNDC1): a Promising Biomarker and Potential Therapeutic Target of Human Diseases." *Genes & Diseases* 8, no. 5: 640–54.https://doi.org/10.1016/j.gendis.2020.08.011

Zhang, Zeng-Li, Na-Na Wang, Qi-Lian Ma, Yang Chen, L. Yao, L. Zhang, Qiu-Shi Li, Min-Hua Shi, Hong-Feng Wang, and Zheng Ying. 2020 Jan. "Somatic and Germline Mutations in the Tumor Suppressor Gene PARK2 Impair PINK1/Parkin-Mediated Mitophagy in Lung Cancer Cells." *Acta Pharmacologica Sinica* 41, no. 1: 93–100.

Zhu, Lizhe, Wei Wu, Siyuan Jiang, Shibo Yu, Y. Yan, K. Wang, Jianjun He, Y. Ren, and Bin Wang. 2020. "Pan-cancer Analysis of the Mitophagy-Related Protein PINK1 as a Biomarker for the Immunological and Prognostic Role." *Frontiers in Oncology* 10: 569887.https://doi.org/10.3389/fonc.2020.569887

About the Editors

Dr. Shalini Mani is Associate Professor in the Department of Biotechnology at Jaypee Institute of Information Technology, Noida, India. She has total 20 years of research and 14 years of teaching experience. She completed her PhD from Centre for Cellular and Molecular Biology, Hyderabad India.

She is a Molecular Cell Biologist and has 14 Years of teaching and research experience. Dr Mani is primarily involved in studying the role of mitochondria in different chronic diseases and effect of different environmental factors on mitochondrial metabolism. As a translational researcher, she is also exploring the molecular mechanism for therapeutic benefits of some selected herbs on mitochondrial activity in different diseases conditions such as diabetes and cancer.

She has published more than 50 research and review articles in international peer-reviewed reputed journals and has authored 25 chapters and 02 books.

She is the recipient of a prestigious Endeavour Research Fellowship award Australia in 2015 and Women in Oncology Research by Healthcopeia Foundation, India in 2021.

She is also a Member of Society for Mitochondrial Research and Medicine, Indian genome Variation database, British Neuroscience Association, ISTAART etc.

Dr. Samiksha Wasnik is an accomplished researcher and academic professional with a strong life science and regenerative medicine background. She obtained her Ph.D. in Life Science from Deakin University in Australia in 2014, where her research focused on developing tissue-engineered acellular matrices for the expansion of hematopoietic progenitors. Before this, she earned her MS in Bio-Technology from Guru Ghasidas University in India, where she explored Y-chromosome and mtDNA phylogeny and researched the molecular epidemiology of alcoholism.

Throughout her career, Dr. Wasnik has made significant contributions to the scientific community through her extensive publications. She has authored 27 research articles in esteemed journals such as FASEB Journal, American Journal of Physiology-Cell Physiology, Neoplasia, Science Advances, and JCI Insight. Additionally, she has contributed to the field by writing five book chapters.

In recognition of her innovative work, Dr. Wasnik holds a patent titled "Methods of treatment of a cytokine storm." This patent, in collaboration with other inventors, showcases her expertise in developing novel treatment methods for cytokine storms.

Dr. Samiksha Wasnik held an academic title as an Assistant Research Professor in the Division of Regenerative Medicine at Loma Linda University in Loma Linda, CA, until 2022. In this role, she conducted groundbreaking research and contributed her expertise to advancing regenerative medicine.

Index

E

F

G

H

I

L

M

N

O

P

R

S

T

U

Y